World Sustainability Series

Series editor

Walter Leal Filho, Hamburg, Germany

More information about this series at http://www.springer.com/series/13384

Walter Leal Filho · Diana Mihaela Pociovălişteanu
Paulo Roberto Borges de Brito
Ismar Borges de Lima
Editors

Towards a Sustainable Bioeconomy: Principles, Challenges and Perspectives

Springer

Editors
Walter Leal Filho
Faculty of Life Sciences
HAW Hamburg
Hamburg
Germany

Paulo Roberto Borges de Brito
Department of Management, College of Business
Colorado State University
Fort Collins, CO
USA

Diana Mihaela Pociovălişteanu
"Constantin Brancusi" University of Târgu Jiu
Târgu Jiu
Romania

Ismar Borges de Lima
Institute for Agrarian & Regional Development
Federal University of Southern and Southeastern Pará
Marabá, Pará
Brazil

ISSN 2199-7373 ISSN 2199-7381 (electronic)
World Sustainability Series
ISBN 978-3-319-73027-1 ISBN 978-3-319-73028-8 (eBook)
https://doi.org/10.1007/978-3-319-73028-8

Library of Congress Control Number: 2017962024

Printed on acid-free paper

This Springer imprint is published by Springer Nature
The registered company is Springer International Publishing AG
The registered company address is: Gewerbestrasse 11, 6330 Cham, Switzerland

Preface

As society progresses, we need to rely more on a sustainable bioeconomy. Models of development based on the excessive use of natural resources have proven to be wrong and unsuccessful. We, therefore, need to move towards a bioeconomy. Indeed, moving towards bioeconomy is one of the key policy strategies of the European Union, whose vision includes a more sustainable use of environmental resources, especially—but not only—in the agricultural and forest-based sectors, in the handling of waste streams and in the production of value-added products and bioenergy.

There is a paucity of publications which take an interdisciplinary look at the various elements which integrate the bioeconomy. It is against this background that the book 'Towards a Sustainable Bioeconomy: Principles, Challenges and Perspectives' has been produced.

This book contains a set of papers which serve the purpose of showcasing experiences from research, field projects and best practice in the field of sustainable bioeconomy. Consistent with the need for more cross-sectoral interactions among the various stakeholders working in the field of bioeconomy, this book aims to:

i. provide research institutions, universities, NGOs and enterprises with an opportunity to display and present their works in the field of bioeconomy;
ii. foster the exchange of information, ideas and experiences acquired in the execution of projects, especially successful initiatives and good practice on bioeconomy from across the world;
iii. introduce methodological approaches and experiences derived from case studies and projects, which aim to show how a sustainable bioeconomy may be pursued in practice.

Last but not least, further aim of this book is to document and disseminate the wealth of experiences available today.

This book is structured into six parts. Part I deals with the concepts of bioeconomy, including some fundamental components and new approaches.

Part II handles aspects related to industry, market and financing possibilities for bioeconomy-related activities. Part III handles innovative approaches and technological possibilities. Part IV describes some developments related to agriculture, biopharming and food production. In Part V, papers related to bioenergy and biofuels are presented.

Finally, Part VI presents papers related to bio-based forest productions and biomass.

We thank the authors for their willingness to share their knowledge, know-how and experiences, as well as the many peer reviewers, who have helped us to ensure the quality of the manuscripts.

Enjoy your reading!

Hamburg, Germany — Walter Leal Filho
Târgu Jiu, Romania — Diana Mihaela Pociovălişteanu
Fort Collins, USA — Paulo Roberto Borges de Brito
Marabá, Brazil — Ismar Borges de Lima
Spring 2018

The original version of the book was revised: Corresponding author's corrections have been incorporated and the incorrect author names have been corrected. The erratum to this book is available at https://doi.org/10.1007/978-3-319-73028-8_30

Contents

Part I
Bioeconomy: Concepts, Fundaments and New Approaches

Fostering Sustainable Bioeconomies: The Role of Conscious Consumption

Madhavi Venkatesan

Abstract Sustainability is typically discussed in a siloed fashion in the United States. Cradle-to-cradle production and regulation to curb greenhouse gas emissions are proffered as salves for evidenced degradation, but little attention is directed to how a society can enable sustainability as a cultural norm. Further and related, the role of the individual economic agent as consumer, investor, and government participant is seemingly not acknowledged. To a large extent, the population majority delegates the powers conferred in the three roles to a minority largely through indifferent conveyance posited on trust, leaving outcomes impacting society as a whole dependent on incentives of a few, who may or may not be aligned with the public welfare. Therefore, given the evidence of marketed demand fostered by a consumerism based economy, perhaps the most significant, powerful, and traction-inducing vehicle for instituting sustainability may be found in enabling conscious consumption at the individual level. Arguably, the conduit for conscious consumption would then be education not limited to defining sustainability, but inclusive of the rationale for sustainability, the patience requisite for implementation, and the acceptance of sustainability as a societal norm of behavior. However, the building block for conscious consumption is found in understanding the basis of present consumption decisions, ultimately the values that shape the behaviors that lead to observable economic outcomes (Venkatesan 2015).

Keywords Consumerism based economy · Conscious consumption
Sustainability · Sustainable bioeconomies

M. Venkatesan (✉)
Department of Economics, Bridgewater State University,
131 Summer Street, Bridgewater, MA 02325, USA
e-mail: mvenkatesan@bridgew.edu

W. Leal Filho et al. (eds.), *Towards a Sustainable Bioeconomy: Principles, Challenges and Perspectives*, World Sustainability Series,
https://doi.org/10.1007/978-3-319-73028-8_1

1 Introduction

1.1 Significance of Educating for Sustainability

In the United States, consumption expenditures contribute to over 65% of gross domestic product (GDP), which since the 1940s has been the international metric for economic progress. Given this linkage and the corresponding focus on GDP growth as a proxy for progress, consumption decisions can have a significant ripple effect throughout a single economy as well as the finite global resource base. Consider for example the use of milk cartons. Wax lined, printed paper milk cartons have been created for the transport and preservation of milk from the production to the consumption stage. However, the components of the carton were not developed with waste disposal in mind, rather increasing distribution and sales were the rationale for the carton. As a result, largely related to the focused basis of its creation, the milk carton serves a consumption purpose without consideration of the impact to the environment and potential future human and animal health due to its non-biodegradable or re-usable composition. This illustration on a broader consumption scale provides a simplified perspective to evaluate the underlying values captured in consumption decisions. From this perspective, production for consumption may be expressed as a myopic activity, focused on near-term satiation of a need or want to the exclusion of the evaluation of the impact or ripple effect of the satiation (Day and Aaker 1970).

The values embedded and communicated within demand and supply, determine the manner in which a need or want is attained. To the extent that there is no discussion of the values and behavioral factors assumed and reflected in demand and supply, arguably, implicit values, the values and the subsequent behaviors become endogenous to the economic system. Therefore, explicit awareness of present behavioral assumptions inclusive of the "unlimited wants" of consumers, profit maximization motivations of producers, and the understated resource depletion resulting from externalized costs, offer the potential to modify active and embedded behavior.

An understanding of economics specifically oriented toward enabling the development of rational economic agent behavior, can raise awareness of the significance of consumption behavior as the activity relates to sustainability, where the defining of sustainability is consistent with the economic perspective of resource utilization bounded by intergenerational equity, which by definition considers the needs of the present and the future relative to current resource capacity and regeneration ability. Awareness of resource limitations in conjunction with temporal allocation, in turn fosters the development and implementation of conscious and unconscious reinforcement of sustainability, which are the needed elements in driving a culture of sustainability.

1.2 Economics in Cultural Context

Economics evaluates human behavior relative to wants, needs and resource allocation within a natural environment. By definition, the parameters of the discipline include other life forms and physical resources needed to maintain both life and environmental regeneration. To the extent that a human culture incorporates non-human elements in decision-making, the economic system includes an understanding of the holistic inter-dependence of living and non-living elements of the planet.

Culture is a significant contributor to what is perceived as valuable and is the determining parameter in the designations that ultimately yield to resource allocation within a society. Given that culture is a learned behavior, culture can either promote or diminish any given society's understanding of the interconnectedness of human and planetary life, thereby determining the extent of the anthropocentric, or human-centered, perspective. The United Nations Educational, Scientific and Cultural Organization, UNESCO, defined culture as a significant component to attaining global sustainability.

Culture shapes the way we see the world. It therefore has the capacity to bring about the change of attitudes needed to ensure peace and sustainable development which, we know, form the only possible way forward for life on planet Earth. Today, that goal is still a long way off. A global crisis faces humanity at the dawn of the 21st century, marked by increasing poverty in our asymmetrical world, environmental degradation and shortsightedness in policy-making. Culture is a crucial key to solving this crisis (UNESCO 2000).

The inputs and outputs of economic systems are dependent on the value structures of a society and to the extent that economics explains observable phenomenon and proposes optimal outcomes, the discipline can be both responsible for the maintenance of an economic framework and also the catalyst for a change. Economic outcomes in essence mimic the values of the participants in an economic system.

Evaluating the historical cultural progression of human society can promote a stronger understanding of the economic relationship with resource allocation, both intra- and inter- society, and most importantly provide insights with respect to how perceptions of the world are shaped through cultural frameworks at a given point in time. The pace at which cultural attributes evolve may also provide a deeper understanding of why institutional and social frameworks may be inconsistent with the manifestation of contemporary challenges. Viewing economic thought or philosophy over time reveals the dynamic and cultural elements of society, as well as the basis of economic thought that remains in the principles literature in the present period.

1.3 Reconciling Economic Theory and Historical Context

The cultural attribution of value is a significant and arguably primary differentiator with respect to the variation in the perspective between societies of the quality of life for both human and non-human elements. Examples of surviving written works that provide a foundation or insight with respect to economic activities include Plato's *Republic* and Aristotle's *Politics*. The similarities in economic circumstances as described by the authors are consistent with the phenomenon observable today; however, the evaluation of human behavior as it applied to accumulation of wealth, stratification of society, and the role and impact of gratification were framed within an evaluation and discussion of moral philosophy and ethics, positioning Western economics up to the eighteenth century within the discipline of moral philosophy and politics. The evolution of the discipline continued through the modern era until the discipline formerly separated from moral and political philosophy through iteration as political economy to its present standalone context as economics. The observable mechanics of economic systems were the basis of discussion in conjunction with the human values, whether assumed as innate or culturally inspired. A connection between the qualitative and quantitative aspects of economic outcomes was articulated and addressed as an evolving and dynamic process. From this perspective economics discussions offered both a *normative* and a *positive* perspective, where the former provided opinions and values related to optimization and the latter described observable activity. At the present time, economics in practice has shed the normative element of the discipline opting for a positive attribution as a means to enhance its standing as a science. In essence, the focus on optimization has been to the exclusion of explicit evaluation of prevailing values. Given the significance of embedded values in conscious decision making, the lack of articulation of values may contribute to the implicit value of outcome based decision-making that only considers the optimization of the outcome rather than the impact of the outcome to others and future consumption.

The foundation for current economic thought can be found in the writings of Smith (1791), Bentham (1879), Ricardo (1911) and Marx (2009) along with many others. However, though all of these authors provided insights related to the human behavior contemporary to their time, the context of their writings has often been neglected in lieu of an adoption of an absolute meaning of their opinions. In essence, allowing the commentaries of these authors to embody a universal significance independent of time has arguably enabled the transfer of the theoretical modeling of a society specific from one period to another, independent of any assessment of the temporal evolution of behavior and underlying values.

To a large extent, the economic principles in practice have maintained the theories espoused by the writers and contributors to economic thought contemporary to the Classical period. Mill's *Principles of Political Economy* (2016) provided a summary of the contributions to economic thought by Adam Smith, David Ricardo and other significant thought leaders of the nineteenth century and became a standard text used in the study of economics into the early twentieth century.

However, of note is that the authors including Mill were relaying behaviors perceived in a society contemporary to their life and questioning aspects of the observed progress of the time including poverty, the role of money, and the potential impact of population growth. Their thoughts were debated discussions and their frameworks were not adopted as immutable facts. Additionally, the issues discussed were similar to those of predecessor Western societies and as evidenced in the moral philosophical discourses of Plato and Aristotle, nearly two millennia earlier. The evaluation of the human condition within a given social and economic framework prompted economic commentators to be both positive evaluators from the perspective that positive signifies reporting on observable and factual phenomenon and normative participants, where normative requires an expression of value judgment.

In contrast with the foundations of the discipline, the present instruction of economics, strongly influenced by Marshall, has eliminated the normative aspects of assessment, reducing economics to mathematical relationships that are addressed in absolute terms rather than in alignment with cultural attributions coincident with their development. Further, the seeming lack of attention to values and behavior incorporated within economic assessment has distanced the tangibility of economics, limiting understanding of the explanatory potential of economics and the application of economics as both a cause and a remedy of unsustainable practices. As a result, at present there is a need to promote and foster an understanding of the role of values in economic outcomes and the sustainability of observed outcomes.

1.4 Perception of Resource Value, Market Outcomes and Price

Economics is the social science discipline that evaluates the relationship between human wants and the resources available to satisfy them. In identifying and explaining the relationship between wants and resources, economists use broad generalizations related to human behavior, arguably the most significant of which relates to wants (Nelson 1995).

Wants are based on the premise that individual economic agents, individuals interacting within the general economy, will always seek to have more of desirable goods and services. Desirable goods include both normal goods, which are goods that an individual will continue to purchase as their income increases and luxury goods, which are goods that are not needed but are wanted to support an external display or perception of status or wealth. Not all goods are desirable, for example, inferior goods represent a classification of goods and services that will be reduced or eliminated by consumers as their incomes increase.

The behavior of wanting more, sometimes referenced as unlimited wants, is a social value, consistent with consumerism, which is defined as the focused act of consuming goods and services to improve utility, the economic concept that defines the benefit of consumption (Czech 2000). Insatiability is not representative of an intrinsic human characteristic but rather a learned behavior. This is an important point. If a behavior is learned it can be unlearned and a new behavior can emerge, which in turn can produce a different economic outcome.

1.5 Market Distortions, Externalities and Failure of Market Equilibrium

Market outcomes, price and quantity, are highly dependent on the information that consumers and suppliers have available. Informational asymmetry, where one party has more understanding or knowledge related to a good than another party, can create price and quantity outcomes that may not effectively consider scarcity. This results in market inefficiency, a situation where resource use is not efficiently allocated by the market. This is a significant issue and one that consumers are only beginning to understand. For example, abundance is a relative term but it is not inconsistent with scarcity; all resources are scarce. The perception of abundance without the recognition of inherent scarcity of resources can hasten resource depletion.

Resources are broadly defined as including all the inputs in the production of final goods and services that are ultimately tied to the satisfaction of a want. From this perspective, resources could include teak wood trees in the making of furniture; water in the production of soda; and cattle in the production of food. Typically, resources are classified into one of three groupings, which include: natural resources, human resources, and capital resources. Trees, water, and cattle are all natural resources. Human labor or entrepreneurship define human resources and capital resources consist of man-made objects that can be used to produce goods and services, such as factories and equipment. Regardless of the type of resource, all resources are finite and so by definition can be qualified as scarce (Choi and Ng 2011; O'Hara 1995).

Scarcity in economics essentially captures the relationship between wants and the access and availability of resources. For example, one could want a mango, see it hanging high on a tree but not have a ladder to reach it. The good in question is available but it is not accessible. Alternatively, one could stumble on a farmer's market selling mangos only to find that all the mangoes on display have been purchased. In this case the mangos are accessible but they are not available. Both of these examples highlight the temporal or time sensitivity of scarcity. In the first example, one could borrow or purchase a ladder but this will take time and in the second scenario, one can drive or walk to another market, but again, additional time will be required to satisfy the want.

Looking at time in a slightly different manner, a community could require lumber for the construction of new municipal buildings. The lumber required will result in the deforestation of one hundred acres. In satisfying the want for lumber today, the community limits access and availability of lumber from the one hundred acres over the time period required for the forest to regenerate, creating time based scarcity.

In a market system, access and availability establishes a perceived scarcity embedded within the supply of a good. Ultimately, the supplier's willingness and ability to sell a specified amount of a good at a prevailing price is assumed to capture the costs of production of the good, implicitly including the scarcity of inputs. As a result it is expected that the higher the degree of perceived scarcity of a resource, the higher its price and in the case of an input, the resulting price of the final good.

The production of goods by producers is based on a competitive framework. Additionally, the producer seeks to minimize costs and maximize revenue, to achieve maximum profitability. As a result of the focus on profitability, there is significant incentive for producers to externalize costs of production as a means of cost minimization. Externalizing costs can include pollution discharge, exploitation of regulatory differences between countries, overuse of natural resources, and limited waste disposal and reduction efficiencies. Though in the immediate period this may be beneficial to profitability, it may promote both short-lived unsustainable returns and longer-term environmental and social costs.

Consumers may not be aware of the implicit tradeoffs being made as a result of the production of a good. This informational asymmetry can be attributable to many reasons, including a belief that regulatory agencies guarantee safety, to just simply a lack of diligence when assessing goods. For consumers, reliance on market efficiency without an understanding of the embedded incentives of producers can promote negative externalities. In effect the pursuit of satisfying unlimited wants may include effectively delegating environmental and social stewardship to producers whose incentives may not include the evaluation of these parameters. The end result is most readily seen in natural resources, where under pricing due to lack of inclusion of scarcity can lead to extinction or elimination of a resource's availability.

In a market driven economy, such as in the United States, the market is credited with efficiently determining the price of an item by implicitly incorporating the costs associated with production. When consumers or producers face low prices for consumption and input purchases, respectively, and the underlying belief is that the price being paid is fully reflective of the cost of the item being purchased, there is less of an incentive for efficient use and higher potential for waste. Price effectively becomes a measure of a resource's worth. When asymmetric or incomplete assessment of scarcity is prevalent, price may not properly indicate the cost of the resource being consumed.

In some areas of the world, forested land has been perceived as abundant and the resulting price for land has been limited to the perception of present period abundance. The net result of the perception has been excessive global deforestation, resulting in present period-pronounced scarcity in some regions. Decades will be required to promote regrowth of the same lands. Had prices considered the impact of forest harvesting, or the price of temporal scarcity, demand would have been lessened. Both consumption and production could have promoted efficient market pricing leading to sustainable resource use, all from this simple inclusion.

Demand and supply yield market outcomes that are assumed to represent an efficient allocation of resources. The price at which the quantity demanded equals the quantity supplied is therefore expected to embody the cost associated with the production and consumption of the good or service. However, production and consumption are not limited to the transactional nature of exchange of the final good at the determined market price. In the process of production and consumption there are costs that are not factored that impact the well being of the economy at large and these are referenced as externalities. In essence, externalities arise when an individual or firm engages in activities that influence the well being of others and where no compensation in provided in exchange for the imposition.

Typically externalities are characterized as negative, signifying that the externality yields an adverse outcome. These externalities are referenced as being *negative externalities*. However, there is a potential that a positive outcome could be generated leading to a positive externality. In the discussion of externalities it often assumed that market participants accept the externalities generated by their actions as acceptable due to their focus on immediate gratification of their needs. For the producer this equates to externalizing the cost of disposal of waste products into waterways and the air where no cost is directly borne to adversely impact profits but qualitative costs are assessed that may impact the enjoyment and longevity of multiple life forms and generations of human life. For the consumer the externality can be evaluated in the indifference to waste creation at the point of the consumption decision or even the externalities associated with the production of the good or service being purchased. In the case of the former, the cost of disposal of packaging material is typically marginal to zero, relatively negligible, but disposal creates a negative externality in the landfill, incinerator or recycling plant that could have been avoided with a thoughtful exercise of demand.

At present, the type of internalizing of externalities that has occurred has been limited to quantifying the externality to an overt cost. However, to the extent that the costs may remain unassessed and the market mechanism is not cognizant and focused on the elimination of the externality-based cost, rather the minimization of overall costs, this process has yielded suboptimal outcomes. For example, assume that a firm produces ambient pollution as a result of incineration of waste. If a governmental regulatory body institutes a fee or cost for pollution, effectively charging the firm for the ability to pollute the air, the producer is able to delegate responsibility for environmental stewardship to the price of pollution. Additionally, depending on the price elasticity of demand for the service offered, the producer may be able to not only transfer the costs now associated with polluting activity to the consumer, but may also be able maintain the pollution level. Assuming that the consumer is inelastic, in this example the negative externality related to internalizing the cost has not changed, instead only the responsibility of pollution has been transferred to a cost, revenue to the regulating body has been generated, and the consumer has suffered erosion in their overall disposable income and purchasing power.

The same type of scenario exists with a permit trading program, where in effect permits are issued for a specific amount of externality emission, allowing economic agents to trade and thereby optimize through again cost minimization. However, the cost minimization is founded on the presumption or delegation of the permit system to fostering socially optimal outcomes, again, relieving the economic agent engaged in the creation of the externality form being directly accountable for qualitative actions. Additionally, the trading of permits assumes that optimal financial outcomes equate to optimal environmental and social outcome due to the aggregated assessment of pollution. However, to the extent that pollution is not distributed evenly and certain locations may have a disproportionate concentration, the permit systems fails to generate a socially optimal outcome. This may be compounded by the impact of inelasticity, which may allow for the transfer of costs of implementation of the permit program to the economic agents the program was designed to protect.

Externalities are defined as a type of market failure based on the premise that optimal social outcomes result from individual economic agents acting in self-interest. However, if instead of being a market failure, externalities could be evaluated to assess and develop an optimizing strategy between individual interests and enhanced social outcomes, externalities could be internalized within the market model as a modification of preference. Perhaps externalities only indicate a lack of holistic awareness on the part of the consumer and producer or a cultural bias toward immediate gratification. These characteristics can be potentially modified through education. Optimal and universally acceptable strategies could then be adopted to promote sustainability.

The success of this internalization strategy relies on the development of the educated rational economic agent as a consumer. If consumers are aware of the responsibility inherent in their consumption and are aware of the environmental and social impact of production processes, consumer demand can create the coalescing framework to augment preference to exhibit demand for sustainably produced products (Boran 2006; Schweitzer 1981). The augmentation in demand does not allow for the opportunity of delegation of responsibility of pollution capacity to a cost or alternatively, the incorporation within a cost minimization framework, as a result, the change in preference and subsequent modification in demand promotes the development of market outcomes that are environmentally and socially optimal from the position of what is supplied.

Resources such as air and water have no market price and are considered to be abundant. On the surface, these resources may appear to be unlimited however, increased population pressures along with externalized costs related to production, such as pollution, have diminished the availability of both potable water and clean air. How could this have occurred?

The lack of price, a market model promoting the focus of profit maximization, and promotion and validation of unlimited wants are largely responsible. Consumers have effectively allowed supply to determine demand by not imposing restrictions on how goods can be produced. Producers have focused on short-term profitability in lieu of long-term strategic resource utilization. In the short-run both consumers and producers have benefitted but the cost of consumption and profitability was externalized to other nations, the environment, and future generations. For example, in the seventeenth century, North American coastal waters were described and recorded as being rich in quantity and diversity of fish; the perception of abundance led over time to overfishing and presently many varieties are endangered or at the risk of extinction. The cost of fishing included the human and capital costs not the replenishment costs. This yielded an ability to maintain artificially low prices, greater yields for profitability (over fishing), and waste.

1.6 Market Prices, Values, and Common Goods

An understanding of the perception of scarcity and abundance provides a strong foundation to understanding supply, demand, and market outcomes as these concepts relate to resource allocation and sustainability. To the extent that consumers delegate responsibility for sustainable consumption to producers and producers are focused solely on profit maximization increased understanding of the responsibility inherent in consumption may provide a catalyst for increasing sustainable production, consumption and development. As holistic evaluation of consumption is an assumed behavior of the rational economic agent, strengthening the understanding of the role of consumption may be significant in enabling the development of the rational economic agent.

Supply and demand reflect the amount that producers or suppliers of a good or service are willing and able to sell at a particular price and the amount that consumers of a good or service are willing and able to purchase at a particular price, respectively. Though on the surface the concepts of supply and demand appear simple the characteristics that determine the explicit willingness and ability can be complex. The complications can arise as a result of differences in the preferences, behaviors, cultural values, financial capacity, as well as resource access and availability to the production process as these relate to suppliers. For consumers or demand, the complications can also be attributed to preferences, behaviors, cultural values, financial capacity and wealth perception, as well as the perception of value and price, along with access and availability, of other substitute and complementary goods. Where and how the supply and demand interact with each other define a market. A market is comprised of a group of producers (supply) and consumers (demand) for a specific good or service, who collectively, as part of their exchange process, determine the market price or equilibrium price of a good or service.

Price is the natural outcome of the supply and demand relationship. It is indicative of the value of a good based on a consumer's assessment of the costs and benefits of purchasing the good. As consumers become increasingly aware of the environmental and social costs of production, the prevailing price may be corrected either through regulatory imposition of the costs of externalities within the market mechanism or via consumers, who will opt to purchase goods not on price but related to holistic production costs.

It is important to note that the market relationship is dependent on information and understanding of the limits of duty of care. The outcome of the market relationship, price and quantity, can only reflect the embedded preferences and social values depicted in demand and supply. If the market outcome does not meet expectations, the market model is not to blame; rather the prevailing value structure may be the flaw (Bishop 1993).

Value in this context is related to how resources are valued from the perspective of the quality of care and maintenance we would be willing and able to provide to ensure the protection of the resource. The use of the word "value" is not directly based on market quantification but expresses the hierarchical importance that consumers and producers would attribute to a resource; examples may include the environment, human health, and animal welfare.

Every day consumers make decisions with the collective strength of aggregated individual demand. These decisions influence supply and demand going forward, including the ability of producers to develop new goods and services, as well as resources and technological advances to satisfy both existing demand and projected future demand. Demand is a powerful catalyst in the evolution of market outcomes. However, to a large extent the power of demand is limited both by the fragmentation of consumers due to limited opportunities for coalescing around specific interests and limited consumer understanding of the inherent power of aggregated consumption decisions. From this perspective, understanding how the market functions and the power of consumption in creating sustainable economic outcomes is one aspect of developing into a rational agent.

The values embedded and communicated within demand and supply, determine the manner in which a need or want is attained. To the extent that there is no discussion of the values and behavioral factors assumed and reflected in demand and supply, arguably, implicit values, the values and the subsequent behaviors become endogenous to the economic system. From this perspective, explicit awareness of present behavioral assumptions inclusive of the "unlimited wants" of consumers, profit maximization motivations of producers, and the understated resource depletion resulting from externalized costs, offer the potential to modify active and embedded behavior.

1.7 Conscious Consumption and the Sustainability

The explicit discussion of the embedded assumptions guiding the behavior of the decision-maker is typically not a part of the economic education process. As a result, to the extent that individual economic agents, producers or consumers of a good or service, are bounded by rationality that does not include addressing the impact of externalized or non-quantified costs, the economic discussion does not promote or position the assessment of alternative outcomes. Implicitly and endogenously, the economic discussion establishes and maintains consumption to production circular flow, focusing on the gratification of consumption and profit taking from production, seemingly eliminating assessment of externalities and holistic dynamics. Returning to the milk carton example provided in the Introduction section, the economic discussion would be limited to the utility gained from consuming the milk and the corresponding profit maximization of the producer. Waste would be regarded as an externality rather than an endogenous aspect of the decision making process. Additionally, costs are priced into the product through efficient market assumptions. In net consumers would expect that the purchase price is indicative of the holistic cost of the product and producers would view production costs as being related to market priced inputs not environmental impacts during or as part of the life cycle of the good.

Economics, in present practice, evaluates efficiency with respect to the use of resources to maximize production and consumption, not by the moral desirability of the physical methods and social institutions used to achieve this end. The factors that are included in an economic evaluation are limited to the tangible quantifiable costs and costs are overlooked where either a market or regulatory oversight has not provided a monetary justification. From this perspective, the impact of consumption decisions on the environment, economic disparity, or endangerment of other species is not an issue. The market mechanism disenfranchises the consumer from the welfare of those impacted by his/her consumption and promotes the perception that price alone is indicative of the true cost of a good. The possibility that consumption should be reduced because the act of consumption is not good for the soul, or is not what actually makes people happy, has no place within the economic value system. The underlying assumption is that consumers are driven to want more. As a result, economic modeling assumes that reduction in consumption in the current period is only addressed through the lens of an increase in consumption in a later period. That the assumption of insatiable want may be taught a learned behavior, reinforced through a market model is not even addressed in economics.

A general and seemingly applicable assumption is that consumers and producers maximize the benefit related to the opportunity accessible in their particular circumstance. The desire to reach an optimal outcome for a given point in time, as has been noted before, is subjective and specific to how these economic agents view the concept of maximization, which in turn is likely to be highly correlated with cultural values. For example, in Indigenous societies there is evidence that a balance between present and future periods along with that of the environmental system, as a whole, was included in decision-making and optimization. In present consumerism fostered economies, the cultural values are less likely or unlikely to incorporate environmental and social justice parameters proactively. The focus of observable and marketed consumption is immediate gratification. However, as consumer awareness of both the impact of consumption and the power of consumption to modify and catalyze economic outcomes increases there is growing evidence of a shifting cultural paradigm to one of sustainability.

2 Final Considerations

Markets do fail to produce optimal outcomes. Sometimes this is due to the myopic focus of market participants as in the case of externalities and in other circumstances it can be attributable to the lack of excludability as in the case of common goods. To some extent cultural values dictate the significance of the adversity related to the creation of externalities or abuse of common goods. The use of market models has been the regulatory mechanism to modify socially non-optimal outcomes, but through relying on the market mechanism rather than simultaneously including mechanics to promote cultural change, the majority of regulatory interventions to date have had limited to questionable success (Venkatesan 2016).

A constituency with an understanding of the holistic relationship between consumption and sustainability and having engagement in government are foundational elements in achieving and maintaining sustainability as a cultural norm. For long-term traction, sustainability is dependent upon holistic and routine evaluation of economic and societal frameworks. These frameworks need to be assessed and modified as part of an on-going continuous improvement process. Fundamentally, what may have been viewed as appropriate action at a point in time may no longer serve the same purpose due to changing environmental, social and cultural parameters (Shah 1999). However, the members of a society have to be both empowered and cognizant of the need for this type of evaluation in order for efficiency and ultimately sustainability to be a realized inter- and intra- generational attribute. From this perspective, the deployment of consumer education programs targeted at defining responsible demand, conscious consumption, are a requisite foundation for sustainability.

References

Bentham J (1879) An introduction to the principles of morals and legislation. Clarendon, Oxford
Bishop RC (1993) Economic efficiency, sustainability, and biodiversity. Ambio 22(2/3):69–73
Boran I (2006) Benefits, intentions, and the principle of fairness. Can J Philos 36(1):95–115
Choi S, Ng A (2011) Environmental and economic dimensions of sustainability. J Bus Ethics 104(2):269–282
Czech B (2000) Economic growth as the limiting factor for wildlife conservation. Wildl Soc Bull 28(1):4–15
Day GS, Aaker DA (1970) A guide to consumerism. Journal of Marketing 34(3):12–19
Marx K (2009) Das Kapital, vol 1. Regnerny Publishers, Washington, District of Columbia
Mill JS (2016) In Ashley WJ (ed) Principles of political economy with some of their applications to social philosophy, 7th ed., 1909. Longmans, Green and Co., London
Nelson RH (1995) Sustainability, efficiency, and god: economic values and the sustainability debate. Ann Rev Ecol Syst 26:135–154
O'Hara SU (1995) Sustainability: Social and ecological dimensions. Review of Social Economy 53(4):529–551
Ricardo D (1911) The principles of political economy and taxation. J.M. Dent & Sons, London
Schweitzer A (1981) Social values in economics. Rev Soc Econ 39(3):257–278

Shah M (1999) Synthesis of ecology and economics: towards a new theoretical paradigm. Econ Polit Wky 34(46/47):3293–3298

Smith A (1791) An inquiry into the nature and causes of the wealth of nations. Tourneisen and Legrand, Basil, p 22

UNESCO (Culture Sector) (2000) World culture report. United Nations, Paris

Venkatesan M (2015, March 15) Values, behaviors and economic outcomes. Retrieved from https://www.youtube.com/watch?v=ipq5owxlU0g

Venkatesan M (2016) Economic principles: a primer, a foundation in sustainable practices. Kona Publishing, Mathews, NC

Service-Based Bioeconomy—Multilevel Perspective to Assess the Evolving Bioeconomy with a Service Lens

Päivi Pelli, Jyrki Kangas and Jouni Pykäläinen

Abstract Increasing role of services is often described as the increased share of services sector employment and value added in economy, although research on services addresses also the multiple ways how services are embedded in socio-economic processes and innovation. Integration of products and services is perceived with potential to improve efficiency in manufacturing as well as sustainability of operations. Improving technologies provide new means to organize production and define how value is created, distributed and captured. This paper seeks to assess the blurring borderline between manufacturing and services and its impact on the primary and processing industries within the evolving bioeconomy conceptualizations. An analytical framework is presented based on service research in the marketing discipline and the multi-level perspective to socio-technical changes; 'service-based bioeconomy' highlights the role that the further downstream industries have in defining the future bioeconomy. The analytical framework is illustrated in an empirical study context of the Finnish bioeconomy strategy with three mini-cases where the value propositions of traditional forest based industries, emerging bio-industries and their further downstream customers are analyzed: biorefineries' supply to sustainable textiles, wood-based solutions for sustainable built environment, and forestry solutions for biomass production and beyond.

Keywords Bioeconomy · Services · Servitization · Socio-technical change
Finnish bioeconomy strategy

P. Pelli (✉) · J. Kangas · J. Pykäläinen
School of Forest Sciences, University of Eastern Finland,
PO Box 111 (Yliopistokatu 7), 80101 Joensuu, Finland
e-mail: paivi.pelli@uef.fi

J. Kangas
e-mail: jyrki.kangas@uef.fi

J. Pykäläinen
e-mail: jouni.pykalainen@uef.fi

W. Leal Filho et al. (eds.), *Towards a Sustainable Bioeconomy: Principles, Challenges and Perspectives*, World Sustainability Series,
https://doi.org/10.1007/978-3-319-73028-8_2

1 Introduction

Today the services sector contributes 70–80% of national value added and employment in developed countries. The picture, however, is more diverse, and services can be found embedded in all socio-economic activities (Coombs and Miles 2000; Bryson and Daniels 2010). Services functions are inherent in manufacturing, and 40–50% of manufacturing employment is in fact service-type occupations such as management, accounting, marketing, research and development, legal and other services (Pilat et al. 2006; Manyika et al. 2012). The input-output analyses show that the services sector, i.e., the input acquired from external service providers, contributes to 40% of realization of the manufacturing output (Stehrer et al. 2014; De Backer et al. 2015). Outsourcing of activities from manufacturing to external service providers has made services visible in economic terms but also changed the dynamics of value creating networks; services companies not only provide the services previously produced in the manufacturing companies, but seek to improve their performance for example by adopting new technology.

Furthermore, services are increasingly produced by the manufacturing companies themselves; analyses on international company financial data show that approximately 30% of manufacturers with over 100 employees offer services (Neely et al. 2011). The phenomenon, called as *servitization of manufacturing* (Vandermerwe and Rada 1988), and the modes how companies transform their operations towards services have been investigated in the engineering, management, marketing and design fields. Wider adoption of enabling technologies, such as digital and sensor applications, has been a driver for developing new products and services, making the tangible and intangible part of company offerings ever more intermingled (Chang et al. 2014; Gallouj et al. 2015). From a business-making perspective the increased knowledge of customer or user processes helps to ensure company competitiveness and sustain revenues (Oliva and Kallenberg 2003), but it may also lead to new business models to connect the supply chains to customer resources and to improve reaction to the changing customer needs (Chesbrough and Spohrer 2006; Bitner and Brown 2008; Ostrom et al. 2010). From a wider societal perspective bundling of products and services may improve resource efficiency and sustainability of operations, thus, reducing also the environmental impact (Mont 2002; Tukker 2004).

Instead of addressing the tangible and intangible production separately within the company processes, integration of products and services, or product-service systems, have become analyzed as a systemic change affecting both the company in-house operations and the configurations of resources across its value networks (Baines et al. 2009; Boehm and Thomas 2013). Phrases like 'next production revolution' (OECD 2016) or 'service economy 3.0' (Chang et al. 2014) emphasize the role of information systems as an enabler to rethink how value is created, distributed and captured. From this perspective, a *service(-based) economy* is not an issue of mere increase in the volume of services, but a more profound change how production, distribution and usage can be organized. It is possible to improve the connection between upstream and downstream of production processes, create benefit to individual parties involved in the value networks, as well as support continuous development of processes.

Bioeconomy is promoted as a new production paradigm; shift from a petroleum-based economy to processes using renewable biological resources and new (bio)technological solutions (e.g. EC 2012). Yet several definitions exist for a bioeconomy due to varying emphases by different actors, such as industry federations and NGOs at international and national levels (Schmid et al. 2012; McCormick and Kautto 2013; Staffas et al. 2013) and the different research disciplines active in the field (Pfau et al. 2014; Bugge et al. 2016). The evolving bioeconomy connects between the traditional production sectors, such as agriculture, forestry, fisheries, forest industry and food industry, as well as the industrial activities emerging with new technologies, new products and processes, for example, in the chemical, medicine, energy and textiles industries.

Developments related to the increasing role of services have gained little attention with regard to the bioeconomy development (Pelli et al. 2017). In fact, if the definition of a bioeconomy is strictly limited to the *production and conversion of renewable biological resources* (EC 2012), services remain largely invisible. It is challenging to define the bioeconomy-related services based on the existing statistical metrics, not to say to distill the information about these activities from the economic data. Services can be found in all areas of the evolving bioeconomy, such as, production related maintenance, delivery channels, trade, retail, R&D and expert services, but also nature-based tourism (Rönnlund et al. 2014; Efken et al. 2016). Thus, when focus is mainly on tangible resources (biomass) and (bio-)products, services are support functions rather than opportunities for new businesses or sources of innovation. However, totally new services can be envisaged to emerge due to the technological advancements in bio- and nanotechnologies in a similar way how information and communication technology (ICT) led to emergence of new types of services, and its convergence with other technologies created new operation modes (Chang et al. 2014; Gallouj et al. 2015). Not only better data and metrics are needed about the role of services, but also new conceptualizations to understand the developments ongoing in production at large: services effect on the evolving bioeconomy, but also bioeconomy effects on what will be defined as services or 'as service' in the future.

This paper investigates the increasing role of services in production as a phenomenon developing parallel to the evolving bioeconomy, and seeks for a means to assess the developments and their possible interlinkages. First, transition to a bioeconomy is described as a system change: the multi-level perspective to socio-technical change (Geels 2002, 2005) is applied to structure the analysis. An integrative approach to products and services is based on service-dominant logic by Vargo and Lusch (2004, 2008) which gives an equal importance in value creation to the customer's networks and the producer's supply chain. The analytic framework —multilevel perspective to assess the evolving bioeconomy with a service lens—is illustrated in an empirical study context of the Finnish bioeconomy strategy.

The service lens conceptualizations stem from the marketing discipline but they are also utilized, for example, in the analyses of operations management as well as technology and engineering management; instead of understanding value created in production, added layer by layer to the goods and measured as *value in exchange*, value is assessed as *value in use* from the viewpoint of the beneficiary.

2 An Analytic Framework to Transitions to a Service-Based Bioeconomy

2.1 Multi-level Perspective to System Change Towards a Bioeconomy

Transition to bioeconomy can be understood as result of co-evolution of economic, social, technological, institutional and ecological developments (Bosman and Rotmans 2016). Renewable biological resources as well as (bio)technological solutions can be utilized in novel ways for fulfilling societal functions, such as energy, mobility, communication and housing. Multi-level perspective (MLP) to socio-technical change (Geels 2002, 2005) has become utilized in transition studies as a model to iterate dynamics of socio-technical changes as well as sustainability transitions that stretch over several decades and are difficult to manage or predict their outcome (Elzen et al. 2004; Geels 2010; Smith et al. 2010; Markard et al. 2012). Transitions are described across three nested levels: *niches* where radical innovations take place, *regimes*, i.e., established dominant socio-technical systems that enable but also constrain radical innovation, and *landscape* which represents the overall trends that are outside the influence of the regime and niche players. Examples of the macro level drivers often indicated in the bioeconomy strategies are the increasing uncertainties due to scarce resources and climate change, increasing demands due to growing global population, as well as changes due to technology and scientific discoveries (e.g. in EC 2012).

As described in the introduction, rather than addressing it as a macro-level trend, the increasing role of services is here understood as a phenomenon *within* the existing production modes and organization of societal functions. The phenomenon has evolved over several decades in the manufacturing and services sectors for organizing and managing necessary infrastructures, knowledge and material flows. Landscape-level drivers for these developments are technological development, international trade, and increased interdependencies of the networked operations (see e.g. Chang et al. 2014; de Backer et al. 2015; Gallouj et al. 2015; OECD 2016). The phenomenon is perceivable also in the primary production: forestry is based on data and monitoring of forest resources and on models, for example, to assess the impact of climate change and the developments in the forest products markets on the future use of forests. While these activities have been support services for administration and forestry production, they provide also international business opportunities, such as consultancy or technology-enabled services for management of natural resources and for sourcing products to the global markets (Näyhä et al. 2015; Pelli et al. 2017).

The bioeconomy strategies set demands for the natural resources sectors, management of their processes, as well as increased R&D for new (bio)technologies, new materials and products. These processes extend beyond the traditional industry borderlines and affect the established regimes for example for energy or chemicals sectors. R&D projects, pilot and demonstration projects are niches where radical innovations are tested and experimented. For example, biorefineries, either as new industrial entities or deploying the biorefinery technologies into existing industries, such as pulp mills, call for new types of partnerships (Pätäri et al. 2011; Näyhä and Pesonen 2014; Bauer et al. 2016). New materials or products of biorefineries are not necessarily directly applicable to the further downstream production but require also adjustments in the customer industry processes. Development of solutions necessitates interaction across several levels of the value networks, and deeper understanding of the customer processes along the supply chains. Services, such as necessary resource management tools, knowledge flows or data and monitoring systems, evolve integrated into the tangible production processes, and they affect both the upstream supply chain and the downstream customer processes. Processes become re-configured through resource integration and mutual learning of multiple actors.

2.2 *Service Lens: An Integrated View on Products and Services*

Instead of focusing on the material flows or technology development, research on services seeks to analyze the above mentioned processes as socio-economic transactions: innovation is not only sought by R&D for new technologies or new products, but innovation is understood more processual and embedded in the interactions where products and services are exchanged (Coombs and Miles 2000; Gallouj and Savona 2009). The service-dominant logic (SDL) by Vargo and Lusch (2004, 2008) provides an integrated view on products and services which is applicable both to business-to-business, business-to-consumer and consumer-to-consumer contexts—or as they call it: to any actor-to-actor context.

An industrial logic—or a 'goods-dominant logic' (GDL) by Vargo and Lusch—is based on linear value-added chains, and it does no more fit well to networked business-to-business operations (Håkansson and Snehota 1989), to value production in volatile competitive environments (Normann and Ramírez 1993), or to value creation enabled with digital means and technology engineering (Chesbrough and Spohrer 2006; Maglio and Spohrer 2008). Linear value-added chains are part of the production processes, but they are nested in processes which require more systemic view on the resource integration and re-configurations. SDL defines service as basis of any socio-economic exchange: using one's competences for the benefit of another. Resources are integrated in value co-creation of multiple actors, i.e., producers, customers, users, suppliers, and wider networks, which are nested in institutional arrangements, that is, the macro-level rules of the game (Vargo and Lusch 2011, 2016). Instead of tangible assets such as raw materials, it is knowledge and skills that are the primer resources and sources for competitiveness.

SDL makes explicit the role of customers and users both as beneficiaries and crucial actors in value co-creation; customers' networks are equally important than the producer's supply chain in providing and integrating resources for value creation (Cova and Salle 2008; Payne et al. 2008). Value is proposed from one actor to another: from producer to customer, from customer to their own customers and stakeholders involved in utilizing the company offering. Servitization of manufacturing becomes a more extensive change than the manufacturers merely adding services on their tangible products; it requires changes both in customer interactions as well as organization of the company in-house operations (Smith et al. 2014). Changing the role of the customer from consuming to co-creating value can be a source of innovation in resource use, developing the existing or creating new resources (Michel et al. 2008; Skålen et al. 2015).

2.3 *Value Proposition as a Tool to Connect Micro, Meso and Macro Levels*

Value proposition by a company seeks to differentiate the company offering from its competitors either based on the product or production features or on the benefits it provides within the customer processes (Anderson et al. 2006; Johnson et al. 2008). The company offering can include both goods and services produced in-house or by external service providers, and its benefits consider both the benefits and costs for the customer in using the company offering (summarily in Payne et al. 2017). Thus, value propositions also define what is the role expected from the customer. In a commodity-type business value propositions can be defined by the producer, while solutions-based business prerequisites more interaction; the more integrated the producer and customer processes become, the more input is needed from the customer already for defining the value proposition (Mathieu 2001; Oliva and Kallenberg 2003).

At a generic level value proposition can be described as a *statement of benefits* to the customer (Cova and Salle 2008). Taking this generic stance, value propositions can be defined both at the macro level of a system and at the company level or in a specific customer context. Also bioeconomy visions and strategies contain value propositions: implementation of the EU strategy (EC 2012) contributes *benefits to the society*, such as improving the management of Europe's renewable biological resources, decoupling economic growth from resource depletion and environmental impact, making the transition towards a post-petroleum society and tackling of the challenges associated with climate change, land use and global food security. Biotechnology development promises a means to radical innovation compared with the present-day socio-technical regimes. In primary production sectors, in turn, *services to society* refer to the sustainable management and use of natural resources, contribution to rural employment and income, as well as maintenance the multiple goods and services that the natural ecosystems provide (Schmid et al. 2012; Pelli et al. 2017). In other words, there are different value propositions within the bioeconomy conceptualizations. The statements of expected benefits —value propositions for the beneficiaries—can be utilized to investigate parallel definitions across industrial sectors that contribute to socio-technical transitions, and to zoom in and zoom out between the different levels of analysis, such as an individual company and the bioeconomy vision and strategies. See Table 1 summarily about the analytic framework.

In the following the Finnish bioeconomy strategy is presented with three mini-cases to demonstrate the analytical framework. The mini-cases were selected so that they illustrate development of cross-sectoral solutions. The case descriptions

Table 1 A multilevel perspective to assess the evolving bioeconomy with a service lens

[Landscape level trends and drivers that affect both the meso and micro levels]	
Bioeconomy—profound change in economy utilizing renewable biological resources and processes as well as biotechnology solutions to fulfil the societal needs	Value proposition as the 'benefits to society' that the transition would contribute Sources of data: visions, strategies, funding programmes
[Meso level] Established socio-technical regimes, for example how housing, energy or mobility is organized, enabling but also constraining radical innovation and new operation modes Bioeconomy solutions call for collaboration across sectors and connecting the upstream (biomass) and downstream of value chains (industrial sectors using the new materials, products and processes)	Value propositions of traditional industries, such as primary production ('services to society') and further downstream industries (benefits to the customer industries and to the stakeholders) Sources of data: industrial sector specific visions, strategies, development programmes
[Niche-level] Pilot projects, demonstrations and R&D to develop, test and experiment radical innovations Biotechnology innovations take long time from research to markets; many parallel technologies are at an early stage of development and also competing solutions are developed	Value propositions of the consortium partners (benefits to the funding organization/ strategies; benefits to the consortium partners/ resource integration; benefits to other stakeholders/differentiation from the competing solutions) Sources of data: project documents, participants
Individual companies and organizations	Value propositions of individual companies: benefits and costs to the customer, sharing resources, package of goods and services, differentiation from the competitors' offerings Sources of data: company documents, company representatives

and analyses are based on information available from public sources, such as strategy documents, financing programmes, technical reports and company materials. An entry point for the analyses was the traditional forest sector actors' role in the niche as industrial partners for technology development before market entry (textiles), as a solution provider for market entry of new solutions (construction), and as coordinator of platform development for more efficient processes (forest data). Second, the collaborating partners and their roles in the niche were identified. Information on the funding programmes and partners' own media releases revealed other niches where the partners participate, as well as niches where competing solutions are developed to provide the expected benefit. The mapping aims not to be exhaustive; focus was placed on the collaboration partners from different stages of the upstream and downstream production and their value propositions and parallel niches were identified. The analyses are concluded to discuss the question: how the increasing role of services, as a phenomenon embedded in existing production processes, interlinks with the evolving bioeconomy?

3 Empirical Study Context: The Bioeconomy Strategy in Finland

3.1 Forest-Based Bioeconomy Between the Old and the New

The Finnish bioeconomy strategy defines bioeconomy as an economy that relies on renewable natural resources to produce food, energy, products and services (Ministry of Employment and the Economy 2014). Forests and forest-based industries are emphasized due to the available resources and their role in the economy (Antikainen et al. 2016; Bosman and Rotmans 2016): the pulp and paper and wood products industries have traditionally had very important roles in the economy. Knowledge base, practices and structures for management and use of forest resources have developed over a long period of time in response to the operating environment, for example, the high share of private forest ownership in Finland. The existing industries have a crucial role for ensuring stable functioning of raw material markets: wood from different stages of a forest lifecycle provides raw materials for different types of products, and residues from one production process are typically used for the production of other products, such as from sawmilling to pulp industry and energy production. In other words, changing one element of the traditional forest industry supply chains affects the future conditions both for forestry and for other woody biomass based industries. Promotion of wood construction within the bioeconomy strategy is part of this equation, and contributes on its part to the raw material supply of biorefineries which are being planned in Finland with domestic and foreign investments.

Bearing this starting point in mind, the output of the *Finnish bioeconomy* already currently exceeds 60 billion euros, and more than 300,000 persons are employed by the *bioeconomy sector* (Ministry of Employment and the Economy 2014). These industries bring more than 25% of the total value of exports in Finland. The fact that bioeconomy is nothing new in the forest sector disguises the strategic goal that in the future, besides wood products, paper and energy, the biomasses originating from forests are expected to be increasingly used in textile fibers, medicines, chemicals, functional foods, cosmetics, biofuels, and new bio-product innovations substituting the use of fossil resources, for example in plastics. At the same time forests are seen increasingly important for other purposes, such as biological diversity, climate change mitigation, nature tourism, game and wildlife management, and multiple other ecosystem services.

The Finnish strategy explicitly mentions also services sectors in developing the bioeconomy as well as new business models on ecosystem services, for example in the construction sector and in water and landscape management (Ministry of Employment and the Economy 2014). Although new production modes, servitization or product-service systems are not mentioned, the Natural Resource Strategy for Finland (Sitra 2009) already envisaged that bioeconomy would mean for the forestry sector a shift from high-volume products maximizing economies of scale to specialized products in *fragmentary or integrated production*, and further, to *integrated solutions* where *side benefits*, such as reduced greenhouse gas and nutrient emissions, can be identified, measured and defined as *tradable services* (Sitra 2011). Such change would require new mindset to design and capitalize the new modes of production. The following cases illustrate the questions that arise from the evolving production modes and the sources of tensions within the established regimes. Table 2 summarizes the findings for all three mini cases, and further details for each case are available in the annexed Tables 3, 4 and 5.

3.2 *Biorefineries Supply to Sustainable Textiles*

The new bioproduct mill by Metsä Fibre Ltd. in Äänekoski is the largest investment of the forest industry in Finland, with the total investment of approximately EUR 1.2 billion. It is inviting businesses of various sizes to its 'ecosystem', each specializing in different bioproducts or services at the same mill site. Main product of Metsä Fibre Ltd. itself will still be softwood pulp, but it also produces heat and electricity. Aim is that all side streams from the bioproduct mill are utilized in the ecosystem to greatest extent possible, for example into tall oil, turpentine, cosmetics, different other green chemicals, and bio-composites. One of the possible product development lines is textiles, and the company has announced its collaboration with the Itochu Corporation, a trading house which owns 24.9% of Metsä Fibre Ltd. and is also its pulp marketing agent. The company has participated, together with other large forest industries in Finland, in technology programmes developing new uses of pulp: the FuBio Future Biorefinery programmes

Table 2 Summary of findings from the three mini-cases: the value propositions with regard to sustainable textiles, sustainable built environment and forestry solutions

Landscape level: scarce resources, increasing uncertainties; new technology applications, new production modes; renewable resources, bio-technology as one of the Key Enabling Technologies			
The Finnish bioeconomy strategy (2014): objective to generate new economic growth and new jobs from an increase in the bioeconomy business and from high added value products and services while securing the operating conditions for the nature's ecosystems			
Meso level: established socio-technical regimes	Textiles sector: sustainable processes; consumer protection; brands	Construction industry and the real estate sector: to promote good construction; resource and energy efficiency	Forestry sector: to promote active and sustainable forest management; profitability of forestry
Niche level: R&D/pilot projects	**Ioncell-F—method for cellulosic fibers from ionic liquid solution**	**Wood City—sustainable urban quarter**	**Forest Big Data—next generation forest resource management system**
	Alternative method for viscose production; improved solution of cellulose-based fibers to the increasing textile materials demand	Functional premises in a central location; showcase of urban wood construction	Efficient utilization of forest data to support the bioeconomy development
Examples of individual organizations (key partners)	Universities and technology development agencies: new technologies, new materials, new products and processes	Construction company: consistent and flexible construction project process that provides economic benefits, faster implementation schedule and better equivalence to the client's needs	Coordination by the forest-based sector organizations (both public and private actors): efficient processes, stable raw material supply, improved models
Examples of industrial partners of the forest-based sector	Forest industry companies (pulp mills and biorefineries): trusted partner for raw materials supply, efficient processes, and as needed, also technical services to the customers	Wood products company: wood-based solutions to address high demands with regard to safety, quality, design and sustainability; tools to support adoption of the products	*Technology companies:* easy-to-use technological solutions; precise, accurate information to support the customers' decisions and processes, and to improve the industry profitability
Examples of further downstream customers	Apparel industry and retail: design, quality, fashion; responsible sourcing (cotton), sustainability, circular processes, climate change actions	City of Helsinki (rental apartments): safe and quality living at a reasonable price	*Forest industry companies:* easy to use services and forest revenues available to the private forest owners

(2009–2014) in the Finnish Bioeconomy Cluster FIBIC and the Acel Advanced Cellulose to Novel Products Programme (2014–2017). Technology called Ioncell has been developed in several project stages coordinated by the Aalto University and involving several universities, R&D partners, industrial partners from forest industry, chemicals industry, as well as from the textile industry.

The Ioncell method utilizes an ionized solvent to replace carbon disulfide, a toxic chemical used in the production of viscose. There are other similar technologies developed in collaboration with the same R&D partners, as well as parallel by other teams. A number of technology start-ups (e.g. Spinnova Ltd., The Infinited Fibre Company Ltd.) and new biomill investment plans (e.g. KaiCell Fibres Ltd.) are seeking to move forward with their technology either to a pilot plant or already to an industrial-scale production in an international collaboration. The Ioncell method is still developed for recovery of the solvent during the process, but Metsä Fibre Ltd. has already presented its first pilot garments in order to raise awareness of the public and potential investors to test and scale up the production. In Aalto University the innovative uses of advanced cellulosic materials are developed in collaboration of two schools, the chemical technology and the arts, design and architecture. The projects involve also industrial partners such as the clothing retail company H&M and furniture retail company IKEA. Ioncell method was first tested with dissolving pulp from the Enocell mill of Stora Enso Ltd., and a prototype dress was produced for the Finnish design company Marimekko. At present Stora Enso Ltd. sells dissolving pulp to China as raw material for production of viscose fibers, thus, for traditional textile production processes.

In the analyzed materials the Ioncell partners present several benefits based on the production process. They define specific value propositions of the *wood raw material* as renewable, traceable, ecological and sustainable material that outperforms other raw materials in land use (requires less arable land, less water than cotton). Value propositions of *viscose* as a product from this raw material highlight its benefits in the dyeing process as well as its moisture absorbance in use. The *forthcoming Ioncell material* is defined through the improved process, thus, raw material with a brand. *Ioncell-F method* is an alternative to the use of harmful chemicals in the production of viscose, and it is applicable to different sources of cellulose, thus, also to textile waste or even recycled paper and cardboard.

The company-level value propositions provide an additional angle to understand also the parallel niches where the collaborating partners are involved. The forest industry value propositions illustrate the company position as raw material producer that provides also technical service for their customers. Product and process qualities are emphasized. The technology partners highlight the variety of potentials available: different technologies, different raw materials, and different uses (not only textiles for apparel but also nonwovens, composites etc.). Further downstream customers emphasize many other aspects than the material sustainability: qualities and use of the material, design of textiles, sustainable lifestyles (not just more with less, but 'all in one' multifunctionality), responsible sourcing and production at all stages of the supply chain (sustainably sourced cotton) and distribution (including packaging), and circular processes as an opportunity for more efficiency (less waste to landfills). Thus, the socio-technical change to sustainable textiles includes alternative pathways how to fulfill the future demand for materials and textiles: the benefits of bioeconomy solutions based on renewable raw materials, the increased efficiency of circular processes within textiles industry and retail, and the sustainability design of processes of various raw materials, production, distribution as well as use.

3.3 Wood-Based Solutions for Sustainable Built Environment

Use of wood in multistorey construction in Finland has been supported by government strategic programmes, including development of regulations and standards (Ministry of Employment and the Economy 2015; Ministry of Economic Affairs and Employment 2017). Overall, the use of wood for multistorey construction has increased during the past five years in Finland, which shows in the number of construction projects, and in the number of companies producing engineered wood products, such as cross-laminated timber (CLT) and laminated veneer lumber (LVL), and prefabricated wood elements and modules. Solutions have been developed in several projects, including technological development, pilot and demonstration sites. The Wood City is provided here as an example; the project was initiated in 2011 in collaboration between the construction company SRV Ltd. and Stora Enso Ltd. as provider of the massive wood construction method, although construction started only in the end of 2016. The reasons postponing the realization of the project fall outside the scope of this paper, but the timescale gives an idea of the complexity of such projects.

Large wood product companies, such as Stora Enso Ltd., operate in international markets, and they have developed services to support use of their products, such as building information model tools for architects and engineers, including country specific requirements for example on acoustics and fire safety in the key market areas. Further examples are the Timber Academy, an on-line learning tool for professionals by Metsä Wood Ltd. and the open building system introduced by Stora Enso Ltd. in 2016. Thus, the companies offer the potential customers with knowledge and tools to support adoption and use of their products. Pilot projects are showcases that provide references of wood-based solutions; use of wood competes with the more established solutions of concrete and steel construction, and construction industry has been slow to adopt new methods (Hurmekoski 2016).

The Wood City quarter in Helsinki comprises of office and hotel premises and two residential buildings for the Helsinki city housing developer ATT Ltd. In the analyzed materials the Wood City partners define several benefits based on the offering: the *Wood City quarter* is promoted with central location, maritime surroundings, design of the quarter (including sense of community, near-by services), as well as benefits by the wood material for human wellbeing. For the potential tenants of the office premises *wooden buildings* are connected with the idea of more efficient working, and for the hotel with the idea of the acoustic and visual appeal of wood. Furthermore, value propositions of the *wood material* highlight it being ecological, sustainable, reusable, renewable and cost-competitive (rapid construction time) construction material. Also its contribution to emissions mitigation is emphasized. The external cladding material Accoya® is produced in the Netherlands of Pinus Radiata, a wood species cultivated for example in New Zealand. Also its benefits are highlighted based on the material properties; technology based on wood acetylation makes the material substitute of tropical hardwoods as a durable and dimensionally stable material, thus, enabling low maintenance costs. The SRV Ltd., in turn, points out that both wood and concrete are good materials assessed for their carbon footprints, but wood provides improved adaptability to user needs; the spaces are easier to modify afterwards. The Wood City *space and service concept* was developed in collaboration with a service design agency. The aim was to design the user experience of common premises so that it would support the tenants' businesses. The concept has been presented in the Helsinki Design Week programme, along with other service design initiatives realized in the vicinity.

The company-level value propositions highlight solutions-based thinking of the wood products industries. Companies offer products and necessary services to support the customer processes in using the products. For the further downstream customers information is at the moment available only on the rental housing for the city of Helsinki; the value propositions highlight the location, safety and reasonable living costs. The city of Helsinki decision about the construction site, in turn, emphasizes the sustainable development goals, but necessitates also energy efficient solutions. The construction company value proposition focuses on the project management and approach to satisfy the client's needs. SRV Ltd. highlights also its other construction projects which design infrastructure, services and housing jointly, such as the Low2No concept for energy efficient building and living with The Finnish Innovation Fund Sitra, and REDI concept in Kalasatama for urban planning, smart systems and new digital services. Thus, instead of individual building, its technical efficiency and material choices, focus is more on service platforms that enable sustainable solutions and support the inhabitants' and tenants' activities (cf. Hietanen 2011; PwC and ULI 2016). Sustainable built environment extends beyond thinking the design and production of an individual building to designing the social functions related to the building, infrastructures and services nearby.

Thus, the socio-technical change to sustainable construction includes alternative pathways how to fulfill the future demands: the benefits of bioeconomy solutions based on renewable raw materials, the increased efficiency of technical systems through smart processes, as well as the design of functions and operations of the built environment as a social system supporting the tenants' operations as well as the inhabitants' health and wellbeing.

3.4 Forestry Solutions for Biomass Production and Beyond

Potential of the forest resources, sustainable forest management and available expertise and forest data are one of the corner stones of the Finnish bioeconomy strategy (Ministry of Employment and the Economy 2014; Ministry of Agriculture and Forestry 2015). 'Wood on the move and new products from forests' is a key project of the Finnish government strategic programme (2015), including regulatory measures, incentives for more active and sustainable forest management, R&D support and investments in infrastructure. Efficient utilization of forest data has been targeted with several measures. For example, an online service was introduced in 2011 to open the forest data collected with public funding to the forest owners directly (metsaan.fi)—previously forest owners could gain this data if they used forest expert organizations to prepare a forest management plan for their property. Access to data however requires changes of regulations, which processes take a considerable time. Recently new types of services have been developed, for example, the online wood procurement platform (kuutio.fi) in collaboration of the forest owners and wood procuring industries. The initiatives of an open forest data strive for platform thinking, thus, enabling also new services and new service providers. For example the Natural Resources Institute Finland (LUKE) has investigated possibilities for berry data services based on the forest inventory data, as well as citizen science based networks for harvest observations of forest berries. Also open-access satellite data is tested for online berry services in collaboration between Aalto University and Satellio Oy (www.berrymonitor.com).

The Forest Big Data project (2014–2016) was initiated as part of the Data to Intelligence programme financed by the Finnish Funding Agency for Innovation Tekes. Its aim was to develop technology and solutions for *more efficient and higher quality planning and operations* in the entire wood supply chain. Research partners included universities, The Finnish Geospatial Research Institute (FGI), Forest Research Institute Metla (now: LUKE), VTT Technical Research Centre of Finland and key industrial partners of forest industry, forest machinery and service companies already operating in the forest sector. In 2017 the service platform development for Forest Big Data continues in a project coordinated by Metsäteho Ltd., a limited company owned by the forest industry organisations and companies in Finland. Parallel projects, for example on satellite-based data, laser scanning data or data collected by harvesters, continue to develop data quality and applications. These measures include also European collaboration: the Efficient forestry for sustainable and cost-competitive bio-based industry (EFFORTE 2016–2019) consortium coordinated by LUKE is co-funded from the Bio-Based Industries Joint Undertaking.

In this niche the development work is coordinated by the forest sector organizations themselves. The forest industry companies are customers for the new technology and applications but they also develop their own online services for private forest owners. Value propositions of the technology developers and service providers highlight easy-to-use technological solutions, precision of data as well as the benefits to the customers' decision making and profitability. Forest industry companies emphasize in their value propositions to the private forest owners forestry easy-to-use solutions (wood procurement, related harvesting and after-harvesting services, forestry services and tools for property management) as well as profitability and revenues available from forest. The analyzed materials do not explicate specific value propositions to further downstream, but focus is on wood mobilization, stable raw material flow to the industry, and improved forest growth models and utilization methods. Thus, there is no information on how the collection of more data at more precision level contributes to further downstream needs and requirements, or what kind of data processes would support development of new products and production processes such as those exemplified in the above two cases. This does not mean that such questions were not already raised; rather, if such questions are investigated and new services developed, the information is not public nor the policy measures to direct such activities emphasized.

3.5 Zooming In and Out Between the Mini-cases

In the analysis across the three mini-cases (Table 2), the forest-based sector regime of established forestry and forest industry practices is visible in the value propositions both at meso level (sectoral strategies) and micro level (individual companies and organizations): focus is on efficient processes and an active utilization of the forest biomass potential. The collaboration partners' value propositions and parallel niches reveal also more radical technological and systemic change opportunities. From the viewpoint of the further downstream industries the bioeconomy solutions for material substitution are one option among others for more efficient, innovative and sustainable solutions.

The mini-cases highlight servitization of primary production and processing industries. Services are a means to increase efficiency and optimization of the existing production processes, although measuring these service activities or outputs separately is challenging. In the textiles mini-case, the role of forest industry—or a biorefinery—is that of raw material supplier. However, the analyzed case also illustrates the efforts to support development of the value chain further downstream: making the opportunities visible and finding partners that would develop the production process into a pilot facility and scale it up to an industrial production. Furthermore, the traditional forest sector product, market pulp, is already complemented with technical services to the customer. The construction mini-case illustrates the wood product companies' efforts to support use of wood by providing solutions instead of mere materials or products and by demonstrating the benefits of the wood-based solutions. The forest data, in turn, is developed to make forestry and production processes more efficient with the support of more accurate and precision data for the decision making. New digital applications, technology solutions and service platforms are foreseen to support the implementation of the Finnish bioeconomy strategy.

The mini-cases on sustainable textiles and built environment illustrate that materials substitution to the existing customer industry processes requires understanding of the customer processes; new materials or products are not mere drop-in substitutes to the existing supply chains, but require also development of the customer processes, often including services to support adoption of new solutions. While the raw material supplier's value proposition is to be a trusted partner for a timely delivery in requested quantities and qualities at a competitive price, development of new bio-products and processes already emphasizes interaction with the (potential) customer processes to an increasing degree. These interactions can be fulfilled either by strengthening in-house capacities (e.g. the technical service by pulp producers) or using external service providers (e.g. technological solutions providers).

4 Conclusions and Discussion

We have investigated increasing role of services and the interlinkage(s) of these developments to the evolving bioeconomy conceptualizations with an analytic framework based on service research conceptualizations that stem from the marketing discipline. The exemplified mini-cases do not allow generalizations to whole forest-based sector neither in the context of Finland or internationally, but they illustrate that what is described as servitization in manufacturing is not unfamiliar in the natural resources sectors either. Increasing efficiency of the established regimes, such as forestry and related forest-based industries in Finland, is sought for by technology-enabled services and tools to improve the industry competitiveness and profitability. The evolving bioeconomy in Finland represents in our analysis an incremental rather than radical change of the production processes (also in: Bosman and Rotmans 2016). Contribution of the 'service lens' applied in this research is twofold: it provides an integrated approach to products and services, and extends the analysis from the producer's supply chain to the customer networks. In the value propositions of the further downstream customers phrases like 'experience', 'design' and 'service concept' become perceivable parallel to the material and process qualities. The service research concepts can be used to add also human and social perspective to the bioeconomy transitions; how societal functions can be fulfilled in alternative ways and what constitutes service in the future bioeconomy (cf. Markard et al. 2012). This could open alternative scenarios on not only making the existing processes more efficient, but also rethinking of the processes in a more profound way.

Further analyses could be made for zooming in and zooming out between the three mini-cases: the alternative future developments in the customer industries indicate also the multiple linkages that the forest sector production has to these further downstream processes. For example, wood pulp from biorefineries is developed as a solution to textiles production, but the forest-based sector already connects to other raw materials tested for textile production (paper and cardboard waste), to efficient distribution (packaging materials) and to corporate responsibility of the brand companies (climate change actions). Also for the sustainable built environment, wood as construction material is one solution, but the forest-based sector expertise on forest ecosystem services, biodiversity, climate change mitigation and urban forests could be connected to the value propositions of the customer industries (design of customer experience, health and wellbeing, efficient work environments). The 'services to society' highlighted by the forest sector strategies extend from mere rural livelihoods to urban living and wellbeing. Such de- and re-configurations of resources requires new thinking. Focus on biological resources and efficient biomass processes or on biotechnological development largely conceals these angles to a future bioeconomy. The mini-cases highlight two angles from the further downstream customer industries; circular processes and attention on larger systems (such as responsible processes of textiles retail or design of the built environment to support tenants' or inhabitants' activities).

The paper provides a conceptual design and analyses of three mini-cases at a fairly general level. The contribution of this approach is to raise awareness of the increasing role of services in the bioeconomy context: the question is not about mere number and volume of services, but also about the changes how production, distribution and use can be organized. The bioeconomy concepts have been criticized of being too technology fixed, neglecting existing knowledge and capacities as a basis of social innovation, and paying limited attention to sustainability of activities (Schmid et al. 2012; McCormick and Kautto 2013; Staffas et al. 2013). The service research conceptualizations can provide additional means to study transitions to a bioeconomy: Bioeconomy evolves through interactions representing the visions and ideas of the traditional natural resources sectors, new emerging bio-industries as well as the customer industries that utilize the materials, products and services in their own processes.

Annexes

See Tables 3, 4 and 5.

Table 3 Biorefineries supply to sustainable textiles, examples of value propositions at macro, meso and micro levels

landscape level: scarce resources, increasing uncertainties; new technology applications, new production modes; renewable resources, bio-technology as one of the Key Enabling Technologies		
The Finnish bioeconomy strategy (2014): objective to generate new economic growth and new jobs from an increase in the bioeconomy business and from high added value products and services while securing the operating conditions for the nature's ecosystems		
Meso level: established socio-technical regimes	Textiles sector value propositions, e.g. the Nordic Fashion Association: "sustainable can be as cool, sexy and fashionable as conventional clothes" (www.nordicfashionassociation.com/): Nordic Initiative, Clean and Ethical initiative; Better Cotton Initiative; Sustainable Apparel Coalition) [government strategic objectives (2015) Team Finland to support internationalization and competitiveness of enterprises, and attract investments to Finland; and 'Wood on the move and new products from forests' for bioeconomy and clean solutions]	
Niche level: R&D/ pilot projects	**Ioncell-F—method for cellulosic fibers from ionic liquid solution (several project stages 2009–)** value proposition: An alternative method for the use of harmful chemicals in viscose production; improved solution of cellulose-based fibers to the increasing textile materials demand	
Including firm level/ individual organizations	Value propositions	Parallel niches
	Aalto University, wood chemistry/ department of forest products technology: new ways to harness wood and cellulose	CHEMARTS collaboration between the School of Chemical Technology and The School of Arts, Design and Architecture: new concepts for the future use of cellulose and other biomaterials (chemarts.aalto.fi/)
Examples of industrial partners of the forest-based sector	Metsä Fibre: trusted partner, knowing customers processes and their business and keeping deep cooperation with them	Äänekoski bioproduct mill ecosystem: the next-generation bioproduct mill (bioproductmill.com/)
	Stora Enso Biomaterials: seeking new, innovative ways to utilize the valuable raw material, wood, while simultaneously running existing pulp and by-products businesses as efficiently as possible, based on customers' needs	'Pure by Stora Enso' pulp for production of textile fibers
Examples of further downstream customers	Marimekko: timeless, distinctive and functional designs "bringing joy to everyday moments"; high quality materials (ALLU designer T.Pöyhönen: "all in one" multifunctional design)	Development of the brand and international distribution network; corporate responsibility (sourcing of cotton)
	H&M: fashion and quality at the best price	Sustainability, responsibility (sourcing of cotton), circular processes, climate change action
	IKEA: Affordable solutions for better living	Circular processes, no waste to landfill

Table 4 Wood-based solutions for sustainable built environment, examples of value propositions at macro, meso and micro levels

Landscape level: scarce resources, increasing uncertainties; new technology applications, new production modes; renewable resources, bio-technology as one of the Key Enabling Technologies		
The Finnish bioeconomy strategy (2014): objective to generate new economic growth and new jobs from an increase in the bioeconomy business and from high added value products and services while securing the operating conditions for the nature's ecosystems		
Meso level: established socio-technical regimes	Construction industry and the real estate sector value propositions, e.g.: combatting the grey economy: to promote good construction (www.tilaajavastuu.fi/en); Digitalization of construction and real estate sectors: an open and interoperable information management ecosystem for the built environment (www.kiradigi.fi/en/) [government strategic objectives (2015) for bioeconomy and clean solutions: feasibility of wood-based solutions, improved competitiveness of wood construction both in domestic and international markets]	
Niche level: R&D/pilot projects	**Wood City—sustainable urban quarter (2011–)** Value proposition: Functional premises in a central location; showcase of urban wood construction (office and hotel premises, and residential buildings)	
Including firm level/ individual organizations	Value propositions	Parallel niches
	Construction company SRV: construction projects in open co-operation with clients, driven by the client needs; consistent and flexible process for economic benefits, faster implementation schedule and better equivalence to client's needs "Construction for life"	Low2No quarter design (Sitra collaboration, energy efficiency); REDI Kalasatama smart solutions design, including Living Lab (www.redi.fi/living-lab/)
Examples of industrial partners of the forest-based sector	StoraEnso Building and Living: wood-based solutions to address high demands with regard to safety, quality, design and sustainability … developing processes, methods and tools to deliver added value to customers	Open building system, and online tools
	Accoya® (external cladding material; Novenberg Ltd as distributor): durable material with lower maintenance costs, substitute to tropical timber (sustainable sourcing)	Sourcing of raw material, production facilities also elsewhere than in the Netherlands
Examples of further downstream customers	City of Helsinki: Housing Production Department ATT and Heka rental apartments: safe and quality living at a reasonable price (tenants of the office and hotel premises are negotiated by SRV Ltd.; no information publicly available at the moment)	City of Helsinki vision 2050: a denser city than today, with several district centres connected by public transport; good urban life

Table 5 Forestry solutions for biomass production and beyond, examples of value propositions at macro, meso and micro levels

Landscape level: scarce resources, increasing uncertainties; new technology applications, new production modes; renewable resources, bio-technology as one of the Key Enabling Technologies		
The Finnish bioeconomy strategy (2014): objective to generate new economic growth and new jobs from an increase in the bioeconomy business and from high added value products and services while securing the operating conditions for the nature's ecosystems		
Meso level: established socio-technical regimes	Forest-based sector value propositions, e.g. National Forest Programme 2025: Sustainable forest management is a source of growing welfare—forests offer solutions for human and societal needs, policy creates a setting for a growing forest-based bioeconomy and more diverse welfare Strategic Programme for the Forest sector (2008–2015): renewal of the forest sector, utilization of forest resources and expertise contribute to increasing employment and economic growth Government strategic objectives (2015) 'Wood on the move and new products from forests' for bioeconomy and clean solutions	
Niche level: R&D/pilot projects	**Forest Big Data—next generation forest resource management system** (several measures during the past 10 years) value proposition: Efficient utilization of forest data supports the bioeconomy development	
Including firm level/ individual organizations	Value propositions	Parallel niches
	DIGILE (2014–2016): intelligent tools and methods for managing, refining and utilizing diverse data in order to enable innovative business models and services Metsäteho Oy (Forest Big Data 2017): support for the development of shareholders' wood procurement and wood production operations and improves the operating preconditions for wood supply	metsään.fi online service (Finnish Forest Centre) offers the latest information directly to forest owners on their properties, and makes data available and into more efficient use European EFFORTE (2016–2019) consortium on Efficient forestry for sustainable and cost-competitive bio-based industry (www.luke.fi/efforte/) LUKE projects to utilize the national forest inventory data and develop the forest berry harvest data

(continued)

Table 5 (continued)

Examples of industrial partners *in this case the technology companies are solution providers*	Arbonaut: innovative and easy-to-use technological solutions to provide the most precise and actionable forestry management information available Savcor: Solutions for performance optimization to improve customers' capability to exceed their business profitably. Timbervision: software products in close co-operation with industry to improve industry profitability Trestima: forest inventory system with world's most accurate and efficient tool for measuring standing timber	Satellito: space data based analysis and monitoring services to support customers' decisions and processes (hakkuut.fi customer: Ministry of Agriculture and Forestry); parallel: space data for (www.berrymonitor.com/)
In this case the forest industry companies are the further downstream customers	Metsä Forest serves forest owners and wood utilizing industries: improve profitability of forestry and reliable wood supply	As part of their wood procurement, all three companies have developed their own on-line services for the private forest owners
	Stora Enso for private forest owners: services for making forest owning easy "satisfied forest owner enjoys life and revenues from forest"	
	UPM for private forest owners: forest services for easy income and effortless forestry	

List of materials (all websites last accessed 12.7.2017):

Aalto University (2017) Ioncell-F website bio2.aalto.fi/en/research_groups/biorefineries/ioncell/

Accoya (2017) product website website www.accoya.com/

Arbonaut Ltd. (2017) company website www.arbonaut.com/

ATT (2014) Toimintakertomus 2013. Helsingin Asuntotuotantotoimisto (Annual report 2013 of the Helsinki housing production, in Finnish)

CLIC Innovation (2017) Acel Advanced Cellulose to Novel Products 2014–2017 clicinnovation.fi/activity/acel/

CLIC Innovation Ltd. (2017) company website clicinnovation.fi/theme/bioeconomy/

Confederation of the Finnish Construction Industries (2017) website www.rakennusteollisuus.fi
DIGILE (2016) Forest Big Data website www.datatointelligence.fi/forest-big-data.html
European Apparel and Textile Confederation (2017) website www.euratex.eu/
FIBIC (2014) Future Biorefineries Products from Dissolved Cellulose Programme Report 2011–2014. Finnish Bioeconomy Cluster FIBIC 2013.
Finnish bioeconomy (2017) website www.bioeconomy.fi/
Finnish Forest Association (2017) Wood-based textiles are being developed in several Finnish projects, 11.5.2017 available: http://www.smy.fi/en/products-services/wood-based-textiles-are-being-developed-in-several-finnish-projects/
Finnish Forest Centre (2017) website www.metsakeskus.fi/more-10-million-hectares-mapped-details-privately-owned-finnish-forests-best-world#
H& M (2017) company website about.hm.com/en/sustainability.html
H & M Foundation (2017) website Global Change Awards globalchangeaward.com/
IKEA (2017) company website www.ikea.com/ms/en_US/this-is-ikea/people-and-planet/energy-and-resources/
Hämäläinen, J., Holopainen, M., Hynynen, J., Jyrkilä, J., Rajala, P.T., Ritala, R., Räsänen, T. and Visala, A. (2014) Perusteita seuraavan sukupolven metsävarajärjestelmälle—Forest Big Data hanke (Next generation forest data system—the Forest Big Data project, in Finnish). Metsätieteen aikakauskirja 4/2014, 235–241.
KIRA-digi (2017) website www.kiradigi.fi/en/
Marimekko (2014) Sustainability review 2013. Marimekko Ltd.
Marimekko (2017) company website company.marimekko.com
Metsä Fibre (2017) Metsä Fibre website www.metsafibre.com
Metsä Forest (2017) Metsä Forest website www.metsaforest.com
Metsä Wood (2017) Metsä Wood website www.metsawood.com
Metsäteho (2016) Forest Big Data final seminar, 8.3.2016 available: www.metsateho.fi/forest-big-datan-tulosseminaari-8-3-2016-heurekassa/
Ministry of Agriculture and Forestry (2017) Forest data and digital services website (in Finnish) mmm.fi/metsatieto-ja-sahkoiset-palvelut
New wood initiative (2017) website www.uusipuu.fi/
Nordic Fashion Association (2017) website www.nordicfashionassociation.com/
Satellio Ltd. (2017) company website www.satellio.com
Savcor Ltd. (2017) company website www.savcor.com/
SRV (2012) Tulevaisuuden rakentaja. Katsaus 25 vuotta eteenpäin rakentamisen ja elämisen tulevaisuuteen. SRV 25 vuotta. (25-year anniversary publication of the SRV Ltd.: an outlook to the next 25 years of building and living, in Finnish)
SRV (2013) Miten me asumme vuonna 2033? SRV:n tulevaisuuden asumisen ideariihen loppuraportti. (How we live in 2033? Final report of the SRV Ltd. future housing think tank, in Finnish)
SRV Ltd. (2017) company website www.srv.fi/
REDI Living-Lab website www.redi.fi/en/living-lab/

Stora Enso (2017a) Stora Enso Biomaterials website biomaterials.storaenso.com/products-and-services/for-textile-and-other-applications
Stora Enso (2017b) Building and living website buildingandliving.storaenso.com/
Stora Enso (2017c) Stora Enso Forest website www.storaensometsa.fi/
Sustainable Apparel Coalition (2017) website apparelcoalition.org/
TEKES The Finnish Funding Agency for Innovation (2017) TEKES funding programmes website www.tekes.fi/en/programmes-and-services/tekes-programmes/
Timber Vision Ltd. (2017) company website www.timbervision.fi
Trestima Ltd. (2017) company website www.trestima.com/
UPM (2017) forest owner services website www.metsamaailma.fi/
VTT Technical Research Centre of Finland Ltd. (2017) key technology platforms for bioeconomy website www.vttresearch.com/services/bioeconomy/key-technology-platforms-for-bioeconomy
Wood City (2017) website http://www.woodcity.fi/

References

Anderson JC, Narus JA, van Rossum W (2006) Customer value propositions in business markets. Harvard Bus Rev 84(3):90–99

Antikainen R, Lehtoranta S, Luoma P, Berghäll E, Valve E, Miller T, Larvus L, Pohjola J, Laturi J, Lintunen J, Tamminen S, Seppälä J, Uusivuori J (2016) Biotalous ja cleantech Suomessa – strategioiden arviointi ja toimenpidesuositukset (Bioeconomy and cleantech in Finland—Assessment of Strategies and development suggestions, in Finnish). Publications of the Govenrment's Analysis, Assessment and Research Activities 51/2016

Baines TS, Lightfoot HW, Benedettini O, Kay JM (2009) The servitization of manufacturing: a review of literature and reflection on future challenges. J Manuf Technol Manage 20(5):547–567

Bauer F, Coenen L, Hansen T, McCormick K, Palgan VY (2016) Technological innovation systems for biorefineries: a review of the literature. Biofuels Bioprod Biorefin 11:534–548

Bitner MJ, Brown SW (2008) The service imperative. Bus Horiz 51(1):39–46

Boehm M, Thomas O (2013) Looking beyond the rim of one's teacup: a multidisciplinary literature review of product-service systems in information systems, business management, and engineering & design. J Clean Prod 51:245–260

Bosman R, Rotmans J (2016) Transition governance towards a bioeconomy: a comparison of Finland and The Netherlands. Sustainability 8:1017

Bryson JR, Daniels PW (2010) Service worlds, the services duality and the rise of the manuservice economy. In: Maglio PP, Kieliszewski CA, Spohrer JC (eds) Handbook of service science. Springer, New York, pp 79–104

Bugge MM, Hansen T, Klitkou A (2016) What is the bioeconomy? A review of the literature. Sustainability 8:691

Chang Y-C, Miles I, Hung S-H (2014) Introduction to special issue: managing technology-service convergence in Service Economy 3.0. Technovation 34:499–504

Chesbrough H, Spohrer J (2006) A research manifesto for services science. Commun ACM 49 (7):35–40

Coombs R, Miles I (2000) Innovation, measurement and services: the new problematique. In: Innovation systems in the service economy. Economics of science, technology and innovation, vol 18, pp 85–103

Cova B, Salle R (2008) Marketing solutions in accordance with the S-D logic: co-creating value with customer network actors. Ind Mark Manage 37:270–277

De Backer K, Desnoyers-James I, Moussiegt L (2015) Manufacturing or services—that is (not) the question: the role of manufacturing and services in OECD economies. OECD Science, Technology and Industry Policy Papers, No. 19

EC (2012) Innovating for sustainable growth: a bioeconomy for Europe. Communication from the Commission to the European Parliament, the Council, the European Economic and Social Committee and the Committee of the Regions COM(2012)60 final, Brussels, 13 Feb 2012

Efken J, Walter D, Kreins P, Knechtd M (2016) Measuring the importance of the bioeconomy in Germany: concept and illustration. NJAS – Wageningen J Life Sci 77:9–17

Elzen B, Geels FW, Green K (eds) (2004) System innovation and the transition to sustainability: theory, evidence and policy. Edward Elgar Cheltenham

Gallouj F, Savona M (2009) Innovation in services: a review of the debate and a research agenda. J Evol Econ 19:149–172

Gallouj F, Weber KM, Stare M, Rubalcaba L (2015) The futures of the service economy in Europe: a foresight analysis. Technol Forecast Soc Chang 94:80–96

Geels FW (2002) Technological transitions as evolutionary reconfiguration processes: a multi-level perspective and a case study. Res Policy 31:1257–1274

Geels FW (2005) Processes and patterns in transitions and system innovations: refining the co-evolutionary multi-level perspective. Technol Forecast Soc Chang 72:681–696

Geels FW (2010) Ontologies, socio-technical transitions (to sustainability), and the multi-level perspective. Res Policy 39(4):495–510

Håkansson H, Snehota I (1989) No business is an island: the network concept of business strategy. Scand J Manag 5(3):187–200

Hietanen O (2011) Rakennetun ympäristön tulevaisuus. In: Airaksinen M, Hietanen O, Manninen AP, Reijula K, Vainio T, Nenonen S (ed) Roadmap of the built environment. Final Report 5/2011. Tekes (in Finnish)

Hurmekoski E (2016) Long-term outlook for wood construction in Europe. Dissertationes Forestales 211. University of Eastern Finland, School of Forest Sciences

Johnson MW, Christensen CM, Kagermann H (2008) Reinventing your business model. Harvard Bus Rev 86(12):50–59 + 129

Maglio PP, Spohrer J (2008) Fundamentals of service science. J Acad Mark Sci 36:18–20

Manyika J, Sinclair J, Dobbs R, Strube G, Rassey L, Mischke J, Remes J, Roxburgh C, George K, O'Halloran D, Ramaswamy S (2012) Manufacturing the future: the next era of global growth and innovation. McKinsey Global Institute Report, Nov 2012

Markard J, Raven R, Truffer B (2012) Sustainability transitions: an emerging field of research and its prospects. Res Policy 41:955–967

Mathieu V (2001) Product services: from a service supporting the product to a service supporting the client. J Bus Ind Mark 16(1):39–61

McCormick K, Kautto N (2013) The bioeconomy in Europe: an overview. Sustainability 5 (1):2589–2608

Michel S, Brown SW, Gallan AS (2008) An expanded and strategic view of discontinuous innovations: deploying a service-dominant logic. J Acad Mark Sci 36:54–66

Ministry of Agriculture and Forestry (2015) National Forest Strategy 2025

Ministry of Economic Affairs and Employment (2017) Wood-based bioeconomy solving global challenges

Ministry of Employment and the Economy (2014) The Finnish bioeconomy strategy—sustainable growth from bioeconomy

Ministry of Employment and the Economy (2015) From forests to pioneering bioeconomy. Final report on the Strategic Programme for the Forest Sector (MSO)

Mont O (2002) Clarifying the concept of product-service system. J Clean Prod 10(3):237–245

Näyhä A, Pesonen H-L (2014) Strategic change in the forest industry towards the biorefining business. Technol Forecast Soc Chang 81:259–271

Näyhä A, Pelli P, Hetemäki L (2015) Services in the forest-based sector—unexplored futures. Foresight 17:378–398

Neely A, Benedettini O, Visnjic I (2011) The servitization of manufacturing: further evidence. University of Cambridge, Cambridge Service Alliance. In: 18th European Operations Management Association conference, Cambridge, 3–6 July 2011

Normann R, Ramírez R (1993) From value chain to value constellation: designing interactive strategy. Harvard Bus Rev 71:65–77

OECD (2016) Enabling the next production revolution: the future of manufacturing and services—Interim Report. Meeting of the OECD Council at Ministerial Level, Paris, 1–2 June 2016

Oliva R, Kallenberg R (2003) Managing the transition from products to services. Int J Serv Ind Manage 14:160–172

Ostrom AL, Bitner MJ, Brown SW, Burkhard KA, Goul M, Smith-Daniels V, Demirkan H, Rabinovich E (2010) Moving forward and making a difference: Research priorities for the science of service. J Serv Res 13(1):4–36

Pätäri S, Kyläheiko K, Sandström J (2011) Opening up new strategic options in the pulp and paper industry: Case biorefineries. Forest Policy Econ 13:456–464

Payne A, Frow P, Eggert A (2017) The customer value proposition: evolution, development, and application in marketing. J Acad Mark Sci 45:467–489

Payne A, Storbacka K, Frow P (2008) Managing the co-creation of value. J Acad Mark Sci 36 (1):83–96

Pelli P, Haapala A, Pykäläinen J (2017) Services in the forest-based bioeconomy—analysis of European strategies. Scand J For Res 7:559–567

Pfau FS, Hagens JE, Dankbaar B, Smits AJM (2014) Visions of sustainability in bioeconomy research. Sustainability 6(3):1222–1249

Pilat D, Cimper A, Olsen KB, Webb C (2006) The changing nature of manufacturing in OECD economies. OECD science technology and industry working papers, 2006/09. OECD Publishing, Paris

PwC and ULI (2016) Emerging trends in Real Estate®—beyond the capital. Europe 2016. PwC, Urban Land Institute (ULI)

Rönnlund I, Pursula T, Bröckl M, Hakala L, Luoma P, Aho M, Pathan A, Pallesen BE (2014) Creating value from bioresources—innovation in Nordic bioeconomy. Nordic Innovation Report 2014:01, May 2014. Gaia Consultin, Agrotech Ltd.

Schmid O, Padel S, Levidow L (2012) The bio-economy concept and knowledge base in a public goods and farmer perspective. Bio-Based Appl Econ 1(1):47–63

Sitra (2009) Background report on the natural resource strategy for Finland: natural resources—an opportunity for change. Sitra, the Finnish Innovation Fund

Sitra (2011) Sustainable bio-economy: potential, challenges and opportunities in Finland. PBI Research Institute, Magnus Gustafsson, Robert Stoor, Anastasia Tsvetkova. Sitra Studies 51, Mar 2011

Skålen P, von Gummerus J, Koskull C, Magnusson PR (2015) Exploring value propositions and service innovation: a service-dominant logic study. J Acad Mark Sci 43:137–158

Smith A, Voss JP, Grin J (2010) Innovation studies and sustainability transitions: the allure of the multilevel perspective and its challenges. Res Policy 39(4):435–448

Smith L, Maull R, Ng ICL (2014) Servitization and operations management: a service dominant-logic approach. Int J Oper Prod Manage 34(2):242–269

Staffas L, Gustavsson M, McCormick K (2013) Strategies and policies for the bioeconomy and bio-based economy: an analysis of official national approaches. Sustainability 5(1):2751–2769

Stehrer R, Baker P, Foster-McGregor N, Koenen J, Leitner S, Schricker J, Strobel T, Vieweg H-G, Vermeulen J, Yagafarova A (2014) Study on the relation between industry and services in terms of productivity and value creation. Final report. ECSIP consortium: wiiw, Ifo and Ecorys

Tukker A (2004) Eight types of product–service system: eight ways to sustainability? Experiences from SusProNet. Bus Strategy Environ 13(4):246–260

Vandermerwe S, Rada J (1988) Servitization of Business: adding value by adding services. Eur Manag J 6(4):314–324

Vargo S, Lusch R (2004) Evolving into a new dominant logic for marketing. J Mark 68:1–17
Vargo S, Lusch R (2008) Service-dominant logic: continuing the evolution. J Acad Mark Sci 36 (1):1–10
Vargo SL, Lusch RF (2011) It's all B2B… and beyond: toward a systems perspective of the market. Ind Mark Manage 40(2):181–187
Vargo S, Lusch R (2016) Institutions and axioms: an extension and update of service-dominant logic. J Acad Mark Sci 44:5–23

Environmental Sustainability Indicators for the Bioeconomy

Tiina Pursula, Maija Aho, Ida Rönnlund and Minna Päällysaho

Abstract Bioeconomy is a rapidly evolving field, where novel value chains and concepts are developed and commercialized. In this field, there is a need to address sustainability issues, specific to biobased value chains, such as origin of biomass feedstock and related land use issues, material efficiency and energy and water requirements. There is a place for biobased products, for fossil based products and for mixtures of these, and environmental sustainability indicators should aim at pointing out the most suitable alternative from an environmental point of view, in each unique case. Although biomass is a renewable resource, the production of it is dependent on limited resources, such as fresh water, land, nutrients and in the case of animals, food. Biomass is thus only renewable if these limited resources are used sustainably: avoiding excessive use and ensuring that the impact from production does not affect other issues negatively, which for example emissions and wastes, chemical use, nutrients depletion or overload may do. In this chapter, we summarize current state of environmental sustainability indicator development and use for the bioeconomy. In the light of this, the chapter further addresses the gaps in sustainability assessment and lists the novel attempts to develop assessment methods needed to ensure the sustainability of bioeconomy from a wider perspective, yet maintaining proper alignment with the LCA principles. The use of comprehensive environmental sustainability indicators in bioeconomy value chains is also concretized with two case studies. The case studies cover aspects of environmental sustainability in a holistic manner, including e.g. efficient use of natural resources, including materials, water, and energy, minimization of wastes and emissions and reduction of risks to humans and the environment from use of chemicals and

T. Pursula (✉) · M. Aho · I. Rönnlund · M. Päällysaho
Gaia Consulting Ltd, Bulevardi 6 A, 00120 Helsinki, Finland
e-mail: tiina.pursula@gaia.fi

M. Aho
e-mail: maija.aho@gaia.fi

I. Rönnlund
e-mail: ida.ronnlund@gaia.fi

M. Päällysaho
e-mail: minna.paallysaho@gaia.fi

W. Leal Filho et al. (eds.), *Towards a Sustainable Bioeconomy: Principles, Challenges and Perspectives*, World Sustainability Series,
https://doi.org/10.1007/978-3-319-73028-8_3

disposal of chemicals. Qualitative indicators are utilized when important sustainability aspects cannot be turned into numeric values. Systemic perspectives have the benefit of showing the wider effects of actions, as many effects are not direct, but indirect, and the use of biobased products may have multiple effects. Lastly, benchmarking is applied in the cases to give a meaning to the calculated values and form a bridge between sustainability indicators, the status of the environment and industrial production. From results of the cases the multi-faceted nature of sustainability in bioeconomy can be clearly seen. While the biobased value chain out performs the fossil value chain in some sustainability aspects, the fossil chains are equally good or superior in certain other sustainability aspects. This highlights the need for comprehensive sustainability assessment. In order to develop a sustainable bioeconomy we need to analyze and benchmark the value chains in a systematic manner and develop the methods and availability of reliable data continuously. The objectives of the paper, bridging the gaps of sustainability assessment of bio-based value chains, supports the inherent target of bioeconomy to utilize bio-based resources in a sustainable manner.

Keywords Bioeconomy · Sustainability · Benchmarking · Indicators Bioproducts

1 Introduction

Biobased materials are made from renewable agricultural and forestry feedstock including organic waste streams and aquatic biomass. Traditional uses of biomass include e.g. food, feed, timber, pulp and paper together with energy uses. Nowadays biomass can replace the consumption of fossil resources in many other uses and solutions, such as production of fuels, chemicals, advanced materials, plastics, composite fibers, and pharmaceuticals. Essentially, use of renewable biomass has been promoted as replacement of finite fossil resources. Additionally, bioeconomy is seen as a solution for reducing environmental impacts, such as carbon intensity and global warming. Therefore, several political measures have been taken to promote bioeconomy and a number of national bioeconomy strategies have increased. EU has launched and adopted Bioeconomy Strategy for Europe (EC 2012) and similarly Economic Co-operation and Development (OECD) has published policy agenda to support and encourage transition towards bioeconomy. Sustainability is one important element of these strategies and agendas. There is also a wide number of sectoral certification schemes available for the bioeconomy products, like biofuels.

Even though bioeconomy can be seen as a solution for global environmental challenges and it is a clear strategic priority, it may have adverse effects as well. According to the Global Risk Report, published by World Economy Forum (WEF 2016), the biggest recognized global environmental risks are failure of climate-change mitigation and adaptation, biodiversity loss and ecosystem collapse, and global water and food crises. Growing field of bioeconomy industry may be one solution to climate change mitigation, but when not sustainable managed, negative impacts on biodiversity or ecosystems may occur along the entire value chain. Furthermore, biomass production requires land, water and fertilizers leading to situation where risk for soil degradation and negative impacts on biodiversity and water exists. Also larger amount of biomass, compared to fossil resources, is typically required to deliver the same functional unit (e.g. energy content) meaning that more transportation and handling capacity is required for biomass. Therefore, greenhouse gas emissions emitted during the whole value chain might be significant in biobased value chains as well.

In order to address all these aspects and provide an overview on the environmental sustainability indicators of relevance for the bioeconomy and concretize these with an example, this study summarizes the main environmental concerns related to biobased value chains and presents an indicator framework which addresses these concerns in a holistic manner.

We first summarize current state of environmental sustainability indicator development and use for the bioeconomy. In the light of this, we further address the gaps in sustainability assessment and list the novel attempts to develop assessment methods needed to ensure the sustainability of bioeconomy from a wider perspective, yet maintaining proper alignment with the LCA principles. The use of comprehensive environmental sustainability indicators in bioeconomy value chains is also concretized with two case studies.

The case studies cover aspects of environmental sustainability in a holistic manner, including e.g. efficient use of natural resources, including materials, water, and energy, minimization of wastes and emissions and reduction of risks to humans and the environment from use of chemicals and disposal of chemicals. Qualitative indicators are utilized when important sustainability aspects cannot be turned into numeric values. Systemic perspectives have the benefit of showing the wider effects of actions, as many effects are not direct, but indirect, and the use of biobased products may have multiple effects. Lastly, benchmarking is applied in the cases to give a meaning to the calculated values and form a bridge between sustainability indicators, the status of the environment and industrial production.

1.1 Origin of Biomass and Related Land Use Issues

Currently, an increasing pressure exists to use land effectively as it provides necessary commodities for human life. Land provides food and feed and other raw materials for the bioeconomy, but it also provides crucial ecosystem services.

Ecosystems are complex and multifunctional systems providing direct and indirect contributions to human well-being. Species diversity have a remarkable role in the functional ecosystems, (TEEB 2010) indicating that changes in biodiversity cause changes in services that ecosystem is able to provide. Ecosystem services can be divided to provisioning, regulating and cultural services (TEEB

2010). Generalized, provision services cover food, clean water and materials that ecosystems are able to provide. Instead, regulating services include immaterial services, such as air quality and climate regulation, erosion prevention, biological control and maintenance of soil quality. Land use changes is considered one of the main drivers of biodiversity loss having an impact on the crucial life-supporting ecosystem services.

Another relevant issue related to biomass production is indirect land use change (iLUC). This results when pristine areas, typically grass lands and forests, are transformed into cultivation areas in order to produce biomaterials. Another example of iLUC is when agricultural land is displaced by biomass production and food production is relocated. This is global phenomena since iLUC may happen in nearby area or in other continent. Typically, forests absorb high levels of carbon dioxide and by converting forests to croplands, CO_2-emissions are indirectly increased due to the land use change. If the biomass production is not sustainably managed, it might have negative impact on climate change.

Furthermore, increasing population growth sets more requirements for food and feed production. FAO (2011) has stated that one real risk for food security is biofuels production since agricultural feedstock is used for the production of biofuels. Even though non-food biomass feedstock is used for biofuels production, biomass cultivation may still be competing with food production in terms of land use.

1.2 Energy

Energy requirements for biomass processing is one key element that should be discussed when evaluating value chain sustainability of biobased products. In case of fuel production, biomass have lower energy intensity compared to fossil resources indicating that required amount of biomass is remarkably higher. External, often fossil fuels are required to cultivate and harvest biomass feedstock, transportation of feedstock and products, and processing, which is often energy intensive (Zhang et al. 2015). In order to decrease energy consumption in value chain, location of feedstock farming and production site as well as end users should be optimized.

1.3 Water

Production and processing of biomass is water intensive. According to Gerbens-Leenes et al. (2009), water requirements of high-productivity biomass production is 70–400 times higher than the water requirements of fossil energy carriers. Increasing demand of biobased products will increase the consumption and competition on fresh water. Water withdrawal and especially overuse might damage freshwater natural ecosystem and water quality. Moreover, fresh water resources are not distributed equally and currently increasing threat of water scarcity has direct consequences on food security and general safety and human well-being (UNESCO, WWAP and UN-Water 2012).

1.4 Land Degradation and Nutrient Balance

Biomass cultivation has an impact on biodiversity loss through land use changes. Especially, if not sustainable managed, industrial biomass growing aims to maximize productivity achieved by monocultural cultivation practices requiring large inputs of fertilizers, water and pesticides (TEEB 2010). Monoculture and loss of biodiversity cause soil degradation and decline soil fertility. Nutrients deficiency is one of the main reasons for poor productivity. Poor soils hold poorly nutrients, which are easily leached to waterbodies. Agricultural practices are in a key role in maintaining soil fertility and biodiversity.

Furthermore, biomass contains nutrients that should be recycled back to the ecosystem. The importance of nutrient recycling is increasing due the fact that many fossil mineral sources are becoming scarce and simultaneously demand for food production is increasing. Many biomass production facilities produce side or waste streams with rich nutrient content. Therefore, it is important to identify raw materials, production technologies and production methodologies, which support nutrient recycling.

2 Status of Current Indicator Development and Use for Bioeconomy

Currently, a large number of sustainability frameworks and indicators have been developed to assess environmental sustainability of countries, companies and products. The common feature for various evaluation systems is that they are used for measuring the progress towards sustainability targets. As an example of sustainability metrics for countries, European Union has developed resource efficiency indicators to monitor Member States progress towards objectives and targets for resource efficient Europe initiative. Similarly, OECD has introduced the Green Growth Strategy (OECD 2011) and connected indicators to support countries' to achieve national growth, while at the same time ensure sustainable use of natural resources. Besides, companies have adopted sustainability reporting as an almost common procedure and sustainability reporting guidelines published by Global Reporting Initiative (GRI), has been popular among the companies. However, in this study we concentrate on sustainability assessment methodologies suitable for biobased value chains.

For evaluation of the environmental sustainability of biobased products and their value chains, several methods, tools and frameworks are available. SAMT project ("Sustainability assessment methods and tools to support decision-making in the process industries"), funded by European Commission, is currently evaluating existing sustainability assessment methods and tools used in process industry. The aim of the SAMT projects is to give recommendations about the most potential methods for sustainability evaluation. The evaluation encompasses 51 methods and 38 tools, which indicates the large number of different assessment methods. Many of these approaches are based on LCA-methodology, aiming for a comprehensive view on the environmental impacts covering the whole value chain of a product. Other similar studies have also conducted to clarify this complex fields of different sustainability assessment methodologies. For example, Singh et al. (2012) reviewed and categorized different sustainability indicators frameworks according their purpose. They were able to list 41 sustainability indicators frameworks within 12 different categories. Their study includes many different approaches and only some of them can be utilized for sustainability evaluation for biobased value chains. However, based on these studies, it is obvious that various different approaches to sustainability evaluation exists without unanimous view that how environmental sustainability should be measured.

Nonetheless, there are on-going initiatives for the development of comprehensive and streamlined indicator frameworks for environmental sustainability evaluation. As an example, European Commission has proposed the Product Environmental Footprint (PEF) method as a common procedure for environmental assessment. PEF—methodology is a multi-criteria measure of the environmental performance of a good or service throughout its life cycle. PEF information is produced for the overarching purpose of seeking to reduce the environmental impacts of goods and services taking into account supply chain activities (from extraction of raw materials, through production and use, to final waste management). Default environmental footprint categories included in PEF evaluation covers impacts to climate change and ozone depletion and formation, eco- and human toxicity, ionizing radiation and acidification and eutrophication potential. Also resource depletion of water and fossil resources are included. Land transformation potential is also included in assessment methodology. The PEF guideline lists additional environmental information, which the PEF could include. This non-exhaustive lists covers among the other things information on energy consumption, product recyclability, share of renewable resources and significant impacts on biodiversity on protected areas. Many of these are highly relevant for the bioeconomy, but currently very little used in practice and lack of standardized methods. Furthermore, one of the issues in the development of PEFs is looking at which issues to include in the Product Environmental Footprint Category Rules. In addition, possibilities to benchmark or compare products are under development (EC 2016). The object is to create benchmark for the specific product category in a way that all products that delivers the same function can be compared and benchmarked.

Another example is Environmental Product Declaration (EPD), which has also gained interest, especially in Europe. EPD is LCA-based sustainability reporting format for products. For different product groups, specific guidelines, product category rules (PCRs) have been established. The purpose of the EPDs is to provide transparent and comparable environmental life-cycle information of a product. PCRs define the requirements, rules and verification procedure that must be complied in order to verify that declarations of similar products are comparable. At the moment, no specific product category rules are established for non-food biobased products.

Both, PEF and EPD approaches are LCA-based product-centric assessment methods taking account special characteristics of different industries through the product category rules. In addition, sustainability assessment frameworks are also developed for specific industry sectors and their needs. For example, Institution of Chemical Engineers (IChemE) have developed sustainability metrics to guide industrial operations to enhance sustainable practices within industry (IChemE 2012). The guidelines encourage companies to measure environmental impacts caused by their own operation in order to follow progress towards more sustainable operations. These metrics are developed for companies own use and do not necessarily take into consideration the whole value chain of a product, which is important aspect when evaluation sustainability of biobased products. Also social and economic aspects were included in IChemE metrics as well as other evaluation frameworks. For example, BASF has developed an Eco-Efficiency Analysis tool especially designed for chemical industry. This assessment method utilizes LCA-based approach including also economic aspects of sustainability with the purpose to assist in decision-making processes. The emphasize lies in interpretation of the LCA results in a manner that non-LCA experts are able to easily understand results (Saling et al. 2002). Selected indicators cover energy and material consumption, emissions, and toxicity and risk potential. The results are weighted and normalized to make interpretation easier. However, weighting and normalization procedures make it difficult to compare results with other studies and have an impact on transparency of results.

The importance of specific sustainability evaluation method developed especially for biobased products has been recognized and different approaches are under development. For example, OECD (2012) has recognized the need to develop harmonized system for sustainability assessment of biobased products and encourages its member states to develop and implement assessment frameworks for biobased products by using life cycle approach. Currently, there are also some novel approaches to evaluate environmental sustainability which concentrates alone on biobased value chains.

Sustainability Assessment Unit of European Commission Joint Research Centre has established initiative to assess environmental sustainability of bioeconomy. The purpose of this initiative is to develop environmental sustainability indicators concerning biomass production, logistics and use. In addition, methodological tools for setting the sustainability criteria across the biomaterial value chains are under development. The initiative allows European Commission to assess progress and impact of bioeconomy (EC JRC 2015).

Furthermore, some other novel approaches have been established to evaluate sustainability of bioeconomy. As an example, US Environmental Protection Agency (EPA) has developed a unique tool, EnviroAtlas, to assess sustainability of a national bioeconomy. It is based on geospatial data providing also interactive sustainability evaluation at early stage. The tool includes data of ecosystem services and related impacts on human well-being. One purpose of the EnviroAtlas is to promote businesses to mitigate impacts of increased land use in USA and assist in decision-making and efficient resource management (EPA 2016; Walton et al. 2015). Even though, the tool does not cover the whole value chain of biomass production, it concentrates on benefits of ecosystem services, which is neglected in many sustainability evaluation frameworks.

Another example of novel approaches is S2Biom project (http://www.s2biom.eu/en/), co-funded by the European Commission. S2Biom project aims to "support sustainable delivery of non-food biomass feedstock in Europe". The purpose of the project is to develop a computerized toolset that provides easy access to data on different characteristics of biomass value chains and supports user to find best combination of biomass and conversion technology, measured in terms of technical suitability, economic efficiency and minimization of greenhouse gases. Ultimately, tool aims to analyze feasibility of biomass delivery chains in terms of availability and economic and environmental consequences in Europe.

3 Challenges and Gaps in Sustainability Assessment in Bioeconomy

Environmental sustainability assessment of biomaterial value chain should encompass relevant set of indicators to obtain the full and complete picture of biomaterial value chain and cover all the relevant characteristics and environmental impacts of bioeconomy. Furthermore, sustainability assessment frameworks should address bio-products contributions to global environmental challenges. Results of the sustainability assessment should also be easy to understand, relevant for the industry and identify the improvement potential or "hot spots" of the value chain. In addition, sustainability assessment should be able to deliver relative information for decision-making process.

Several approaches and on-going initiatives are trying to find the optimal solution for environmental sustainability assessment for the bioeconomy. Many of the current approaches utilize LCA-approach covering the whole value chain, but no united or standardized methodology exists yet, although many initiatives aim to standardize or harmonize a methodology. Furthermore, inconsistent scopes, allocations methodologies and different impact assessment methods cause incomparable results. Cristobal et al. (2016) concluded that strong need exists for methodological harmonization and more comprehensive LCA of bioeconomy value chains.

3.1 Complementary Indicators

Many of the existing initiatives are focusing on including complementary indicators to assessment framework to gain more comprehensive picture of environmental impacts. However, some special characteristics of bioeconomy are difficult to measure solely by LCA methods. As discussed earlier, bioeconomy has impacts on food and water security, loss of biodiversity, environmental impacts caused by land use changes, and ability to provide necessary ecosystem services.

Increasing demand for biomass utilization for non-food purposes may compete with food crops production. This is due to the fact that the land needed for food production may be diverted to biomass production if not sustainable managed. However, many of the sustainability assessment methodologies fail to include this aspect into evaluation framework. Beuchel et al. (2015) stated that even though environmental and social standards exist for biomass production, practical approach towards food security is missing in sustainability assessment frameworks.

In addition, many assessment methodologies concentrate on measuring the environmental impacts in terms of global warming, acidification, and eutrophication potential, which may have substantial impact on ecosystem services. Although, for example Burkhard et al. (2009, 2014) have developed a methodology to evaluate ecosystem service potential, this aspect is rarely included in sustainability assessment methodologies.

Moreover, methods that measure only direct GHG—emissions through the value chain fail to include impacts of indirect land use changes (iLUC) on GHG emissions. According to De Rosa et al. (2014), iLUC assessment methods are very versatile and no uniform methodology exists for evaluation for the iLUC.

3.2 Qualitative Methods

Typically, environmental sustainability indicators derived from LCA calculation are quantitative. Numeric values are easy to compare but sometimes environmental impacts might be difficult to express as a number and essential information can remain hidden. This is the case especially in bioeconomy, when biomass production influences on biodiversity and ecosystem services with indirect impacts for example for food and water security. In many times, these issues are also location specific, and also this characteristic should be able to include in assessment methodology. Furthermore, many biomass certification systems (e.g. FSC certification for forests) are based on qualitative or descriptive evaluation and similar approach can be used in sustainability evaluation when product is checked against predefined sustainability criteria. By this approach, for example agricultural or forest management practices can be included into evaluation. Although qualitative methods exist for assessing non-quantitative sustainability aspects of the bioeconomy, using of quantitative and qualitative methods in parallel requires careful scoping and consideration.

3.3 Benchmarking

Typically, the results of the sustainability assessment are expressed as quantitative figures, but without benchmarking it does not necessarily express anything about the performance level of sustainability. Comparing, or benchmarking help to understand how the product compares to other similar products; which are the main benefits or challenges in relation to other products. In addition, results of the sustainability assessment should be easy to understand and provide essential information for decision-makers.

There are different methods how the benchmarking can be conducted; companies can have their internal goals for which they compare achieved environmental performance or technology providers can benchmark their technology to competitors' in terms of environmental performance. Benchmarking can also compare different products/value chains that serve the same end use purpose. However, when discussing on sustainability, the assessment and benchmarking should cover the whole value chain from cradle to grave. Moreover, sustainability benchmarking of a product through its value chain is not straightforward. Typically, the main challenge is gathering the relevant reference data for benchmarking purposes.

4 Application of Environmental Sustainability Indicators in Bioeconomy: Case Studies

4.1 Benchmarking the Environmental Sustainability of Bio-Based Materials, Chemicals and Fuels

Environmental sustainability framework Gaia Biorefiner is developed for addressing the earlier described challenges in existing bioeconomy sustainability assessments and providing means for sustainability benchmarking. The indicator methodology has been applied also in the metallurgical industry (Rönnlund et al. 2016a, b, in press). Gaia Biorefiner methodology benchmarks the environmental sustainability of products and value chains. The assessment takes account of global challenges, such as climate change, raw material consumption and the environmental impacts of raw material processing. In addition, the assessment covers local factors, such as regional water scarcity and sustainable land use. Method of evaluating the sustainability of specific value chain is to study the value chain both with respect to its geographical location and the inputs and outputs of the value chain. Reference values are used to understand how the value chain compares to other similar products. Based on the results, alternative value chains can be easily benchmarked.

Gaia Biorefiner methodology is based on ten indicator groups (Fig. 1). The most important sustainability indicators are included providing a comprehensive and useful view on the most critical sustainability issues in the value chains. Both

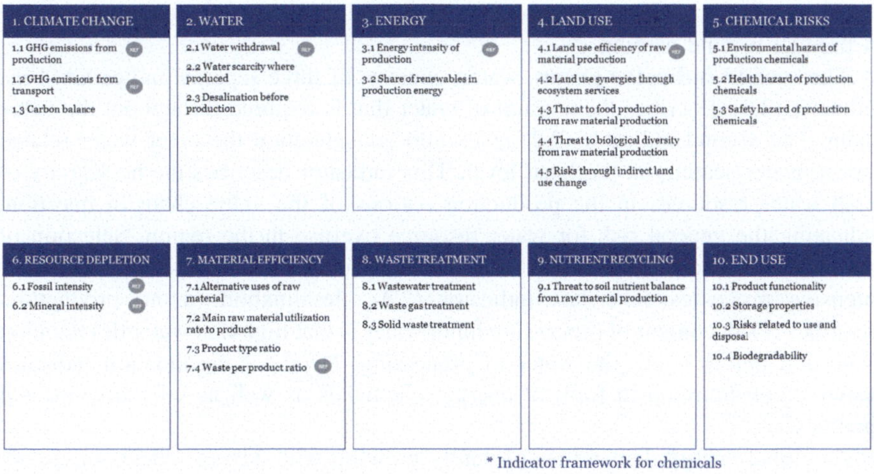

Fig. 1 Ten indicator groups and 30 indicators of Gaia Biorefiner benchmarking tool

potential competitive advantages and risks are identified. The value of each indicator is evaluated based on raw material, process, location and transport data. In the quantitative indicators, the result is benchmarked against a reference group and in the qualitative indicators against indicator specific alert lists and sustainability criteria. The indicator can give three levels of results: green indicating a possible competitive edge, red indicating a possible alert, and yellow being between these classes.

Three main product categories are included in the framework: bio-based materials, chemicals and fuels. Each product category has subcategories and each subcategory is connected to a specific reference group. Issues such as energy intensity, water intensity, fossil intensity and raw material intensity increase with the degree of refinement and value adding steps. By comparing products that belong to a certain product subcategory, it can be ensured that the products are competing in the right league. Reference date is updated regularly reflecting to the technology development.

The first indicator group, climate change, addresses the value chain contribution to greenhouse gases and carbon balance. First and second (1.1–1.2) indicators in this group measures GHG emissions from production and transportations as two separate indicators due to the fact that improvement measures are different. By this approach, improvement potential of processing and transportation are possible to manage separately. The GHG emissions from production covers direct emissions from production and indirect emissions from energy consumption throughout value chain. Emissions from transportation covers all the logistics, accounting both transportation means and distances. Greenhouse gas emissions can be reduced by utilizing as local raw materials and production concepts as possible.

The last indicator (1.3) of the group examines carbon balance indicating the ability of product to bind carbon.

The second indicator group, water, consists of three separate indicators. First indicator (2.1) describes the amount of water that is required throughout the value chain. The second indicator (2.2) takes into consideration the other water related aspect, water scarcity at the local level. This indicator describes the he scarcity of fresh water resources in the production regions of the value chain in question, indicating the general risk for water resource overuse in the region. Selection of production locations with low risk of water scarcity is crucial especially for water intensive processes. The last indicator (2.3), desalination before production, describes if salt content of processing input water is that high that water desalination is needed before using the water in processing. Need for desalination indicates resource consumption in form of energy, chemicals as well as infrastructure and facilities.

As global energy demand is constantly growing with adverse effects of energy production, energy efficiency and sustainable renewable energy resources are therefore one of the key design criteria of production concepts. Therefore, **the third indicator group** covers energy intensity of production as well as share of renewable energy resources (3.1–3.2) covering procurement and all the processing stages of value chain concerned. By this approach, this indicator guides towards using BAT or better as well as more sustainability energy forms.

Increasing pressure exists to use land effectively as land provides crucial ecosystem services, including food and feed production. **In the indicator group four**, impacts of the value chain to land use are measured with five indicators covering different aspects of land use. The first indicator (4.1) measures the land use efficiency, which indicates that how efficiently raw material is produced. Second indicator (4.2) describes land use synergies through ecosystem services, indicating whether the land area used for primary raw material production is simultaneously beneficial for other synergistic purposes, such as providing of ecosystem services like nutrient cycling, air purification, and recreation. The third indicator (4.3), threat to food production from raw material production, describes whether the raw material production is a threat to the food production at the local scale. Indicators 4.4 and 4.5 describes the threat to biological diversity and risks through indirect land use change, respectively. Threat to biodiversity describes if raw material is produced in areas that are identified as rich in biodiversity indicating potential threat e.g. through directly destroying such ecosystems or indirectly impacting their status and development. Risks through indirect land change addresses emissions and threat to food production and biodiversity from indirect land use change.

The fifth group of indicators concentrates on risk related to chemicals that are used during the production. Chemicals are used in processing and our everyday life in numerous ways for human benefit, however they also have negative properties that may result in a risk to the environment, health and safety. Minimization of chemical risks is also one aspect to take into account when benchmarking alternative processing technologies. Indicators 5.1–5.3 describes hazardous properties of used chemical for environment, health and safety, respectively. Risk estimates are based on the seriousness of the hazardous properties of the chemicals.

The **sixth indicator group** covers two resource depletion issues; fossil intensity (6.1) and mineral intensity (6.2). Fossil intensity describes the amount of fossil resources required to produce a certain product, whereas mineral intensity indicator describes whether mineral intensive steps are included in the value chain. Both raw material production (fertilizer use in cultivation) and processing (direct use of minerals and use of mineral-derived process chemicals) are included in the analysis. Mineral intensity can be minimized through selection of raw material and processing technology.

Material efficiency and related issues are included in **seventh indicators group**. The more efficiently the resources can be utilized, the fewer primary resources are required in producing the product. Often material efficiency is beneficial also in minimizing of adverse environmental impacts. The first indicator of the group (7.1) describes whether there are alternative uses for the raw material, e.g. in food production, which is prioritized use. Established alternative uses must be taken into account as well when assessing overall justification for raw material utilization. If there are no other established uses, and the raw material would otherwise be waste, the overall impact on the material efficiency is positive. The second indicator (7.2) indicates the utilization rate of materials. In material efficient production concepts nearly all fractions are utilized for sellable products, indicating efficient utilization of production side streams. Third indicator (7.3) describes the produced waste material and the last indicator of the group compares the share of material products versus energy products. This approach encompasses the general prioritized status of material use.

Also waste treatment is covered in **the indicator group 8.** Typically waste treatment cause negative environmental aspects and require resources and facilities/ infrastructure for processing purposes. Therefore, indicators 8.1–8.3 indicates how much unwanted liquid, gaseous, and solid output requires processing, respectively.

Ninth indicator group encompasses issues related nutrient recycling, which has become increasingly important as many fossil mineral sources are becoming scarce, global demand for food grows and nutrients are currently utilized inefficiently in many value chains. This is measured in one indicator (9.1) that describes the threat to soil nutrient balance from raw material production, indicating potential local adverse changes in nutrient balance in the environment.

The last group of indicators covers the end use of the product after it has been finished. The characteristics and functionality of the product, the life span and recyclability and the end of life treatment requirements are all important sustainability issues. The first indicator of a group (10.1) indicates if the functionality of the product is better, similar, or worse than comparable end product(s) for similar end uses. Product functionality is linked to sustainability e.g. through resource efficiency and environmental impacts. The second indicator (10.2), storage properties, points out the special storage requirements for the material, which are linked to sustainability e.g. via increased resource or energy requirements. Third indicator

(10.3) describes the hazards that are related to the product properties indicating risks related to end use. The last indicator (10.4) describes biodegradability, which indicates the disposal properties of a product.

4.2 Case Example: Biochemicals

4.2.1 Sustainability Benefits and Risks of Crude Tall Oil Based Tall Oil Fatty Acids

The sustainability benefits and risks of tall oil fatty acids (TOFA), produced from crude tall oil, and soybean oil fatty acids, produced from soybean oil were analyzed with Gaia Biorefiner indicator framework. Fatty acids from vegetable oils can be used for example for printing ink, oil paints and surfactants production. Gaia Biorefiner evaluates the quantitative indicators against a product specific reference group. Of the five reference product subcategories of chemicals (Main bulk chemicals, Intermediate bulk chemicals, Fine Chemicals and Pharmaceuticals), Intermediate Bulk Chemicals is chosen as the sub-category corresponding to the end products TOFA and soybean oil fatty acids. The reference group for Intermediate Bulk Chemicals is made of 20 common intermediate bulk chemicals (biomass based and petrochemical) and five oleochemicals. The results of the sustainability benchmarking are shown in Fig. 2.

The indicator based sustainability analysis shows that the value chain for the production of TOFA has several advantages from an environmental point of view. The only potential risk found in the analysis is related to solid waste treatment from the raw material production step, sulphate pulping. The production of Soybean Oil Fatty Acids has somewhat fewer advantages and also includes many potential risks, mainly related to raw material production and land use. The climate change indicators indicate that both value chains have a low impact on climate change due to GHG emissions from processing and raw material production. The production of TOFA occurs closer to the market, which results in lower GHG emissions from transport and gives competitive edge compared to Soybean Oil Fatty Acids. The production of TOFA's raw material crude tall oil is however water and energy intensive. Nevertheless, high use of renewable energy sources gives potential competitive edge and the high water intensity is not critical at the specific production locations as fresh water is not scarce in the production areas. The value chain for TOFA production raises no land use alerts, while these are raised in the value chain for Soybean oil production. Chemicals used in TOFA production have moderate or no hazards for environment, health and safety, while high risk chemicals are used in Soybean oil production. Use of fossil materials is low for both cases, but mineral intensity is higher in the value chain for Soybean oil production due to intensive fertilizer usage. TOFA production is very material efficient and the raw material crude tall oil is non-edible. On the contrary the raw material Soybean is edible and thus an alert is raised in the alternative uses of raw material indicator.

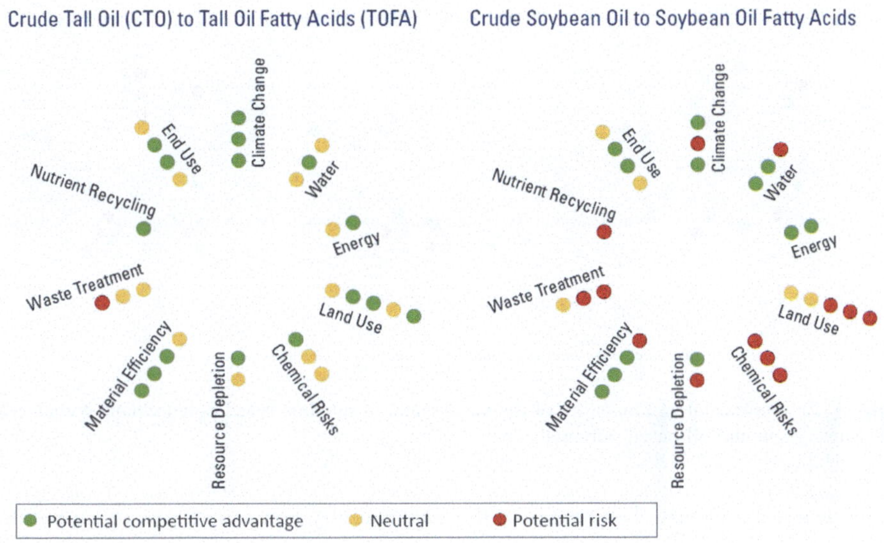

Fig. 2 Environmental sustainability of the value chain of TOFA compared to the value chain of soybean oil fatty acids

Waste treatments raise alert in both value chains. Regarding nutrient balance raw material for TOFA is produced from unfertilized forests and does not change the nutrient balance of N, P and K as much as agricultural production of soybean oil. Soybeans for soybean oil production are produced in areas where nearby waters are eutrophicated and soil is nutrient depleted.

4.3 *Case Example: Biomaterials*

4.3.1 Benchmarking the Environmental Sustainability of Three Raw Materials Used in Construction Insulation

The sustainability benefits and risks of sawdust based polyurethane and wheat straw and oil based polyurethane were analyzed with Gaia Biorefiner indicator framework. Polyurethane is used as insulation material in construction industry. Polyurethane is currently manufactured from crude oil based adipic acid as well as wheat straw based succinid acid. Option to produce succinic acid from sawdust is in research stage and the potential environmental sustainability aspects (positive and negative) were focus of this study. Even though the end product polyurethane is an insulation material, the reference group was built from bulk chemicals in order to better benchmark the differences between selected three raw materials and adipic as

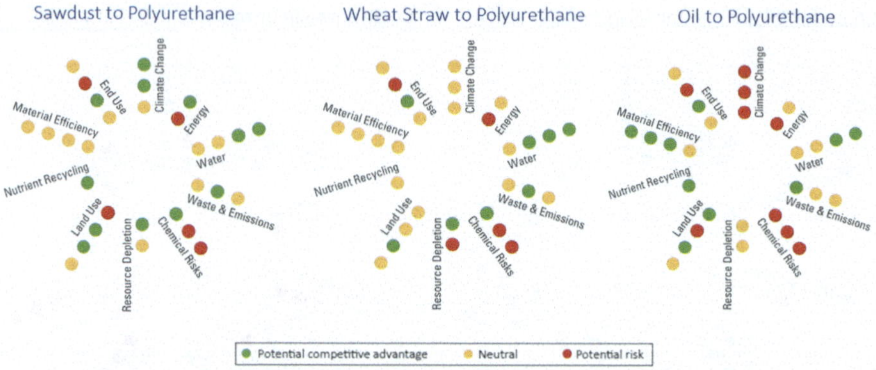

Fig. 3 Environmental sustainability of the value chain of sawdust based polyurethane compared to wheat straw and oil based polyurethane

well as succinid acids. The results of the sustainability benchmarking are shown in Fig. 3.

The indicator based sustainability analysis shows that sawdust value chain has several sustainability advantages as well as disadvantages from an environmental point of view. The potential competitive sustainability advantage of the sawdust to polyurethane value chain lies in its minimal impact on climate change, land use and resource depletion. This is due to high share of renewable resources as a raw material and energy source in the value chain. The potential environmental sustainability risk lies in material-efficiency. Oil-based polyurethane performs better in almost all material efficiency indicators. All three value chains are energy intensive compared to a reference group of other bulk chemicals. This is a potential risk inherent in all benchmarked value chains.

5 Conclusions

Nowadays, utilization of biomass for different purposes is encouraged and supported by national and global bioeconomy policies and strategies. As bioeconomy can be seen as a solution for these global challenges, it may have also adverse impacts on environment, in terms of land degradation, loss of biodiversity and decreased ability to deliver ecosystem services. Therefore, comprehensive evaluation methods covering the special characteristics of environmental sustainability should be employed for evaluation of the environmental impacts of biobased products through the entire value chain.

Currently, numerous different methods and frameworks exist for sustainability evaluation. LCA-based assessment is needed as it enables comprehensive view covering all the processing stages from raw material production to end-of-life procedures. Many existing sustainability assessment frameworks utilize LCA-based methods, however, more comprehensive approaches and harmonized methodologies are needed. Furthermore, qualitative methods are required when important aspects cannot be turned into numerical values.

The Gaia Biorefiner indicator methodology was introduced as an example of a comprehensive approach combining the qualitative and quantitative sustainability indicators for bioeconomy value chains. The case studies presented on benchmarking of tall oil fatty acid and saw dust-based succinic acid with comparable value chains confirm that the holistic approach towards sustainability evaluation for biobased value chain is required. As seen in the case studies, in some sustainability impact categories, such as climate change, the biobased value chains can have potential sustainability advantages while simultaneously comparable fossil value chains perform equally or even better in certain other sustainability aspects, like material efficiency. It is important to be able to identify also the possible sustainability trade-offs of bio-based products compared to relevant reference products. With this information bio economy development can be focused into the applications with best overall sustainability impact against reference products, and the challenging impact targets of the bioeconomy development can thus better be reached. The case studies also show the benefits of the benchmarking as it provides a meaning to calculated values and makes the interpretation of results easier. However, it should be noted that data availability and reliability should be developed further to conduct more comprehensive and transparency evaluation methods.

By this approach, the field of bioeconomy as a whole can be develop towards more sustainable operations and practices. For policy makers the presented methods and similar indicator based systems give a tool to analyze externalities and wider sustainability impacts of policy choices. In bioeconomy regulation plays an important role and holistic analysis of impacts of the policy choices i.e. on incentives for bioeconomy is the pre-requisite for informed decisions. For companies and investors these methods enable holistic benchmarking of various biomass sources, production and product options. Sustainability risks can be highlighted and positive externalities identified, which support decision making on tomorrow's bioeconomy value chains. In the future we see that these approaches can be applied to a wider amount of case products in order to gain more comprehensive pool of test cases.

The limitations of the indicator methods are in its focused applicability. As the method addresses individual value chains, the results cannot be scaled to wider level, addressing biomass use at large and sustainability aspects in the system level. In addition, benchmarking between different value chains is only possible if system scopes can be defined similarly.

References

Beuchelt T, Mohr A, Schneider R, Virchow D (2015) Integrating food security aspects in biomass sustainability standards and certifications through rights-based criteria. Global Bioeconomy Summit, Berlin, 25–26 Nov 2015

Bukhard B, Kroll F, Muller F, Windhorst W (2009) Landscapes' capacities to provide ecosystem services—a concept for land-cover based assessments. Landscape 15:1–22. https://doi.org/10.3097/LO.200915

Bukhard B, Kandziora M, Hou Y, Muller F (2014) Ecosystem service potentials, flows and demands—concepts for spatial localisation, indication and quantification. Landscape Online 34:1–32. https://doi.org/10.3097/LO.201434

Cristóbal J, Matos CT, Aurambout J-P, Manfredi S, Kavalov B (2016) Environmental sustainability assessment of bioeconomy value chains, biomass and bioenergy. Available online 12 Feb 2016 (in press)

De Rosa M, Schmidt J, Trydeman Knudsen M, Hermansen JE (2014) Methodologies accounting for indirect land use change (iLUC): assessment and future development. In Proceedings of the 9th international conference on life cycle assessment in the agri-food sector (LCA Food 2014), San Francisco, USA. ACLCA, Vashon, WA, USA, 8–10 Oct 2014

EPA (2016) United States Environmental Protection Agency. EnviroAtlas—ecosystem services in enviroatlas. https://www.epa.gov/enviroatlas/ecosystem-services-enviroatlas. Accessed 13 Apr 2016

European Commission (EC) (2012) The bioeconomy strategy. http://ec.europa.eu/research/bioeconomy/index.cfm?pg=policy&lib=strategy. Accessed 14 Apr 2016

European Commission (EC) (2016) Product environmental footprint pilot guidance—guidance for the implementation of the EU Product Environmental Footprint (PEF) during the Environmental Footprint (PEF) pilot phase, Version 5.2. http://ec.europa.eu/environment/eussd/smgp/pdf/Guidance_products.pdf. Accessed 14 Apr 2016

European Commission Joint Research Centre (EC JRC) (2015) Bioeconomy information system and observatory (BISO)—environmental sustainability assessment—progress workshop. https://biobs.jrc.ec.europa.eu/sites/default/files/generated/files/documents/1%20%20BISO%202015%20Workshop%20Summary_151002%20FINAL.pdf. Accessed 14 Apr 2016

FAO (2011) Looking ahead in world food and agriculture—perspectives to 2050. In Conforti P (ed). http://www.fao.org/docrep/014/i2280e/i2280e00.htm. Accessed 13 Apr 2016

Gerbens-Leenes PW, Hoekstra AY, van der Meer Th (2009) The water footprint of energy from biomass: a quantitative assessment and consequences of an increasing share of bio-energy in energy supply. Ecol Econ 68:1052–1060

IChemE (2012) The Institution of Chemical Engineers, the sustainability metrics—sustainable development progress metrics recommended for use in the process industries. Available online https://www.icheme.org/communities/subject_groups/sustainability/resources//~/media/Documents/Subject%20Groups/Sustainability/Newsletters/Sustainability%20Metrics.pdf. 14 Apr 2016

OECD (2011) Towards Green Growth. OECD Publishing, Paris, OECD Green Growth Studies. https://doi.org/10.1787/9789264111318-en

OECD (2012) Recommendation of the council on assessing the sustainability of biobased products. Available online 14 Apr 2016

Rönnlund I, Reuter M, Horn S, Aho J, Aho M, Päällysaho M, Ylimäki L, Pursula T (2016a) Eco-efficiency indicator framework implemented in the metallurgical industry: part 1—a comprehensive view and benchmark. The International Journal of Life Cycle Assessment, accepted (in press)

Rönnlund I, Reuter M, Horn S, Aho J, Aho M, Päällysaho M, Ylimäki L, Pursula T (2016b) Eco-efficiency indicator framework implemented in the metallurgical industry: part 2—a case study from the copper industry. Int J Life Cycle Assess, accepted (in press)

Saling P, Kicherer A, Dittrich-Krämer B, Wittlinger W, Zombik W, Schmidt I, Schrott W, Schmidt S (2002) Eco-efficiency analysis by BASF: the method. Int J LCA 7(4):203–218

Singh RK, Murty HR, Gupta SK, Dikshit AK (2012) An overview of sustainability assessment methodologies. Ecol Ind 15:281–299

TEEB (2010) The economics of ecosystems and biodiversity ecological and economic foundations. In Kumar P (ed) Earthscan, London and Washington

UNESCO, WWAP and UN-Water (2012) Managing water under uncertainty and risk—The United Nations world water development report 4: executive summary. Accessed http://www.zaragoza.es/contenidos/medioambiente/onu//newsletter12/789-eng-sum-ed4.pdf. 18 Apr 2016

Walton BT, Perla D, Smith ER (2015) Sustainability assessment of a national bioeconomy: approaches and tools to evaluate emergent properties. SETAC North America 36th annual meeting

World Economic Forum (WEF) (2016) The global risk report 2016, 11th edition. http://www3.weforum.org/docs/GRR/WEF_GRR16.pdf. Accessed 12 Apr 2016

Zhang F, Johnson DM, Wang J (2015) Life-cycle energy and GHG emissions of forest biomass harvest and transport for biofuel production in Michigan. Energies 8:3258–3271

A Macroeconomic Perspective on Green Growth

Mounaim Sebastian Rhozyel and Jolanta Žalpytė

Abstract Prevailing political economy is failing to maintain environmental, social, political and economic coherence. A fundamental shift towards a new economic model is needed 'where the acknowledged priority is to sustain human and natural communities' (GTI in Beyond the growth paradigm: creating a unified progressive politics, 2011: 1). Therefore, it is widely accepted that the current linear so-called "take-make-dispose" economy is not sustainable and that solutions need to be found in order to decrease both the input of limited resources as well as the output of human waste in any form. This paper examines how far the ideas of green growth are capable of handling this problem as they promise material welfare while reducing the impacts on the environment. Even though the underlying approaches like a circular economy, zero-emissions economy or Factor X seem to be desirable, there is reasonable scepticism in how far advances in resource efficiency can ensure a sustainable future while industrialised societies are constantly raising living standards. We show that there is an intrinsic contradiction within the desire for green growth to be an integral step into a sustainable future.

Keywords Circular economy · Material throughput · Green economy Green growth · Sustainable development

1 Introduction

The following paper points out that the current patterns of generating economic welfare do not necessarily contribute to a total reduction of the material throughput even after the resource efficiency was significantly raised. In order to stop this trend,

M. S. Rhozyel
Tamen GmbH, Berlin, Germany
e-mail: mounaim.rhozyel@outlook.de

J. Žalpytė (✉)
Car2go Deutschland GmbH, Berlin, Germany
e-mail: j.zalpyte@gmail.com

W. Leal Filho et al. (eds.), *Towards a Sustainable Bioeconomy: Principles, Challenges and Perspectives*, World Sustainability Series,
https://doi.org/10.1007/978-3-319-73028-8_4

different terminology and concepts emerged to describe an economic system that goes hand in hand with nature (green economy, low carbon economy, circular economy, bioeconomy etc.).

It is very often being propagated that world development is driven by growth. Subject to sustainable development, it is said that this growth should be "greener". We will give a short overview of these green visions that all claim to ensure both economic prosperity and a reduction of the material throughput. Economists agree that economic growth must be decoupled from the material throughput but the current trend does not demonstrate a promising future. Consequently, the voices which argue that a renunciation of the growth paradigm will become inevitable in order to take pressure off the environmental capacities are getting louder. This provides a source of controversy not only in the scientific world. A shift from quantitative growth to qualitative change is needed.

This discussion only slowly finds its way to the political arena so far. Governments and businesses consider eco-efficient technologies as a potential solution for resource scarcity as well as a driving force for innovations, and of course growth. Nevertheless, there are movements like "degrowth" that are gaining more and more attention in the general public. Also, science is starting to offer rationales why savings of the material throughput on one end often lead to emissions on the other. The so called rebound effect (e.g., to buy a more efficient car but drive more).

This paper gives an insight into this intense discussion and addresses the question whether green growth is a conceivable reality or a perspective for sustainable development. In doing so, we propose a simple model that describes a crucial contradiction within the concept of green growth on a macroeconomic level.

2 Conventional Industrial Metabolism

The functionality of an economy or economic activity in general can be best described in the most abstract but plausible way: natural resources get extracted from the biosphere, are then either used directly or transformed into certain consumption goods and thereafter sold to an end-consumer. At the end of this process is the release of waste, emissions and unnecessary resource losses. So the biosphere provides the source of the material on one hand and the sink for waste on the other in order to allow economies to run (Paech 2005). In this sense the process of production and consumption is linear and stresses both, the capability of the biosphere to offer natural resources as well as its capability to serve as a sink for waste in any form. In this regard, mankind interrupts the natural flow of matter and energy.

The global resource extraction of the year 1980 is likely to double by 2020, which reflects the growing demand for natural resources (OECD 2008). Even though drastic technological improvements in energy and resource efficiency were implemented over the last 3 decades, they have been compensated by a massive economic growth. Global carbon emissions from energy use have increased by 40% since 1990 (Jackson 2009a). There is no doubt that the total amount of economic activity dictates both the resource input as well as the emissions output. The Russian Federation is the only large economy (beside Germany) which reduced its carbon emissions extensively since 1990, as a consequence of the breakdown of its heavy industries. This illustrates how strongly the Gross Domestic Product (GDP) correlates with the emissions emitted into the atmosphere and with the material throughput in general. According to Hoffmann (2011), not only the technologically leading countries would have to endure a similar economic breakdown increase threefold in order to limit the rise of global temperature to a maximum of 3 °C.

In this regard, the degree of sustainability of material welfare corresponds to the amount of physical materials circulating throughout the global economy: the material throughput. This notion describes the relationship between the physical size of the global economy and the ecosystems it is influencing. The aforementioned functionality of any economic activity describes the chain of actions impacting the physical world, depending on energy and at the same time degrading ecosystems. Consequently, the Gross World Product (GWP) is the measure of an economic activity and at the same time a measure of the material throughput: the amount of matter and energy used in each production-consumption process (Daly and Farley 2004). Increasing population and consumption result in a higher material throughput. Hence, "understanding throughput is central to understanding sustainability" (Santa-Barbara et al. 2005: 2) because the global economy simply would not work without natural resources and ecosystem services. Further environmental degradation can be limited if the absolute limits to the material throughput are set up (Santa-Barbara et al. 2005).

Moreover, this paper proceeds with a vision of a green economy. It presents options to avoid natural resource scarcity and nature degeneration in order to achieve a balance where nature and economics meet sustainability targets.

3 Vision of the Green Economy

The rising demand for resources and steadily increasing emissions of human waste stress the limits of the biosphere. 'Earth's regenerative capacity can no longer keep up with demand—people are turning resources into waste faster than nature can turn waste back into resources' (WWF 2006: 2). New implications are necessary in order to transform the economic system towards a stable and sustainable combination which in the best case scenario covers the needs human beings have. Senge et al. (2001) predict a "Next Industrial Revolution" that aims at creating the ideal sustainable bio-based economy. The main conditions for the vision of such economy are a provision of sufficient and healthy global food supplies as well as a production of high quality products from renewable raw materials (BMBF 2011).

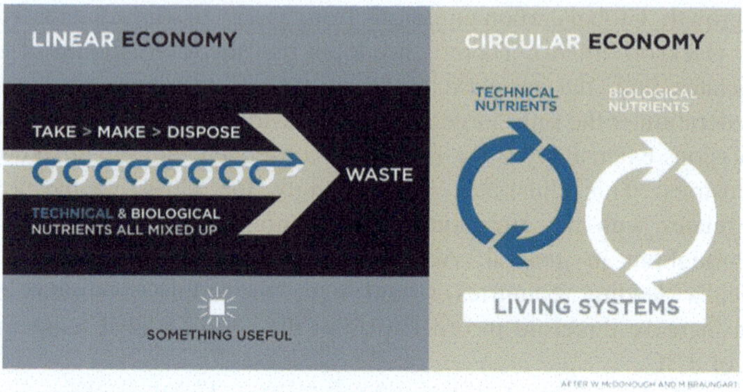

Fig. 1 End of the line or full circle? (EMF 2012)

Such a visionary economy would be characterized by having no stressing impact on ecosystem services: neither on the use of natural resources nor on the sink functions that ecosystems provide, regardless if the underlying approach is called circular economy (e.g., Preston 2012; UNEP 2006), zero-emissions economy (e.g., Ayres 2004; Baumgartner and Zielowski 2007) or Factor X (e.g., Beyers 2005; von Weizsäcker et al. 1995). All these concepts demand a so-called decoupling of economic activity from its material throughput, so that the economic system goes conjointly with the natural system.

Since nature works in perfectly closed cycles without producing any waste that is not further reused (Senge et al. 2001), the current linear pattern of production and consumption should accordingly be 'replaced by systems that reuse resources and conserve energy' (Preston 2012: 2). Following the example of nature, such a circular economy would use old products to make new ones and reintroduce biological ingredients back into the biosphere (see Fig. 1).

The demand for energy to further process unsuitable nutrients for the biosphere such as metals or plastics is generated by renewable energy sources. Wind, sun and biomass are abundant, the technologies either already exist or will be further developed (Paech 2010). A fix amount of resources would constantly provide material welfare to mankind. The previously mentioned conception idealizes the illusion of a superlative world.

4 Towards the Green Economy

For shaping the progress towards a perfectly green economy, several key elements of current production and consumption patterns need to be transformed. It will be inevitable to reorganise the structure of industrial systems both on a local and on a global scale. Material flows must be coordinated in order 'to use the "waste" streams from one factory as a resource for other companies and consumers'

(Preston 2012: 5). Using industrial waste heat to provide heating for households would be a well-known example on the local scale in this regard. Following this idea, several so-called eco-industrial parks (e.g., in Kalundborg, Denmark) were constructed in which the facilities provide each other with residual materials and energy and thus, generate diverse synergies (Preston 2012). On a large scale, industrial material flows need to be massively intensified in a way that they once may function in a closed cycle.

The construction of products must be revolutionised so they can either be recycled and reused or biodegraded. Therefore, it is crucial to explore technologies which aim at the material separation of goods so that waste becomes a new source of raw materials (Baumgartner and Zielowski 2007). A so called recycling concept of the "cradle-to-cradle" production does not minimise the total material flows but rather transforms them into a more cyclical functioning (Högner et al. 2012). It is argued that this would form 'a supportive relationship with ecological systems and future economic growth' (Braungart et al. 2006: 1338).

As both the eco-efficient construction of products as well as the reorganization of industrial systems aim at the production-side, it is important to mention that the consumption-side holds a large potential to "green up" economic systems. There are several concepts aiming at higher resource efficiency by changing consumption patterns. Many of them propose a shift away from conventional ownership towards a consumption system that is characterized by 'sharing, bartering, lending, trading, renting and gifting' (Preston 2012: 10). This social innovation requires less physical materials to satisfy the people's needs. More importantly, it results in more sustainable consumer behaviour (EEA 2016). In how far material welfare can contribute to humanity's perceived wellbeing is another debate that is not part of this work but should always be kept in mind.

Meanwhile, the necessity of "greening up" the patterns of production and consumption are especially accepted in large parts of the industrialised world. It is even assumed that innovative green technologies might be key to push economic prosperity. A new dogma is born—the "green growth imperative".

The idea is that this immense transformation can be carried out by forces which are already shaping our current economic system because 'change through market-driven innovation is the type of change our society understands best' (Senge et al. 2001: 26). Green growth implies an intrinsic faith that the relief of the biosphere itself can serve as a driver of further increasing material wealth (Paech 2010) which can create a win-win scenario. In the following part we analyse if the reduction of the material throughput fits into the concept of economic growth.

5 A Critical Discussion on Green Growth

Green growth can be a perspective for creating new growth notions with a reduced negative environmental impact while still promoting related technological and structural change. But can an increase of economic activity be a strategy to decrease environmental degradation? Aren't the "economic goods" creating "environmental bads"? How to resource the world?

Many questions arise regarding this dubiousness. Moreover, it is a question of income equality and population growth: increased Green House Gas (GHG) emissions and consumption is not related to increased population as long as the majority of the people remain poor. But what about one of the main Millennium Development Goals: "Eradicate extreme poverty and hunger"? There are many questions to answer when debating about economic prosperity in terms of sustainable development.

However, this paper focuses strictly on the macroeconomic dynamics which go along with the concept of green growth and its ability to guide mankind into a sustainable future.

6 The Impact of "Green Growth" on the Material Throughput

In order to deal with the strain on both the sink as well as the resource capacity nature provides to human beings it is important to clarify in which way this should be measured. The relevant number in this context can only be the absolute resource use and the absolute emissions since nature itself works in strictly in absolute terms as well. In order to understand green growth ideas it is important to make a clear distinction between the absolute values of resource in- and emissions output as well as the relative resource in- or emissions output per unit of GDP (Behrens et al. 2007). Jackson (2009b) uses the term "ecological intensity" in this context. This contains indicators with regard to waste disposal such as the carbon intensity along with the resource use serving as the fossil fuel intensity.

During the last decades, several improvements were achieved referring to ecological expanses. For example, the global carbon intensity was reduced from about 1 kg CO_2 per $US in 1980 to less than 770 g of CO_2 per $US in 2006 (Jackson 2009b). Nevertheless, the absolute CO_2 emissions continued to grow. Hence, the improvements in material efficiency (in this case the carbon efficiency) were simply offset by economic growth.

Meanwhile the question arises how to set the concept of green growth into this frame. Primarily, it requires to consider the effects of the process of "greening" an economy on the resource in-and/or the emissions output. The aforesaid deviation from an economy with a high material throughput to an economy with a downsized one certainly diminishes the resource use in absolute terms. It is therefore greatly commendable and indispensable.

Accordingly, green growth has the potential to influence the relative dematerialization positively whereas growth seems to have an increased impact on the absolute resource use. Indeed, this is a given as long as the economic performance is positively coupled with the material throughput even if it is assumed to be close to zero. In other words, additional economic welfare that is generated with the aid of natural resources must by definition raise the absolute use of resources, regardless to the underlying relative dematerialization. Thus, economic growth without an increased impact on the absolute resource use can only be achieved in case of a complete decoupling of the economic performance and resource use, for example, in the case of the perfect circular economy.

What does that tell us? Obviously there is a clear contradiction within the idea of green growth itself. As previously mentioned, the "green" part of the term implies the reduction of the ecological intensity while in contrast growth implies a strict increase of the material throughput. Therefore, an environmental policy can only be successful if the effect of lowering the environmental intensity succeeds the effect deriving from growing economic activities, meaning that 'the decrease in material, energy and pollution intensity is higher than the economic growth rate' (Giljum et al. 2005: 33).

7 Condition to Achieve an Absolute Reduction in Material Throughput

In order to understand the correlation between ecological intensity and economic growth, we propose a simple model presented in Box 1. It shows in a comprehensible way that a total reduction of material throughput can only be achieved if condition (1) is fulfilled. Furthermore, the absolute material throughput reduces above average in case both the ecological intensity and the change in GDP become negative (being below 1). This is because the correlation is not linear, so that in theory a shrinking economy that reduces its ecological intensity is more efficient in taking pressure off the biosphere than an economy with a stable GDP which reduces its ecological intensity even more. The underlying policy implication deriving from this could be stated as both "go green" and "de-growth" at once.

Box 1: A Simple Model

To visualize the interrelation between the material throughput and the economic activity (expressed as the GDP), we simply set up the following equation, where E_t is the absolute material throughput (measured, e.g., by CO_2 emissions or resource use in tons or any equivalent that is of interest for measuring the material throughput) given for a certain period. It is expanded by the absolute GDP as well given in period t. Of course this is also valid for any previous and following periods, so that

$$E_t = \frac{\mathrm{E}_t}{\mathrm{GDP}_t} \times GDP_t \quad \text{and} \quad E_{t+1} = \frac{\mathrm{E}_{t+1}}{\mathrm{GDP}_{t+1}} \times GDP_{t+1}$$

are given. An absolute increase of the material throughput is avoided in case

$$E_{t+1} \leq E_t$$

is fulfilled and respectively

$$\frac{E_{t+1}}{GDP_{t+1}} \times GDP_{t+1} \leq \frac{E_t}{GDP_t} \times GDP_t.$$

Since $\frac{E_t}{GDP_t}$ and $\frac{E_{t+1}}{GDP_{t+1}}$ simply express a certain ecological intensity, $(e_t \text{ and } e_{t+1})$ we simplify the equation to

$$\frac{e_{t+1}}{e_t} \leq \frac{GDP_t}{GDP_{t+1}}$$

and consequently to

$$\Delta e \leq \frac{1}{\Delta GDP} \quad (1)$$

It is interpretable as the condition that must be fulfilled so that an absolute increase in the mateiial throughput is avoided in the observed period of time.

An additional finding which is expressed in condition (1) is the possibility to reduce the absolute material throughput while the GDP increases. In this case, as previously mentioned, the reduction of the ecological intensity must exceed the effect caused by growth.

In conclusion, we developed a very simple but significant condition that policy makers should take into account if sustainable development, as a goal for the environmental and economic policy, is persistently considered. Not expanding the economic activity is a potentially strong instrument for achieving a sustainable pathway of development.

8 The Growth Dilemma

At this point it is necessary to raise the question in how far it is possible for an economy to "de-grow". Without going into detail, some core principles should be mentioned.

Modern economies are based on constant advances in technology related to efficiency gains. This leads to a downgrading of labour since less work is needed to produce the same amount of goods and services (Jackson 2009b). Thus, in order to prevent unemployment and a resulting drop in demand for means, such economies compensate this by the expansion of their economic activity: they grow. The problem is not realizing what the main goal of the growth is. Shouldn't this goal be to eliminate poverty?

Another reason for growth being inevitable lies within the monetary system and in the nature of debts. Making debts implies that you are forced to pay them back, including an additional amount of money deriving from a certain interest rate. This difference requires an additional generation of money. A state in debt can compensate this either by printing additional money (resulting in inflation) or generating economic growth (O'Neill et al. 2010).

In case of a shrinking economy, these described core requirements cannot be fulfilled anymore and the economy might enter into a spiral of recession which entails a risk of system collapse. The crisis in Southern Europe, especially in Greece, demonstrated this dilemma in a very tragic way. The fact that Greece was forced to retrench their spending made the Greek economy shrink. And this lead to a dramatic increase of the unemployment rate, political instability and thereby disrupted the Hellenic nation. Accordingly, economic growth was and still is urgently needed. Jackson (2009b) describes this dilemma to the point where he states that 'growth may be unsustainable, but „de-growth"appears to be unstable' (Jackson 2009b: 8).

9 Conclusion

What we can learn from this paper is that the massive increase of economic activity is closely related to both the input of extracted natural resources as well as the output of harmful human waste. Mankind stresses the limits of what the biosphere is able to handle and thus threatens its own livelihood. The risks deriving from the current linear production and consumption patterns are widely recognised, especially in those parts of the world that are mainly responsible for the huge throughput of natural resources. It means that our current lifestyle is not sustainable and natural resources are under threat.

For this reason solutions are needed but only compromises were found so far. Several visions were developed that all have the common aim of an economic system which ensures material welfare and prosperity without having negative impacts on the environmental dynamic. The "growth imperative" gets continuously converted into a "green growth imperative". Green technologies offer a promising future for growth.

On a macroeconomic level we have shown that there is a clear contradiction within the concept of green growth since economic growth strictly implies a higher level of the material throughput. Abating ecological intensity can only have an advert effect on this. An absolute reduction of the material throughput can solely be achieved if the decreasing effect of the ecological intensity exceeds the increasing effect deriving from growth. Nevertheless, the strategy of reducing economic activity while being more resource- and energy efficient would be above the average efficiency. Unfortunately, policies aiming at generating a lower GDP are not very popular because this would confront the structure of current market economies. This can be seen the best in the developing countries because it is not easy to promote a de-growth strategy where growth is the main macroeconomic incentive. This growth dilemma will become more famous the closer humanity gets to the limit of the world's biosphere system.

References

Ayres RU (2004) On the life cycle metaphor: where ecology and economics diverge. Ecol Econ 48:425–438

Bartlett AA (1978) Forgotten fundamentals of the energy crisis. Am J Phys 46(9):879–880

Baumgartner RJ, Zielowski C (2007) Analyzing zero emission strategies regarding impact on organizational culture and contribution to sustainable development. J Clean Prod 15:1321–1327

Behrens A, Giljum S, Kovanda J, Niza S (2007) The material basis of the global economy. Worldwide patterns of natural resource extraction and their implications for sustainable resource use policies. Ecol Econ 10 (in press)

Beyers B (2005) Wirtschaftswachstum und Wohlstand entkoppeln. FACTOR X – Grundlagen, Aachener Stiftung Kathy Beys, p 9

Braungart M, McDonough W, Bollinger A (2006) Cradle-to-cradle design: creating healthy emissions—a strategy for eco-effective product and system design. J Clean Prod 15:1337–1348

Daly HE, Farley J (2004) Ecological economics: principles and applications, 2nd edn. Island Press, Washington, D.C., p 544

Ellen Macarthur Foundation (EMF) (2012) "Good rather than less bad": the circular approach. Circular Economy. The Principles

European Environment Agency (EEA) (2016) Circular economy in Europe. Developing the knowledge base, No. 2. Available online 10 Mar 2016: file:///C:/Users/jz/Downloads/Circular%20economy%20in%20Europe.pdf

Federal Ministry of Education and Research (BMBF) (2011) National research strategy BioEconomy 2030, our route towards a biobased economy

Giljum S, Hak T, Hinterberger F, Kovanda J (2005) Environmental governance in the European Union: strategies and instruments for absolute decoupling. Int J Sustain Develop 8(½):31–46

Great Transition Initiative (GTI) (2011) Beyond the growth paradigm: creating a unified progressive politics

Hoffmann U (2011) Some reflections on climate change, green growth illusions and development space. United Nations Conference on Trade and Development (UNCTAD) Discussion Paper, No. 205

Högner S, Freitag O, Hube G (2012) Ökologische Nachhaltigkeit in klein- und mittelständischen Betrieben, Schriftenreihe der IHK Würzburg-Schweinfurt, No. 31/2012. Available online 18 Mar 2016. https://www.wuerzburg.ihk.de/fileadmin/user_upload/downloads/Presse/Schriftenreihe/Oekologische_Nachhaltigkeit.pdf

Jackson T (2009a) Industrial ecology in Europe. Beyond the growth economy. J Ind Ecol 13 (4):487–490

Jackson T (2009b) Prosperity without growth? The Transition to a sustainable economy

McDonough W, Braungart M (2002) Cradle to cradle: remaking the way we make things, 1st edn. North Point Press, New York

OECD (2008) Measuring material flows and resource productivity. Synthesis report. Available online 10 Mar 2016: http://www.oecd.org/env/indicators-modelling-outlooks/MFA-Synthesis.pdf

O'Neill DW, Dietz R, Jones N (2010) Enough is enough: ideas for a sustainable economy in a world of finite resources. In: Steady state economy conference. Centre for the Advancement of the Steady State Economy and Economic Justice for All, Leeds, UK

Paech N (2005) Nachhaltigkeit zwischen ökologischer Konsistenz und Dematerialisierung: Hat sich die Wachstumsfrage erledigt? Natur und Kultur 6(1):52–72

Paech N (2010) Vom grünen Wachstumsmzthos zur Postwachstumsökonomie

Preston F (2012) A global redesign? Shaping the circular economy. The energy, environment and resource governance research area (EERG) at Chatham House, London

Santa-Barbara J, Czech B, Daly HE, Farley J, Malghan D (eds) (2005) Sustainable scale in environmental education: three rules, two perspectives, one overriding policy objective, and six cultural shifts. In: Proceedings of the centre for environment education 2005 conference, Ahemedabad, Gujurat, India

Senge PM, Carstedt G, Porter PL (2001) Innovating our way to the next industrial revolution. MIT Sloan Manage Rev 42(2):24

United Nations Environment Programme (UNEP) (2006) Circular economy. An alternative model for economic development

Von Weizsäcker EU, Lovins AB, Lovins LH (1995) Faktor Vier: doppelter Wohlstand – halbierter Naturverbrauch: der neue Bericht an den Club of Rome. Droemer Knaur, München

WWF (2006) Living planet report. Available online at http://assets.panda.org/downloads/living_planet_report.pdf

Part II
Bioeconomy: Industry, Market and Financing Possibilities

The Biodiversity Finance Initiative: An Approach to Identify and Implement Biodiversity-Centered Finance Solutions for Sustainable Development

Marco Arlaud, Tracey Cumming, Ian Dickie, Marlon Flores, Onno van den Heuvel, David Meyers, Massimiliano Riva, Andrew Seidl and Annabelle Trinidad

Abstract Global biodiversity finance needs are on the order of $150 to $440 billion per year. The Biodiversity Finance Initiative—BIOFIN (www.biodiversityfinance.net)—is a global partnership addressing the biodiversity finance challenge in a comprehensive manner in order to identify and implement biodiversity-centered sustainable economic development solutions. BIOFIN provides an innovative accounting and finance approach, enabling countries to measure their current biodiversity expenditures, assess their financial needs and identify the most suitable finance solutions to achieve their stated national biodiversity targets and green,

This chapter is derived from the following report: UNDP (2016). The 2016 BIOFIN Workbook: Mobilising resources for biodiversity sustainable development. The Biodiversity Finance Initiative. United Nations Development Programme: New York. http://www.biodiversityfinance.net.

M. Arlaud · T. Cumming · M. Flores · O. van den Heuvel · D. Meyers · M. Riva · A. Seidl (✉) · A. Trinidad
Biodiversity Finance Initiative, United Nations Development Programme, New York, USA
e-mail: Andrew.seidl@colostate.edu

M. Arlaud
e-mail: marco.arlaud@undp.org

T. Cumming
e-mail: tracey.cumming@undp.org

M. Flores
e-mail: marlon.flores@undp.org

O. van den Heuvel
e-mail: onno.heuvel@undp.org

D. Meyers
e-mail: david.meyers@undp.org

M. Riva
e-mail: Massimiliano.riva@undp.org

W. Leal Filho et al. (eds.), *Towards a Sustainable Bioeconomy: Principles, Challenges and Perspectives*, World Sustainability Series,
https://doi.org/10.1007/978-3-319-73028-8_5

bio-based, sustainable economic development goals. The BIOFIN approach currently is piloted in 31 countries worldwide, including ten of the 17 most biodiverse countries in the world BIOFIN works primarily in support of strengthening the national biodiversity finance framework and closing the national finance gap for the conservation and sustainable use of biodiversity. This chapter reviews the BIOFIN approach, including motivations and the business case for biodiversity investment, our theory of change, highlights some early country results, and the way forward for mainstreaming biodiversity into national policy and decision making and private sector engagement toward sustainable economic development goals.

1 The Wealth of Nature Is the Wealth of Nations

Biodiversity is "Nature"—life on Earth. Biodiversity includes living organisms and ecosystems which underpin human well-being and economies by providing the essentials to healthy and productive human life like clean air, food security and fresh water.[1] Investments in biodiversity, including green economic policy and private sector bioeconomic approaches, are investments in sustainable development, contributing directly to poverty reduction, economic sustainability and the full range of Sustainable Development Goals (SDGs).[2]

International consensus has been reached over the importance of biodiversity in the 17 Sustainable Development Goals (SDGs). Biodiversity is the direct focus of two SDGs: (14) Life Below Water and (15) Life on Land,[3] and contributes to many other goals. For example, in 2013, fish provided 3.1 billion people with almost 20%

A. Trinidad
e-mail: annibelle.trinidad@undp.org

I. Dickie
Eftec (Economics for the Environment Consultancy) Ltd., London, UK
e-mail: ian@eftec.co.uk

A. Seidl
Department of Agricultural and Resource Economics, Colorado State University, Fort Collins, CO, USA

[1]For a technical definition of biodiversity see Article 2 of the Convention on Biological Diversity: www.cbd.int/convention/articles/default.shtml?a=cbd-02.

[2]United Nations Sustainable Development Goals. http://www.un.org/sustainabledevelopment/sustainable-development-goals.

[3]In 2015 The United Nations announced new 17 Sustainable Development Goals with 169 associated targets which are integrated and indivisible. The targets will guide the decisions taken over the next 15 years.

https://sustainabledevelopment.un.org/content/documents/21252030%20Agenda%20for%20Sustainable%20Development.

of their average per capita intake of animal protein and is the most valuable agricultural commodity traded internationally.[4] Forestry provides formal or informal employment for about 50 million people worldwide and accounts for more than 10% of GDP in many of the world's poorest countries.[5]

Additional examples of how biodiversity can contribute to the SDGs include:

- Well-managed forests can provide long-term water security and can serve as emergency stores of energy during times of energy crisis.
- Protected and restored wetland ecosystems can buffer coastal and lowland communities against the impacts of floods, and can provide critical water filtration services. This natural infrastructure can reduce the need for built water-treatment infrastructure.
- A national protected area system can provide tax revenue and support local jobs and livelihoods.
- Agricultural genetic diversity can help to ensure long-term national food security and guard against catastrophic losses, particularly for species that are well adapted to climate extremes, such as flood, drought and excessive heat.
- Sustainable management practices of agriculture, forestry and aquaculture will ensure the sustainable flow of goods and services and can reduce supply chain disruption and price fluctuation.
- Identifying, preventing and eradicating invasive alien species increases the productivity of natural ecosystems and decreases risks from natural disasters.
- Coastal ecosystem protection and restoration efforts can buffer poor and vulnerable coastal communities from the impacts of climate change.
- Sustainable use of biomass holds the promise to contribute to decoupling economic development from fossil fuel use and increase national scale energy independence.
- Well-managed ecosystems can provide a storehouse of medicinal resources that can be critical for maintaining health in rural areas.
- The protection and restoration of coral reefs can ensure the long-term health of fisheries, providing both critical nutrition and livelihoods to millions.

Although the societal benefits of biodiversity are quite clear, they seldom are well reflected in markets. Efforts to quantify the benefits described above in economic terms (i.e. monetary value) can increase awareness and potentially facilitate improved decision making around investments in and management of biodiversity and ecosystem services (or 'natural capital' more broadly). For example, the total annual economic value of the world's renewable natural assets was estimated at US$24 trillion, or US$ 4266 per person on average in 2005,[6] and

[4]FAO (2014).

[5]FAO (2016).

[6]http://www.worldbank.org/en/news/press-release/2011/01/20/countries-manage-natural-wealth-long-term-move-up-development-ladder-says-new-world-bank-book.

Costanza et al. (2014)[7] estimated the world's natural wealth at about US$125 trillion per year. For perspective, the cumulative global Gross Domestic Product (GDP) stood at US$73 trillion in 2015.[8]

Economic valuation of biodiversity and ecosystem services perhaps has greater policy relevance at the sectoral scale. For example:

- Pollinator dependent crops contribute to 35% of global crop production volume, and an annual market value of US$235–577 billion (in 2015) worldwide is directly attributable to animal pollination.[9]
- Global benefits from coral reefs including tourism, fisheries and coastal protection are estimated at US$30 billion per year.[10]
- The market for Chinese herbal medicine was estimated at US$83 billion in 2012.[11]
- The total economic value of the world's renewable resources was estimated at US$24 trillion, or US$ 4266 per person on average.[12]
- The estimated global financial risks from unpriced natural capital to primary production (agriculture, forestry, fisheries, mining, oil and gas exploration, utilities) and processing (cement, steel, pulp and paper, petrochemicals) total US $7.3 trillion, or 13% of GDP in 2009.[13]

2 Nature's Bank Account Is in the Red

By maintaining biodiversity and ecosystems, we retain the ability of the planet to sustain our prosperity. Unfortunately, the value we derive from nature has not resulted in our thoughtful management of the natural wealth we depend on. News of widespread coral bleaching, unintended ecological-economic consequences of agricultural, forest, fisheries and bio-based fuel policies, and catastrophic pollinator deaths stand as testament to our mismanagement. In 2005, the Millennium Ecosystem Assessment (MEA) found that all ecosystems have been negatively

[7]Costanza et al. (2014).

[8]World Bank national accounts data. Available from: http://data.worldbank.org/indicator/NY.GDP.MKTP.CD.

[9]Potts et al. (2016).

[10]World Meteorological Organization (WMO) (2010).

[11]World Health Organization (WHO) (2010).

[12]Based on data in The Changing Wealth of Nations: Measuring Sustainable Development in the New Millennium, World Bank (2011) and includes the estimated value (in 2005) of Crop Lands, Pasture Lands, Forests/ Timber, Forests/Nontimber and Protected Areas in the 152 countries where the data was available. See page 47 Table 2A.2 for more information. https://openknowledge.worldbank.org/bitstream/handle/10986/2252/588470PUB0Weal101public10BOX353816B.pdf?sequence=1&isAllowed=y.

[13]Russi et al. (2013).

affected by human actions. The MEA cites as evidence, for example, a 35% decline in mangroves, 20% loss of coral reefs and about 50% loss in tropical forests.[14]

In 2013, the IUCN Red List named 19,817 threatened species, including: 41% of amphibians, 33% of reef-building corals, 25% of mammals, 13% of birds, and 30% of conifers.[15] The average rate of vertebrate species loss over the last century is up to 100 times higher than the natural rate.[16] Persistent overfishing has a severe impact on marine biodiversity and reduced the total biomass of predator fish species by 52% between 1970 and 2000.[17]

Moreover, extinction rates are unequally distributed around the world. They are significantly higher on islands, with 95% of the world's bird extinctions, 90% of reptile extinctions, 69% of mammal extinctions and 68% of plant extinctions.[18] Small Island Developing States (SIDS) are typically highly dependent upon their natural capital and are recognized as uniquely prone to increasing risk of catastrophic (human, economic and environmental) loss due to climate change.[19]

3 Current Levels of Investment in Biodiversity Are Inadequate

Recognition that biodiversity is the basis of global wealth and increased information about its decline has not been enough to result in adequately financed, sound management for healthy, sustainable ecosystems. Clearly, additional investment is needed.

In benefit-cost analysis of biodiversity investments, at least three pieces of information would be essential: how much is invested in biodiversity currently; how much is needed; and what is the return on that investment. Budgeting and spending are difficult to track because there is no budget line for biodiversity in national and corporate accounts. However, there are several emerging public and private sector efforts to better integrate natural capital accounting and circular economy tracking approaches.[20] In addition, there tend to be no clear investment strategies and no direct measures of performance of investments against biodiversity indicators or targets. Specific information about how much we are currently

[14]Millennium Ecosystem Assessment (2005).

[15]IUCN Red List Committee (2013).

[16]Ceballos et al. (2015).

[17]Secretariat of the Convention on Biological Diversity (CBD) (2014).

[18]CBD (2010).

[19]UNFCCC (2005).

[20]Natural Capital Coalition (https://naturalcapitalcoalition.org/), World Business Council on Sustainable Development. http://www.wbcsd.org/, UN System for Environmental-Economic Accounting (https://unstats.un.org/unsd/envaccounting/seea.asp), The Prince's Accounting for Sustainability Project: https://www.accountingforsustainability.org/en/index.html, Ellen Macarthur Foundation: https://www.ellenmacarthurfoundation.org/, among others.

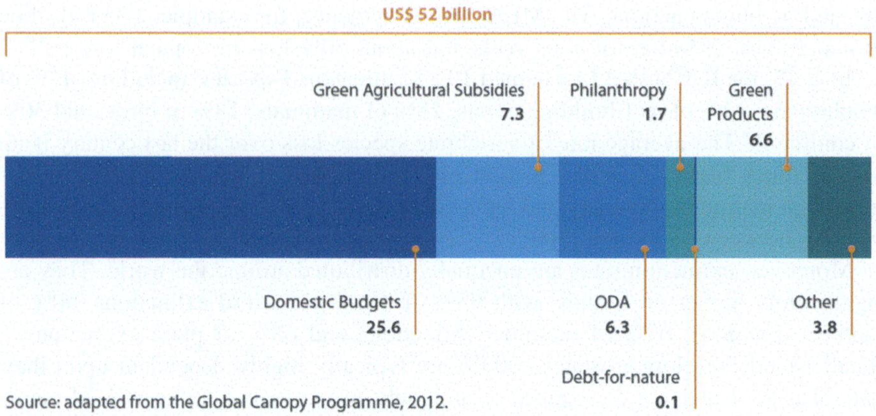

Fig. 1 Estimated global historical annual biodiversity finance investments

spending, on what and to what end is sorely lacking, but global best guesses can provide a start.

Parker et al. (2012) writing on behalf of the Global Canopy Network estimate that at least US$52 billion is spent on biodiversity per year globally.[21] About half of all biodiversity expenditures are thought to come from domestic budgets and fully half of the remainder is from other sources of public funding (Fig. 1). In a world where the private sector dominates biodiversity dependent industries (e.g., agriculture, fisheries, forestry, tourism), private sector investment is either very small or difficult to quantify.

The UN Convention on Biological Diversity convened an expert ('High-Level') panel to provide a top-down estimate of the global biodiversity finance needed to fulfill the 2020 Strategic Plan. In 2012, their best estimate was an annual investment need between US$150 and US$440 billion, implying commitments on the order of 0.08–0.25% of global GDP.[22] Parker et al. (2012) generated similar estimates. The financing gap, evident in both developed and developing countries, is a major challenge, hampering the achievement of both the Convention on Biological Diversity's (CBD) Strategic Plan and the Sustainable Development Goals. However, estimates of the global financial commitment required to meet the SDGs represents between just 0.2 and 0.6% of the estimated US$73 trillion of global GDP.[23]

These broad global investment and need estimates, taken in conjunction with the World Bank's conservative estimate of the total economic value of renewable resources, imply a global return on potential investment on the order of 10–1 (Fig. 2).

[21] Parker et al. (2012).

[22] Convention on Biological Diversity (CBD) (2012).

[23] World Bank (2015).

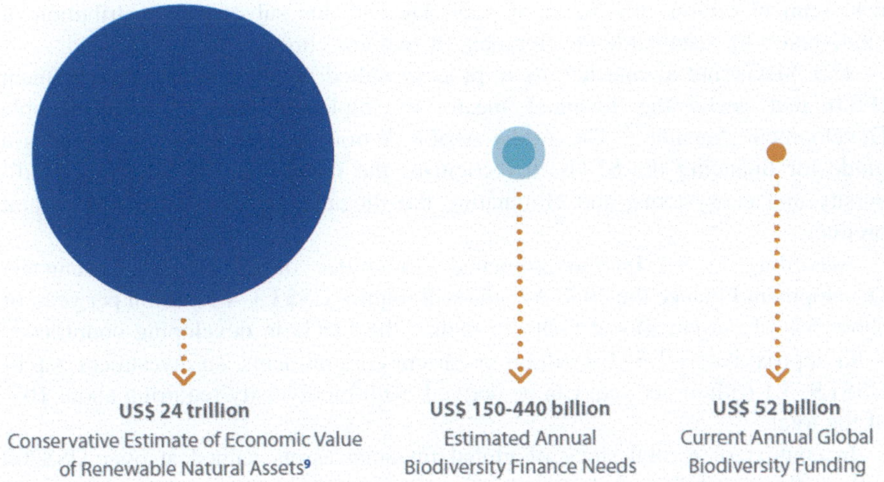

Fig. 2 Return on Investment: Biodiversity asset value versus annual maintenance costs

However, most biodiversity investments are not made at the global level. Such investments are planned, prioritized, financed and implemented at the national level and below. As a result, data tracking biodiversity investments and actions are best collected and tracked at national and subnational levels. Subsequent to the High Level Panel estimate, the CBD recommended supporting national efforts to collect and report biodiversity finance information to contribute to better understanding of the biodiversity investment sources, levels, gaps, and needs and to identify different means for biodiversity finance resource mobilization.

4 Biodiversity Finance Is Part of the Broader Sustainable Development Finance (SDF) Challenge

Many of the 2.7 billion people who survive on less than US$2 a day, depend directly on biodiversity and healthy ecosystems.[24] "The potentially catastrophic changes to biodiversity will have major consequences for people living in poverty who disproportionately rely on biodiversity for their subsistence."[25] For example, up to 70% of the energy in Africa comes from wood fuel.[26] Global annual net forest loss—of the order of 3.3 million ha in 2010–2015[27]—directly limits sustainable

[24]World Bank (2014).

[25]Intergovernmental Committee of Experts on Sustainable Development Finance (2015).

[26]Mead (2001).

[27]Food and Agriculture Organization (FAO) (2015).

development options for the rural poor. Despite this substantial contribution of biodiversity to sustainable development, it remains chronically underfunded.

The SDGs are connected to a process called Financing for Development (FfD) that seeks the financial means to implement the 2030 Sustainable Development Agenda.[28] The Addis Ababa Action Agenda (AAAA) provides a guide for financing the SDGs. It recognizes the importance of protecting biodiversity and ecosystems, and eliminating the illegal trade of species and natural products.

According to the Intergovernmental Committee of Experts on Sustainable Development Finance the 2030 Agenda will require US$3.3–4.5 trillion per year, or about 5% of current global GDP, to realize the SDGs in developing countries.[29] With approximately US$1.4 trillion in current commitments, an investment gap of US$1.9–3.1 trillion per year can be derived, with biodiversity requiring about 10% of the total.

In context of a total stock of global financial assets valued at over US$200 trillion, the possibility of closing this finance gap is within reach. However, most countries are not investing adequately in biodiversity or sustainable development despite there being no shortage of liquidity in the world. A shift is required towards new investment and fiscal policy paradigms that better incorporate the economic value and financial benefits of biodiversity and sustainable development.

Biodiversity finance is the practice of raising and managing capital and using financial incentives to support sustainable biodiversity management.[30] There is increasing interest and demand to use biodiversity finance tools for public and private investments in biodiversity and ecosystem services. Biodiversity finance and investments in biodiversity focused economic development policy aligns biodiversity conservation and ecosystem service management with sustainable economic development plans and goals.

Finance and economics provide critical perspectives to make a compelling 'business' case for investments in biodiversity. The range of available finance solutions is increasing, and the ways in which resources are both mobilized and spent have become progressively diversified. "Blended" finance approaches benefiting from collaborations among public, philanthropic and private actors have become common. The value of green finance markets is booming in part due to the development of green bonds and of more innovative forms of venture capitalism and "impact investing".[31]

---

[28]Addis Ababa Action Agenda: http://www.un.org/esa/ffd/wp-content/uploads/2015/08/AAAA_Outcome.pdf.

[29]Intergovernmental Committee of Experts on Sustainable Development Finance (2015).

[30]Adapted from Clark (2007).

[31]Conservation Finance. From Niche to Mainstream: The Building of an Institutional Asset Class. Credit Suisse Group AG and McKinsey Center for Business and Environment (2016).

5 Repairing and Upgrading Our Broken Compass

The Biodiversity Finance Initiative (BIOFIN) is part of the solution. BIOFIN was developed in response to the 10th Conference of the Parties (COP-10) of the Convention on Biological Diversity (CBD) which identified the need for better information on current expenditures and financing needs, and for a comprehensive approach to develop resource mobilization strategies. BIOFIN is considered an important support for the CBD Strategic Plan for 2011–2020 and responds directly to Aichi Target 20 on resource mobilization and supports the other 19 targets to facilitate the delivery of the National Biodiversity Strategies and Action Plans (NBSAP).[32]

BIOFIN was launched at the COP-11 in 2012, under an initial grant from the EU, and to date has received additional financial support from Germany, Norway, Switzerland and Flanders. BIOFIN is today a UNDP-managed global partnership that supports countries to enhance their financial management for biodiversity and ecosystems: 31 countries have initiated a national BIOFIN process, including ten of the 17 megadiverse countries on the planet (Fig. 3).[33,34] An increasing number of countries are using part of the approach and implementing BIOFIN with other donor money and support.

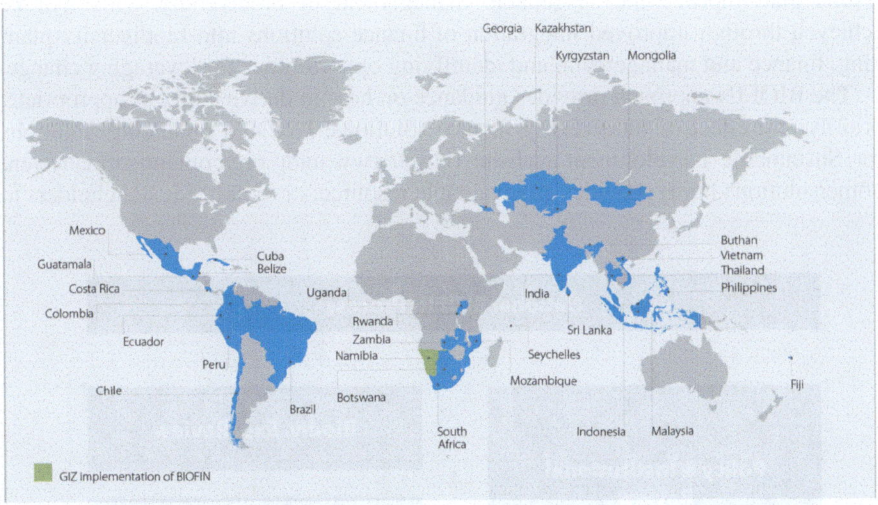

Fig. 3 The 31 countries of the Biodiversity Finance Initiative

[32]Convention on Biological Diversity National Biodiversity Strategies and Action Plan: https://www.cbd.int/nbsap/.

[33]Mittermeier et al. (1997).

[34]BIOFIN: www.biodiversityfinance.net.

BIOFIN makes use of three detailed country-level assessments to develop a biodiversity finance plan, drawing on qualitative and quantitative data, innovative methodologies, and global and national expert input. The BIOFIN approach provides an innovative, stepwise and adaptable process that enables countries to:

- Assess the policy, institutional, and economic context for biodiversity finance (Biodiversity Finance Policy and Institutional Review; PIR);
- Measure and analyse current biodiversity expenditures, from the public and private sectors, donors and NGOs (Biodiversity Expenditure Review; BER);
- Make a reliable estimate of the finance needed to achieve a country's biodiversity goals, and compare this to current biodiversity expenditures and other resources available (Financial Needs Assessment; FNA); and
- Develop a biodiversity finance plan that identifies and mobilizes the resources and policies required to implement the most suitable finance solutions (Biodiversity Finance Plan; BFP) (Figs. 4 and 5).

Thorough assessments are being produced by countries to build an evidence base from which to identify, prioritize and implement different finance solutions to improve the sustainable management of biodiversity and the contribution of biodiversity to sustainable development. Biodiversity finance solutions are ways of using one or more finance mechanism or instrument (e.g. taxes and subsidies) in a particular context (e.g. finance sources and agencies/institutions involved), targeting results that improve the sustainable management of biodiversity. This will be achieved through improved integration of finance solutions into biodiversity planning, finance and management, and identifying opportunities for leveraging change.

The BIOFIN approach provides guidance on how to derive a mix of appropriate, priority and effective biodiversity finance solutions. UNDP's Financing Solutions for Sustainable Development website (http://www.undp.org/content/sdfinance/en/home/solutions.html) provides an invaluable resource, or toolkit, for stakeholders to

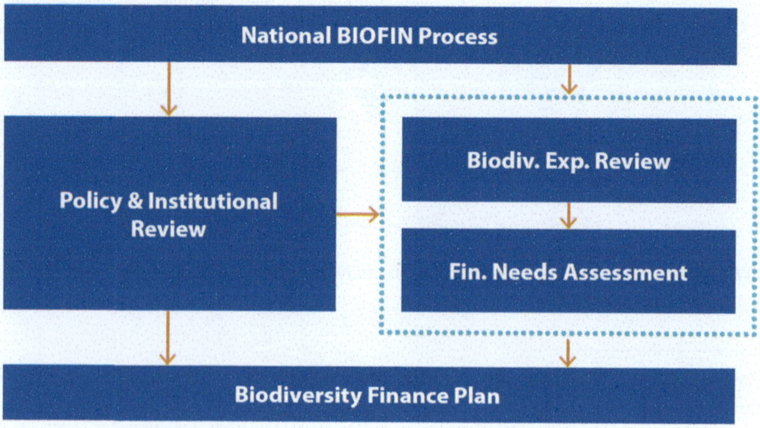

Fig. 4 The National BIOFIN process overview

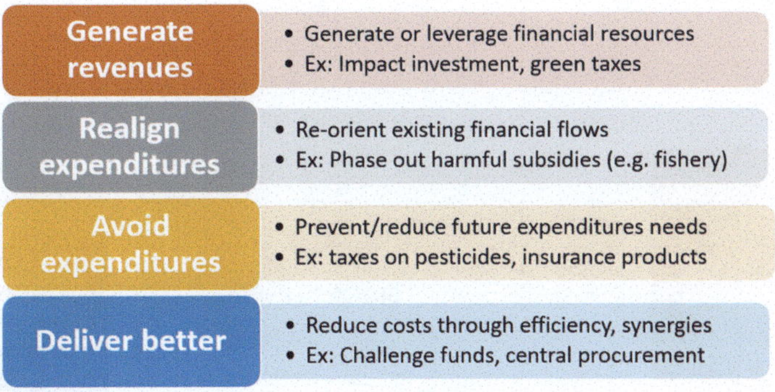

Fig. 5 How biodiversity finance solutions can achieve their objectives

explore potential biodiversity finance mechanisms (tools, instruments, and strategies) and to craft feasible finance solutions. The website provides an overview of the mechanism, typical uses, revenue potential, risks and feasibility. Tools can be sorted by type, sectoral application, or SDG. Finance solutions can achieve their desired impact through (Fig. 5):

- Generating new revenues targeted towards biodiversity;
- Reorienting or realigning existing financing to reduce negative impacts and improve outcomes;
- Avoiding future expenditures through strategic investment and policy;
- Delivering better conservation through improved effectiveness, efficiency and synergies (Fig. 6).

BIOFIN also seeks to increase its effectiveness by promoting synergies with related international and national programmes, such as:

- Programming on Conservation Finance by a large variety of organizations,[35] often includes work on protected area finance and innovative finance solutions, which should be closely involved when developing the finance plan;
- Global programmes working to expand environmental accounting practices: The United Nations System of Environmental-Economic Accounting (UN SEEA)[36] is an international standard for including environmental data in national statistical reports and especially in national accounts.
- The World Bank's Wealth Accounting and Valuation of Ecosystems (WAVES and WAVES+) helps countries to establish Natural Capital Accounts and carry out the required economic valuation studies.[37] In the medium and long term,

[35]UNDP-Global Environment Facility (GEF), and non-government organizations such as World Wildlife Fund, Conservation International, Wildlife Conservation Society, The Nature Conservancy and Birdlife International.

[36]United Nations (2016).

[37]World Bank Group (2016).

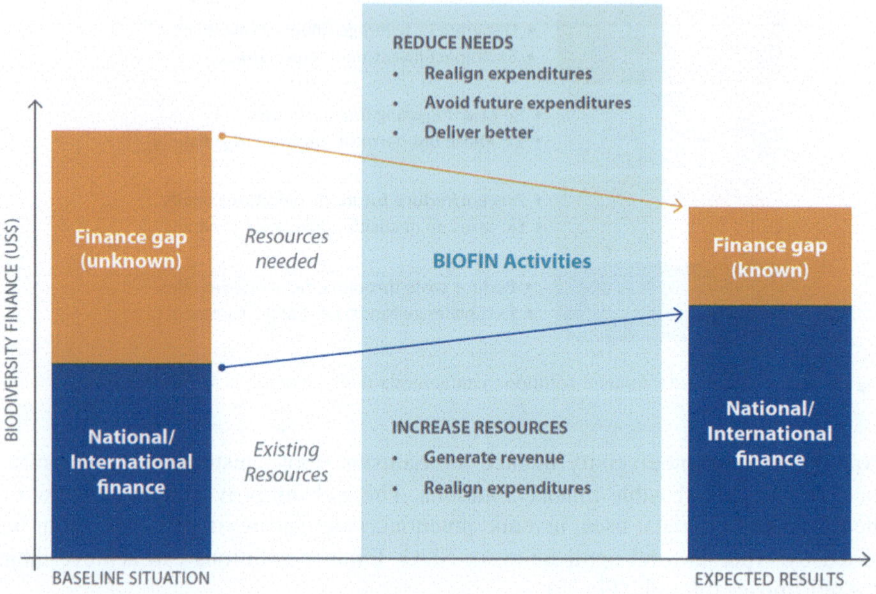

Fig. 6 The National BIOFIN approach and outcomes

tracking stocks and flows of environmental assets can contribute to the documentation of impacts from investments in biodiversity;

- The United Nations Environment Programme's The Economics of Ecosystems and Biodiversity (TEEB)[38] works in many countries assessing, summarizing and deepening understanding of how economic valuation studies can provide insight and guidance on biodiversity policy and planning.

The BIOFIN process actively seeks buy-in from finance and environmental stakeholders and decision-makers (e.g. ministries of finance, business organizations, ministries of environment, NGOs) to identify and mobilize policies, resources and institutional capacities to implement biodiversity finance solutions. This ensures biodiversity finance solutions are:

- Politically realistic, drawing on knowledge of relevant institutions and fiscal policy;
- Financially sound, showing the returns on biodiversity investments, backed by an economic case considering the distribution of the costs and benefits; and
- Integrated into the wider sustainable development agenda, contributing to more effective, efficient, and equitable sustainable biodiversity management and development (Fig. 7).

[38]The Economics of Ecosystems and Biodiversity (TEEB) (2016).

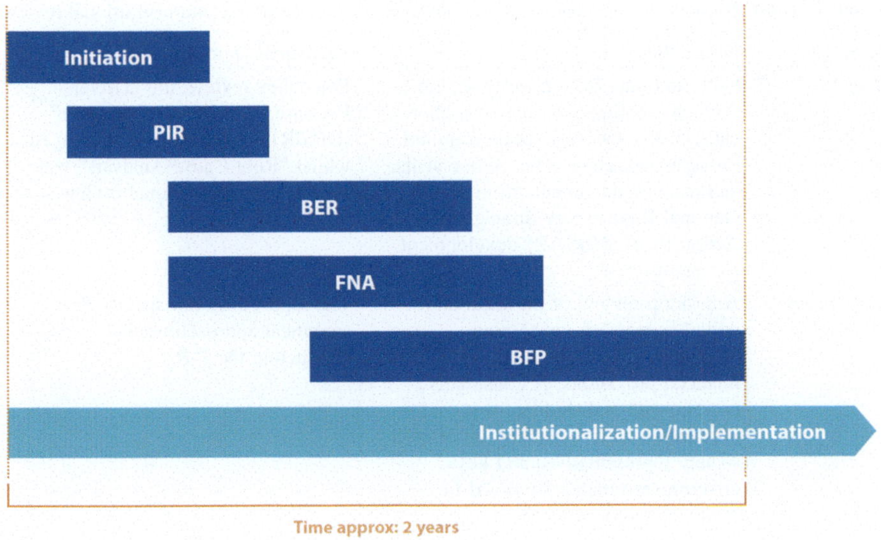

Fig. 7 BIOFIN timeline and path to institutionalisation

BIOFIN aims to be integrated into relevant country level processes in order to influence change.

It is important to ensure that sufficient capacity is developed to sustain finance solutions into the future. The uptake and successful implementation of finance solutions will be strengthened by a convincing business case for investing in biodiversity, aimed at both the public and private sectors. Using the language of finance and economics, the value of biodiversity for specific stakeholder groups needs to be articulated effectively. Investing to support the conservation and sustainable use of biodiversity is a fundamental component of achieving the Sustainable Development Goals—the global intention to harness collective power to achieve social, economic and environmental targets.

6 A Roadmap for Global Biodiversity Finance Solutions: A Sampling of Illustrative Results

The Biodiversity Finance Policy and Institutional Review (PIR) analysis establishes what will be analysed within the National BIOFIN study (e.g. which biodiversity targets) and the context for the intended change in financing. The PIR has strong parallels with similar processes countries may undertake for climate, water, forests, transportation and/or health, for example. The PIR includes an assessment of the NBSAP, principal (biodiversity-dependent) industrial sectors, the national budget cycle, and the laws and policies that affect performance, principal stakeholders and actors, and existing biodiversity finance solutions (Table 1).

Table 1 BIOFIN countries' approaches to the biodiversity finance policy and institutional review

Country	Data collection	Analytical approaches
Uganda	Collected data from priority sectors through meetings, discussions, phone calls, and research. Reports acquired through the web searches, office visits, and existing documents from previous National Biodiversity Strategies and Action Plans (NBSAP} development processes	For policy review, the "Driver-Pressure-State-Impact-Response" (DPSIR) framework was employed, whilst "Root Cause Analysis" was used for the institutional review
Seychelles	A desktop review of the legal and policy framework concerning biodiversity which focused on financial provisions. A consultative process with stakeholders was established with the setting up of civil society representatives and keys government officials involved in biodiversity conservation	The report does not specify the analytical approach used for conducting the PIR
Philippines	Independent desk review of biodiversity-related laws with focus on financial provisions. A consultative and collaborative process with stakeholders from various sectors were held through PIR workshops	The report does not specify the analytical approach used for conducting the PIR
South Africa	Existing data from the 5th Country Report to the Convention on Biological Diversity and the National Biodiversity Assessment was used to identify 15 key trends, supplemented by peer review discussions with a reference group	The "Pressure-State-Response" model was employed for the PIR
Botswana	A desktop review, prioritization and analysis of biodiversity conservation-related policy and legal instruments. Then a prioritization task of stakeholder and national development planning priorities was conducted	The report does not specify the analytical approach used for conducting the PIR
Georgia	A desktop review of NBSAP, relevant laws and policies, national reports, research papers was conducted for the PIR	The "Pressure-Status-Response" model and "root-cause analysis" were adopted for the PIR
Thailand	A desktop review of NBSAP, National reports, and Master Plan on Biodiversity Resource Conservation	The "Pressure-Status-Response" model and "root-cause analysis" were adopted for the PIR

(continued)

Table 1 (continued)

Country	Data collection	Analytical approaches
Fiji	A desktop review of biodiversity-related laws and policies was carried out focusing on its impacts on the status and trends of biodiversity and the financial provisions. National reports, papers, manuscripts were used for the review. A series of consultations and workshops were held with stakeholders from various sectors	The report does not specify the analytical approach used for conducting the PIR

For example, in South Africa, the PIR reveals a national vision for biodiversity stewardship and economic development that creates opportunities to highlight and mainstream the role of biodiversity in national development plans (NDP). The NBSAP outlines a roadmap to ensure the management of biodiversity assets and ecological infrastructure to support South Africa's economic development. The vision of the NBSAP is to "Conserve, manage and sustainably use biodiversity to ensure benefits to the people of South Africa, now and in the future." The NDP affirms that South Africa "needs to protect the natural environment in all respects, leaving subsequent generations with an endowment of at least equal value," a clear nod to the essential tenets of sustainable development. The NDP deals extensively with natural resources and biodiversity focused on tourism, agriculture and rural development, economic infrastructure (water) and human settlements (spatial planning).[39]

The Biodiversity Expenditure Review (BER) explores public and private expenditures benefitting biodiversity to establish past and projected biodiversity expenditures. In the BER, national BIOFIN teams compare, for example, optimistic, pessimistic, and expected 'business-as-usual' budget scenarios for the short and medium term planning horizons, often now looking to 2030 in line with the SDG reporting period. Since there is no budget line or code for biodiversity in most national, sectoral or private sector accounting systems, the BER crucially establishes a systematic approach to tagging budget items and attributing the percentage of the budget item that is relevant to biodiversity (OECD[40] and SEEA[41] have systems that may help national efforts). Analyses can focus on NBSAP or other national reporting categories, industrial sectors, line ministries, budget allocation vs expenditures, as well as leverage of public domestic expenditures with international and/or private spending (Table 2).

[39]http://www.gov.za/issues/national-development-plan-2030.

[40]http://unstats.un.org/unsd/envaccounting/seea.asp.

[41]http://unstats.un.org/unsd/envaccounting/seeaRev/SEEA_CF_Final_en.pdf.

Table 2 Key economic sectors identified as highly relevant to biodiversity-based economic development policies and finance solutions

Countries	Key sectors of biodiversity management
Uganda	Water and environment; agriculture; tourism; wildlife and antiquities; energy and mineral development; and works and transport sectors
Seychelles	Land use and housing; building and construction; utilities and waste management; transport; trade and communication; agriculture; fisheries; tourism; and protected areas
Philippines	Industrial Manufacturing and Processing; Forestry and Forest-related Activities; Protected Areas; Cave and Cave Resources; Agriculture and Agro-biodiversity; Tourism; Energy; Transportation and Infrastructure; Water; Fisheries; Mining; Human Settlements; and Wild Management
South Africa	Mining; infrastructure development; agriculture; land use; water; pollution and waste; and protected areas
Botswana	Water; energy; agriculture; biodiversity; infrastructure; tourism; wildlife management; and land degradation/rehabilitation
Georgia	Energy; agriculture; forestry; tourism; and mining
Thailand	Coastal and marine; terrestrial (forest and mountain); wetlands and rivers; and urban biodiversity
Fiji	Tourism; manufacturing; commerce; agriculture; forestry; fisheries; land resources development and management; water and waste management; and housing and urban development

For example, in Kazakhstan, an expenditure attribution system based on the "impact" a project has on biodiversity and the Aichi targets of the CBD was developed through expert opinion. This is captured by an attribution score of 0–100%, with 100% reflecting activities which have a "direct" influence on biodiversity conservation, 90–5% reflecting activities with an increasingly "indirect" influence on biodiversity and 0% meaning no impact on biodiversity. This system was applied to the Kazakhstan national budgets from 2008 to 2014 and to develop historical spending patterns and alternative budget scenarios based on historical experience and future expectations.

The Financial Needs Assessment (FNA) estimates the financing required to deliver national biodiversity plans, targets and results, and then assesses the financing gap between this and the projected expenditures. Typically, the FNA will use the NBSAP as its guidance document to calculate biodiversity investment needs. Some countries, however, will use their national planning document for this purpose. The FNA produces a detailed budget against each "costable action," result, outcome or project, the scale of which depends on the specificity of the guidance documents. The FNA will typically use the same tagging and attribution system used in the BER, undertake a prioritization exercise, and produce a detailed estimate of the finance gap between business-as-usual expenditures and the financial needs of the national (biodiversity) plan.

For example, the Philippines developed a detailed costing analysis of its NBSAP (there called the Philippines Biodiversity Strategy and Action Plan—PBSAP) using an iterative process involving the main stakeholders, key experts and government officials. Figure 8 shows estimated costs (in US$2015) for each year from 2015 to 2028 categorized by the main PBSAP categories. Initial costs were high due to investments associated with protected areas (PAs) establishment efforts to relocated human occupants of the PAs to areas of lower biodiversity value. Figure 9 then illustrates the gap between business as usual expected budget scenarios and investment need.

Finally, the Biodiversity Finance Plan (BFP) prioritizes financing solutions that will close the financing gap by optimizing current and expanding future investments (public, private, national, international, traditional and innovative) in biodiversity management, and develops the business case for the best options. The BFP aims to present a coherent and comprehensive national approach to biodiversity finance that encompasses a full suite of finance solutions, well beyond the mobilization of new and additional resources. Finance solutions in planning or implementation stages to date include:

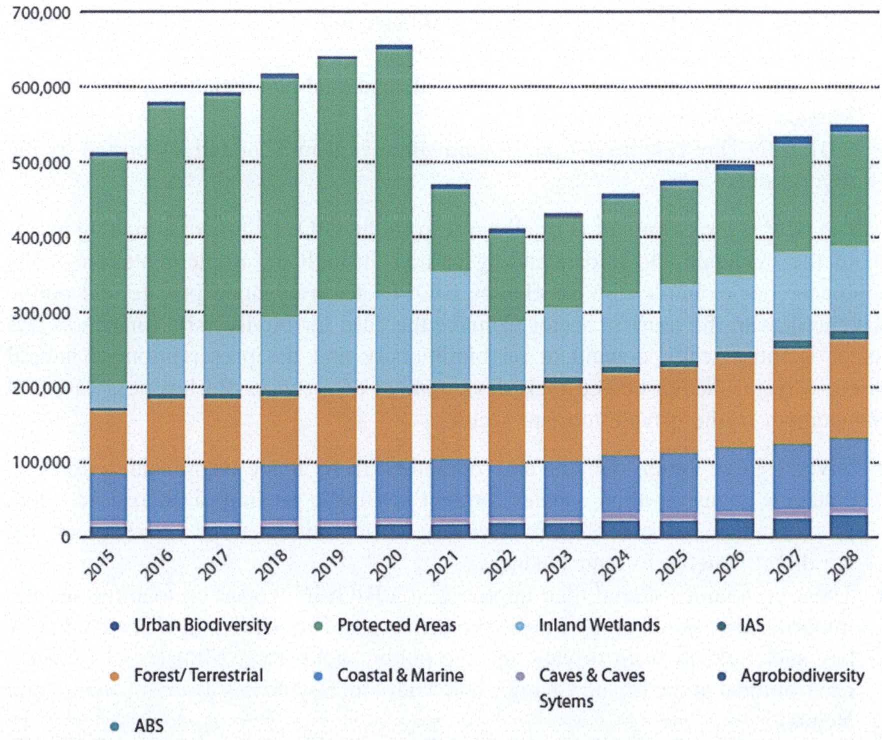

Fig. 8 Timeline of investment needs for implementing the Philippines NBSAP, 2015–2028 (thousands of US$2015)

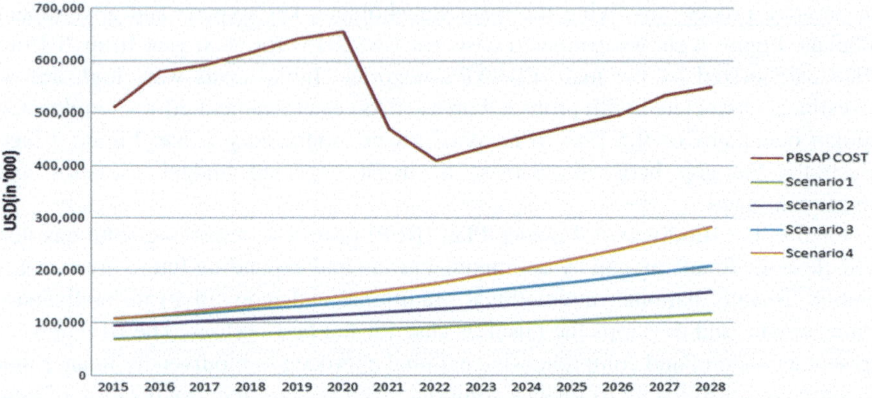

Fig. 9 Philippines National Biodiversity Financing Gap, 2015–2028, (US$2015)

- exploring a biodiversity investment portfolio within the Islamic Investment Fund in Indonesia;
- legislation to incentivize private wildlife preserves in South Africa and Belize;
- utilizing climate funds for biodiversity finance in Fiji and several other BIOFIN countries through the Global Climate Fund;
- payment for Ecosystem Services proposals and enabling legislation in Malaysia, Mexico and Peru; and
- a BIOFIN Day celebration and fundraising event in Thailand supported by the royal family.

The BFP is the culmination of all work from the BIOFIN process, making use of all of the evidence and understanding gained through the implementation of the approach. For example, the Seychelles BIOFIN team engaged private and public stakeholders in the tourism sector to make the case for biodiversity finance as this sector is intrinsically dependent on biodiversity and the preservation of natural assets. Several factors helped their development of a successful business case and engagement of the private tourism sector:

- Several large hotel owners and operators have made significant efforts to eradicate invasive alien species, protect sea turtle nesting sites, restore coral, protect mangroves and other biodiversity conservation actions as they recognize the direct benefits to their business.
- These companies shared their approach at a BIOFIN workshop to showcase the importance of biodiversity for the private sector. The workshop was attended by key stakeholders from private and the public sector (e.g. Ministry of Finance, Environment and Climate Change, and Tourism, Seychelles Tourism Board, and NGOs).
- The hotels actively financing biodiversity understand that the opportunity cost of biodiversity degradation is very high to them as individual enterprises and at

a national scale. Investing in biodiversity provides competitive advantages to the hotel companies and creates positive market externalities thanks to cooperation and strategic efforts.
- All businesses are required to pay a mandatory Corporate Social Responsibility Tax of 0.5% of sales. The workshop provided an opportunity for the private sector and government to discuss the use of these tax revenues to fund biodiversity programmes.

7 The Way Forward and Concluding Remarks

The Biodiversity Finance Initiative anticipates closure of its pilot phase in 2018. Expected outputs of BIOFIN by the end of 2018 include:

- The 30 pilot countries publish their reports and share their lessons learnt in creating biodiversity finance plans and implementing biodiversity finance solutions in their countries using the BIOFIN approach.
- Statistical analysis of the 30 pilot countries' financing information create biodiversity investment project information for other countries.
- E-learning modules and formal curricular development are delivered to increase global capacity in biodiversity finance approaches, to continue to support BIOFIN pilot countries, and other countries interested in the BIOFIN approach.
- A final revision of the BIOFIN approach incorporating the lessons learnt from the pilot phase and a fully integrated database tool will be published.
- Three regional nodes (Panama, Turkey and Thailand) will provide technical support for countries interested in implementing the BIOFIN approach, but that were not involved in the pilot phase, using self-generated or alternative sources of financial support.

For the BIOFIN process to be effective the initiative should not focus solely on the Aichi Targets, the CBD and the NBSAP. There are other important national and international initiatives that can be integrated into the BIOFIN process in each country. Just as national teams build collaboration by working closely with a variety of government partners, civil society, and other national stakeholders and experts, the BIOFIN process seeks to connect with a wide range of related initiatives.

These initiatives can include a range of national and donor driven projects and programmes, related global initiatives, other related conventions, research initiatives, and more. For example, BIOFIN seeks to identify and utilize economic analyses of changes to biodiversity and ecosystems, particularly on the value of ecosystem services. BIOFIN assessments and plans will probably include strategies that are derived from other Conventions such as: the Convention on Migratory

Species; the Convention on International Trade of Endangered Species; the RAMSAR Convention on Wetlands; the United Nations Educational, Scientific and Cultural Organization World Heritage Convention; the United Nations Convention to Combat Desertification; and the United Nations Framework Convention on Climate Change.

In the long run BIOFIN aims to lay the groundwork for full institutionalization of the biodiversity finance plan and ongoing biodiversity mainstreaming and monitoring efforts piloted in the BIOFIN process. Early steps toward institutionalization have included:

- the generation and adoption of new information on spending and investment needs;
- new policies to create the enabling condition or the removal of inhibiting conditions to biodiversity positive investments;
- piloting of new budget tagging, attribution and accounting systems across sustainable development finance objectives; identification of new sources of public and private investment strategies.

Typically, this requires strong ownership by the Ministry of Finance and ongoing engagement by civil society private and public sector partners. Sufficient human and financial resources must be in place for implementation, effective planning and implementation must follow.

Nature conservation, restoration and sustainable management provide us practical, cost-effective, no-regret solutions to simultaneously benefit human wellbeing and safeguard biodiversity. A wide range of finance solutions, including environmental fiscal reform, biodiversity offsets, environmental trust funds, debt-for-nature-swaps, impact investing, markets for green and bio-based products, and streamlining biodiversity income options into financial planning processes of government among many other finance alternatives present opportunities to wed economic development and biodiversity conservation for a prosperous future. Together, through effective biodiversity finance solutions, we can deliver better policy outcomes, reduce future finance needs, realign expenditures, and generate more revenue to bridge our investment gap and work toward a better, more resilient, tomorrow for our world.

Acknowledgements The authors would like to thank the following individuals for their contributions to the 2016 Workbook from which this chapter was derived: Lucia Motaung, Blerina Gjeka, Oscar Huertas, Csoban Somodi, Semiray Emeksiz, Georgina Wilde, Alessandra Alfieri, Herve Barois, Bertrand Rassool, Mariana Bellot, Juan Bester, A. Damodaran, Gail Hurley, Helen Ding, Devika Iyer, Bruno Mweemba, Amath Pathe Sene, Carlos Manuel Rodriguez, Ainur Shalakhanova, Anton Steurer, Monika Wozowczyk, Guillermo Zuniga, Jamie Ervin, Usman Iftikhar, Nik Sekhran, Yves de Soye, Caroline Petersen, Adriana Dinu, Midori Paxton, James Maiden, and Ray Victurine. In addition, we would like to thank our partners for their support: The European Union, the Governments of Germany, Switzerland, Norway and Flanders. The governments, partners, and consultants from the 30 BIOFIN countries also contributed ideas and insights into this approach. All errors remain the responsibility of the authors.

References

CBD (2010) Biodiversity scenarios: projections of 21st century change in biodiversity and associated ecosystem services. https://www.cbd.int/doc/publications/cbd-ts-50-en.pdf

Ceballos G et al (2015) Accelerated modern human-induced species losses: entering the sixth mass extinction. Sci Adv. http://advances.sciencemag.org/content/1/5/e1400253.full

Clark S (2007) A field guide to conservation finance. Island Press, Washington, DC

Convention on Biological Diversity (CBD) (2012) Resourcing the aichi biodiversity targets: a first assessment of the resources required for implementing the strategic plan for biodiversity 2011–2020. https://www.cbd.int/doc/meetings/fin/hlpgar-sp-01/official/hlpgar-sp-01-01-report-en.pdf

Costanza R et al (2014) Changes in the global value of ecosystem services. Global Environmental Change, pp 152–168

Credit Suisse Group AG and McKinsey Center for Business and Environment (2016) Conservation finance. From Niche to mainstream: the building of an institutional asset class. https://www.credit-suisse.com/media/assets/corporate/docs/about-us/responsibility/banking/conservation-finance-en.pdf

FAO (2014) The state of world fisheries and aquaculture. http://www.fao.org/3/a-i3720e.pdf

Food and Agriculture Organization (FAO) (2015) Global forest resources assessment 2015. http://www.fao.org

FAO (2016) 2016 state of the world's forests. http://www.fao.org/3/a-i5588e.pdf

Intergovernmental Committee of Experts on Sustainable Development Finance (2015) https://sustainabledevelopment.un.org/intergovernmental/financecommittee

IUCN Red List Committee (2013) The IUCN red list of threatened species strategic plan 2013–2020. http://cmsdocs.s3.amazonaws.com/keydocuments/red_list_strategic_plan_2013_2020.pdf

Mead D (2001) Plantations and wood energy. Food and Agriculture Organization (FAO). http://www.fao.org/3/a-ac125e.pdf

Millennium Ecosystem Assessment (2005) http://www.millenniumassessment.org/en/index.html

Mittermeier RA, Robles-Gil P, Mittermeier CG (eds) (1997) Megadiversity. Earth's biologically wealthiest nations. CEMEX/Agrupaciaon Sierra Madre, Mexico City

Parker C et al (eds) (2012) The little biodiversity finance book. Global Canopy Programme, Oxford. http://globalcanopy.org/publications/little-biodiversity-finance-book-3rd-edition-2012

Potts S et al (2016) Summary for policymakers of the assessment report of the Intergovernmental Science-Policy Platform on Biodiversity and Ecosystem Services on pollinators, pollination and food production. http://www.ipbes.net/sites/default/files/downloads/Pollination_Summary%20for%20policymakers_EN_.pdf

Russi D et al (2013) The economics of ecosystems and biodiversity for water and wetlands. http://doc.teebweb.org/wp-content/uploads/2013/04/TEEB_WaterWetlands_Report_2013.pdf

Secretariat of the Convention on Biological Diversity (CBD) (2014) Global biodiversity outlook 4. https://www.cbd.int/gbo/gbo4/publication/gbo4-en-hr.pdf

The Economics of Ecosystems and Biodiversity (TEEB) (2016) The economics of ecosystems and biodiversity. www.teebweb.org

UNDP (2016) The 2016 BIOFIN workbook. http://www.biodiversityfinance.net

UNFCCC (2005) Climate change, small island developing states. Issued by the CLIMATE CHANGE SECRETARIAT (UNFCCC), Bonn, Germany

United Nations (2016) System of environmental-economic accounting (SEEA) contains the internationally agreed standard concepts, definitions, classifications, accounting rules and tables for producing internationally comparable statistics on the environment and its relationship with the economy. http://unstats.un.org/unsd/envaccounting/seea.asp

World Bank (2014) Investing in natural capital for eradicating extreme poverty and boosting shared prosperity: a biodiversity roadmap for the WBG. https://openknowledge.worldbank.org/bitstream/handle/10986/19998/887530REVISED000NC0385272B00PUBLIC0.pdf;sequence=1

World Bank (2015) World Bank national accounts data. http://data.worldbank.org/indicator/NY.GDP.MKTP.CD

World Bank Group (2016) Wealth accounting and valuation of ecosystem services (WAVES). https://www.wavespartnership.org/
World Health Organization (WHO) (2010) WHO traditional medicine strategy: 2014–2023. http://apps.who.int/iris/bitstream/10665/92455/1/9789241506090_eng.pdf?ua=1
World Meteorological Organization (WMO) (2010) Climate, carbon and coral reefs. http://www.wmo.int/pages/prog/wcp/agm/publications/documents/Climate_Carbon_CoralReefs.pdf

Bioeconomy Opportunities in the Danube Region

Miklós Gyalai-Korpos, Zoltán Szabó, Miklós Hollósy, Bence Dávid, Kinga Pencz, Csaba Fehér and Zsolt Barta

Abstract The Danube Region is an area of Europe with high potentials of biomass resources, and many studies point out the comparatively high biomass potential in the Central European region in the forms of different forestry and agricultural residues. However, in order to exploit these resources, considerable investments are needed to build biorefineries, trigger market demand for bio-based products and guarantee the security and sustainability of the long-term biomass supply requiring active involvement of farmers. These challenges are sizeable, and call for a holistic approach, a transition period and harmonized policy support. However, the authors believe that the transition can speed up with the involvement of industry players, building on existing value chains and infrastructures, all placed in an enabling and business supporting environment. In this chapter, the authors aim to introduce the Danube Region and its potentials in sustainable biomass resources, providing an overview of main supply chains and players as well as of policy instruments and

M. Gyalai-Korpos (✉) · M. Hollósy
PANNON Pro Innovations Ltd., Budapest, Hungary
e-mail: miklos.gyalai@ppis.hu

M. Hollósy
e-mail: dr.hollosy.miklos@gmail.com

B. Dávid · K. Pencz · C. Fehér · Z. Barta
Department of Applied Biotechnology and Food Science, Faculty of Chemical Technology and Biotechnology, Budapest University of Technology and Economics, Budapest, Hungary
e-mail: david.bence93@gmail.com

K. Pencz
e-mail: pekiki91@gmail.com

C. Fehér
e-mail: csaba_feher@mail.bme.hu

Z. Barta
e-mail: zsolt_barta@mail.bme.hu

Z. Szabó
LátensDimenzió Consultancy, Budakeszi, Hungary
e-mail: zoltan.szabo@latensdimenzio.com

W. Leal Filho et al. (eds.), *Towards a Sustainable Bioeconomy: Principles, Challenges and Perspectives*, World Sustainability Series,
https://doi.org/10.1007/978-3-319-73028-8_6

constraints that are currently slowing the development of the bioeconomy. Based on the current situation and its critical analysis, the authors propose a strategy that can move forward the bioeconomy in the Danube Region and probably multiplied also in other geographical areas. In order to realize the multiple benefits and decrease the financial and market risks, in this chapter, we will introduce circular economy and sustainable intensification approaches that build on industry involvement in order to obtain working business models. This approach could facilitate the bioeconomy transition by merging novel technologies into working value chains to valorize by-products and could make stakeholders interested by providing an additional revenue stream instead of waste management fee.

Keywords Danube region · Bioeconomy · Biomass · Industrial symbiosis Sustainability

1 Introduction

In 2010 the European Commission put forward an ambitious European Strategy for the Danube Region (European Commission 2010), accompanied by a detailed Action Plan (European Parliament 2010a). The Council of the European Union (EU) endorsed both of these documents in early 2011 (Council of the European Union 2011). The Danube Strategy, which builds upon the experience from the first EU macro-region initiative (the EU Strategy for the Baltic Sea Region), aims to develop the significant economic potential and improve environmental conditions in the region. With its focus on sustainable growth, it also makes an important contribution to achieving the Europe 2020 goals. The strategy, encompassing 115 million inhabitants of 14 countries (Danube-region.eu 2016), seeks synergies and trade-offs, better alignment and greater coherence of policies and funding, and overcoming fragmentation.

The EU Strategy of Danube Region (EUSDR) is implemented with the support of the European Commission Directorate General for Regional Policy, which is also responsible for coordinating the Strategy at the policy level with the assistance of the High Level Group on macro-regional strategies made up of official representatives from all EU Member States. The EUSDR addresses a wide range of topics, divided into 11 Priority Areas. The work of each Priority Area is jointly coordinated by two participating countries that work in consultation with the Commission, relevant EU agencies and regional bodies (Danube-region.eu 2016). Activities at national level are steered by National Coordinators, who are also responsible for informing stakeholders of the developments and for promoting the EUSDR. A Danube Strategy Point was established in 2015 to improve the implementation process and facilitate information flow among priority areas and national coordinators.

The countries of the Danube Region have already recognized the potential lying in biomass utilization by publishing and promoting the Danube Region Biomass Action Plan (Hujber and Szilágyi 2014). The aim of this action plan was to summarize, showcase and analyze the state of play of the Danube Region regarding biomass use from the point of view of the EU Biomass Action Plan of 2005 (European Parliament 2005) and the National Renewable Energy Action Plans of Danube EU Member States and similar initiatives in non-EU Danube countries. Based on the findings, the study aimed to identify concrete priorities for further improvement to promote the development of a common position on the use of biomass in the countries in the region. It also aimed to formulate concrete recommendations at Danube macro-regional and national level. However, this document provides a restricted and specific techno-economic snapshot that does not look either at the regional specifics (Nomenclature of Territorial Units for Statistics—NUTS—levels 1, 2 or 3), or at the broader, integrated sustainability (socio-economic, environmental, apart from climate change) implications of bioenergy.

Stakeholders, as well as the EU policy makers, are aware of the complexity of the Danube Region that is characterized by extremely diversified political, economic and social contexts, e.g. the political will to meeting "20-20-20 targets", the ongoing EU enlargement policy, the significant academic capacities, the presence of skilled workforce, the importance of agriculture and forestry in national economics. Given this context, it is crucial to learn about the expectations of stakeholders and how they can be motivated to work together for the bioeconomy considering also building on traditional and well established sectors, such as the agriculture and other relevant industries processing biomass/organic material.

The Climate-Knowledge and Innovation Communities (KIC) Biohorizons project—with the involvement of PANNON Pro Innovations Ltd.—focused on consulting stakeholders across the bioeconomy. This was carried out by an online survey with the purpose to ask how to support development of the bioeconomy in the best way, what are the main barriers and what interventions are needed to overcome them. In 2014, this survey received nearly 500 credible responses. The results highlight the need to ensure continuity of policy in order to build investor confidence in bioeconomy, as well as to mobilize feedstock and human and financial resources, all of which are critical factors to working supply chains (Hodgson et al. 2016). Hence, establishment of new supply chains and involvement of biomass producers to secure feedstock supply for long-term operation and/or practice industrial symbiosis approaches to integrate into existing industries (business to business—B2B) are of crucial importance.

Despite the significant biomass potential, the Biohorizons project resulted in a very low number of answers from the Danube Region, leading to the conclusion that the stakeholders may not be aware of each other and lack the necessary capacities. In order to enhance sustainable bioenergy and advanced biomass use, the authors now propose a strategy to involve industry partners from agriculture and process industries. In line with the survey results, the proposed approach aims to create more B2B cooperation in order to learn the market for bio-based products, overcome knowledge gaps, and create supply chains.

Creating novel supply chains can have a series of challenges (Fig. 1), as well as dynamic feedback along the whole supply chain and with the entire system should

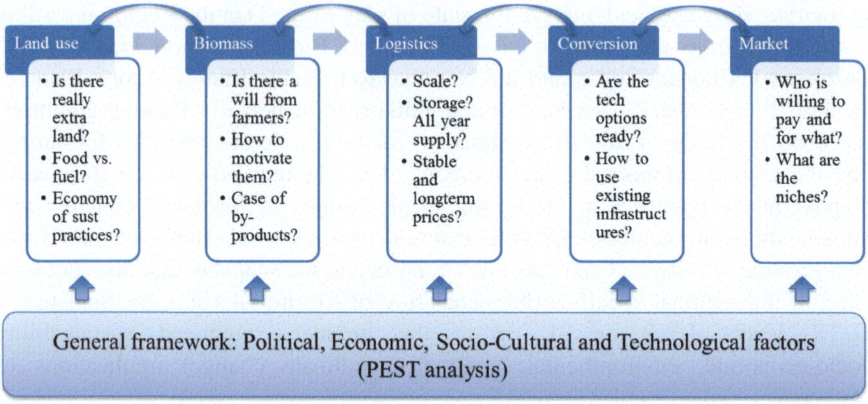

Fig. 1 Challenges of novel supply chains

be considered. Mobilizing biomass feedstocks in a sustainable and resource-efficient manner remains an issue of vital importance for Europe. First of all, for meeting the targets of the Renewable Energy Directive (European Parliament 2009a) and Fuel Quality Directive (European Parliament 2009b) for a low carbon energy system in 2020 and beyond. It also plays an important role in fostering the sustainable development of the European bio-based economy and in strengthening rural areas.

The outcomes of the Biohorizons survey (Hodgson et al. 2016) points out that the main bottlenecks in creating the bioeconomy and its biomass supply chains are related to resource mobilization combined with financial constraints and uncertain policy framework. Additionally, personal experiences of the authors and yet unpublished findings of a similar mapping survey focusing on the Danube Region underpin that the perception about the main bottlenecks is similar by stakeholders of the Danube countries. Thus, our approach for this paper is to propose policy and strategy elements that can facilitate mobilization of the significant biomass potential of the Danube Region while involving existing industries. If implemented as a joint effort of industry, policy and academia cooperation, then it can prompt also the mobilization of other resources such as skills and financials. This theoretical background is supported by the methodology on how existing technologies and policies can be in support of biomass mobilization and the bioeconomy transition, in general.

Our methodology focuses on two fields; sustainable intensification of biomass and circular economy, as two potential tools for efficient mobilization of biomass. The first one is a set of technology solutions and approaches for sustainable enhancement of agricultural production making it possible to have more biomass from the same fields. The second one is a novel direction for the economy urging the switch from linear production chains to circular ones. While this is a general vision, it can also apply to biomass by-products from different industries, as another source of bioeconomy feedstock and key to mobilize them. The reason for choosing those routes is due to our approach: common in both that policy decisions and directions (by means of financing instruments and legal framework) can significantly influence their implementation, and thus impacting the bioeconomy, as well as both build on the engagement of existing industries that are needed to realize innovation in supply chains.

The connection between sustainable intensification and circular economy is that both target the biomass mobilization part of the biomass supply chain by exploiting novel ways to expand the potential sources of biomass. By their basic principles, these routes offer sustainable and resource efficient mobilization of biomass, thus support the "more from less" approach in bioeconomy. With the involvement of current industries and offering them financially promising approaches, both can make current stakeholders interested in the bioeconomy transition. This approach also represents that the bioeconomy, as a very complex and overarching vision, cannot be managed by a sole Priority Area of the Danube Region Strategy (as it happened for the Biomass Action Plan investigated by Priority Area 2, Sustainable Energy) but rather require cooperation between relevant Priority Areas to accelerate in all aspects from agriculture to business creation.

Hence, this chapter focuses on the biomass sourcing and mobilization part of the supply chain, building on involvement of existing industry to support the bioeconomy transition. The approach starts from introducing the promising biomass potential of the Danube Region followed by presenting the above two ways to exploit this sustainably, as well as suggests a strategy to link together industry players and benefiting existing infrastructures. This approach is then underpinned by showcasing three case studies of the Danube Region that already apply or intends to apply these principles. The chapter is closed by conclusions to define measures that can move forward the bioeconomy in this sense. The key is to improve resource efficiency and close production loops (between potential and realistic yields, and between industries processing biomass), which could fast track bioeconomy development in the Danube Region, and can also work in different locations.

2 Biomass in the Danube Region

The Danube Region is well recognized for its biomass potential. Analysis of the National Renewable Action Plans submitted to the European Commission as a consequence of the implementation of Renewable Energy Directive (SEC 2005)—targeting that at least 20% of the total energy need of the EU is based on renewables by 2020—shows that in the Danube Region biomass plays a crucial role in reaching these targets as outlined in Table 1 (Beurskens and Hekkenberg 2011).

While the biomass ratio in renewable electricity is the most diversified one, at least two-third of the 2020 renewable targets in heating and transport is to be supplied by biomass. These ambitious targets favour the use of biomass for energy production. While the governments are legally obligated, the farmers and other

Table 1 Biomass ratios in renewable electricity, heating and transport, biomass ratio in 2020 targets and 2020 renewable targets of some countries of the Danube region (Beurskens and Hekkenberg 2011)

Country	Biomass ratio in renewable electricity (%)	Biomass ratio in renewable heating (%)	Biomass ratio in renewable transport (%)	Biomass ratio in 2020 targets (%)	2020 renewable targets (% of total energy mix)
Hungary	59.4	68.8	95.5	72.2	13.0
Bulgaria	11.4	97.3	95.0	83.5	16.0
Germany	22.8	78.7	n.d.	54.1	18.0
Austria	9.8	86.3	68.2	50.0	34.0
Romania	9.2	96.0	92.3	63.5	24.0
Slovenia	11.0	84.0	94.6	57.7	25.0
Slovakia	21.4	84.1	91.8	65.4	14.0

players in the supply chain, who practically are responsible for the production and mobilization of this large amount of biomass may be not aware of the sustainability practices, and how to tackle the "food vs. fuel" debate, as well as land use change conflicts. The strategy presented in this chapter can foster a comprehensive approach to help players in the supply chain to adapt and apply sustainability practices, and facilitate market uptake of biomass-based solutions and implementation of the policies of European Commission.

There are different types of biomass resources with specific characteristics and potential to become sustainable feedstock for bioeconomy purposes in the Danube Region. Even though many potential estimations are available for this wide spectrum of feedstock, there is consensus that agricultural and forestry residues hold an enormous unused potential by means of volume and energy content despite the different methodologies, boundaries, input data and frameworks resulting in relatively large deviation. Therefore, it is absolutely necessary to comply and assess their properties and availability, acknowledging the fact that their importance might vary depending on the ecological circumstances. Moreover, along with the biomass supply chain, there are conflicting interests and effects, e.g. food supply, natural conservation, which can limit feedstock availability.

In many countries, like Austria, Bosnia and Herzegovina, Croatia, Montenegro, Romania, Slovakia and Slovenia forestry products and by-products contribute to the biomass potential to a large extent due to the high ratio of forest cover. Use of by-products is also in advanced situation in some cases, e.g. in Austria in 2009 wood chips, bark and sawmill by-products supplied more primary energy than fuel wood did. Wood chips and sawmill by-products are primarily used in the sawmill and wood processing industry, as well as in cogeneration and district heating plants, whereas pellets are mainly used in growing quantity in domestic heating systems (Austrian Energy Agency 2012). In Slovakia, the residues of the wood processing factories are usually utilized on site to cover the energy demand of the plant (Oravec and Slamka 2013).

Countries of high potential of agricultural residues are Bulgaria, Czech Republic, Hungary, Ukraine and Serbia, where the forestry products are complementary. In the Czech Republic, the residual agricultural biomass—straw and hay—has the greatest potential, however, this potential is currently used only to a minimum extent (Tluka and Stupavský 2011). While in Serbia 45% of the wheat residue and 61% of the barley residue have been already utilized for gaining bioenergy due to an effective policy, as more than 60% of the green-energy comes from utilization of biomass (Stojadinovic 2009).

Potential of agricultural residues in the Danube Region is also outstanding in European perspective. Monforti et al. (2012) carried out the geographical assessment of potential bioenergy production in the countries of EU (EU-27) from residues available from eight crops (wheat, barley, rye, oat, maize, rice, rapeseed and sunflower). The method applied also considered competitive uses from farming and the environmental constraints by retaining a minimum ratio of residues in soils. As results, it was found that on average (EU-27) 42% of produced residues could be sustainably collected. There are also geographical differences in Europe: the highest ratio of the residues collectable sustainably was found for Hungary (46%) followed by Italy, France, Germany, Austria and Poland. Considering the competitive uses after collection—as ratio of the collected amount—on average 83% would be remained as available amount; in the case of Hungary this was found to be 96%, also ranked as the highest.

Although a great number of studies reviewed the biomass potential in the Danube Region, the data are scattered throughout the literature, and usually consist only of the calorific value of the different types of biomass. Danube Biomass Action Plan (Hujber and Szilágyi 2014) also reinforces that there is lack of available biomass related data in the Danube Region, however underpins that biomass has the greatest significance among renewable energy sources in the Danube Region. These facts emphasize the importance of the topics, however, also raise the lack of regional and common approaches for estimating the potential and assessing the technological and economic competing interests.

Calorific values give little information on the actual type and property of the biomass and the suitability for advanced bio-based application. Additionally, these potential estimations do not consider the aspects of practical utilization such all-year round availability, storage, large radius collection area and the diversity of farmers owning the production fields and their interests.

3 Sustainable Intensification of Biomass Supply

As shown above there is a potential in biomass supply considered not to compete with food and feed supplies, however, its mobilization and creation of new supply chains can be of significant difficulties due to different reasons. An opportunity for the Danube Region, which is often overlooked and not considered in practice and in policy to a sufficient degree, is the sustainable intensification of current farming practices. The sustainable intensification of crop production concept aims to address the significant environmental impact (soil, water bodies, global climate, local biodiversity and human health) of croplands, which is the primary land use in Europe. Besides that, there are large areas of unused land and fragmented properties in the Danube Region, of which utilization and cooperation between farmers could enhance the biomass potential.

Crop yields in Western Europe are the highest in the world, while those in Eastern Europe are low. The yield gaps showing the difference between actual and potential crop yields are particularly large in Central and Eastern Europe and neighbouring countries (may be over 10 tonnes per hectare, Wicke et al. 2015), and yield contest results illustrate that average cereal yields could be at least doubled. This represents a wasted opportunity in meeting crop demand with minimal environmental impact (Tilman et al. 2011). If land were more efficiently utilized, more food, feed, fibre as well as fuel (energy) could be produced on the same amount of land. As an example, average annual yield increase of 2.5% per hectare of land in Western Europe has been achieved in the production of arable crops since the late 1990s, while the total production area has decreased. Consequently, an average increase of 2.7 million tonnes of biomass is realized annually from the same crop area creating virtual hectares without real area expansion (Langeveld et al. 2014).

A resource efficient, technologically smart pathway exists also for the Danube Region for the intensification of conventional arable farming with substantial untapped climate change mitigation potential. This pathway needs policy support and financing, and also careful implementation in order to avoid possible side effects on the environment and economy. The additional biomass produced by improved cropland management may be used for bioenergy production or bioeconomy purposes, also leading to climate change mitigation.

Sustainable intensification means enabling farmers to produce more food, feed, fibre, and fuel, while less water, land, energy, and other inputs are used, thus improving resource efficiency in farming with the help of smart technologies. Food and Agriculture Organizations of the United Nations presents sustainable crop production intensification as a new paradigm (FAO 2011). Smith (2013) defines sustainable intensification as "the process of delivering more safe, nutritious food per unit of input resource, whilst allowing the current generation to meet its needs without compromising the ability of future generations to meet their own needs". Hence, intensification should not be confused with intensive farming; sustainable intensification of crop production does not assume a shift from less to more intensive modes, or vice versa (Siplatform.org.uk 2016). Instead, arable land is managed to maximize outcomes across economic, environmental and social dimensions.

Despite the fact that sustainable intensification holds great promise, the concept currently has had little coverage in the policy domain. Discussions in general about the nexus of farming and bioenergy insufficiently include actual ways how to utilize the potential. Nevertheless, there is large potential in Europe to produce additional bioenergy and bioeconomy feedstocks as well as food, feed and fibre in a sustainable way considering only arable land that is cultivated by "intensive farming technologies" possibly subject to sustainable intensification. This does not apply to all agricultural lands in Europe, since small scale organic arable farming, extensive grasslands and agricultural land with ecosystem services or non-commodity outputs cannot be the subject of this approach. In general, the already intensively cultivated arable fields with unsustainable practices may benefit the most from sustainable intensification. The focus of sustainable intensification is the cropping systems, where unnecessarily high input (such as fertilizer) use is associated with emissions and relatively low yields. Assessing agricultural productivity and climate change implications of closing the yield gap, Valin et al. (2013) found that sustainable land intensification would increase greenhouse gas (GHG) savings by one-third when compared with a fertilizer intensive pathway.

Sustainable intensification is a technological way to close the yield gap, hence it is of importance for the Danube Region. Mueller et al. (2013) found that the changes of management practices vary considerably by region and current intensity.

Two of the major ways to sustainably increase productivity of land are by increasing yields above baseline and applying double-cropping. Yield increases can be achieved through improved fertilizer applications, mechanization, better seeds, precision farming, modern irrigation techniques, application of drones, etc. The increased production of feedstock does not imply the application of more inputs, thus the burden on the environment is not increased, and might be even reduced. Double or multi-cropping is a technique, where an extra crop is cultivated in a given year on the same plot. Typically, a short rotation cycle energy crop (e.g. sorghum) is produced before or after the regular crop cultivation. Double cropping increasingly allows farmers to increase the harvested area on shrinking agricultural areas (Langeveld et al. 2013). The spread of multi-cropping techniques, such as the establishment of a winter cover crop, can bring additional environmental benefits in terms of soil erosion and soil GHG emission profile of cropping.

Data-enabled agriculture allows for better decision-making through more specific knowledge. By precisely applying fertilizer tailored to the specific need of each plant, leaching is reduced, thus abating soil and water pollution. Precision farming technologies has been around for some time, and now latest technologies, application of drones and processing of big data will further stimulate resource efficiency, and mitigate emissions. In our interpretation precision farming techniques complemented by drones and "big data" processing at landscape level comprises sustainable intensification, and, perhaps, it may bring us to the next stage in crop production development.

Technologies are readily available and rapidly evolving. The use of global positioning system (GPS) and geographic information system (GIS) technologies and modern farming machinery enables site specific application of the right amount of input materials (e.g. fertilizer) at the right time. The application of drones is a promising innovation in farming, with multiple benefits. Data from drones assessing e.g. the level of moisture in the topsoil, the chlorophyll content of the crop may be used for adjusting the spread of nitrogen fertilizer to the optimal level required for every part of the farm. In this way, yields are raised, nitrogen leaching is reduced, and nitrification and/or eutrophication are mitigated. Likewise, sensors mapping the field can detect the greenness of the plants, and as plants having insufficient nitrogen tend to turn pale, it can guide the amounts of nitrogen to be applied.

4 Circular Economy Approach

Waste has always been a part of human living, however, with the advent of consumption society the types and volumes of waste create a real challenge and threat to environment. Hence, as a result of the increasingly strict environmental regulations and social pressure a number of companies apply new means to enhance their environmental performance. Many have already realized that this is the way to remain competitive in an era with provisional resource scarcity and rising price volatility. Such new tools are the life cycle assessment of products, environmental management systems and industrial symbiosis, whereas the new concept referred to as circular economy. The industrial symbiosis devises cyclical processes similar to natural ecosystems, and strives to domesticate the circulation of material and energy in industrial processes as found in natural ecosystems. While industrial symbiosis is a typical end-of-life approach in waste management, i.e. it offers solutions to waste streams, the circular economy concept provides a wider approach advocating solutions not only according to the waste hierarchy but more general in resource efficiency by advanced production chains, technology and business models, as well as in communication to raise social awareness. As a result, improved resource efficiency, which is the overarching aim of the circular economy, will lead to economic growth (Heck and Rodgers 2014).

The European Commission and policy landscape also recognize the environmental, climate and economic impact of the circular economy and the substantial value of wastes landfilled or incinerated without new policy framework. In 2011 only a limited share (40%) of the municipal waste generated in the EU was recycled, with the rest being landfilled (37%) or incinerated (23%), of which around 500 million tonnes could have been otherwise recycled or reused (European Commission 2014). In this sense, in December 2015 the European Commission adopted an ambitious new Circular Economy Package that foresees measures to encourage both industrial symbiosis and bio-based economy (European Commission 2015).

The circular economy approach also offers opportunities to develop the bioeconomy in the Danube Region. Industries producing large volumes of high organic content wastewaters and other biowastes may cooperate with technology providers in order to turn by-products into value added output and optimize processes. Due to feedbacks, linear production processes from the raw material to the end-product are complemented by cyclical processes, and production loops close. A symbiosis is born, where manufacturers of different industrial sectors cooperate, and where the waste of a company is used by the other as raw material. Hence, industrial symbiosis is a mutually beneficial partnership of businesses, a cooperation of companies, where the objective is to reduce the total volume of used raw material and generated waste, and by the exchange of materials and energy all participants get a competitive advantage. Thus, one party of the synergy is a company wishing to get rid of its waste or resource in an environmentally conscious and economic way, while the other is its partner, who is able to use that resource (e.g. by-product, energy) in its own industrial production or enterprise. Waste is diverted from the landfill, while both the expenses of the company and the environmental burden have decreased. Government and other third parties can support this symbiosis by providing matchmaking support and enabling legal framework for waste upcycling, additionally the conscious consumers can also demand more sustainable products affecting corporate social responsibility policies.

In order to raise industry awareness, as well as to stimulate the market for bio-based products, more effective involvement of current biomass processing industry, e.g. agro-food processing, pulp and paper industry, forestry-based industry, fibres, feed, and cosmetics sectors, is inevitable. In those sectors, biomass input is considerably high, and generates large quantity of residues and biowastes (solid by-products and high organic content wastewater). Handling and disposal of these streams could be problematic, and could result in high environmental burden and additional financial implications for the company. Nevertheless, these streams available on-site can also be the basis for processes and added value products by guaranteeing constant feedstock supply. Hence, advanced technologies could be integrated into existing infrastructures and value chains utilizing by-products and entering niche markets. This approach could facilitate the bioeconomy transition by involving current stakeholders and merging novel technologies into working value chains. Hence, by-products can be valorized, and additional revenue streams can be obtained while avoiding waste management fees. The resource efficiency of existing biomass value chains may be enhanced this way while relying on existing infrastructures and networks. This approach can be especially successful in the Danube Region, because of the presence of traditional industries processing different types of biomass. However, in order to exploit these resources considerable investments are needed to integrate biorefineries, reveal market demand for bio-based products and guarantee the security and sustainability of the long-term biomass supply requiring active involvement of farmers and/or other industries. These challenges are sizeable, and call for a holistic approach, a transition period and harmonized policy support.

Financing considerations also prefer integration into current or proposed factories by side stream optimization. From a technological point of view, side streams and/or by-products of many biomass related industries (biofuel, food, paper, etc.) can be processed further resulting in value added compounds. On the one hand, this way the industrial stakeholders can be involved, and be aware of the new solutions and their economics, they can broaden their product spectrum, which may result in economic advantages and potential entries into niche markets. On the other hand, by utilizing the existing infrastructures, the up-front cost can be decreased significantly by avoiding building basic connections and common process elements (biomass storage, downstream processing etc.). Hence, the biorefinery concept can guarantee additional revenue streams for facilities including biofuel plants, which makes steps towards the market-based operation. From policy and economic points of view, lock-in situations could be avoided by motivating current industry players to lead the transition by implementing other technologies while holding their market position with their existing production. Hence, paths to develop bio-based value chains should consider integrating current infrastructures, which can result in win-win situations.

5 Case Studies

This section highlights a few projects and initiatives from the Danube Region, in which the industrial partners apply the approaches introduced above.

5.1 *Case 1: Wastewater Integrated Algae and Other Biorefinery Options*

Municipal wastewater estimated to be more than 330 km^3 globally provides a basically unlimited potential market and feedstock source, and it is forecasted to grow further (Drechsel et al. 2015). In the Danube Region the number of wastewater treatment plants has grown rapidly due to the progressive implementation of the Urban Wastewater Treatment Directive (European Communities 1991) and its later amendment (European Commission 1998). With the advent of circular economy practices and policies, and the demand for a more resource efficient operation, wastewater may be considered more as a secondary raw material. Beside removing hazardous substances below the legal threshold, the treatment also aims to recover materials and energy, which can generate additional revenue, and/or can reduce operational cost of the whole treatment.

The municipal wastewater value chain has been considered to be strictly linear using an end of pipe approach. Once the sewage network delivered the wastewater to the treatment plant, the water line provides the clean water, and the sludge line manages the removed organic content in the form of sludge. Sludge is used to produce biogas, and then the digested sludge is separated to sewage sludge (usually landfilled) and leachate containing nutrients above the threshold of discharge, thus usually returned to the start of the treatment. This otherwise problematic leachate, available in most of the wastewater treatment plants, is a potential feedstock for bioeconomy purposes. In the case of industrial processes the wastewater with specific origin and stable composition is usually treated on-site, which could also create opportunities for upgrading.

The Climate-KIC co-funded Microalgae Biorefinery 2.0 project aims to demonstrate the technical and economic viability of integrating algae production into wastewater treatment, and targets wastewater treatment companies demanding a solution, which can make their operation more efficient and sustainable. This solution can also reduce emissions, and provide additional revenues at the same time. Algae can remove the excess amount of nutrients from the leachate reducing the load of return flow while producing biomass that creates added value products and revenue streams. Budapest Sewage Works Ltd. as industrial partner supplies the necessary feedstock, and hosts the demonstration plant at the North Budapest Wastewater Treatment Plant. The demonstrated solution turns leachate into added value algae biomass by recovering of nutrients and capturing CO_2 of biogas plant. Furthermore, it helps the company to develop commitment to sustainability values and enhance the 'green' image.

In a broader picture, wastewater is more often considered as resource instead of waste due to legislation and public concerns, hence algae technology may be substituted or complemented with other wastewater valorization solutions, such as bioplastics production (Eggplant.it 2016) and implementation of microbial fuel cell (Dimou et al. 2014). Therefore, the circular bio-based economy should consider integration into wastewater treatment that offers the benefit of existing infrastructures and all-year, concentrated supply with foreseen supportive changes in legislation.

5.2 *Case 2: Turning Bioethanol Plant into Biorefinery by Capturing CO_2*

First-generation bioethanol plants utilizing starchy grains, mostly feed corn, can also be the basis of the circular economy and benefit from sustainable intensification, and thus can function as biorefinery and in turn become drivers of sustainable biomass production. Pannonia Ethanol is the largest ethanol plant in Central and Eastern Europe, and after its expansion in 2016 is the second largest plant in Europe.

Pannonia Ethanol is located in Dunaföldvár, Tolna County, Hungary. The plant started operating in March 2012. The facility currently utilizes roughly 1.1 million tonnes of corn annually, and it produces about 450 million liters of ethanol and nearly 300,000 tonnes of distiller's dried grains with solubles (DDGS, a high protein animal feed), and 10,000 tonnes of corn oil, an animal feed ingredient. Animal feed co-production in biofuel plants saves million hectares of crop production in the EU (Penlington et al. 2015).

An assessment based on the methodology of Renewable Energy Directive of the EU showed that the pathway of Pannonia Ethanol achieves 65% reduction in GHG reductions compared to fossil fuels, which is beyond the minimum GHG reduction threshold applicable for biofuels defined in the Renewable Energy Directive.

The production process in the plant uses a closed system, minimizing waste as well as purifying and reusing water involved in the production.

Pannonia Ethanol has been considering various opportunities for cutting carbon emission. These directions focusing on resource efficiency and industrial symbiosis include capture and utilization of CO_2 produced during fermentation, and investigation of cellulose-based ethanol production. The approach of Pannonia Ethanol represents the integration possibilities into a factory running on existing and well-established technology. The integration of other technologies can facilitate innovation while contributing also to the improvement of the original process by enhancing efficiency, optimizing by-product streams and reducing wastes and fossil fuel use. Pannonia Ethanol also helps the suppliers to switch to more effective farming practices to sustainably enhance yield and deliver savings on life cycle basis.

5.3 *Case 3: Paper Recycling and Biorefining*

LC FUEL s.r.o (Lcfuel.sk 2016) realized a project entitled "Research of technology for the production of bioethanol from lignocellulosic raw materials in industrial size with continuous operation" between 2014 and 2015. The aims were to design a suitable technology solution for the production of biofuels from secondary materials, and the project delivered the following results: (1) old corrugated cardboard (OCC) was found to be a suitable feedstock; (2) a processing technology was proposed using the existing facilities of the former pulp and paper mill in the industrial park of Štúrovo; (3) the by-products of the technology were identified, and a suitable processing option was developed; (4) the economic analysis showed market potential. The technology supplier of the project was the firm of SAL-TECH s.r.o. (Saltech.sk 2016).

In the first step the plant will convert 30,000 tonnes of OCC dry weight annually. The collected OCC is pretreated in a hydropulper, and then the cellulose fraction is broken down into glucose in enzymatic hydrolysis carried out by cellulase enzymes. The hydropulper and the tanks used for enzymatic hydrolysis and fermentation of ethanol are those used in the former paper mill. Regarding the cellulase enzymes there are two options: either they will be purchased, or they will be fermented on site using the OCC as carbon source. The latter may decrease the GHG emission of the process significantly (Hong et al. 2013). The fermentation will be carried out by ordinary baker's yeast, and the ethanol will be purified by distillation and molecular sieve adsorption. The stillage will be separated, and the liquid fraction will be evaporated. Then the obtained syrup will be mixed with the dried solid fraction, and they will be incinerated together to produce heat and electricity for the process. As the plant will be located in the industrial park of Štúrovo, and the process is energy-demanding, waste heat available in other factories will also be used. This further improves the environmental performance of the process, and creates industrial symbiosis. Currently the technology is in early implementation phase on a pilot scale.

The project also includes extension options, in which other biorefinery alternatives will be developed. Beside bioethanol organic acids can also be fermented from the sugars obtained in the enzymatic hydrolysis, and some of them, which already have a developed market, will also be considered as products resulting in a multi-product biorefinery.

6 Conclusions

The chapter discussed the approaches of sustainable intensification and circular economy to increase the available biomass base for bioeconomy purposes in the Danube Region while not competing with food and respecting the principles of sustainability. Additionally, both approaches build on the involvement of current players of the biomass value chain, hence facilitating the bioeconomy transition through well-functioning infrastructures and networks and bringing economic and additional benefits.

Sustainable intensification and resource efficiency allow to increase the availability of biomass without increasing the burden on the environment. There is a large scope for increasing biomass availability in Europe, and particularly in the Danube Region in a sustainable way, hence fostering bioenergy production. Technologies including precision farming are available, their uptake needs to be supported, however, it will not happen autonomously. In addition, evidence shows that the yield gap persists in many countries (Wicke et al. 2015), and sustainable intensification could be a solution. Why has it not happened on a large scale yet? Perhaps, market incentives were missing. With bioenergy production, the necessary market push may arise. Bioenergy and bioeconomy may be the catalyst for moving towards sustainable intensification. The dialogue between demand and supply sides can facilitate joint development strategies. In this process the public sector role is also paramount for ensuring the appropriate enabling framework, e.g. by creation of certification bodies, financial instruments and support schemes to innovation and mechanization in agriculture. The support to biomass production can move forward to equitable pricing and control over sustainability requirements, certification and stable supply. The unused lands in the Danube region would also help to obtain the important investments in agricultural mechanization and education.

Widening the product portfolio, optimizing processes and eliminating the by-products streams could be attractive for companies processing biomass and owning organic waste or wastewater. Experiences confirm that bioeconomy stakeholders need to be involved and interested. A vehicle to catalyze this is the industrial symbiosis that links together industries to share existing infrastructures, knowledge, biomass streams and business models in order to create successful B2B cooperation. Nevertheless, as in most cases the primary feedstock comes from the agriculture, the established relations to the agricultural sector is essential to guarantee reasonable price, good quality and long-term stable supply of feedstock. In turn biomass processors need to encourage producers to switch to sustainable practices that also improve yields.

In both approaches the knowledge economy will also be strengthened, as sustainable intensification essentially links to the application of "big data", while circular economy implies process optimization and networks. Decision-makers will need to acknowledge the potential benefits, and enabling and/or supporting policies need to be designed and implemented. An enabling policy framework in the Danube Region needs to provide perspective and maximize benefits. As a result, biorefineries working in strong cooperation between agriculture and/or industries would provide a variety of products delivering social, environmental and economic benefits while decreasing fossil fuel dependence.

References

Austrian Energy Agency (2012) Basic data bioenergy. [online] Available at http://www.energyagency.at/fileadmin/dam/pdf/publikationen/berichteBroschueren/bioenergy-basicdata.pdf. Accessed 19 Apr 2016

Beurskens LWM, Hekkenberg M (2011) Renewable energy projections as published in the National Renewable Energy Action Plans of the European Member States [online]. Available at http://www.ecn.nl/docs/library/report/2010/e10069.pdf. Accessed 19 Apr 2016

Council of the European Union (2011) 3083rd general affairs council meeting, general affairs [online] Available at http://www.consilium.europa.eu/uedocs/cms_data/docs/pressdata/EN/genaff/121513.pdf. Accessed 19 Apr 2016

Danube-region.eu (2016) The Danube region's website [online]. Available at http://www.danube-region.eu/about/the-danube-region. Accessed 19 Apr 2016

Dimou O, Andresen J, Feodorovich V, Goryanin I, Harper A, Simpson D (2014) Optimisation of scale-up of microbial fuel cell for sustainable wastewater treatment with positive net energy generation. New Biotechnol 31:213

Drechsel P, Qadir M, Wichelns D (2015) Wastewater, economic asset in an urbanizing world, 1st edn. Springer, The Netherlands. Available at http://www.springer.com/us/book/9789401795449. Accessed 19 Apr 2016

Eggplant.it (2016) EggPlant's official website [online]. Available at http://www.eggplant.it/. Accessed 19 Apr 2016

European Commission (1998) Commission Directive 98/15/EC. Off J Eur Commun [online] 41. Available at: http://eur-lex.europa.eu/legal-content/EN/TXT/?uri=OJ:L:1998:067:TOC. Accessed 19 Apr 2016

European Commission (2010) European Union strategy for Danube Region [online]. European Commission, Brussels. Available at http://eur-lex.europa.eu/LexUriServ/LexUriServ.do?uri=COM:2010:0715:FIN:EN:PDF. Accessed 19 Apr 2016

European Commission (2014) Proposal for a directive of the European Parliament and of Council [online]. European Commission, Brussels. Available at http://ec.europa.eu/environment/waste/pdf/Legal%20proposal%20review%20targets.pdf. Accessed 19 Apr 2016]

European Commission (2015) Closing the loop: commission adopts ambitious new Circular Economy Package to boost competitiveness, create jobs and generate sustainable growth [online]. Available at: http://europa.eu/rapid/press-release_IP-15-6203_en.htm. Accessed 19. Apr. 2016

European Communities (1991) Council directive. Off J Eur Commun [online] 34. Available at: http://eur-lex.europa.eu/legal-content/EN/TXT/PDF/?uri=OJ:L:1991:135:FULL&from=EN. Accessed 19 Apr 2016

European Parliament (2005) SEC(2005) 1573 /* COM/2005/0628 final. Off J Eur Commun [online] 46. Available at http://eur-lex.europa.eu/legal-content/EN/TXT/?uri=celex:52005DC0628. Accessed 19 Apr 2016

European Parliament (2009a) Directive 2009/28/EC. Off J Eur Commun [online] 52. Available at http://eur-lex.europa.eu/legal-content/EN/TXT/?uri=uriserv:OJ.L_.2009.140.01.0016.01.ENG. Accessed 19 Apr 2016

European Parliament (2009b) Directive 2009/30/EC. Off J Eur Commun [online] 52. Available at http://eur-lex.europa.eu/legal-content/EN/TXT/PDF/?uri=CELEX:32009L0030&from=EN. Accessed 19 Apr 2016

European Parliament (2010a) SEC(2010) 1489. Off J Eur Commun [online] 54. Available at http://eur-lex.europa.eu/legal-content/EN/NOT/?uri=CELEX:52010SC1489. Accessed 19 Apr 2016

FAO (2011) Save and grow—a policymaker's guide to the sustainable intensification of smallholder crop production [online]. FAO, Rome, pp 1–116. Available at http://www.fao.org/docrep/014/i2215e/i2215e.pdf. Accessed 19 Apr 2016

Heck S, Rodgers M (2014) How resource scarcity is driving the third industrial revolution. Available at http://www.mckinsey.com/business-functions/sustainability-and-resource-productivity/our-insights/how-resource-scarcity-is-driving-the-third-industrial-revolution. Accessed 19 Apr 2016

Hodgson E, Ruiz-Molina M-E, Marazza D, Pogrebnyakova E, Burns C, Higson A, Gyalai-Korpos M, Rehberger M, Hiete M, Di Lucia L, Noël Y, Woods J, Gallagher J (2016) Horizon scanning the European bio-based economy: a novel approach identification of barriers and key policy interventions from stakeholders in multiple sectors and regions. Biofuels Bioproducts Biorefining 10(5): 508–522. Accessed 3 Jan 2017

Hong Y, Nizami A, Bafrani MP, Saville BA, MacLean HL (2013) Impact of cellulase production on environmental and financial metrics for lignocellulosic ethanol. Biofuels Bioproducts Biorefining [online] 7(3), 303–313. Accessed 19 Apr 2016

Hujber D, Szilágyi T (2014) Danube Region biomass action plan, 1st edn.. Ministry of Foreign Affairs of Hungary, Budapest, pp 2–132. Available at http://danubebiomass.eu/documents/Danube_Region_Biomass_Action_Plan_03191001.pdf. Accessed 19 Apr 2016

Langeveld H, Dixon J, van Keulen H (2014) Biofuel cropping systems: carbon, land and food. Earthscan, Oxford, p 271

Langeveld JWA, Dixon J, van Keulen H, Quist-Wessel PMF (2013) Analyzing the effect of biofuel expansion on land use in major producing countries: evidence of increased multiple cropping. Biofuels Bioproducts Biorefining [online] 8(1):49–58. Available at http://onlinelibrary.wiley.com/doi/10.1002/bbb.1432/abstract. Accessed 19 Apr 2016

Lcfuel.sk (2016) LC FUEL s.r.o.'s official website [online]. Available at http://www.lcfuel.sk/. Accessed 19 Apr 2016

Monforti F, Bódiss K, Scarlat N, Dallemand J (2012) The possible contribution of agricultural crop residues to renewable energy targets in Europe: a spatially explicit study. Renew Sustain Energy Rev [online] 19:666–677. Available at http://www.sciencedirect.com/science/article/pii/S1364032112006740. Accessed 19 Apr 2016

Mueller ND, Gerber JS, Johnston M, Ray DK, Ramankutty N, Foley JA (2013) Closing yield gaps through nutrient and water management. Nature [online] 490:254–257. Available at http://www.nature.com/nature/journal/v490/n7419/abs/nature11420.html. Accessed 19 Apr 2016

Oravec M, Slamka M (2013) Regional profile of the biomass sector in Slovakia [online]. Zvolen, Foropa, pp 2–29. Available at http://www.foropa.eu/files/country_reports/country%20profile%20slovakia.pdf. Accessed 19 Apr 2016

Penlington N, Rabbich S, Gibbons J, Keatinge R, Vickers M et al (2015) Environmental and nutritional benefits of bioethanol co-products. ENBBIO Project Report No. LK0697 (in press)

Saltech.sk (2016) SAL-TECH s.r.o.'s official website [online]. Available at http://www.saltech.sk/. Accessed 19 Apr 2016

Siplatform.org.uk (2016) The sustainable intensification research platform's official website [online]. Available at http://www.siplatform.org.uk/. Accessed 19 Apr 2016

Smith P (2013) Delivering food security without increasing pressure on land. Global Food Secur [online] 2(1):18–23. Available at http://www.sciencedirect.com/science/article/pii/S2211912412000363. Accessed 19 Apr 2016

Stojadinovic D (2009) Serbia—biomass for energy [online], pp 1–16. Available at http://www.eclareon.eu/sites/default/files/biomass_in_serbia_stojadinovic.pdf. Accessed 19 Apr 2016

Tilman D, Balzer C, Hill J, Befort BL (2011) Global food demand and the sustainable intensification of agriculture. Proc Natl Acad Sci USA [online] 108(50):20260–20264. Available at http://www.pnas.org/content/108/50/20260.short. Accessed 19 Apr 2016

Tluka P, Stupavský V (2011) Study on biomass trade in the Czech Republic [online]. Available at http://www.central2013.eu/fileadmin/user_upload/Downloads/outputlib/4Biomass_Trade_study_Czech_Republic_uploaded.pdf. Accessed 19 Apr 2016

Valin H, Havlík P, Mosnier A, Herrero M, Schmid E, Obersteiner M (2013) Agricultural productivity and greenhouse gas emissions: trade-offs or synergies between mitigation and food security? Environ Res Lett [online] 8(3). Available at http://iopscience.iop.org/article/10.1088/1748-9326/8/3/035019/pdf. Accessed 19 Apr 2016

Wicke B, Faaij A, Brinkman M, Gerssen-Gondelach S, van der Laan C (2015) ILUC prevention strategies for sustainable biofuels [online], Utrecht, pp 1–61. Available at: www.uu.nl/en/file/10451/download?token=3Ewt93Ks. Accessed 19 Apr 2016

Barriers to Acceptance of Bio-Based Substitutes: How Schema Incongruity Can Hinder the Purchase of Bio-Based Products

Katja Rudolph

Abstract Bioeconomy encompasses the production of renewable biological resources and their conversion into bio-based products with less environmental impact (e.g. bioplastics). To boost the bioeconomy, it is crucial that companies not only develop these products, but also successfully market them. However, companies are often confronted with poor product acceptance by end-consumers. Depending on the product category, different barriers to acceptance seem to prevail but these are not yet fully understood. This paper seeks to contribute to understanding and overcoming these barriers by analysing product acceptance using the schema theory. It is hypothesized that if associations with the core product attributes do not match the associations with environmentally friendly products, consumers are likely to experience schema incongruity. This rather unpleasant cognitive state can, as a consequence, induce negative a product evaluation. Therefore, adding the label 'green' or 'environmentally friendly' to inappropriate products might lead to a more widely-held negative perception of bio-based products. Consequently, this lowers consumer acceptance of the products and thus impacts upon market success. To analyse this effect, people's associations with environmentally friendly products were tested by an Implicit Association Test. Additionally, a bio-based substitute suffering from schema incongruity was identified and examined using the empirical setting of an experiment and a questionnaire. The results show that environmentally friendly products are somewhat more strongly associated with 'gentle' than with 'strong'. There was also a significant effect of congruity in the product presentation on the perception of stability and environmental friendliness. Thus, to market bio-based substitutes, companies should weaken core product characteristics like strength as little as possible by emphasizing environmental friendliness. Hence, the product presentation should focus on information regarding core attributes and market sustainability as an additional product benefit.

K. Rudolph (✉)
SVI-Endowed Chair of Marketing, esp. E-Commerce and Cross-Media Management,
HHL Leipzig Graduate School of Management, Jahnallee 59, 04109 Leipzig, Germany
e-mail: katja.rudolph@hhl.de

W. Leal Filho et al. (eds.), *Towards a Sustainable Bioeconomy: Principles, Challenges and Perspectives*, World Sustainability Series,
https://doi.org/10.1007/978-3-319-73028-8_7

Keywords Acceptance barriers · Bio-based substitutes · Implicit association test Schema theory · Barriers to sustainable consumption

1 Introduction

The bioeconomy is expected to expand worldwide, not least due to the support it receives through policy measures such as national bio-economy strategies (e.g. in Germany, Finland, South Africa, India) and cross-national initiatives such as the European Union's Bioeconomy Strategy (Bioökonomierat 2015). In this context, companies and researchers are developing new production methods based on renewable biological resources to manufacture products with significantly less environmental impact. Besides completely new products, a lot of companies are enlarging their product portfolios with bio-based substitutes. These are bio-based products that satisfy the same needs as conventional products but that have less environmental impact (e.g. bioplastics).

Nonetheless, technological innovation alone is not sufficient to foster sustainable development. To boost the bioeconomy, it is essential that companies successfully market their bio-based products. This is important in order to make the business models profitable, to incentivize companies to invest in the development of sustainable products and to expand their activities within the field of bioeconomy. In addition, the successful marketing of bioeconomy products will further increase the number of customers purchasing sustainable products and, in consequence, lower the total environmental impact (Thøgersen 1999; Jansson 2011; Balderjahn et al. 2013). Thus, bio-based substitutes can at the same time help solve ecological problems while fostering company growth and contributing to a country's economic development (Beise and Rennings 2005).

However, bio-based substitutes compete with established conventional products in the market place and companies selling bio-based products often encounter barriers at the customer level that hinder the acceptance of environmental friendly products (Gleim et al. 2013). Although most customers claim to be concerned about the environment, sustainable products only represent a small percentage of overall demand (Kalafatis et al. 1999; Bruhn and Kirchgeorg 2007; Esch and Brunner 2010; Luchs et al. 2010; Gleim et al. 2013; Tate et al. 2014). Moreover, the stated willingness to buy sustainable products is far higher than is evidenced through the real consumption behaviour (Luchs et al. 2010; Carrington et al. 2010). In the literature, this phenomenon is discussed as the attitude-behaviour gap (Vantomme et al. 2006; Pickett-Baker and Ozaki 2008; Gleim et al. 2013). It is one of the most consistent findings within research on sustainability and consumer behaviour, yet it is poorly understood (Belz and Peattie 2009; Carrington et al. 2010). As customers buy less products than they indicate they would, the investments made by firms to develop sustainable products are not viewed as being rewarded as expected by investors, business plans etc. (Prothero et al. 2011; Haws et al. 2014). If this gap is left unaddressed, producers of sustainable product alternatives might not meet the economic goals or return-on-investment targets associated with these investment expenditures and therefore reduce their commitment to sustainability. This would limit the offer of sustainable products and impede development towards more sustainable consumption (Prothero et al. 2011).

Although consumption barriers for sustainable products are of increasing importance to firms, most research still concentrates on why consumers purchase sustainable products instead of examining potential consumption barriers (Tanner and Wölfing Kast 2003; Smith and Brower 2012; Gleim et al. 2013). In order to help understand and overcome customer resistance to sustainable consumption, this paper investigates consumption barriers, notably by considering customer's associations with environmental friendly products and how they influence product acceptance. In the literature review, relevant barriers hindering the consumption of sustainable products are discussed, especially focussing on the influence of consumer associations on the perception of environmentally friendly products. To explain this influence, the schema theory is introduced and applied to the examination of the perception of environmentally friendly products. Finally, an empirical study based on an Implicit Association Test (IAT) and an experiment was conducted. Thereafter follows a discussion of the results and the practical implications.

2 Literature Review

2.1 Barriers to Sustainable Consumption

The research findings indicate that one reason why consumers claim to be pro-environmental but purchase few sustainable products might be that their general attitudes are not specific enough to induce pro-environmental behaviour. Thus, pro-environmental actions are hindered by specific subjective interests (Pickett-Baker and Ozaki 2008). For example, a customer intends to buy more environmental friendly products but then buys a cheaper product instead to save money for new shoes that are subjectively more important. Asking customers why they do not buy sustainable products most often reveals obstacles including the higher price, reduced availability, perceived uncertainty as to whether the product is really better for the environment and the required behavioural change (Ottman 1998; Rehfeld et al. 2007; Faltins 2010; Ozaki 2011; Gleim et al. 2013; Umweltbundesamt 2013). Another major barrier to sustainable consumption are concerns regarding product quality and performance (Pickett-Baker and Ozaki 2008; Chang 2011) and a lack of customer expertise with respect to sustainable products (Gleim et al. 2013). This implies that many consumers have little trust in sustainable products to be as good as conventional products or they feel unsure due to a lack of experience with the product. Even if some consumers with high pro-environmental values believe that sustainable products perform as well as conventional ones, the consumer majority is less environmentally oriented and remains sceptical (Pickett-Baker and Ozaki

2008). The development of sustainable substitutes is of little use if consumers do not adopt them. Hence, consumers play a central role in reducing environmental impact and promoting sustainable substitutes (Kotler 2011; Umweltbundesamt 2013). A study by Belz and Schmidt-Riediger (2009) confirms that consumers are one of the main drivers of companies applying sustainability marketing strategies. So, for business and environmental reasons, it is crucial to better understand green consumer behaviour (Belz and Peattie 2009; Jansson et al. 2010). This presents a significant problem for firms that needs to be addressed by research on how to overcome consumers' buying resistance.

Interestingly, looking at the percentage of sustainable products sold per product category, consumption barriers seem to affect products differently (Esch and Brunner 2010; Smits et al. 2014; Luchs et al. 2010; Umweltbundesamt 2013). For instance, Esch and Brunner (2009) found that consumers attach greater importance to environmentally friendly production with regard to milk products than with regard to body care products. Remarkably, according to this study, quality is by far the most important buying criteria within these two product categories and in the top three of other product categories like clothing and cleaning products (Brunner 2014). Thus, another factor seems to influence the mentioned consumption barriers and, consequently, pro-environmental purchase behaviour. Hence, to investigate these barriers, product category specific factors have to be taken into consideration.

2.2 Relevance of Acceptance in the Adoption Process

In order to comprehend the individual decision making of customers, it is essential to analyse the adoption process. Adoption means that a new product leads to a positive purchase decision and a continued use of the product (Robertson 1971; Binsack 2003). The adoption process can be understood as a mental process where an individual makes a decision with respect to a new product. It includes the process from information intake, through information processing, to the memorizing of information (Weiber and Rosendahl 1997; Rogers 2003; Binsack 2003). In the end, the adoption process determines whether a new product will be used or not. Hence, it is critical to the success with consumers of new bio-based substitutes. The acceptance of a product is an important element within the adoption process, especially in the initial phase (Nabih et al. 1997). Consumers' product acceptance is defined as a positive attitude in conjunction with a willingness to use the product, which in turn leads to receptiveness towards the new product. Hence, acceptance can be understood as a determinant of the adoption process and has a positive influence on the purchasing and usage behaviour (Binsack 2003). Therefore, in marketing science, the acceptance construct is often used to explain the failure or success of new products, as it is only if new bio-based substitutes are accepted by customers that they will be tried and, in the case that customers are convinced, ultimately be used continuously (Nabih et al. 1997).

Regarding sustainable products, research shows that their acceptance will heavily depend on the expectations that consumers form towards products from certain product categories (Luchs et al. 2010). Research indicates that if products are associated with 'gentle' characteristics (e.g. food, cosmetics), the supplement 'green' fosters a positive product evaluation and hence acceptance, whereas the effect is reverse if the products are associated with strength and performance (e.g. car tires). Luchs et al. (2010) show that, for example, the preference for a car shampoo offered by a sustainable brand is significantly lower than for a baby shampoo likewise offered from a sustainable brand. Thus, whether sustainability is an advantage for a product or not depends on the benefits that are valued in a certain product category.

Luchs et al. (2010) see the reason for this in the expectations which an individual forms towards certain products. The attribute 'sustainable' is more strongly associated with 'gentle' whereas 'not sustainable' is associated with 'getting the job done'. Van Doorn and Verhoef (2011) confirm these results by finding that the quality perception of organic products was lower for 'vice products' than for 'virtue products'. The authors argue that the wholesomeness signalled by organic claims might reduce the pleasure of a vice product. In both studies, the different evaluations of sustainable and non-sustainable products among product categories were attributed to the associations with the product itself, the concept of sustainability, and whether the product category and sustainable attributes match. Still, current research lacks insights on how this can be explained against the background of information processing within the decision making process. Hence, additional research is required to better understand the underlying psychological processes.

2.3 Schema Theory and Evaluation of Environmental Friendly Products

Numerous studies have shown that consumers evaluate new products using previous knowledge (Binsack 2003). More precisely, they evaluate new products in comparison with a reference point, which is often represented by products they already know. Improvements are rated as profits and deteriorations are seen as losses, whereby losses influence consumer behaviour more than profits (Gourville 2006). Furthermore, if consumers cannot intensively evaluate a new product due to the limits of the human data processing capacity, their present product knowledge is the most important valuation basis (Binsack 2003). This means that the present cognition of an individual strongly influences the evaluation of new products (Sujan 1985; Binsack 2003). It is well known from cognitive psychology that schemata play an important role within information processing (Goldstein 2011; Anderson 2015). A schema 'is a person's knowledge about some aspects of the environment'(Goldstein 2011). It can be understood as a cluster of knowledge describing typical aspects of a concept it represents (Thorndyke and Yekovich

1980). For example, a schema of a dowel could consist of the following elements: 'made of plastic', 'small', 'helpful', 'strong', 'putting up pictures', 'DIY' etc. Schemata are a main component in human knowledge processing, as they steer perception, simplify mental processes, and organize information memory (Kroeber-Riel and Gröppel-Klein 2013). They influence the interpretation of perceived information as well as expectations and conclusions (Fiske and Linville 1980). Hence, schemata affect the attitude people have towards new products and consequently are a crucial factor for the acceptance of bio-based substitutes (Ram and Sheth 1989). Thus, it can be assumed that schemata influence the acceptance of bio-based substitutes.

Mandler (1982) proposes that the schema of a product category influences the information processing with respect to a product that should be evaluated, especially the congruity between the schema and the product. He theorizes that the level of congruity between an object and the category schema influences how a new object is processed and hence how it is evaluated. He distinguishes between two main options: congruity and incongruity. In the case of schema congruity, the new object fits the schema and meets the individuals' expectations. This results in a positive but not ecstatic evaluation. The second option comprises incongruity and describes a case where new objects, in our case new bio-based substitutes, do not fit to the schema and are incompatible with a person's existing knowledge. In case the incongruity is severe and cannot be solved, a strong negative evaluation is to be expected. However, if consumers are able to resolve the incongruity by assimilation, alternate schemata, or accommodation, the evaluation is positive (Meyers-Levy and Tybout 1989).

In the context of bio-based substitutes, this means that consumers may experience incongruity if the associations with the core product attributes (e.g. strength) do not match the associations with 'bio-based' (e.g. gentle) products (Luchs et al. 2010). If the incongruity cannot be overcome, a rather unpleasant cognitive state can, as a consequence, induce a negative product evaluation. Thus, the greater the discrepancy between what a new product is and the expectation, the higher the risk that the product will not be accepted and, as a result, be unsuccessful (Jhang et al. 2012). Adding the label 'green' or 'bio-based' to products that are associated with strength, consequently, might lead to a lower performance perception and, accordingly, to lower consumer acceptance which, as a result, hinders market success. But, if incongruity can be resolved, it would present a substantial opportunity for bio-based substitutes, as they could expect a highly positive evaluation. Thus, it is hypothesized that consumers associate bio-based products more with attributes related to gentleness than with attributes related to strength. Furthermore, it is assumed that in the case of a very incongruent sustainable product presentation of a strength-related product, the acceptance is going to be lower than in the case of a congruent conventional product presentation. In the case of a moderately incongruent sustainable product presentation of a strength-related product, the level of acceptance is expected to be higher than with respect to a very incongruent product presentation.

3 Methodology

3.1 Implicit Association Test (IAT)

To explore the influence of schema incongruity on the acceptance of bio-based substitutes, an online survey (that included an Implicit Association Test (IAT)), an experiment, and a questionnaire were conducted. The IAT, developed by Greenwald et al. (1998), is an established tool from the field of psychology, which is used to measure implicit associations between concepts. Even if the IATs validity has been challenged critically (Fiedler et al. 2006), it has consistently been shown that the IAT has better reliability and validity than other implicit measures (Egloff et al. 2005; Teige et al. 2004).

For this study, an IAT was used to demonstrate associations between conventional and environmentally friendly products with attributes of strength and gentleness, respectively. The IAT measured the response latency and compared the difference between two experimental settings (critical test blocks). The reaction time reflects the strength of the associations and how they are interconnected in the memory, and thus sharing the characteristics that are ascribed to the schemata (Segal 1988). According to the established procedure proposed by Greenwald et al. (1998), participants run through seven blocks where they sort, as quickly as possible, pictures and words presented in the middle of the screen to two categories on the top of the screen. For the present study, the categories were 'environmentally friendly products' and 'conventional products' and the two attributes were 'strong' and 'gentle'. The stimulus material consisted of pictures showing conventional products and respective bio-based alternatives (e.g. detergents or glue) and words similar to 'strong' and 'gentle'. In the first two blocks, participants practiced sorting pictures or words to the categories. In block 3, the categories were combined (see Table 1) and practiced again. Block 4, then, was the first block to test the associations. In the following, the categories were changed (see Table 1) and, after two test blocks, a second critical test block was conducted. To control for order effects, participants were randomly assigned to one of two groups, which differed regarding the block order and thus if they started with the congruent or incongruent category combination. Table 1 shows the block sequence for group A. The participants of group B were presented an IAT where block 1 and 5, block 3 and 6, and block 4 and 7 were swapped. For the data analysis, both groups were combined into one group, as the order variable did not show any significant main or interaction effect.

3.2 Experimental Setting

To test the assumptions, a bio-based substitute, hypothetically suffering from consumer schema incongruity, was identified through explorative discussions with consumers and firms. As a result, a new bio-based dowel was selected which was

Table 1 IAT block sequence for group A

Block	Function	Left category	Right category
Part 1: test of reaction times for congruent condition			
1	Practice (category 1)	Conventional products	Environmentally friendly products
2	Practice (category 2)	Strong	Gentle
3	Practice (category 1 + 2)	Conventional products/ strong	Environmentally friendly products/gentle
4	**Test block**	Conventional products/ strong	Environmentally friendly products/gentle
Part 2: test of reaction times for incongruent condition			
5	Practice (category 1 switched sides)	Environmentally friendly products	Conventional product
6	Practice (category 1 + 2)	Environmentally friendly products/strong	Conventional product/ gentle
7	**Test block**	Environmentally friendly products/strong	Conventional product/ gentle

only briefly on the market and thus little known by consumers. Within the online survey, the participants started with the IAT followed by a question meant to elicit their free associations with a dowel in order to measure the concrete associations with the product used in the setting. For the experimental setting, the participants were split into three groups. The participants of each group were shown one of three different product presentations. The groups differed in terms of the incongruity induced by the shown product: group 1 (high schema incongruity): 40%; group 2 (moderate schema incongruity): 40%; and group 3, as the control group (no incongruity): 20%. Each product presentation consisted of a picture of a package of dowels and a short product description. Table 2 gives an overview of how the product presentations were manipulated for the experimental setting.

At the end of the survey, the participants were asked questions regarding their attitude and willingness to buy and use the presented product. For the evaluation, items measuring the attitude toward a product (e.g. good vs. bad; useful vs. useless) were combined into one variable. Items measuring the willingness to buy and use the product (e.g. possible vs. impossible, very likely vs. very unlikely) were also combined into one variable.[1]

The participants were invited to participate in the survey through a panel, personal contacts, shares in social networks, and a newsletter. In total, 985 respondents participated in the survey, 656 finished it and answered all the questions (dropout rate: 33%). Table 3 gives an overview of the socio-demographic characteristics of

[1]Prior to this, a factor analysis was conducted that confirmed that the attitude items could be combined into one single variable (Cronbach's alpha = .93). The items measuring willingness to buy and use the product could be combined to one variable as well (Cronbach's alpha = .92). Items were measured on a bipolar scale (semantic differential).

Table 2 Characteristics of the product presentations

	Product presentation	Dowel colour	Packaging	Label	Product description
Group 1	Very incongruent	green	green with leafs	→Bio-based →Nylon quality →Non-halogen	Focusing only on sustainability aspects
Group 2	Moderately incongruent	green	simple, green	→Nylon quality →Bio-based →Non-halogen	Focusing on sustainability aspects, technological development, security, durability and functionality
Group 3	Congruent	grey	red	→Nylon quality →Non-halogen	Focusing only on security, durability and functionality

Table 3 Sample overview

	Group 1 (%)	Group 2 (%)	Group 3 (%)	Total (n)
Gender				
Female	50.8	58.6	55.8	340
Male	49.2	41.4	44.2	279
Age				
14–24 years	14.0	11.8	8.4	78
25–34 years	33.3	33.8	34.4	220
35–49 years	24.0	26.2	25.2	164
50–64 years	22.5	21.3	22.1	143
65 years and older	6.2	6.8	9.9	47
Household size				
1 person	26.0	20.6	26.7	156
2 persons	34.1	40.8	38.2	245
3 persons	22.5	18.3	21.4	134
4 persons or more	17.4	20.2	13.7	116
Professional status				
Pupil	0	.4	0	1
Vocational trainee/military or civilian servant	0	1.5	.8	5
Student	23.7	22.3	21.4	148
Full-time employee	42.0	41.3	41.2	271
Part-time employee (part-time/by the hours/from time to time)	11.7	15.9	10.7	86
Housewife/househusband	5.1	3.0	4.6	27
Currently not employed	3.9	1.5	.8	15
Retired	8.9	9.5	13.7	66
Self-employed	4.7	4.5	6.9	33

(continued)

Table 3 (continued)

	Group 1 (%)	Group 2 (%)	Group 3 (%)	Total (n)
Financial situation				
I do not have to restrict myself in any way	3.5	3.5	6.3	26
I am well provided for and I am able to afford various things	29.5	32.0	43.3	214
By and large, I can manage	45.3	42.5	35.4	272
I can barely make ends meet	16.3	17.4	12.6	103
I cannot make ends meet	5.4	4.6	2.4	29

the sample. The sample covers all the important socio-demographic groups of German consumers that are responsible for the main purchases of their household above 16 years of age, but it is not absolutely representative for the German population. The participants' demographic distribution is similar across each group so that no distorting effects are expected.

4 Results

4.1 Implicit Association Test (IAT)

Based on the theoretical reflections, it was assumed that participants classify the stimuli more quickly if the paired categories are congruent (conventional products + strong vs. environmentally friendly products + gentle) rather than incongruent (environmentally friendly products + strong vs. conventional products + gentle). The IAT data was analysed by means of the improved scoring algorithm (D score) proposed by Greenwald et al. (2003). The scoring procedure compares the performance between the congruent blocks (practice and test block) and the incongruent blocks (practice and test block). The mean reaction times between the congruent and incongruent blocks were divided by the overall standard deviation. The D score gives information about the size and direction of the associations: A positive D score indicates, in this context, that the participants respond more quickly if the category combination is congruent than if it is incongruent. A negative score implies a reverse pattern and a D score of close to zero indicates that there is no effect. The D score showed a positive medium effect (D = .569). As expected, the average reaction time for the congruent setting was significantly shorter than for the incongruent category combination (M_{con} = 947.07 ms; SD_{con} = 570.10; M_{inc} = 1252.03 ms; SD_{inc} = 1478.54; $t(653) = 5.04$, $p = .000$). Thus, in the study, environmentally friendly products are relatively more strongly associated with 'gentle' than with 'strong', as the reaction time for the category 'environmentally friendly products' with 'gentle' was shorter than with 'strong'. The reaction times for the category 'conventional products' with

Table 4 Most frequently stated characteristics of a dowel

Characteristics[a]	In (%)	n
Stability	25.9	170
Durability	24.1	158
Strength	21.3	140
Safe	6.6	43
Plastic	6.4	42
Robustness	6.3	41
Useful	5.9	39
Small	5.8	38
Grey	3.8	25
Sustainability	0.3	2

[a]Words with the same word stem were aggregated to one category (e.g. stable was added to stability)

words similar to 'strong' indicate a stronger association of conventional products with 'strong' than with 'gentle'.

4.2 *Pretest for the Experiment (Product Associations)*

To test the assumption of the experiment that dowels are typically associated with strength, the survey participants were asked to write down three characteristics they associate with a dowel. The ten most frequently stated characteristics are summarized in Table 4. Attributes like 'stability' (25.9%), 'durability' (24.1%) and 'strength' (21.3%) were most often named by the participants. Thus, it can be assumed that a dowel is a product associated with attributes like stability and strength and thus that it can be used for the experiment to test incongruity with environmental friendliness.

4.3 *Experiment*

After the participants had seen the product presentation, they were asked to evaluate the product regarding its stability and environmental friendliness. When looking at the overall MANOVA results, the perception of stability and environmental friendliness differed significantly between groups; in other words, depending on the product presentation: $\Lambda = .74$, $F(4,\ 1296) = 53.58$, $p = .000$. Furthermore, the

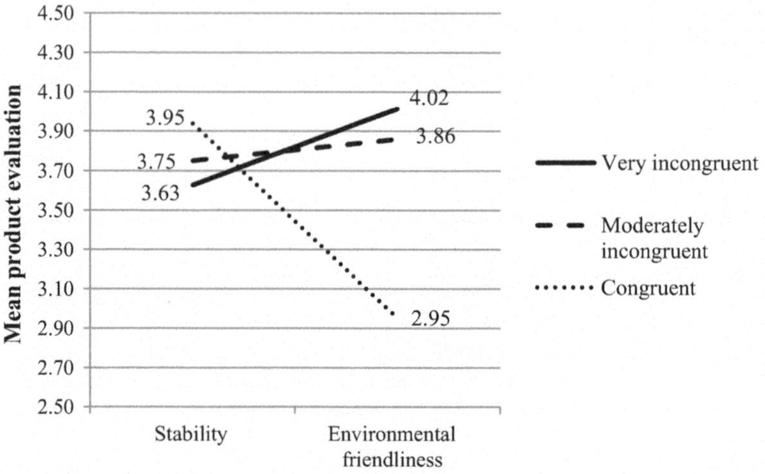

Fig. 1 Product evaluation regarding stability and environmental friendliness depending on congruity in the product presentation (*Legend* 1 = not stable at all, 5 = very stable; 1 = not environmental friendly at all, 5 = very environmental friendly)

analysis revealed a significant effect of product presentation on perceived stability, $F(2,649) = 6.18$, $p = .002$ and on perceived environmental friendliness, $F(2,649) = 75.98$, $p = .000$. Figure 1 illustrates these effects.

A Hochberg's GT2 post hoc test showed a significant difference with respect to the perception of stability between the group with the very incongruent product presentation and the group that saw the congruent product presentation ($p = .001$). However, no significant difference with respect to the perception of stability could be found comparing the group with the moderately incongruent product presentation to the very incongruent ($p = .258$) and to the congruent group ($p = .091$). Looking at environmental friendliness, the test results reveal a significant difference between the group with the moderately incongruent product presentation and the group that saw the congruent product presentation ($p = .000$). There was also a significant difference between the group with the very incongruent product presentation compared with the group that saw the congruent product presentation ($p = .000$). The only groups that did not differ significantly were the group with the very incongruent and the group with the moderately incongruent product presentation ($p = .103$).

Using MANOVA, no significant effect of congruity in the product presentation on attitude towards and on willingness to buy and use the product was found, $\Lambda = .998$, $F(4, 1278) = .347$, $p = .846$; see also Fig. 2).

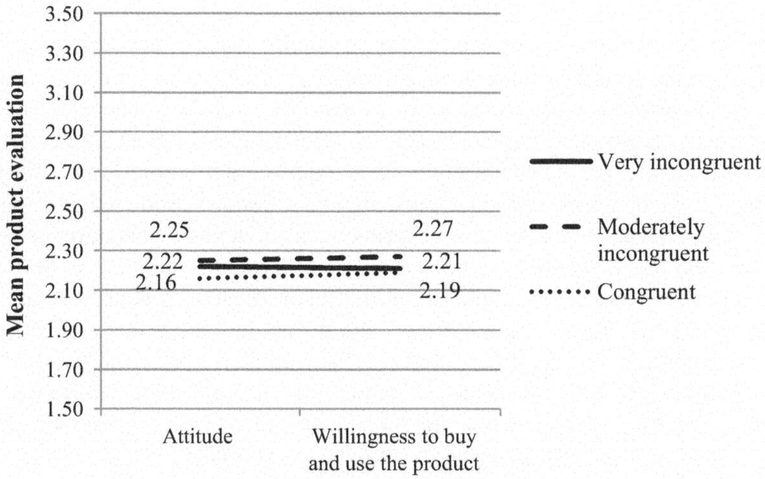

Fig. 2 Product evaluation regarding attitude and willingness to buy or use the product depending on the congruity in the product presentation (*Legend* 1 = positive, 5 = negative; 1 = high willingness to buy and use the product, 5 = low willingness to buy and use the product)

5 Discussion

This paper aimed to analyse barriers to sustainable consumption looking on schema incongruity in product presentation. It was assumed that schema incongruity can hinder the acceptance of a new product and consequently have a negative impact on the adoption process of environmental friendly product innovations, which may be a driver of the attitude-behaviour gap. The results of the IAT suggest that the association between ‘environmentally friendly products’ and ‘gentle’ were somewhat stronger than between ‘environmentally friendly products’ and ‘strong’. This supports the assumption that consumers associate bio-based products more with attributes related to gentleness than with attributes related to strength. According to the schema theory, consumers may experience incongruity if the core product attributes do not match the attributes associated with a product being ‘environmentally friendly’. To investigate this, we choose a dowel as an example to test the effect of incongruity in an experiment. The association test confirmed that the main characteristics consumers associate with this product were stability, durability, and strength. Thus, the product could be used for the experimental setting in order to create incongruity by presenting it as a primarily environmentally friendly product.

The results show that there is a different perception of stability and environmental friendliness between the three groups that saw either a strongly incongruent, moderately incongruent, or congruent product presentation of the same dowel. As expected, *environmental friendliness* was evaluated highest for the product presentation focusing highly on the environmental friendliness of the product (very incongruent condition), whereas the dowel in the control group was evaluated as being least environmentally friendly. The moderately incongruent as well as the very incongruent dowel were perceived as significantly more environmentally friendly than the dowel in the control group.

The evaluations also differed in terms of *perceived stability,* where the dowel in the congruent condition was perceived most strongly. While both moderately and very incongruently presented products were rated as less strong, only the group with the very incongruent product presentation differed significantly to the congruent group. Thus, based on this experiment, a moderately incongruent presentation of a bio-based substitute emphasizing stability and functionality, on the one hand, and environmental friendliness, on the other, leads to a better evaluation of environmental friendliness, but not to a significant loss with respect to the perception of stability compared to a conventional product. This is important, as the survey participants rated stability as being significantly more important than environmental friendliness when buying a dowel ($M_{stab} = 4.76$; $SD_{stab} = .6$; $M_{EF} = 3.25$; $SD_{EF} = 1.211$; $t(643) = -28.678$, $p = .000$).

For marketers, this implies that a presentation focussing solely on the environmental friendliness of a product can have a negative influence on the perceived performance if strength and stability represent important product attributes. Thus, marketing should focus on minimizing the perception of a loss of strength by using a moderately incongruent product presentation, emphasizing both environmental and performance aspects. The focus of the product presentation should still be on core product attributes, while presenting environmental friendliness as an additional asset. In general, marketers should keep in mind which attributes are associated with the product category—for some categories (e.g. baby food, cosmetics), the core associations might be more in line with environmental friendliness and the incongruity effect might be less severe.

Even though the results regarding the perception of stability and environmental friendliness support the assumption that incongruity in the product presentation influences product evaluation, the hypothesis cannot be fully confirmed. No differences could be shown between the three congruity conditions regarding product acceptance, consisting of the attitude towards the product and the willingness to buy and use it. Apparently, the different product presentations and the different perceptions of stability and environment friendliness seem to have no effect on the acceptance of a bio-based substitute. However, this contradictory finding might be explained by the fact that the participants indicated in their comments at the end of the survey that they had difficulties assessing their overall attitude towards the product because the dowel is a product that is rarely used. Therefore, the stated purchase likelihood might be low for all groups, independently of the product presentation. Thus, it would be interesting to replicate this study using a bio-based everyday product with strength as an important product attribute and testing the effect of schema incongruity in the product presentation on product acceptance again. Furthermore, research on the interaction of schema incongruity, price, and perceived risk with regard to bio-based substitutes could have important implications for the marketing of sustainable products.

6 Conclusion

This study indicates that environmentally friendly products are more strongly associated with attributes pertaining to gentleness than to strength. Consumer perception of a bio-based dowel differed significantly depending on the way in which its environmental friendliness was highlighted in detriment of core product attributes like strength and stability. The results indicate that incongruity between core product attributes and environmental friendliness in the product presentation has a significant influence on the perception of products. This can hinder the acceptance of environmental friendly product innovations and thus can be an additional reason for the attitude-behaviour gap.

However, even if the results of the IAT and the assessment of the product attributes suggest such an effect, schema incongruity did not show a significant effect on the acceptance of bio-based products in the experiment. Thus, further research to investigate this barrier to sustainable consumption is necessary, e.g. by focussing on every day products that consumers can evaluate more easily than a dowel.

For the marketing of environmentally friendly products, the results underline the importance of analysing the core product attributes that are important to customers and to accordingly adapt the communication of environmental friendliness to this. Core product characteristics not closely connected to environmental friendliness, like strength or stability, should be diluted as little as possible by highlighting the product's environmental friendliness. Thus, information regarding core attributes should be focused on in the product presentation and sustainability issues should be presented as an additional product benefit.

References

Anderson JR (2015) Cognitive psychology and its implications, 8th edn. Worth, New York

Balderjahn I, Buerke A, Kirchgeorg M, Peyer M, Seegebarth B, Wiedmann K-P (2013) Consciousness for sustainable consumption: scale development and new insights in the economic dimension of consumers' sustainability. AMS Review 3(4):181–192

Beise M, Rennings K (2005) Lead markets and regulation: a framework for analyzing the international diffusion of environmental innovations. Ecol Econ 52(1):5–17

Belz F-M, Peattie K (2009) Sustainability marketing: a global perspective. Wiley, Chichester, West Sussex, U.K., Hoboken, NJ

Belz F-M, Schmidt-Riediger B (2009) Marketing strategies in the age of sustainable development: evidence from the food industry. Bus Strategy Environ 19(7):401–416

Binsack M (2003) Akzeptanz neuer Produkte: Vorwissen als Determinante des Innovationserfolgs, Gabler Edition Wissenschaft Forschungsgruppe Konsum und Verhalten, 1st edn. Dt. Univ.-Verl, Wiesbaden

Bioökonomierat (2015) Auf dem Weg zur biobasierten Wirtschaft. Available at: http://www.biooekonomierat.de/fileadmin/Publikationen/Sonstige/Der_Biooekonomierat.pdf. Accessed 27 Jan 2015

Bruhn M, Kirchgeorg M (2007) Marktorientierte Führung im Wandel von Umweltbewusstsein und ökologischen Rahmenbedingungen. In: Bruhn M, Kirchgeorg M, Meier J (eds) Marktorientierte Führung im wirtschaftlichen und gesellschaftlichen Wandel. Gabler Verlag, Wiesbaden, pp 83–114

Brunner CB (2014) Corporate Social Responsibility in der Markenkommunikation nutzen. In: Esch F-R, Tomczak T, Kernstock J, Langner T, Redler J (eds) Corporate brand management. Springer Fachmedien Wiesbaden, Wiesbaden, pp 345–370

Carrington MJ, Neville BA, Whitwell GJ (2010) Why ethical consumers don't walk their talk: towards a framework for understanding the gap between the ethical purchase intentions and actual buying behaviour of ethically minded consumers. J Bus Ethics 97(1):139–158

Chang C (2011) Feeling ambivalent about going green. J Advert 40(4):19–32

Egloff B, Schwerdtfeger A, Schmukle SC (2005) Temporal stability of the implicit association test-anxiety. J Pers Assess 84(1):82–88

Esch F-R, Brunner CB (2009) "Schein und Sein", *Wirtschaftswoche*, pp 40–42

Esch F-R, Brunner CB (2010) "Mehr als nur leere Worte?", *absatzwirtschaft*, No. 4, pp 75–77

Faltins R (2010) Bio-Lebensmittel in Deutschland: Kaufbarrieren und Vermarktung, Reihe Nachhaltigkeit, vol 31. Diplomica-Verl, Hamburg

Fiedler K, Messner C, Bluemke M (2006) Unresolved problems with the "I", the "A", and the "T". A logical and psychometric critique of the implicit association test (IAT). European Rev Soc Psychol 17(1):74–147

Fiske ST, Linville PW (1980) What does the schema concept buy us? Pers Soc Psychol Bull 6(4):543–557

Gleim MR, Smith JS, Andrews D, Cronin JJ (2013) Against the green: a multi-method examination of the barriers to green consumption. J Retail 89(1):44–61

Goldstein EB (2011) Cognitive psychology, 3. ed., international student ed., special ed. Wadsworth Cengage Learning, Belmont, Calif

Gourville JT (2006) Wann Kunden neue Produkte kaufen. Harvard Bus Manag 28(8):44–57

Greenwald AG, McGhee DE, Schwartz JLK (1998) Measuring individual differences in implicit cognition: the implicit association test. J Pers Soc Psychol 74(6):1464–1480

Greenwald AG, Nosek BA, Banaji MR (2003) Understanding and using the implicit association test: I. An improved scoring algorithm. J Pers Soc Psychol 85(2):197–216

Haws KL, Winterich KP, Naylor RW (2014) Seeing the world through GREEN-tinted glasses: green consumption values and responses to environmentally friendly products. J Consum Psychol 24(3):336–354

Jansson J (2011) Consumer eco-innovation adoption: assessing attitudinal factors and perceived product characteristics. Bus Strategy Environ 20(3):192–210

Jansson J, Marell A, Nordlund A (2010) Green consumer behavior: determinants of curtailment and eco-innovation adoption. J Consum Market 27(4):358–370

Jhang JH, Grant SJ, Campbell MC (2012) Get it? Got it. Good! Enhancing new product acceptance by facilitating resolution of extreme incongruity. J Mark Res 49(2):247–259

Kalafatis SP, Pollard M, East R, Tsogas MH (1999) Green marketing and Ajzen's theory of planned behaviour: a cross-market examination. J Consum Market 16(5):441–460

Kotler P (2011) Reinventing marketing to manage the environmental imperative. J Market 75 (4):132–135

Kroeber-Riel W, Gröppel-Klein A (2013) Konsumentenverhalten, 10th edn. Vahlen, München

Luchs MG, Naylor RW, Irwin JR, Raghunathan R (2010) The sustainability liability: potential negative effects of ethicality on product preference. J Market 74(5):18–31

Mandler G (1982) The structure of value: accounting for taste. In: Clark MS, Fiske ST (eds) Affect and cognition: the 17th annual Carnegie symposium. Erlbaum, Hillsdale, pp 3–36

Meyers-Levy J, Tybout AM (1989) Schema congruity as a basis for product evaluation. J Consum Res 16(1):39–54

Nabih M, Bloem SG, Poisez TBC (1997) Conceptual issue in the study of innovation adoption behavior. Adv Consum Res 24(1):190–196

Ottman JA (1998) Green marketing: opportunity for innovation, 2nd edn. NTC Business Books, Lincolnwood, Ill
Ozaki R (2011) Adopting sustainable innovation: what makes consumers sign up to green electricity? Bus Strategy Environ 20(1):1–17
Pickett-Baker J, Ozaki R (2008) Pro-environmental products: marketing influence on consumer purchase decision. J Consum Market 25(5):281–293
Prothero A, Dobscha S, Freund J, Kilbourne WE, Luchs MG, Ozanne LK, Thøgersen J (2011) Sustainable consumption: opportunities for consumer research and public policy. J Public Policy Market 30(1):31–38
Ram S, Sheth JN (1989) Consumer resistance to innovations: the marketing problem and its solutions. J Consum Market 6(2):5–14
Rehfeld K-M, Rennings K, Ziegler A (2007) Integrated product policy and environmental product innovations: an empirical analysis. Ecol Econ 61(1):91–100
Robertson TS (1971) Innovative behavior and communication, Editors' series in marketing. Holt, Rinehart and Winston, New York
Rogers EM (2003) Diffusion of innovations, 5th edn. Free Press, New York
Segal ZV (1988) Appraisal of the self-schema construct in cognitive models of depression. Psychol Bull 103(2):147–162
Smith KT, Brower TR (2012) Longitudinal study of green marketing strategies that influence Millennials. J Strateg Market 20(6):535–551
Smits M, Wald D, Vismans D, Huet E (2014) When social responsibility leads to growth No. 6, Boston
Sujan M (1985) Consumer knowledge: effects on evaluation strategies mediating consumer judgments. J Consum Res 12(1):31
Tanner C, Wölfing Kast S (2003) Promoting sustainable consumption: determinants of green purchases by Swiss consumers. Psychol Market 20(10):883–902
Tate K, Stewart AJ, Daly M (2014) Influencing green behaviour through environmental goal priming: the mediating role of automatic evaluation. J Environ Psychol 38:225–232
Teige S, Schnabel K, Banse R, Asendorpf JB (2004) Assessment of multiple implicit self-concept dimensions using the Extrinsic Affective Simon Task(EAST). Eur J Pers 18(6):495–520
Thøgersen J (1999) The ethical consumer. Moral norms and packaging choice. J Consum Policy 22(4):439–460
Thorndyke PW, Yekovich FR (1980) A critique of schema-based theories of human story memory. Poetics 9(1–3):23–49
Umweltbundesamt (2013) Green products in Germany 2014. Status Quo and Trends. Available at: http://www.umweltbundesamt.de/sites/default/files/medien/378/publikationen/uba_green_in_germany_2014.pdf. Accessed 8 Apr 2015
van Doorn J, Verhoef PC (2011) Willingness to pay for organic products: differences between virtue and vice foods. Int J Res Mark 28(3):167–180
Vantomme D, Geuens M, DeHouwer J, DePelsmacker P (2006) Explicit and implicit determinants of fair-trade buying behaviour. Adv Consum Res 33:699–703
Weiber R, Rosendahl T (1997) Anwendungsprobleme der Conjoint-Analyse. Die Eignung conjointanalytischer Untersuchungsansätze zur Abbildung realer Entscheidungsprozesse. Mark ZFP 19(2):107–118

Key Factors for the Successful Implementation of Payments for Environmental Services and Offsets for Biodiversity Management and Sustainable Development

P. Puydarrieux and A. L. Mésenge

Abstract Production and consumption affect land use and management. Pollution and habitat loss, including degradation and fragmentation, resulting from non-sustainable production and consumption are important drivers of biodiversity and natural resources loss globally. Economic mechanisms, such as payments for environmental services (PES) and mitigation banking for biodiversity offsets, can be used to support the transition toward more sustainable practices, to avoid net loss of biodiversity and to advance biodiversity net gain. Understanding the key factors of success for efficient implementation of such mechanisms is crucial for business and governments. Based on new institutional economics, this article explores the organizational form and the transaction cost structure of these economic mechanisms. PES and mitigation banking can be seen as hybrid organizational forms between markets and hierarchies. The properties of natural assets and particularly biodiversity depend largely on their location and their unique structural and functional characteristics. Analysis suggests that this great specificity of natural assets is a key element to take into account when designing these economic mechanisms. This approach helps us to identify two distinct ways to implementing PES and offsets: land acquisition and strong intermediation. These conceptual elements are empirically tested in a number of case studies in the paper, including biodiversity offsets and PES schemes implemented to ensure the production of water in watersheds and climate change mitigation. The study provides replicable lessons, specifically the key role of regulation, the importance of the generic form of governance of these mechanisms, the structure of transaction costs and their potential return on investment.

P. Puydarrieux (✉)
Lead Natural Resource Economist, IUCN (International Union for the Conservation of Nature), Gland, Switzerland
e-mail: philippe.puydarrieux@iucn.org

A. L. Mésenge
Economist ECOFOR, Paris, France
e-mail: anne-laure.mesenge@gip-ecofor.org

W. Leal Filho et al. (eds.), *Towards a Sustainable Bioeconomy: Principles, Challenges and Perspectives*, World Sustainability Series,
https://doi.org/10.1007/978-3-319-73028-8_8

Keywords Economic instruments · Biodiversity offsets · Mitigation banks Payments for environmental services (PES) · New institutional economics

1 Introduction

Planetary boundaries defined as a safe operating space for humanity is a recent approach to global sustainability (Rockström et al. 2009a, b). Biodiversity, including ecosystems, is one of the four boundaries which are already exceeded today. And, the Red List Index (IUCN 2017), an indicator of the changing state of global biodiversity, clearly shows that the rate of biodiversity loss is still increasing. Such boundaries are placed before the thresholds, allowing time to react and to come back within the planet limits. Exceeding the boundaries can be seen as a debt that human society has to pay off and which is yet costing more than USD 750 billion annually in lost ecosystem services (GCP 2012).

According to estimations, annual global conservation needs are estimated to be USD 300–400 billion (Gutman 2010; James et al. 2001; Berry 2007). The Global Canopy Programme (GCP 2012) estimates that current flows of funds to conservation are around USD 52 billion per year, the greatest part of which is domestic government budgetary spending. Assuming current governmental and philanthropic conservation efforts were to roughly double to USD 100 billion per year, a gap of USD 200–300 billion would remain that is to say 20–30 times greater than today investable cash flows from conservation projects (Huwyler et al. 2014). Public sector finance and philanthropic capital alone are not sufficient to address this estimated gap.

More problematic, world's most biodiversity-rich countries remain highly underfunded. For instance, the 40 most underfunded countries shelter 32% of all threatened mammalian species (Waldron et al. 2013). Scaling up sustainable investment models and raising awareness of the potential importance of private capital to conservation are crucial.

The planetary boundaries approach is compatible with production, consumption and economic growth, if efforts are concentrated towards the transition to sustainable development and reducing human pressures on nature. Certain economic mechanisms can be mobilized to contribute to transition toward sustainable development in diverse areas. This article focusses on payments for environmental services (PES) and mitigation banks which are among the best known economic mechanisms[1] for conservation finance (OECD 2013; Fétiveau et al. 2014). These mechanisms can be used to integrate the values of biodiversity and ecosystem services into decision-making processes as intended by Aichi target 2[2] (CBD 2010).

[1]The terminology used here refers to OECD (2013).

[2]The Strategic Plan 2011–2020 of the Convention on Biological Diversity is comprised of 5 strategic goals and 20 targets, collectively known as the Aichi Targets.

PESs are considered as appropriate instruments to support changes of practices in agriculture, forestry, and even fisheries. Mitigation banks have been developed in relation with "no net loss" policies to compensate loss of biodiversity generated by land take, i.e. turning natural land into converted areas as a result of urbanization.

A good understanding of key factors for their successful implementation is necessary in order to scale up their development and to increase financing for biodiversity. As PES and mitigation banks involve both public and private sectors, this article analyses the role of government and regulation, and identifies key parameters for the development of profitable solutions by corporations.

This article addresses the question of the design and the implementation of these two mechanisms and brings some answers based on New Institutional Economics. The analysis focuses on governance forms and transaction costs. The study of transaction costs structure (*ex-ante* and *ex-post* costs) and their distribution over time is of importance to design these mechanisms.

2 Regulation: The Starting Point for Innovative Financing Mechanisms

2.1 The Tragic Question of the Sustainable Use of Public Goods

On September 25th 2015, countries adopted the 2030 Agenda for Sustainable Development and its 17 Sustainable Development Goals (SDGs) that came into effect on January 2016, with a targeted fulfilment date of 2030. Built upon the Millennium Development Goals, and expanding their reach and scope, the SDGs have the central objective to promote global economic growth and development that are sustainable, for humanity and for the planet. The adoption of these SDGs is based on the observation that resources required to sustain our current consumption and lifestyle demand are more than the planet can regenerate (Global Footprint Network 2016). This is as if we were living on credit with a risk of cease of payment. Reason forces us to change our lifestyle and consumption patterns.

However, this objective faces the well-known "tragedy of the commons" (Hardin 1968), evocating over-grazing of common land. In these conditions, without proper regulation and management, a common, freely accessible limited resource will inevitably be overexploited (Bureau and Bonnet 2015).

Natural resources and biodiversity could also be largely qualified as public goods. Public goods are non-rivalrous (Consumption of this good by anyone does not reduce the quantity available to other agents) and non-excludable (It is impossible to prevent anyone from consuming that good). Due to its heterogeneity and context-dependence, biodiversity could even appear as a series of public goods that overlap from the local to global levels. From an individualistic perspective, everyone wishes to consume public goods but no-one has an interest to pay for it or to invest for their maintenance. Indeed, the risk is in being the only party to pay and in seeing the other consumers derive a benefit from an investment in which they did not contribute. The problem is a lack of incentive for individuals to invest in them and an inaccurate price signal in the market. Ultimately, the result is underinvestment.

The underlying economic issue is a problem of "externality", that is a cost or a benefit that affects a party who did not choose to incur that cost or benefit. As public goods, natural resources and biodiversity cannot be substituted by forms of capital produced according to incentives provided by the market. The provisions to be introduced to solve a problem of externalities of public goods are all those that introduce exclusivity and rivalry. These provisions such as contracts, controls, and access restrictions are intended to transform the public goods into club goods or into private goods.

By resorting to uniform and inflexible standards we are failing to acknowledge the heterogeneity and the context-dependence of natural assets regarding the opportunities to reduce their use, which generates additional costs and eventually leads to the reduction of environmental ambitions.

By relying only on voluntary approaches, we ignore that economic agents are, above all, guided by their private interests (Bureau and Bonnet 2015). Therefore, we have to bring these private interests into line with the public interest.

2.2 *Economic Mechanisms to Support the Implementation of Changes*

Several economic mechanisms can be mobilized to raise finance for the maintenance of biodiversity and natural resources. In this perspective, those instruments contribute to align private interests with public interest, in particular investment in natural assets. Payment for environmental services[3] (PES) and mitigation banking[4] are some of the best known economic instruments for the finance of nature conservation (OECD 2013). These mechanisms both integrate a part of the value of natural assets in their design and can support the implementation of change of practices.

A PES is considered as a cost-effective means to achieve conservation goals by rewarding specific actors for their efforts in providing ecosystem services, such as watershed protection, soil stabilization, carbon sequestration, etc. This instrument recognizes the need to bridge the interests of landowners, land users and outsiders and has been defined (Wunder 2014) as:

[3]Also referred to as payment for ecosystem services.

[4]Also referred to as conservation banking, species banking, habitat banking or biobanking.

(1) voluntary transactions
(2) between service users
(3) and service providers
(4) that are conditional on agreed rules of natural resource management
(5) for generating offsite services.

PES schemes have various potential sources of funding, which are sometimes combined.

- Payment by final users of the ecosystem service;
- Payment by a public authority as in the case of agri-environmental measures of the European Common Agricultural Policy;
- Payment by operators which are required to offset their impacts on natural assets.

The buyer of the service satisfies a specific need by contracting a PES on a voluntary basis. The valuation of the payment is supposed to be based on the opportunity cost imposed by the preservation of ecosystems providing the services, in realty generally lower than the social value created. This concept applies the logic of Coase Theorem (1960), which shows that in a model where transaction costs are zero and property rights are clearly defined, the free play of negotiation between parties affected by an externality leads at an optimum independent on the initial allocation of rights. The concept of PES relies on a beneficiary pays approach, as opposed to a polluter pays approach. However, because the assumptions of the Coase theorem are not usually satisfied, the need or demand is mostly defined by the community and environmental regulation appears as a catalyser to increase the development of this instrument.

The Vittel mineral water company (France) pays farmers to make sure that their agricultural practices do not affect the quality of the water resource. But if the company decided to develop a PES scheme, it was principally because of a legal constraint. Legally, in France, mineral natural water is defined by its nature characterized by its "content in minerals, trace-elements, and other components", its "original purity", these "characteristics having to be conserved intact because of the underground origin of the water that has to be preserved of risk of pollution".[5] When in 1987, the company observed a change in the content of nitrate in the water feeding the source of mineral water, this legal non-conformity had to be solved (Barbier 2011). In fact, PES can be seen by actors as a solution to fulfil their legal obligations or their collective commitments.

Biodiversity offsets are "measurable conservation outcomes resulting from actions designed to compensate for significant residual adverse biodiversity impacts arising from project development after appropriate prevention and mitigation measures have been taken" (BBOP 2009, 2012). Their goal is to achieve no net loss and preferably a net gain of biodiversity on the ground with respect to species composition, habitat, structure and ecosystem functions and people's use and

[5]Cf. Art. 2—décret n°89–336, 6 Jun 1989.

cultural values associated with biodiversity. They are used to allow certain developments within an overall objective of no net loss and preferably net gain of biodiversity. Biodiversity offsets happen during the final step of the mitigation hierarchy "avoid, minimize, restore and offset", appear as a last resort and are only applied to the residual impacts. Even though voluntary offsetting is possible, biodiversity offsets mostly occur in the framework of environmental public regulations. For instance, this is typically the case for schemes implemented in United States thanks to the Clean Water Act (1977) and the Endangered Species Act (1973) or in France thanks to the Law of August 8, 2016, on the recovery of biodiversity, nature and landscapes.[6] The implementation of biodiversity offsets usually takes one of the three most common following forms (OECD 2013):

1. One-off approach: the biodiversity offset is carried out by the developer or a subcontractor;
2. In-lieu arrangement: the developer pays a fee to a third party to compensate for residual biodiversity impacts;
3. Mitigation banking.

A mitigation bank refers to a pool of offset credits representing a gain in biodiversity resulting from actions to restore, establish, enhance, and/or preserve biodiversity. Offset credits can be sold to developers for paying off their ecological debt resulting from a project.

Regulation is a prerequisite to the creation and the development of PES and mitigation bank in particular by defining the conditions and the limits for their design. With this understanding, and taking into account the risks related to these economic instruments, the establishment of guidelines, thresholds and safeguards was required. To this end, IUCN developed the first-ever global Policy on Biodiversity Offsets (IUCN 2016), which was adopted by IUCN Members at the World Conservation Congress in September 2016 and is now offering a practical support for national action in favour of regulation design. The development of the IUCN policy involved a robust stakeholder process allowing a strong acceptance from governments, conservation organizations and companies.

PES and Biodiversity offsets are both able to mobilize significant financial flows from both the public and the private sector. For instance, in 2011 biodiversity offsets programmes were estimated to have mobilized at the global scale between USD 2.4 and USD 4 billion (OECD 2013). PES programmes were estimated to have mobilized more than USD 9 billion in 2008 alone for watershed services (OECD 2013). However, for now, these resources remain insufficient to bridge the funding gap in conservation.

Due to their significant potential to finance the nature conservation, it is urgent to better figure out the key factors for their successful implementation. The New Institutional Economics gives a framework for the study of the transactions related

[6]Cf. Art. 69—Law No. 2016–1087 of 8 Aug 2016.

to natural assets and underline at least two essential connected criteria: asset specificity and transaction costs.

3 New Hybrid Governance Required to Reduce Transaction Costs

3.1 Specificity of Natural Assets: A Key Characteristic to Consider for Transactions Relating to Natural Capital

"Natural capital" is a term proposed by Pearce et al. (1989), as "a metaphor to shed light on the role of nature in supporting the economy and human welfare" (ten Brink 2015). Following the metaphor, natural capital consists of natural assets including both biotic and abiotic elements and delivering flows of goods and services to human societies. Natural assets are in particular ecosystems, landscapes, seascapes, habitats, species, and genetic resources. Human societies derive multiple benefits from goods and services delivered by ecosystems and biodiversity, such as climate regulation, natural hazard mitigation, contribution to human health, food security, etc. (MEA 2005).

These benefits depend essentially on the functions of ecosystems and the interactions between nature and human activities. Investment in natural capital, for instance by adopting sustainable consumption and lifestyle or by conservation actions, can enhance the benefits for human societies. Inversely, nature degradation has a negative impact on human society, escalating global crises from conflicts to food security (Rüttinger 2015). Investing in natural assets consists in implementing actions that aim to maintain or increase the capacity of this natural resource to deliver goods and services. Various categories of actions could serve this purpose: protection of natural areas and natural resources, improvement of the quality of a resource, restoration of degraded sites and populations (Levrel et al. 2015).

PES and biodiversity offsets are both economic mechanisms allowing investment in natural assets by maintaining or increasing the quality of ecosystem services or by generating a gain of biodiversity in a specific geographic area. In all cases, these instruments involve transactions between actors relating to natural assets. The study of transactions helps to define the conditions of a better efficiency of these mechanisms, particularly by reducing the transaction costs.

The economics of transaction costs was introduced by Ronald Coase in his article "The Nature of the Firm" (Coase 1937). For him, "*an economist thinks of the economic system as being co-ordinated by the price mechanism and society becomes not an organisation but an organism.*" By these words, Coase criticizes the understanding of the economic system as an automatic process regulated by the market. This idea, which questions the cost-effectiveness of markets, is reinforced by the existence of the firm, another organizational arrangement in which regulatory control is high. Thus the author shows that the price mechanism is not the "required coordination" of the economic system, agents being able to choose from among organizational alternatives which could appear as second best. In various conditions, the firm can replace the market because there is a cost of using the price mechanism, which Coase calls "market transaction costs".

Oliver Williamson investigated the economics of transactions by pursuing two objectives. The first is an operational conceptualization of transaction costs, while the second explains the distribution of transactions among existing governance forms based on human factors such as opportunistic behaviour and limited rationality (Williamson 1981).

Williamson defines the transaction as the transfer of a good or service through a technologically separable interface. Transaction costs can be defined as the costs of gathering information and analysis, time for negotiations, development of contractual guarantees, and the monitoring of the performance of the commitments (Levrel et al. 2015). The Williamson's model of transaction costs relies on three criteria: degree of uncertainty, frequency of transactions and asset specificity. Uncertainty and frequency are important in determining what type of contract should be implemented in order to minimize transaction costs. Frequency is characterized by the repetition of similar transactions and uncertainty is defined as "the degree to which the future state of the world cannot be anticipated". The concept of "asset specificity" is more difficult to grasp. This is why Williamson has greatly developed this notion to increase its practicality.

An asset is specific when its use value by a given partner, is higher than its market value. The market value of this asset, apart from the transaction for which it is dedicated, is low or zero. Asset specificity becomes an issue because of opportunism.

The multidimensional property of asset specificity is well acknowledged. The five following dimensions are commonly quoted (Williamson 1983; Malone et al. 1987):

- Site specificity;
- Physical asset specificity;
- Human asset specificity;
- Dedicated assets;
- Time specificity.

Natural assets, in particular biodiversity, are mostly specific assets. Their site specificity is often very high, it is to say that a biological resource can be available at a certain location but movable only at great cost. Two areas are never exactly ecologically identical. Designing offsets to compensate impacts of projects requires assessment of how to achieve biodiversity gain at offset sites that are equivalent to biodiversity loss at the impact site. This means to ensure the "ecological equivalence" which, in the context of biodiversity offsets, is synonymous with the concept of 'like for like' and refers to areas with highly comparable biodiversity components (species diversity, functional diversity and composition, ecological integrity or condition, landscape context and ecosystem services) (BBOP 2012; IUCN 2016).

Biodiversity is defined by "*the variability among living organisms from all sources including, inter alia, terrestrial, marine and other aquatic ecosystems and the ecological complexes of which they are part; this includes diversity within species, between species and of ecosystems*" (Convention on Biological Diversity 1992, Art. 2). In this context, ecological functions reflect the complexity of the quality and diversity of genetic resources, species and ecosystems and the interactions between living and non-living components of nature. This complexity characterised the great physical asset specificity of biodiversity.

The more specific the natural assets, the higher the transaction costs to be met. As biodiversity and more generally natural assets have very high asset specificity, the transaction costs concerning investments in natural capital are very high. This could appear as a strong constraint to the development of financial mechanisms for biodiversity.

3.2 Organisational Innovations Required for the Successful Design of PES and Mitigation Banks

Two central behavioural assumptions influence the level of transaction costs: the limited rationality and the opportunism of the actors. The level of this behavioural uncertainty determines the choice of a governance structure to frame a transaction.

Williamson distinguishes three governance forms: market, firm and hybrid. Each of these structures is characterized by a type of contract and a degree of asset specificity. The firm and the market form the two extremes of the hierarchy of modes of coordination, while the hybrid form is an intermediate level.

The market is the least hierarchical form of governance. Contracts involve non-specific assets and are generally complete, as it is easy to specify future contingencies. The autonomy of the contractors is maintained and the risks of opportunistic behaviour are low because of the substitutability of the sellers. The incentives to follow contractual commitments are maximum.

The hybrid form establishes a bilateral dependency between the contractors when the conventional contract of the market is no longer appropriate. It takes shape when assets are specific and generate a quasi-rent. The risk of opportunistic behaviour increases, which is why elastic contracts are put in place to ensure continuity of the long-term relationship while allowing *ex-post* adjustments to the terms of the contract. However, if the degrees of specificity and uncertainty of the assets become too high, a third level of governance is required.

The firm or hierarchy is the most integrated form of governance. Adaptation between agents is no longer autonomous but coordinated in the framework of an institutional arrangement. Because of the specificity of the assets and the uncertainty, the contracts are very incomplete but very elastic. These are contracts of subordination because the coordination and control of opportunistic behaviours rest on hierarchy and authority. *Ex-post* modifications to the terms of the contract and the requested tasks are possible, whereas on the market a contract has to be renewed.

In short, the hybrid form offers more incentives than the firm and more authority than the market. If the risks of opportunistic behaviour are low, the market will be the privileged mode of governance, while if the risks of opportunistic behaviour increase, contracts will become increasingly incomplete and the organizational forms chosen will be increasingly hierarchical. Figure 1 shows transaction costs as a function of asset specificity in three cases of governance: market, hierarchies (or firm) and hybrid forms (Williamson 1996).

The market appears as relevant governance for transaction regarding little specific assets. To the contrary, when concerning high asset specificity transactions, appropriate governance are hierarchies or vertical integration.

Regarding carbon emissions, their specificity is low and the market could be the most relevant governance form to be used for transactions on carbon credits. Due to the high asset-specificity of biodiversity, market is not the most appropriate governance to manage transactions concerning most of environmental services or biodiversity offsets. Firm or vertical integration could refer to situations when land acquisition is the solution for an actor to comply with his environmental obligations. But as shown by Fig. 1 this governance form is not the most cost-effective. Hybrid forms of governance appear to be more appropriate alternatives.

Actually, analysis of mitigation banks and PES schemes in various countries shows that hybrid form is a convergent choice to manage those transactions. Levrel et al. (2015) mentioned "it consequently seems to us that the mitigation banking system cannot be considered to be a market instrument in the strict sense of the

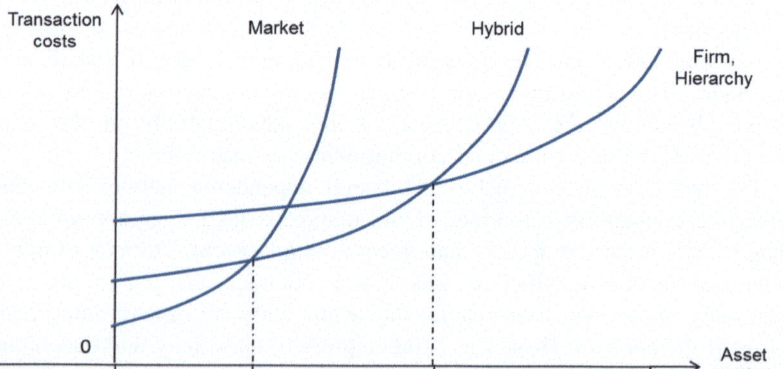

Natural asset specificity	Low	Medium	High
Examples	Carbon	Broad natural habitats, Ecosystem services	Endemic species
Most cost-effective governance	Market	Hybrid	Firm, Hierarchy, Vertical integration

(Adapted from Williamson 1996)

Fig. 1 Transaction costs as a function of asset specificity Adapted from Williamson (1996)

term. It rather appears to be a hybrid form that combines commercial, hierarchical and community characteristics".

4 How to Design Profitable Instruments for Sustainable Development?

4.1 Typology of Transaction Costs in the Implementation of PES and Mitigation Banking

The notion of "transaction" refers to the period of time in which the exchange takes place. An exchange often takes place over time in particular to solve problems due to the heterogeneity and asymmetry of information between the contracting parties. Transaction costs take place during this period. According to Williamson, two types of transaction costs are usefully distinguished: *ex-ante* costs and *ex-post* costs (Williamson 1985).

Ex-ante costs are not costs that occur before the start of the relationship but prior to signing the contract. These costs are associated with the drafting of the agreement, its negotiation and its guarantee.

The drafting of the agreement requires the search for information, the search for future partners and the drawing up of specifications. Contracts or documents binding the contracting parties may be drafted with numerous guarantees or, on the contrary, with a lack of voluntary precaution. The incompleteness of the contracts is linked to the limited rationality of the agents who cannot anticipate all the possible future contingencies. This allows subsequent adjustments, linked to unanticipated events. The costs of drafting the contract increase in particular with asset specificity in order to secure the commitments.

Ex-post costs refer to costs incurred after contracting. They are partially correlated with *ex-ante* costs and are strongly influenced by uncertainty about the future status of the contractual relationship. Four of these costs were mentioned by Williamson:

- The "maladaptation costs" are due to contractual misalignments.
- The "haggling costs" are incurred when *ex-post* misalignments are to be resolved by bilateral efforts.
- The "setup and running costs associated with the governance structures to which the disputes are referred" when the contractors have not found solutions to their *ex-post* misalignments. Opportunistic behaviour is the main cause of this form of *ex-post* costs.
- The "bonding costs" of effecting secure commitments represent the costs of predicting procedures to secure these commitments. These procedures may, for example, take the form of penalties or guarantees.

Other forms of transaction costs have been acknowledged such as *ex-post* evaluation costs, costs of using a second opinion (Bowen and Jones 1986).

4.2 Managing a Scenario of Financial Flows in the Implementation of PES and Mitigation Banking

The implementation of PES and mitigation banks refers to transactions concerning highly specific assets, and often requires the establishment of hybrid forms of governance. The design of these mechanisms involves specific transactions between various actors and intermediaries concerning precise items, especially the definition of ecological targets and the means to achieve them. The relation between the means and the target has to rely on sound science and this could require ad hoc research work. For instance, in the case of Vittel PES scheme, a partnership between the Water Company and INRA (*the French National Institute for Agricultural Research*) has been set up in order to define some of the PES characteristics (Barbier 2008).

Analysis of various PES and mitigation banks shows that *ex-ante* costs are often very high and could cause the failure of a project. This was the case with a project of mitigation banking aiming to offset the Black-bellied Hamster (*Cricetus cricetus*) habitat destruction in Alsace (France). In this case, this mechanism was expected to deal with a very specific asset. This increased the difficulty to develop a hybrid form for the transaction such as a mitigation bank. A hierarchy (firm integration) seemed to be the most appropriate solution in this case.

Usually *ex-ante* and *ex-post* transaction costs could be summarized as following (See Table 1).

Table 1 Different forms of transaction costs involved in the design of PES and mitigation banks

Category of transaction costs	Nature of transaction costs
Ex-ante	Scientific analysis
	Design of ecological targets
	Design of the means (environmental service, restoration measures, etc.)
	Contract negotiation
	Contract establishment
	Guarantees
Ex-post	Maladaptation costs
	Haggling costs
	Setup and running costs
	Bonding costs

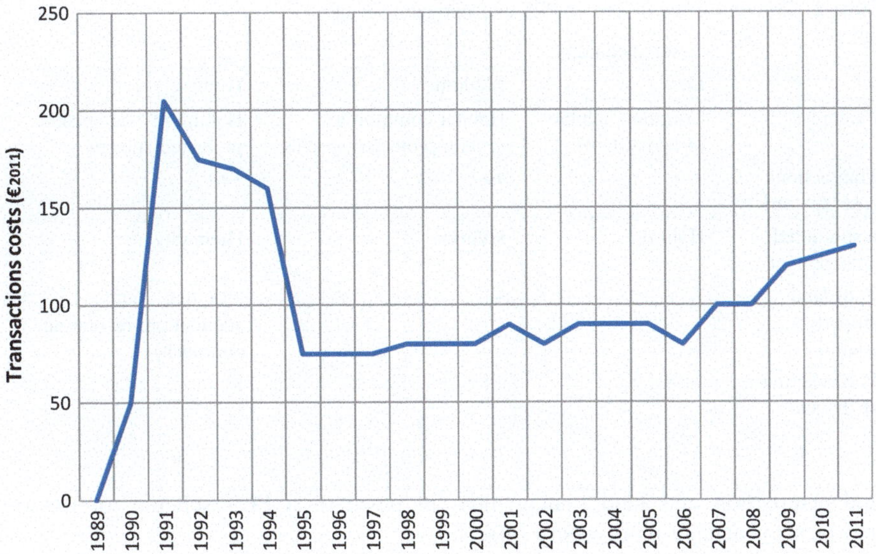

Fig. 2 Temporal structure of transaction costs (Adapted from Defrance 2011. Evolution of transaction costs related to the creation and implementation of a PES scheme in Evian watershed)

The level of *ex-ante* costs of transaction is generally significant but conditions the control of *ex-post* costs.

Figure 2 shows clearly that *ex-ante* transaction costs are higher than *ex-post* ones in the specific case of the implementation of a PES scheme in Evian watershed. This example illustrates how significant financial resources are required to launch a project investing in natural assets.

4.3 *Investing in Natural Capital: Not Only a Cost but also Earnings*

A cost-effective design and management of a PES or a mitigation bank allows investors to expect realistic commercial earnings. The key parameters to take into account are:

- The asset specificity,
- The size of the project,
- The internal rate of return (IRR) of the project.

As mentioned previously, asset specificity conditions the governance form chosen for the transaction. Given that PES and mitigation banks can be considered as hybrid forms involving various stakeholders, if asset specificity is too high, a more hierarchical form of governance will be more appropriate. This could mean

Table 2 Criteria for the design of PES and mitigation banks

	Asset specificity		
	Low	Medium	High
Examples	Common habitats or ecosystems	Habitat common to several protected species	Habitat of one specific protected species
Transactions costs (*ex-ante*)	+	++	+++
Institutional arrangement	Hybrid	Hybrid	Hierarchy
% of land property control	+	++	+++ (e.g.: land acquisition, or one-off approach)
Maximal size of the site	+++	++	+

land acquisition and land management in place of a PES scheme, or one-off approach to manage biodiversity offsets.

As mitigation banks refers to a repository of existing offsets credits, the higher the specificity of site is, the harder it could be for the manager of the bank to sell the credits. The development of mitigation banks seems to be less risky for limited geographic areas than for large service areas.

Expected IRR is also an appropriate input to design business models of PES and mitigation banks, allowing sizing several options of the project (such as guarantees, etc.) (See Table 2).

Taking into account that the *ex-ante* costs of transaction, and more generally the establishment costs, are very high and that sales-revenues happen mostly later, the cost of capital invested in the project has to be strongly controlled and reduced. The public or common nature of goods and services delivered by Natural Capital justify the use of public funding in order to reduce the cost of capital invested in Natural Capital and sustainable development. Several facilities have been developed by public institutions (e.g.: the Natural Capital Financing Facility developed by the European Bank of Investment and the *Investments for the Future Programme* in France) in order to reduce the weighted average cost of capital (WACC) invested in such kinds of projects. These financing facilities could take the form of grants or loans.

Due to the public-good nature of Natural Capital, the measure of Return on Investment is not necessarily evaluated in terms of financial earnings. Actually, the results of such investment are usually a net gain in delivery of bundles of ecosystem services, whose benefits can be assessed by their social value. Furthermore, the purpose of these investments is to maintain and restore biodiversity and the Return on Investment might also be measured in terms of net gain concerning biodiversity and ecosystems.

5 Conclusion

PES and mitigation banks are considered as two appropriate economic mechanisms to improve the financing of Natural Capital. Understanding how to successfully design and implement them is urgently needed to address the funding gap to avoid the loss of biodiversity. Regulation is a key starting point for their development because by relying only on voluntary approaches, we ignore the fact that economic agents are, above all, guided by their private interests.

Because of the great specificity of natural assets, the market cannot be the most cost-effective institutional arrangement to deal with transactions concerning Natural Capital and neither are hierarchies such as land acquisition and land management. Therefore, we have to advocate for hybrid forms that is to say more complex institutional organizations.

The development of PES and mitigation banks requires high *ex-ante* costs of transaction in order to precisely define the ecological targets and the necessary means to reach these goals. These initial costs could be seen as an obstacle to the scaling-up of such mechanisms. Public funds could be used to reduce the cost of the capital needed to impulse investment in Natural Capital. Operators willing to develop a commercial offer of services including biodiversity offsets and PES should consider less specific assets and limiting the size of the site implicated in order to reduce their financial risks.

The role of governments appears central for facilitating investments in natural capital by establishing appropriate regulations and financing facilities.

References

Barbier M (2011) L'eau minérale naturelle ou le paradoxe de la pureté originelle. In: Sophie H, Olivier T (eds) Humains, non-humains. Comment repeupler les sciences sociales. La Découverte, Paris, pp 22–29

Barbier M (2008) «Water in bottles, farmers in green. The sociotechnical and managerial construction of a "dispositif" for underground water quality protection». Int J Agric Res Gov Ecol IJARGE 7(1/2):174–197

BBOP (Business and Biodiversity Offsets Programme) (2009) business biodiversity offsets and BBOP an overview. BBOP, Washington, DC

BBOP (Business and Biodiversity Offsets Programme) (2012) Glossary, 2nd updated edn. BBOP, Washington, D.C.

Berry P (2007) Adaptation options on natural ecosystems. A report to the UNFCCC secretariat financial and technical support division. Retrieved from ipcc-wg2.gov/njlite_download.php?id=5755

Bowen DE, Jones GR (1986) Transaction cost analysis of service organization-customer exchange. Acad Manag Rev 11:428–441

Bureau D, Bonnet X (2015) How can we remedy the failings of «the invisible hand»? CGDD (2015), Nature and the Wealth of Nations, Ministère de l'Ecologie. du Développement Durable et de l'Energie, Paris, La Revue du CGDD, pp 133–140

CBD (2010) Strategic plan for biodiversity 2011–2020 and the Aichi targets "living in harmony with nature". Secretariat of the Convention on Biological Diversity, Montréal

Coase R (1937) The nature of the firm. Econometrica 4:386–405

Coase R (1960) The problem of social cost. J Law Econ 3:1

Defrance P (2011) Evaluating economic policy instruments for sustainable water management in Europe. Financial compensation for environmental services: the case of Evian Natural Mineral Water, WP3 EX-POST Case studies, 44p

Fétiveau J, Karsenty A, Guingand A, Castellanet A (2014) Etude relative aux initiatives innovantes pour le financement de la biodiversité et l'identification des mécanismes à fort potentiel, rapport final, février 2014. Ministère des affaires étrangères et du développement international - France, Paris

Global Canopy Programme (GCP) (2012) The little biodiversity finance book, 3rd edn. Retrieved from http://www.globalcanopy.org/materials/little-biodiversity-finance-book

Global Footprint Network (2016) National Footprint Accounts, 2016 edn. www.footprintnetwork.org/en/index.php/GFN/blog/national_footprint_accounts_2016_carbon_makes_up_60_of_worlds_footprint

Gutman P (2010) World Conservation Magazine. The magazine of the international union for conservation of nature. Jul 2010 edn. Retrieved from http://cmsdata.iucn.org/downloads/101252_uicn_24pp.pdf

Hardin G (1968) The tragedy of the commons. Science 162(3859):1243–1248. https://doi.org/10.1126/science.162.3859.1243.PMID5699198

Huwyler F, Käppeli J, Serafimova K, Swanson E, Tobin J (2014) Conservation finance. Moving beyond donor funding toward an investor-driven approach Credit Suisse. McKinsey & Company, WWF

IUCN (2016) IUCN Policy on Biodiversity Offsets. WCC-2016-Res-059-EN

IUCN (2017) IUCN red list of threatened species—version 2016.3 www.iucnredlist.org

James A, Gaston K, Balmford A (2001) Can we afford to conserve biodiversity? Bioscience 51:43–52. Retrieved from https://www.cbd.int/doc/articles/2002-/A-00486.pdf

Levrel H, Hassan F, Scemama P, Vaissière AC (2015) Institutional and organisational innovations for increasing investment in natural capital. In CGDD (2015), Nature and the Wealth of Nations, Ministère de l'Ecologie, du Développement Durable et de l'Energie, La Revue du CGDD, Paris, pp 145–158

Malone TW, Yates J, Benjamin RI (1987) Electronic markets and electronic hierarchies. Commun ACM 30(6):484–497

MEA (Millennium Ecosystem Assessment) (2005) Ecosystems and human well-being: synthesis. Island Press, Washington, DC

OECD (2013) Scaling-up finance mechanisms for biodiversity. OECD Publishing, Paris

Pearce DG, Markandya A, Barbier E (1989) Blueprint for a green economy. Earthscan, London

Rockström J et al (2009a) Planetary boundaries: exploring the safe operating space for humanity. Ecology and Society 14(2):32

Rockström J et al (2009b) A safe operating space for humanity. Nature 461:472–475. https://doi.org/10.1038/461472a

Rüttinger L, Stang G, Smith D, Tänzler D, Vivekananda J et al (2015) A new climate for peace: taking action on climate and fragility risks. In An independent report commissioned by members of the G7. Adelphi, International Alert, Woodrow Wilson International Center for Scholars, European Union Institute for Security Studies

ten Brink P (2015) In CGDD (2015), Natural capital—an old concept with a new life. CGDD (2015), Nature and the wealth of nations. Ministère de l'Ecologie, du Développement Durable et de l'Energie, Paris, La Revue du CGDD, pp 41–50

Waldron A, Mooers AO, Miller DC, Nibbelink N, Redding D, Kuhn TS, Roberts JT, Gittleman JL (2013) Targeting global conservation funding to limit immediate biodiversity declines. Proc Natl Acad Sci USA 110(29):12144–12148

Williamson O (1981) The modern corporation: origins, evolution, attributes. J Econ Literature XIX:1537–1568

Williamson OE (1983) Credible commitments: using hostages to support exchange. Am Econ Rev 73(4):519–538

Williamson O (1985) The Economic Institution of capitalism: firms, markets and relational contracting. The Free Press, a division of Macmillan Inc, New-York, p 452

Williamson O (1996) The mechanisms of governance, Oxford university Press Ed., New-York, 429p

Wunder S (2014) Revisiting the concept of payments for environmental services. Econ Ecol. https://doi.org/10.1016/j.ecolecon.2014.08.016

Part III
Bioeconomy: Innovative Approaches and Technological Possibilities

Development of Blue Revolution Through Integrated Bio-cycles System on Tropical Natural Resources Management

Cahyono Agus

Abstract The high net primary production in tropical ecosystem were more supported by the rapid organic-cycling than their low fertility weathered acid soil, due to the high temperature, rainfall, moisture and light intensity along a year. Tropical natural resources have a high biomass productivity but still less economical values. New paradigm from extraction to empowerment of natural resource will give new challenge to shift from red- & green economic to blue economic concept that should be more smart, global, focus, and futuristic for sustainable development. Integrated Bio-cycle System (IBS) is a closed-to-nature ecosystem on landscape ecological management to manage land resource (soil, mineral, water, air, microclimate), biological resources (flora, fauna, human) and their interaction to have more high added value in environment, economic, socio-culture and health aspect. The bio-economic chain should be managed through 9A (Agro-production, -technology, -industry, -business, -distribution, -marketing, -infrastructure, -management, -tourism) with 9R (reuse, reduce, recycle, refill, replace, repair, replant, rebuild, reward). The system has multifunction and multi-product, that will meet with the expected basic need for daily-, monthly-, yearly- and decade's income at short-, medium- and long- term periods for small, medium, and big stakeholder. IBFS can produce "gold of life", such as: yellow gold (food, rice, corn), green gold (vegetables), brown gold (plantation wood), red gold (meat), white gold (milk, fish), black gold (organic fertilizer), transparent gold (water), gas gold (oxygen), blue gold (biogas, biomass energy, bio fuel), king gold (herbal medicine), prosperity gold (tourism), inner gold (mystic). IBS with ABCG (academic, business, community, government) networking has a good prospect for sustainable environment and life.

Keywords Bio-economic chain · Blue revolution · Integrated bio-cycles system
Gold of live · Tropical ecosystem

C. Agus (✉)
Faculty of Forestry, Universitas Gadjah Mada, Bulaksumur,
Yogyakarta 55281, Indonesia
e-mail: cahyonoagus@gadjahmada.edu
URL: http://www.acahyono.staff.ugm.ac.id

W. Leal Filho et al. (eds.), *Towards a Sustainable Bioeconomy: Principles, Challenges and Perspectives*, World Sustainability Series,
https://doi.org/10.1007/978-3-319-73028-8_9

1 Introduction

1.1 Tropical Forest

The decrease in forest area around the world is becoming a major problem, and tropical forests have been disappearing with dramatic speed. In 1980, the forest area decrease amounted to 7.5 million ha of closed forest and 3.9 million ha of open forest (Pancel 1993). Lee and Park (2001) reported that tropical forests (50% of the world's forest) are disappearing at a rate of 10–16 million ha yr^{-1}, whereas forest cover in Asia was reduced by about 3.4 million ha in 1995 alone. In Indonesia, it is estimated that forest will be lost at a rate of approximately 1.6 million ha yr^{-1} during the 2000s (Baplan 2000). Tropical forest area will be reduced 500–800 million ha by 2050. In 1995, forests covered an area of more than 3.4 billion ha, or about 26.6% of our geosphere land (Freezailah 1998).

It is understood that world's forests perform many functions. However, tropical forests are different from temperate and boreal forests because of their (a) complex ecosystems (b) highly efficient in total biomass production, (c) subject to property problems and population pressures, (d) storehouses of the world's genetic resources and medicinal plants, and (e) disastrous forest fires (Freezailah 1998).

Deforestation and degradation of tropical forests has been accelerated, causing valuable natural resources to be lost in recent years. Timbers from natural forests in tropical countries have been regarded as a free gift bestowed by God. Natural forests are not able to supply enough timber for world demand in a short period. The demand for intensive silviculture technology will increase with the increasing global interest in rehabilitating degraded forests in many countries, including Indonesia. Intensive silviculture will become a promising choice for creating productive and sustained forests.

Regarding to tropical deforestation and degradation, forest plantation in tropical countries requires careful consideration (Freezailah 1998). This plantation must be given greater emphasis in work and activities for the following reasons: (a) may be converted into highly productive forest, (b) contribute to the restoration of our planet's ecological health, (c) may supply the demand for domestic timber consumption and international trade, (d) will help to relieve pressure on, and to save our remaining precious tropical forests and their rich biodiversity.

One way to replenish forests after logging is through carrying out enrichment planting and reforestation of cleared areas. In recent years, deforestation of tropical regions has accelerated and valuable natural resources have been lost. Many natural forests in tropical regions of the world have been converted into cultivated land, degraded secondary forests, artificial plantations and grasslands. The main causes of forest destruction are illegal logging, slash-and-burn cultivation, extensive cattle grazing and natural forest fires (Ohta et al. 2000). The destructions of tropical forestland are also responsible for major impacts on weather, affecting rainfall and temperatures locally and even globally. Understanding the soil characteristics of tropical forest ecosystems is important not only for maintaining land for sustainable production, but also for estimating effects of forest change on global environmental change.

1.2 Short Rotation Plantation

The conservation and preservation of natural forest will certainly become more difficult if timber suppliers dependent upon plantations are not established in an appropriate manner and scale. Short rotation plantations are mostly monoculture ecosystems, and so have certain limitations in species diversity and long-term productivity when compared to polyculture systems and natural forests. Nutrient uptake in natural forests is relatively constant with nutrients being rapidly recycled in the ecosystem, whereas in plantation forests nutrients are recycled quickly and the amount of nutrient uptake varies with age of the stand (Agus et al. 2004).

Natural forests cannot supply the world timber demand, so intensively managed plantations may be an important means for a productive and sustainable strategy to meet needs for woods. Plantation forests can also fulfill a number of functions; in many locations, plantations have been established for environmental rehabilitation as well as for soil and water conservation. In anticipation of the growing role of plantation forestry, Indonesia is planning to establish 6.2 million ha of industrial plantation forests (Baplan 2000). Nevertheless, conversion of natural forests to large-scale plantations of fast growing species introduces the possibility of various risks such as damage caused by diseases and insects, and deterioration in sustained forest productivity of tropical soils due to the susceptibility of forest monocultures. Furthermore, forest soil in tropical regions is generally nutrient-poor; over half of all tropical soils are highly weathered and leached (Jordan 1985), and attempts to rehabilitate tropical forest will face difficulty.

1.3 Legume Cover Crop

Beneficial effects of leguminous plants on accompanying or subsequent crops have been recognized for thousands of years in various cultures. The use of cover crops is particularly relevant in the humid tropics because high rainfall has often depleted nutrients in soil, especially nitrogen (AICAF 1995). Soil's organic N is continually lost in natural or man-made ecosystems through plant removal and loss processes such as leaching, denitrification, and ammonia volatilization. Biological N_2 fixation helps replenish the soil N pool. Sustainable forests require compensation of nutrients lost by the harvesting of products.

In tropical regions, legumes have been utilized frequently to fulfill the following functions (AICAF 1995):

a. Preservation of soil fertility and effects on other crops in multiple cropping systems
b. Contribution to soil structure improvement through tap roots of the root system
c. Protection and mellowing by covering of soil surface with legume plant canopy
d. Fixation of soil particles by root system and prevention of soil erosion by water and wind
e. Weed control through soil covering with canopy
f. Ease of selection and use for multiple cropping systems due to diversification in species, varieties and ecotypes of legumes.

Soil fertility declines rapidly through erosion, leaching and insulation due to the exposure of soil resulting from clear cutting in tropical regions. During the establishment phase, some nutrient supplements are usually necessary to enhance growth or to ensure the survival of planted forests in degraded land. However, in maintaining a sustainable forest soil productivity program, priority should be given to employing organic matter and symbiotic microbe methods (ITTO 1993). Legumes play an important role in improving the productive capacity of the soil as a source of N_2–fixation from the atmosphere, as well as improving recycling of nutrients, protection from soil erosion, the build up and maintenance of soil organic matter and the suppression of weeds (Hairiah et al. 1992; Skerman et al. 1988). The use of leguminous plants as 'cover crop' is one of the oldest agricultural practices for improving soil fertility, however, no reports of the practice in tropical plantation forests are known to the authors.

1.4 Sustainable Blue Earth

Our blue planet consists of a blue ocean of 72% and a blue sky over 95%, should be supported by blue earth. A red-economy that is merely economic-oriented has resulted in environmental and life damage. A green economy oriented towards environmental and healthy value, although actually was expensive and dangerous. The concept of the blue economy was developed by Gunter Pauli of ZERI Foundation in 2009 (Agus 2016) through the acceleration of natural cycle processes with the empowerment of land resources (land, water, and mineral), biological (plants, animals, and humans) with added value of economy, environment, socio-culture, technology, sustainable management. The blue economy offers efficient investment, increased innovation, increased funding, job creation, social capital development, entrepreneurial stimulation. Performed by the utilization of waste and abandoned goods to be food, energy and employment, thus turning poverty to be sustainable development, and scarcity to be availability. The blue economy has provided new and innovative new creative and innovative opportunities, clean and dignified.

2 Materials and Methods

2.1 Site of Study

The study sites used are located in East Kalimantan, Indonesia at an altitude of 20–60 m above sea level. Climate of the study area is wet tropical rain forest (Köppen classification) characterized by heat and moisture throughout the year. Soils are classified as *Typic Hapludults* (USDA classification). In general, the climate of the research sites (Sebulu and Kenangan sites) is characteristic of a tropical rainforest region; hot and wet throughout the year. The average annual rainfall from 1991 to 2000 was 1937 $\pm$ 567 mm. The total monthly rainfall varied from year to year, and the minimum and maximum amount of annual precipitation recorded in the research sites was 985 mm in 1998 and 2725 mm in 2000, respectively (Agus 2003).

The monthly distribution of rainfall shows twin peaks of over 200 mm month^{-1} in both April and December, caused by east and west monsoon effects. The minimum and maximum average monthly rainfalls were 78 $\pm$ 38 mm in September and 238 $\pm$ 92 mm in December, respectively (Agus 2003). Unexpectedly moist and/or dry months occurred during some years in Kalimantan. There were consecutive drastically dry years in 1997 and 1998. This phenomenon was linked to particularly strong El Nino-Southern Oscillation (ENSO) and forest fire events (Agus et al. 2004).

The average annual evaporation from 1988 to 1991 was 1273 mm per year, and the minimum and maximum of average monthly evaporation rates were 75 mm in November and 145 mm in February, respectively (Agus 2003). The monthly rainfall was almost always higher than rate of evaporation, except for the dry month of September.

The mean diurnal air temperature was 27.6 $\pm$ 1.2 °C. The mean maximum air temperatures were 32 °C and the mean minimum air temperature was 23 °C. The mean diurnal air temperature was identical to air temperature at 10 AM. The mean air temperature differed little between months. Diurnal ranges of air temperature were approximately 19 °C for clear days, and the changes in daily maximum and daily minimum temperature from month to month were smaller than the daily range of change. The mean relative air humidity was 81.9 $\pm$ 5.0%, with maximum of 99% at the night and minimum of 40% at the afternoon (Agus 2003). The average daily range of relative humidity was larger than the range between months.

The pattern of fluctuation in soil temperature followed those of air temperature. However, the values and degree of variations of hourly soil temperature were relatively lower than those of air temperature. Furthermore, the value and variations at 10 cm soil depth were lower than those in the upper layer. Diurnal soil temperature at 0 cm soil depth was 27.1 $\pm$ 0.8 °C, and at 10 cm soil depth was 27.0 $\pm$ 0.5 °C (Agus 2003).

2.2 Focus of Study

Studies on sustainable site productivity in tropical regions have been carried out by the following items (Agus 2003):

a. Effects of land use change (such as from natural forest (NF) to secondary forest (SF), degraded forest (DF), slash-cut & burn areas (CB), alang-alang (*Imperata cylindrica*), first rotation of yemane (Ga) and coppice (Co) on the chemical properties of forest soil.
b. Establishment of site index and nutrient dynamics in a short rotation plantation of yemane in good, moderate and poor sites. Items researched were: plant growth, plant biomass (aboveground, understorey and belowground biomass), litterfall, soil chemical properties, forest floor, decomposition rates of leaf-litter, characteristics of N-mineralization in soil and ion movement in soil.
c. Effects of ages on the biogeochemical cycles in 2, 4, 6 and 8-year-old stands of yemane on the moderate site at the first rotation. Items researched were: plant growth, plant biomass (aboveground, understory and belowground biomass), litterfall, soil chemical properties, forest floor, characteristics of N-mineralization in soil and ion movement in soil.
d. Effects of LCC on soil amelioration after clear cutting in short rotation plantation by planting of *Mucuna anagyroides* (CA), *Crotalaria chochuchinensis* (MC) and *Calopogonium caeruleum* (CC). Biomass of aboveground, belowground and soil chemical properties were investigated. Furthermore, N_2-fixation rate of *Styloxanthes quionensis* (CIAT-184), *Flamengia congesta* (FC), *Mucuna chochuchinensis* (MC) and *Mimosa pudica* (MP) were also studied.
e. Development of Integrated bio-cycles System.

3 Results and Discussions

3.1 Effects of Tropical Land Use Changes

The soil pH(H_2O) in plot of Cut & Burn forest was slightly higher than that of all other sites (Agus 2003). This phenomenon was almost identical to results reported by Ohta et al. (2000) that the pH(H_2O) in primary forests was lower than those in over-burned areas of degraded ecosystems, particularly alang-alang grasslands. The widespread forest fire in the 1980s and 1990s and the land preparation system using fire in the other land use sites were suggested to be the cause for the increase in soil-pH. The burning of forests and the residues that accompany a large input of ash to the soil surface in CB may have caused the rise in pH(H_2O), CEC, BS, T-N and exchangeable base cations, especially exchangeable K, Ca and Mg. The land use change from NF to another land use caused a decrease in reduction-oxidation potential (Agus et al. 2004). The Eh value in CB was lowest among the other land uses (Agus 2003).

The small disturbance of forest stands through the conversion to SF by selective cutting from NF increased the T-N and exchangeable nutrients (K, Ca, and Mg) in the topsoil drastically (Agus 2003). The selective cutting in NF had a good effect on soil chemical properties, because of the higher amount of litter's decomposition. The land use change to DF tended to cause a decrease in organic matter, exchangeable base cations (K and Mg), because of less nutrient cycles in the DF ecosystem. NF had a greater carbon stock in topsoil in tropical region and NF decreased sharply following conversion to other land uses. The land use change to CB seemed to have stimulated an increase of soil chemical properties and nutrient amounts over a short time because of the supply of decomposed materials. The yemane plantations (Ga and Co) caused an increase in T-N, exchangeable-Ca and Mg but seemed to be problematic with regard to the decrease in EC, CEC and organic matter. Nutrient lost from forest ecosystem by harvesting and an in-efficiency or die-back of old roots correlated to the slower growth of yemane coppice at the second rotation rather than those of first rotation. The land use changes influenced soil chemical properties, especially in the topsoil (0–10 cm). It is needed to preserve nutrient availability in the topsoil and nutrient cycles in forest ecosystem during land use change in tropical region (Agus et al. 2004).

Natural forests acted as the greatest T-C sinks in tropical forest topsoil (0–10 cm soil depth), but this capacity in lower soil layers (10–30 cm) was almost identical in all plots (10–15 Mg ha^{-1}) (Agus 2003). Total C stock in the topsoil of NF (70 Mg ha^{-1}) decreased sharply to approximately half (20–30 Mg ha^{-1}) following conversion to other land uses. The closed cycling in the mature natural forest holds a large amount of litter in the O horizon. With the high C/N ratio, therefore C stock in NF seemed to be difficult to be decomposed by microorganism and tend to be accumulated in the topsoil. When natural forests are logged, the organic matter is decomposed rapidly and inorganic nutrients are lost through leaching and erosion. The supply of organic material from forest fires and harvesting in CB could not increase the T-C amount in forest soil as high as in NF yet (Agus et al. 2004). The new establishment of a closed cycling ecosystem in Ga and Co didn't show a different amount of T-C than in DF, but was still lower than in NF. The amounts of T-N in the SF, CB, DF, Ga and Co land uses were slightly greater than NF for topsoil (0–10 cm) and subsoil (10–30 cm). The disturbances during the change of land use supplied more easily decomposable organic matter by soil microorganism, therefore total N tend to be increased.

3.2 Growth and Nutrient Dynamics by Site Class in Gmelina Arborea *Plantations*

The growth of yemane was significantly influenced by site-quality. The differences in stand height were 3 m between the site classes (poor, moderate and good) at the 6 years and differences in stand volume were approximately 40 $m^3 ha^{-1}$ (Agus 2003).

Aboveground biomass production at the end of the rotation (6-year-old plantations) differed significantly among the site classes. Biomass on the good, moderate and poor sites was 40 ± 20, 86 ± 17 and 120 ± 12 Mg ha^{-1}, respectively. Halenda (1993) reported the aboveground biomass of the yemane in Sarawak to be 85 Mg ha^{-1}, which is similar to productivity on the moderate site in this field. Chijioke (1980) reported total aboveground biomass for roughly 6-year-old yemane to be 60–122 Mg ha^{-1} in Brazil and 63–137 Mg ha^{-1} in Nigeria. These tendencies implied that production can be increased by choosing suitable sites (Agus et al. 2004). In general, better soil and site class would produce more biomass and minimize the risks that may cause land degradation.

FAO (2001) reported that the worldwide average of aboveground woody biomass in forests was 109 Mg ha^{-1}. Although the physicochemical properties of soil in tropical regions is poorer than in temperate regions, biomass productivity in tropical regions is higher than in temperate regions, possibly due by the higher temperature, rainfall, humidity, soil microorganisms and growth periods in tropical regions over the years.

The nutrient amounts in the aboveground biomass at the end of the rotation (6-year-old) among the three site class were 280–840 for N, 40–120 for P, 200–590 for K, 110–320 for Ca and 15–45 for Mg (Agus 2003). The species and understorey biomass is affected by the site quality and degree of closing capacity of the canopy, which has direct relation to the level of relative light intensity (RLI) on ground surface. We found some pioneer plants as indicator plants of site quality in tropical region, such as: kreyu (*Euphatorium odoratum*), Macaranga (*Macaranga* sp.), *Prihinium* sp in the good site; anggrung (*Trema orientalis*), mikania (*mikania* sp.), macia, sirih-sirihan (*Piper adencium*) in the moderate site; and alang-alang (*Imperata cylindrica*) in the poor site (Agus et al. 2004).

Soil pH(H_2O) and pH(KCl) showed a relatively acidic constant value irrespective of soil depth in the poor and moderate sites, but pH value in the good site was clearly higher in the topsoil (Agus 2003). A more neutral pH in the good site would provide better conditions for Eh and CEC. Robson (1989) reported that soil acidity decreased plant growth through many causes (e.g. deficiencies of P, Mo and Ca, and toxicity of Al, Mn and H^+ ions).

Concentrations of exchangeable Ca, exchangeable Mg, T-C, T-N and T-Ca increased toward the top, due to decomposition of litterfall by microorganism above the topsoil and its percolation. Concentration of exchangeable Al increased in lower layers of soil depth, because of the vertical shift of the low molecular weight of organic substances, together with sesquioxides (Fe, Al, Mn) in the podzolization process (Zech 1993). The concentrations of exchangeable Fe, T-P, T-K, T-Mg, T-Fe and T-Al showed a relatively constant value irrespective of soil depth (Agus 2003).

The amounts of exchangeable Al were greater than those of exchangeable Ca, therefore soil acidity and toxicity were high. The ratios of available/exchangeable/free nutrient amounts to total amounts on the three sites were trace for N, 1–3% for P, K and Fe, 10–30% for Mg and Al and 40% for Ca (Agus 2003). The nutrient management should be considered for good supporting in tree growth. The good site provided greater amounts of T-P, T-Ca, T-Mg, available-P, exchangeable-Ca and Mg in soil which could be circulated in the ecosystem than that of the moderate and poor sites. Free Fe and exchangeable Al amounts in 0–10 cm soil depth of the good site were smaller than those of the moderate and poor sites. The high concentration of toxic exchangeable Al in the topsoil tended to decrease yemane growth (Agus et al. 2004).

The amounts of MBC in 0–30 cm soil depth in the poor site were approximately 300 kg ha^{-1} and in the good and moderate sites were approximately 800 kg ha^{-1} (Agus 2003). The amount of MBN in poor site was 130 kg ha^{-1} and in the good and moderate sites were 300–370 kg ha^{-1}. MBC and MBN in the good and moderate sites were two to three times the amount of the poor site. A part of inorganic-C and N were dissolved from microbial biomass into soil water every year, because of the short lifespan (2–3 years) of microorganism. The role of microorganism on nutrient cycles and site productivity in tropical region was very important (Agus et al. 2004).

The litterfall of yemane plantation was distributed almost evenly throughout all months of the year, although it is categorized as a deciduous tree. This phenomenon was influenced by the evenly high temperature and amount of rainfall throughout a year in a wet tropical rain forest region. Leaf litter was the major contributor (80%) to total litter production. Total annual litterfall in the moderate and good sites was approximately 8 Mg ha^{-1} yr^{-1} and 12 Mg ha^{-1} yr^{-1}, respectively (Agus et al. 2004). The annual litterfall was affected by site productivity.

The nutrient input through litterfall in the good site was approximately 50–100% higher than that of the moderate site (Table 1). The nutrient input from litterfall in yemane plantations was considerably higher compared to other fast growing species of *Eucalyptus grandis* in subtropical regions (Turner and Lambert 1983) and natural tropical forests (Vitousek 1984). Nutrients released from leaf litter in the good site were approximately 30–80% higher than those in the moderate site (Table 1).

The initial concentration of NO_3-N during their N-mineralization in 0–5 cm and 5–10 cm soil depth were 20 and 2 mg kg^{-1}, respectively regardless the site class (Agus 2003). The initial concentrations of NH_4-N were trace in all soil layer and site class. N-mineralization in yemane plantations was rapidly converted to NO_3-N and the production of NH_4-N was trace. Ammonium-N produced by microorganisms was converted into NO_3-N by nitrification bacteria (e.g. *Nitrosomonas* or *Nitrobacter*). The accumulation of NO_3^-N in tropical forest soils caused the low soil pH (Agus et al. 2004). It is necessary to investigate the biomass and activities of microorganism.

Annual N-mineralization in 0–10 cm soil depth contributes 25% of annual uptake by yemane in the good site, and 28% in the moderate site (Table 1). The inorganic-N amount in soil was very low (12–13 kg ha^{-1} in 0–10 cm soil depth), therefore the amounts of N-mineralization was very valuable for site productivity. Annual N-mineralization amount was 4–6 times of their inorganic-N (NO_3^- + NH_4^+) in the soil (Agus et al. 2004).

Harvesting removes approximately of 60% of stems from the forest ecosystem. The stem contribution to the total aboveground biomass was about 85%, higher

Table 1 Annual nutrient dynamics at the end of the rotation (6-year-old forest) of yemane plantation in the moderate sites

	Dried matter	C	N	P	K	Ca	Mg
	Mg ha^{-1}		kg ha^{-1}				
Biomass							
Aboveground	86	46	599	84	417	229	32
Understorey	3	2	39	2	30	14	1
Belowground	22	10	173	–	81	93	14
Forest floor	14	3	105	22	54	36	4
Soil (0–10 cm) exchangeable*		12	3	17	187	18	
(0–100 cm) exchangeable*		16	35	131	453	131	
Soil (total)		131	25,652	1445	11,219	1054	657
Total forest		192	*932	*143	*713	*825	*182
Harvesting	43	23	300	42	209	115	16
Fluxes (1 year)							
Wood increment (WI)	14	8	100	14	70	38	5
Litterfall (Lf)	8	5	86	7	35	89	5
Uptake (WI + Lf)		–	186	21	104	127	10
Leaves decomposition	6	3	88	–	50	73	4
N-mineralization (0–10 cm)			53				
Nutrient movement 0 cm			9	–	87	17	8
10 cm			6	–	15	8	6
30 cm			5	–	6	3	6
50 cm			5	–	2	3	6
70 cm			4	–	6	4	7
Turn-over time (Ao/Lf)	1.8	0.5	1.2	3.1	1.5	0.4	0.9
Below/aboveground	0.3	0.2	0.3	–	0.2	0.4	0.4
Litterfall/uptake			0.46	0.33	0.33	0.70	0.50
Uptake/exch-soil			11.6	0.6	0.8	0.3	tr
Increment/exch-soil			6.3	0.4	0.5	0.1	tr
Harvest/exch-soil			18.8	1.2	1.6	0.25	0.1
Aboveground/ exchangeable-soil			37.5	2.4	3.2	0.5	0.2

*Exchangeable-nutrient, not for total-nutrient

than that in Sarawak (77%, Halenda 1993). Approximately 50% of the aboveground biomass and nutrient amounts will be lost from the forest ecosystem as harvested materials (Table 1). The nutrient amounts lost in the good site were approximately 30–50% higher than those in the moderate site. the contribution of nutrient loss by harvesting to soil nutrient availability is higher in tropical forests than in temperate forests.

Whole tree harvesting of yemane on both the moderate and good sites removed 18.8–22.2 times of available N, 1.2–1.4 times of available P, 1.6–3.0 times of exchangeable K, 0.3–0.4 times of exchangeable Ca and 0.1 times of exchangeable Mg in 0–100 cm soil depth (Table 1). Therefore, an over-exploitation of nutrients on the low soil fertility will lead to land degradation in tropical regions. Nitrogen and P will be a major caused of land degradation. Potassium may not became a major problem because of the supply (55–80 kg ha^{-1} yr^{-1} in tropical region) from rainfall and throughfall (Tanaka et al. 2000). Amount of lost nutrients by harvest should be added in order to sustain productivity. Without proper management, the speed of the nutrient cycle in the next rotation of a tropical forest will slow down because of land degradation due to erosion and decrease of organic matter (Agus et al. 2004).

The annual nutrient uptakes were approximately 11–17 times of available N, 0.3 times of available P, 1.1 times of exchangeable K, 0.9 of exchangeable Ca and 0.05 times of exchangeable Mg in 0–100 cm soil depth (Table 1). The nutrient uptake in higher amounts than their availability in soil (especially N) will become the nutrient limiting for growth during the next rotation. A part of annual uptake was returned to the soil through litterfall at approximately 50% for N and Mg, 35–40% for P and K, and 70% for Ca. Furthermore, the high decomposition rate of their litter will increase their availability through the nutrient cycles.

Yemane plantation forest was supported more by nutrient cycling rate than by amount of nutrient availability in the soil. The high temperature, rainfall and humidity throughout the year in tropical forests results the high rates of tree growth, microbial activity and nutrient cycling. The good site had better productivity and more rapid nutrient circulation than that of the moderate site.

3.3 The Role of Legume Cover Crops

The aboveground biomass at 4 MAP was 2.1–2.9 Mg. ha^{-1} (Agus 2003). The belowground (root) biomass of all observed LCC were close to a similar amount of 1.7 ± 0.4 Mg ha^{-1}. A similar aboveground biomass was also reported for other related species of LCC on the agricultural land of South Sumatra until 3 or 4 MAP (Hairiah et al. 1992). Because their life span is about 6 months, the legumes observed in this study can be planted twice a year in this research field. Thus, the LCC could supply organic matter of over 8–10 Mg ha^{-1} yr^{-1}. Therefore, the amount of organic matter produced by the LCC that could be decomposed and supplied to the soil was greater than its total biomass.

Nitrogen amounts of aboveground and belowground biomass were respectively 29–41 and 36–38 kg ha^{-1} (Agus 2003). The amounts found through this investigation were lower compare to the report of Hairiah et al. (1992) that LCC as a green manure might contribute about 70 kg ha^{-1} of N from the total biomass in one cycle. The amounts of cations in the total biomass of the LCC at 4 MAP were 27–33 for K, 15–23 for Ca and 2–3 kg ha^{-1} for Mg, respectively.

Skerman et al. (1988) reported that LCC plays an important role in the supply of organic matter and the stimulation of biological activity in soil. Kita et al. (2005) estimated that N_2-fixation by some kinds of LCC were 10–60 kg ha^{-1} yr^{-1} in low productivity soil at the same research site. Planting of LCC is a good method for maintaining land productivity and sustainability of a short rotation plantation such as a yemane plantation.

The total (aboveground plus belowground) biomass of LCC at 6 MAP was 13.8 Mg ha^{-1} in SQ, 9.3 Mg ha^{-1} in FC, 7.7 Mg ha^{-1} in MC and 7.3 Mg ha^{-1} in MP (Agus 2003). There were significant differences in total biomass between SQ and the other observed LCCs. The life spans of SQ and FC were more than 2 years, whereas those of MC and MP were 6 months and can be cut and planted twice per year in the research field. In the case of the coppice stand of yemane, SQ was able to be cultivated for 2 years after clear cutting of yemane. Smaller amounts of biomass were found for the other legume species (3.8–4.7 Mg ha^{-1} at 4 MAP) on the poorer productivity site, as well as the other species on the agricultural land (5–6 Mg ha^{-1} at 6 MAP) in South Sumatra (Hairiah et al. 1992).

The total biomass amount of N (140–460 kg ha^{-1}) in moderate-class sites was higher than in poor sites (70–80 kgN ha^{-1}) (Agus 2003), and in other reports (30–110 kgN ha^{-1} yr^{-1}, Hairiah et al. 1992). The cycles of N and other nutrients in the legumes would double if the legumes were planted twice per year. One cycle (6 MAP) of LCC biomass included amounts of 54–65 kg ha^{-1} for K, 25–61 kg ha^{-1} for Ca, 12–19 kg ha^{-1} for Mg. Calcium amounts in SQ were more than that in MP. Stem harvesting in a short rotation plantation of yemane on the moderate site removed approximately 300 kg ha^{-1} for N, 25 kg ha^{-1} for P, 210 kg ha^{-1} for K, 115 kg ha^{-1} for Ca and 15 kg ha^{-1} for Mg. This means that SQ and FC can circulate their lost amounts within one cycle, whereas MC and MP take two cycles, if all the organic matter decomposes completely.

Diurnal changes in the N_2 fixation rate of the investigated legume's nodule ranged from 0.1–1.3 μmolesN_2 g^{-1} nodule h^{-1} (Agus 2003). The average rate of N_2 fixation in FC seemed to be the highest followed by SQ, and MP and MC had the same rate. The rate of N_2 fixation of each legume changed with time during the day. The N_2 fixation rate during the daytime was higher than that in the nighttime, with the exception of MC.

The degree in response of N_2 fixation rate to temperature change differed among species. There was a positive correlation between soil temperature and N_2 fixation by FC. FC had a high response to changes in soil temperature. There was positive relationship found between soil temperature and N_2 fixation rate in FC, SQ and MP plots between 26 and 30 °C. The optimum temperature range for N_2 fixation of legumes in tropical regions seemed to be narrower and higher than in other regions. Since the temperature in tropical regions is relatively constant year-round, the seasonal changes of N_2 fixation rates seemed to be less sensitive in comparison with other regions.

Daily N_2 fixation rate was 24 μmolesN_2 g^{-1} d^{-1} in FC, 14 μmolesN_2 g^{-1} d^{-1} in SQ, 10 μmolesN_2 g^{-1} d^{-1} in MC and 8 μmolesN_2 g^{-1} d^{-1} in MP (Agus 2003). N_2 fixation rate in FC was clearly higher than that in MC and MP. The quantity of N_2 fixation by legume nodules varied with the type of legume host and the *Rhizobium* strain. Nodule biomass in FC was 235 kg ha^{-1}, MP was 161 kg ha^{-1}, SQ was 84 kg ha^{-1} and MC was 80 kg ha^{-1}. FC had a greater biomass, a bigger nodule size and seemed to have more effective root nodules than the other three legumes. Skerman et al. (1988) reported that the total nodule biomass per plant is constant but differs with plant host. The effectiveness of the nodule also affected the N_2 fixation rate, depending on the number of bacteroids present in the nodule (Akkermans and Houwers 1983). This study did not consider the relationship between nodule size, nodule color and their effectiveness. Therefore, in the future it will be necessary to measure N_2 fixation rate based on each plant host, *Rhizobium* strain, nodule size and nodule color.

Annual N_2 fixation by LCC was 12 kgN ha^{-1} yr^{-1} in SQ, 57 kgN ha^{-1} yr^{-1} in FC, 9 kgN ha^{-1} yr^{-1} in MC and 13 kgN ha^{-1} yr^{-1} in MP (Agus 2003). These figures were about half the N_2 fixation amount (20–120 kgN ha^{-1} yr^{-1}) by legumes in semi-arid Mediterranean areas (Papastylinou 1988). Legumes planted in a better quality site in tropical forestland seemed to have lower N_2 fixation rates, but had a greater biomass (Agus et al. 2004; Skerman et al. 1988). The contribution of atmospheric N_2 fixation to amounts of N in legumes in tropical forestland was 5–20%. This value was almost identical to the results (5–20%) of AICAF (1995), but was lower than those of Mediterranean areas (40–80%) using ^{15}N methodology (Papastylinou 1988).

3.4 Integrated Bio-cycle System

Tropical bio-geo-resource has a high biomass productivity but still less economical values. Integrated Bio-cycle Farming System (IBFS) is an alternative system which harmoniously combines agricultural sectors, and non-agricultural aspects, on landscape ecological management. The cycle of energy, organic matter and carbon, water, nutrient, production, crop, money was managed through 9R (reuse, reduce, recycle, refill, replace, repair, replant, rebuild, reward) to get optimal benefits for the farmer, community, agriculture and global environment. The system has multi-function and multi-product (Food, Feed, Fuel, Fiber, Fertilizer, Pharmacy, Edutainment, Eco-tourism). They will meet with the expected basic need for daily-, monthly-, yearly- and decade's income at short-, medium- and long- term periods. IBFS was expected to provide additional benefits for farmers with small, medium and big capital, through the recycling of organic waste into renewable resources to produce high-value production, such as organic fertilizer (liquid and solid), animal feed, and sources of bio-gas energy.

Integrated Bio-cycle Farming System (IBFS) is an alternative system of agriculture which harmoniously combines agricultural sectors, such as agriculture, horticulture, plantation, animal husbandry, fisheries, forestry with non-agricultural aspects, such as settlements, agro-industry, tourism, industry which are managed based on landscape ecological management under one integrated area. IBFS was developed by UGM through ICM (Integrated Crop Management), INM (Integrated Nutrient Management), IMM (Integrated Soil Moisture Management) and IPM (Integrated Pest Management). The system should collaborate and develop networking system between ABCG (Academic, Business, Community and Government) with economic, environmental and socio-cultural approach as a characteristic of Education for Sustainable Development (Agus 2010). This model facilitates the learning needed to maintain and improve our quality of life and the quality of life for generations. It is about equipping individuals, communities, groups, businesses and government to live and act sustainably; as well as giving them an understanding of the environmental, social and economic issues involved. Integrated farming could support for better sustainable life and environment.

The IBFS have comprehensive characteristics compare to other integrated farming system, namely: Low input farming, Organic Farming, Bio-dynamic Farming and Agroforestry system (Table 2). The key characteristics of IBFS developed in UGM University Farm are (i) an integration of agriculture and non-agriculture sector, (ii) value of environment, esthetics and economics, (iii) rotation and diversity of plants, (iv) artificial and functional bio-technology, nanotechnology, pro-biotic, (v) management of closed organic cycle and integration in an integrated area among ICM, IPM, IMM, INM, IVM, (vi) management of integrated bio-protection and ecosystem health management, (vii) landscape ecological management, agro-politan concept, (viii) specific management of plant and (ix) holistic and integrated system (Agus 2010).

IBFS is expected to be one alternative solution for improving land productivity, program development and environmental conservation and rural development in an integrated management. They will meet with the expected basic need at short-, medium- and long- term for food, clothing and shelter. Thus, IBFS could provide income at daily-, monthly-, yearly- and decade's term for farmers.

The role of micro-, meso- and macro-organisms on biogeochemical and nutrient cycling in increasing of land productivity is very important. Microorganisms are able to provide essential nutrients to plants through both mutualistic symbiotic and non-symbiotic. Agus et al. (2004) showed that the ability of N mineralization in tropical soil is 3–5 times higher than that available in the soil. Meanwhile, legume cover crops could to supply nitrogen 9–27 times higher than that available in the soil (Agus et al. 2003). Biotechnology including bio-artificial and functional nanotechnology will greatly enhance the success of integrated bio-cycles farming in tropical region.

IBFS was expected to provide additional benefits for farmers with small, medium and big capital, through the recycling of organic waste into renewable resources to produce high-value production, such as organic fertilizer (liquid and solid), animal feed, and sources of bio-gas energy (Agus et al. 2011). That will be a good prospect that organic farming can provide sustainable economic, environment and socio-cultural aspect. IBFS can produce "gold of life", such as: yellow gold (food, rice, corn), green gold (vegetables), brown gold (plantation wood), red gold (meat), white gold (milk, fish), black gold (organic fertilizer), transparent gold (water), gas gold (oxygen), blue gold (biogas, biomass energy, bio fuel), king gold (herbal medicine), prosperity gold (tourism), inner gold (mystic) (Agus 2016).

Table 2 Key characteristics of various types of sustainable agricultural system

Low input/ integrated	Organic farming	Bio-dynamic	Agroforestry	Integrated bio-cycle
Integration of advantageous natural process	Integration of land, environment and health of human beings	Management of organism which optimize quality of land, plants, animal and human health	Intégrate of wood and herbal plantes	Integration of agriculture and non-agriculture sector
Adding environmental values	Natural fertilizer. Environmental values	Economic values	Environmental values	Value of environment, esthetics and economics
Plant rotation	Plant rotation, diversification and ideal space	Plant rotation, diversification and ideal space	Spatial diversitas tipe crop	Rotation and diversity of plants
Impact of minimum land management	Adequacy of N through fixation-N	Adequacy of N through fixation N, special preparation for improvement of land quality and living plants	Plant variation and pastoral system	Artificial and functional bio-technology, nanotechnology, pro-biotic
The use of chemical fertilizer	Prohibition on treatment of plants and fertilizer	Prohibition on treatment of plants and fertilizer	Fertilization on agricultural plants, the use of cycle in forest plants	Management of closed organic cycle and integration in an integrated area among ICM, IPM, IMM, INM, IVM
The use of pesticide	Management of traditional animal	Management of traditional animal		Management of integrated bio-protection and ecosystem health management
General principle	Principle of grouping units	Principle of grouping units	General principle	Landscape ecological management, agro-politan concept
Specific management of plants	Specific management of plants	Specific management of plants	Specific management of plants	Specific management of plant
Semi-traditonal	Natural	Integrated	Traditional	Holistic and integrated
Stockdale and Cookson (2003), Chan (2006)	IFOAM (1998)	Koepf et al. (1976)	Stockdale and Cookson (2003)	Agus (2006, 2010)

4 Conclusions

The high net primary production in tropical ecosystem were more supported by the rapid organic-cycling than their low fertility weathered acid soil, due to the high temperature, rainfall, moisture and light intensity along a year. Tropical natural resources have a high biomass productivity but still less economical values. New paradigm from extraction to empowerment of natural resource will give new challenge to shift from red- & green economic to blue economic concept that should be more smart, global, focus, and futuristic for sustainable development. Integrated Bio-cycle System (IBS) is a closed-to-nature ecosystem on landscape ecological management to manage land resource (soil, mineral, water, air, microclimate), biological resources (flora, fauna, human) and their interaction to have more high added value in environment, economic, socio-culture and health. The bio-economic chain should be managed through 9A (Agro-production, -technology, -industry, -business, -distribution, -marketing, -infrastructure, -management, -tourism) with 9R (reuse, reduce, recycle, refill, replace, repair, replant, rebuild, reward).The system has multifunction and multi-product, that will meet with the expected basic need for daily-, monthly-, yearly- and decade's income at short-, medium- and long-term periods for small, medium and big stakeholder. IBFS can produce gold of life, such as: yellow gold (food, rice, corn), green gold (vegetables), brown gold (plantation wood), red gold (meat), white gold (milk, fish), black gold (organic fertilizer), transparent gold (water), gas gold (oxygen), blue gold (biogas, biomass energy, bio fuel), king gold (herbal medicine), prosperity gold (tourism), inner gold (mystic). IBS with ABCG (academic, business, community, government) networking has a good prospect for sustainable environment and life.

References

Agus C (2003) Sustainable site productivity and nutrient management of short rotation plantation of Gmelina arborea Roxb. plantation forest at tropical region. Dissertation at Tokyo University of Agriculture & Technology. Tokyo Japan

Agus C, Karyanto O, Hardiwinoto S, Haibara K, Kita S, Toda H (2003) Legume cover crop as a soil amendment in short rotation plantation of tropical forest. J For Env 45(1):13–19

Agus C, Karyanto O, Kita S, Haibara K, Toda H, Hardiwinoto S, Supriyo H, Na'iem M, Wardana W, Sipayung M, Khomsatun, Wijoyo S (2004) Sustainable site productivity and nutrient management in a short rotation Gmelina arborea plantation in East Kalimantan, Indonesia. New For J 28:277–285

Agus C (2006) Integrated Bio-cycles Farming System (IBFS). Research Update. Research Cluster of Agro-Science. Institute of Research and Community Services UGM, Yogyakarta

Agus C (2010) Organic matter management in UGM University farm for sustainable food, feed, fiber and fertilizer. In Proceeding seminar on the 16th annual international sustainable development research conference. Hong Kong

Agus C, Sunarminto BH, Suhartanto B, Pertiwiningrum A, Wiratni Setiawan I, Pudjowadi D (2011) Integrated Bio-cycles farming system for production of Bio-gas through GAMA DIGESTER, GAMA PURIFICATION AND GAMA COMPRESSING. J Jpn Inst Energy 90(11):1086–1090

Agus C (2016) Revolution of sustainable blue Earth, In: Agus C, Suratman, Panjono (eds) Blue Earth Rahayu: sustainable environment and life. UGM Press, Yogyakarta (in Indonesian)
AICAF (1995) Handbook of tropical legume cultivation. Association for International Cooperation of Agriculture & Forestry, Tokyo, p 109
Akkermans ADL, Houwers A (1983) Morphology of nitrogen fixers in forest ecosystems. In. Gordon JC, Wheeler CT (eds) Biological nitrogen fixation in forest ecosystems: foundation and applications. Martinus Nijhoff/Dr. W. Junk Pub. (Kluwer Academic Pub. grup), The Hague, pp 7–53
Baplan (2000) Indonesian forest statistics. Ministry of Forest & Estate Crop., Republic of Indonesian, p 233
Chan GL, (2006) Integrated farming system. http://www.scizerinm.org/chanarticle.html. Access 5 Sept 2012
Chijioke EO (1980) Impact on soils of fast growing species in lowland tropics. F.A.O. Forestry Paper No. 21, Rome, p 16
FAO (2001) State of the world's forests. FAO of UN, Rome, p 181
Freezailah BCY (1998) General lecture by executive director of International Tropical Timber Organization (ITTO) at the University of Gadjah Mada, Yogyakarta on 10 Aug 1998, p 10
Hairiah K, Utomo WH Van der Heide J (1992) Biomass production and performance of leguminous cover crops on an Ultisol in Lampung. Agrivita 15(1):39–43
Halenda CJ (1993) Aboveground biomass production and nutrient accumulation of a *Gmelina arborea* plantation in Sarawak, Malaysia. J Trop For Sci 5(4):429–439
Hardiyanto EB, Ryantoko A, Anshori S (1999) Effects of site management in *Acacia mangium* plantation at PT Musi Hutan Persada, South Sumatra, Indonesia. In: Nambiar EKS, Tiarks A, Cossalter C, Ranger J (eds) Site management and productivity in tropical plantation forests. Workshop proceedings, CIFOR, Bogor, pp 41–49
International Federation of Organic Agriculture Movements, IFOAM (1998) Basic standards for organic production and processing. IFOAM. Tholey-Theley, Germany
ITTO (1993) International Tropical Timber Trade Organization (ITTO) policy development 4. ITTO guidelines for the establishment and sustainable management of planted tropical forests. ITTO, Yokohama, p 38
Jordan CF (1985) Nutrient cycling in tropical ecosystems. Wiley, Chichester, p 190
Kita S, Agus C, Toda H, Haibara K (2005) Effects of cutting and regeneration on microbial activity of natural forest in East Kalimantan. J Jpn For Soc 87:217–224
Koepf HH, Pettersson BD, Schaumann W (1976) Biodynamic Agriculture. Anthroposophic Press, Spring Valley, New York
Lee DK, Park YK (2001) Degradation issues in the Southeast and Northeast Asia. In: Proceeding of 10th international workshop of BIO-REFOR, The University of Tokyo, Tokyo, pp 3–24
Ohta S, Morisada K, Tanaka N, Kiyono Y, Effendi S (2000) Are soils in degraded dipterocarp forest ecosystems deteriorated? A comparison of Imperata grasslands, degraded secondary forests and primary forests. In: Guhardja E, Fatawi M, Sutisna M, Mori T, Ohta S (2000) Rainforest ecosystems of East Kalimantan El. Nino, drought, fire and human impacts. Springer, Tokyo, pp 49–57
Pancel L (1993) Tropical forestry handbook. Springer, Berlin, p 810
Papastylinou I (1988) The role of legumes in the farming systems of Cyprus. In Osman AE, Ibrahim MH, Jones MA (eds) The role of legumes in the farming systems of the Mediterranean areas. Developments in plant and soil sciences. Kluwer Academic Publishers, Dordrecht, pp 39–50
Ritchie GS (1989) The chemical behavior of aluminium hydrogen and manganese in acid soil. In: Robson AD (ed) Soil acidity and plant growth. Academic Press, Harcourt Brace Jovanovich Publisher, Sydney, pp 1–60
Robson AD (1989) Soil acidity and plant growth. Academic press, Sydney, p 306
Skerman, PJ, Cameron DG, Riveros F (1988) Tropical forage legume. 2nd ed., FAO-UN, Rome, p 692

Soerianegara SS, Lemmens RHMJ (1993) Plant resources of South-East Asia No 5(1). Timber trees: major commercial timbers. Prosea, Bogor, pp 215–220
Stockdale EA, Cookson WR (2003) Sustainable farming systems and their impact on soil biological fertility-some case studies. In: Abbott LK, Murphy DV (eds) Soil biological fertility. A key to sustainable land use in agriculture. Kluwer Academic Publishers, Dordrecht, pp 225–239
Tanaka N, Aksa D, Effendi S, Ohta S (2000) Vertical transportation of chemical elements through water movement in different vegetation. In: Guhardja E, Fatawi M, Sutisna M, Mori T, Ohta S (eds) Ecological studies, rainforest ecosystems of East Kalimantan. El Nino, drought, fire and human impacts, vol 140. Springer, Tokyo, pp 59–67
Turner J, Lambert MJ (1983) Nutrient cycling within a 27-year-old Eucalyptus grandis plantation in New South Wales. For Ecol Manage 6:155–186
Vitousek PM (1984) Litterfall, nutrient cycling and nutrient limitation in tropical forests. Ecology 65:285–298
Zech W (1993) Geology and soils. In: Pancel L (ed) Tropical forestry handbook. Springer, Berlin

Local Knowledge and Resources as Driving Forces of Sustainable Bioeconomy

Maris Klavins and Vaira Obuka

Abstract A major driving force to promote the idea of sustainable bioeconomy could be local experiences, skills and knowledge in respect to the use of local and natural materials (at first, biomaterials). Sustainable bioeconomy is a concept under development, and as such it requires argumentation and demonstration of efficiency. The aim of this chapter is to study the local knowledge of the Baltic region in terms of the applicability of local biomaterials in production. In the context of bioeconomy, there is an evident need to identify the possibilities for the use of natural and local materials as well as the knowledge to manage these resources. Natural materials of the Baltic region, such as hemp, straw, timber, grain processing products (husk), reeds, moss and flax, will be studied in the historical context and in the use for innovations in modern bioeconomy. In addition, such resources as clay, organic lake sediments (sapropel), peat, sludge, ash, coal and biochar will be evaluated as potential source materials for the manufacture of innovative products. Regarding the use of natural resources, different sectors will be analysed, for example, agriculture and construction. The obtained results will give an insight into the knowledge and traditions of the Baltic region concerning the use of natural materials as a key for sustainability.

Keywords Sustainable development · Bioeconomy · Natural materials
Building materials · Knowledge integration

M. Klavins (✉) · V. Obuka
Department of Environmental Science, University of Latvia, Riga, Latvia
e-mail: maris.klavins@lu.lv

V. Obuka
e-mail: vaira.obuka@lu.lv

W. Leal Filho et al. (eds.), *Towards a Sustainable Bioeconomy: Principles, Challenges and Perspectives*, World Sustainability Series,
https://doi.org/10.1007/978-3-319-73028-8_10

1 Introduction

Bioeconomy in our time is considered as one of the key concepts supporting development. In European Union, bioeconomy is an essential element included in the development strategy and supporting sustainable development. The position is firmly stated in the European Commission's "Strategy for a sustainable bioeconomy to ensure smart green growth in Europe" (European Commission 2012) as well as in the "European strategy and action plan towards a sustainable bio-based economy by 2020" (European Commission 2010). As a new concept, the term "bioeconomy" or "bio-based economy" requires some clarification, and discussions on the proper understanding of the content of the term continue.

However, in general, the European Commission's definition can serve as the starting point for understanding the concept: "The term 'Bioeconomy' means an economy using biological resources from the land and sea as well as waste, including food wastes, as inputs to industry and energy production. It also covers the use of bio-based process for sustainable industries" (European Commission 2012). Work on the definition of the concept bioeconomy continues, and there is a consensus that it should have a much broader scope, considering the need in developing future systems of the use of natural resources as well as delivery of societal benefits and public goods (Schmid et al. 2012). This definition reveals not only the planned outcomes of the implementation of the concept of bioeconomy and aims of such implementation but also helps to understand the need to reorganise the existing basic principles of economic development, in particular, to reduce the dependence on fossil resources (including fossil fuels) and their unlimited exploitation, taking into account the quantitative limits of non-renewable resources.

The implementation of the concept of bioeconomy can significantly contribute to reaching the UN Sustainable Development Goals (2015). Bioeconomy is considered to be capable of meeting the major challenges that Europe and the world are facing: (a) increasing populations that must be fed; (b) depletion of natural resources; (c) impacts of ever increasing environmental pressures; and (d) climate change. The goal of bioeconomy is to move towards a more innovative and low-emissions economy, meeting demands for sustainable agriculture and fisheries, food security and the sustainable use of renewable biological resources for industrial purposes. The indirect aims of bioeconomy promotion includes environmental aims (reduction of greenhouse gas emissions, reduction of pollution due to mining of non-renewable resources, reduction of waste streams), new breath in economic developments (creation of new products, markets, jobs) and political aims (reduction of dependence on countries rich in resources, for example, Middle Eastern countries, Russia, China and others). At the same time, bioeconomy will support innovation, improve the quality and safety of food, support the development of rural and special costal communities and improve the efficiency of agricultural, food and general industrial production and distribution systems. Bioeconomy can promote building of low-carbon societies and thus achieve far-reaching aims.

Development of bioeconomy is urgent for EU member States with often limited national availability of non-renewable resources and to a similar extent also for developing countries. The widest positive response to the need to implement this concept in the reorganisation of economic production this concept is received in EU and Japan (Priefer et al. 2017). Several EU Member States (e.g., Latvia) have defined bioeconomy as one of the directions of national development and as part of smart specialisation plans.

The EU Bioeconomy Action Plan is based on three pillars (European Commission 2012):

1. Strengthening of research in biosciences and life sciences and supporting innovations by providing significant EU and national funding;
2. Stakeholders' involvement and development of synergies between different sectors of economies and society;
3. Enhancement of markets and supporting competitive activities as well as improving resource efficiency.

Largely, developments in bioeconomy are associated with research progress, focusing on agricultural and food technologies, biomass processing and biotechnology, while omitting the need to achieve noticeable progress in transforming waste streams into valuables, recovery of essential elements (for example, phosphorus compounds) and other directions of activities. On the one hand, the existing progress is largely based on research achievements: major progress in biosciences, development of new technologies and their implementation. Further aims are related to the development of innovation capacities. However, traditional knowledge and historical approaches relevant in pre-industrial era can also be a source of new knowledge.

In the context of bioeconomy, there is an evident need to identify the possibilities for the use of natural and local materials as well as the knowledge to manage these resources. The aim of this paper is to study the local knowledge of the Baltic region in terms of the applicability of local biomaterials for production in the context of developments in the field of bioeconomy. Natural materials of the Baltic region, such as hemp, straw, timber, grain processing products (husk), reeds, moss and flax, will be studied in the historical context and in the use for innovations in modern bioeconomy.

1. Natural and traditional use of biomaterials for bioeconomy

From a historical perspective, natural materials are the only ones available for humans. A well-known example of the use of natural materials is medicine, where natural, biologically active substances are used as drugs. More than half of the existing active substances on the market are either directly obtained from plants and animals or inspired by nature. One of the major branches in current biopharmacy—phytochemistry—is engaged in search of new active substances by analysing plants and their composition (Shah 2009). Not only active, isolated substances but also extracts of plants and animal source materials are directly used in healthcare, cosmetics and other applications. Considering the size of the biopharmaceutical market and its relation to large-scale chemical industry, the significance of natural substances and their applications are remarkable, and this segment of industry constantly grows and also makes way into bioeconomy. Scientific research in this field plays an important role. In addition, it is important to emphasize that cooperation between the state, business and academia plays a key role in developing of well-founded, high value-added products and technologies.Another major field of revival of traditional knowledge to support the development of bioeconomy is agriculture. Even most advanced agricultural technologies are largely relying on traditional knowledge and experiences and historical traditions. In these days, traditional knowledge is more and more often considered as a source of knowledge to advance contemporary agriculture globally, to support the development of new technologies in agriculture and to disseminate the traditional, local, knowledge globally. An example of such approach is the use of biochar to improve fertility of soils and reclaim poor, degraded soils (Lehmann and Joseph 2012). The origins of the use of biochar are in the traditional, native South American cultures. Biochar, a product of low-temperature pyrolysis of waste biomass, can slow down carbon turnover and thus enhance carbon sequestration in soils, restore ecosystem services and increase the fertility (and thus also productivity) of the soil. Further, biochar increases the water and nutrient holding capacity of the soil and is therefore beneficial for stressed soils. At the same time, biochar in soils is much more refractory in respect to mineralisation than conventional compost and can aid in the utilisation of organic wastes by returning the organic material and fertilisers to production as well as to the carbon cycle. So, the use of biochar is a telling example of a sustainable approach where traditional knowledge helps boosting bioeconomy.

Next, building materials are also examplary in this respect. The use of natural and local materials as well as traditional approaches is embodied in the concept of vernacular architecture and materials and their application in the construction industry. Vernacular materials can be considered as an alternative for sustainable construction, as they have significantly lower environmental impacts in comparison to industrially-produced ones (Oliver 2006; Sassi 2006). One of the main challenges for vernacular building is related to energy savings during the production of building materials, as the traditional construction industry is one of the largest energy-consuming sectors of the economy, being responsible for almost one third of all carbon emissions (Urge-Vorsatz et al. 2007). The use of natural materials, avoiding energy-intensive technologies, can help significantly reduce the greenhouse gas emissions.

2. Natural materials for bioeconomy: use of sapropel

A prospective material with a widely diverse application potential is the sedimentary deposit in waterbodies—sapropel. Sapropel is formed due to sedimentation of organic substances produced by living organisms, and thus a major source of sapropel organic matter are the decay products of algae, macrophytes as well as planktonic organisms (Leonova et al. 2011; Liužinas et al. 2005). At the same time, the contribution of mineral substances, like alumosilicates, silica and iron oxides entering from the waterbody basin is also significant. Sapropel occurs all over the world, but most intensive accumulation takes place in temperate climatic zones in Asia and Europe. Largest amounts of sapropel are found in Russia and also in the Baltic countries. Just in Latvia, the estimated sapropel resources are about 2 billion m^3. On the one hand, considering the permanently ongoing sedimentation processes in waterbodies, sapropel can be considerd as a renewable resource, especially in eutrophic waterbodies. On other hand, one of the key steps of waterbody recultivation is the removal of sediment mass. Thus, the extraction of sapropel is essential in achieving the recovery aims of eutrophic waterbodies, and sapropel mining is a means to restore the quality of water. In terms of the sapropel properties application, significant studies have been carried out in Latvia (Stankevia et al. 2016; Stankeviča 2012).

The main components of sapropel are organic materials, inorganic (mineral) substances and living organisms. The colour of sapropel varies from black to brownish, dark green to even purple, and the colour indicates the humification degree of organic material (humic substances are black to dark brown) as well as the presence of plant pigments (chlorophyll remaims can make sapropel greenish). The organic substances in sapropel are most valuable for the majority of applications, and their amount varies in a wide range—from 15% up to 90% and more (Stankeviča and Kļaviņš 2013). The organic matter of sapropel consists of C 50.8–59.2, O 27.9–35.2, H 6.7–7.4, N 4.7–5.4 and S 0.6–1.4%. Depending on the formation conditions, sapropel can contain groups of organic substances, such as humic substances 5.1–61.9%, hemicellulose 9.8–52.5%, aminoacids 9.8–17.8%, celulose 0.4–6.0%, bitumens and waxes 6.8–15.2%. Sapropel also contains many trace elements: Si, Al, Fe, Ca, Be, Sr, Mg, Ti, Na, K, V, Cr, Mi, Ag, Mo, Ga, Pb, As, Sn, P, S, Na, Sc, Ni, As, Rb, Y, V, I, Zr, Nb, Mo, Cd, Cs, Ba, La, Ce, Hf and Th (Leonova et al. 2011).

One of the most important properties of moist sapropel is its colloidal suspended phase structure, which determines the ability of the organic colloidal particles of sapropel to absorb large quantities of water. So, it has a high moisture absorbtion capacity—from 70 to 97%—and a low filtration rate (Liužinas et al. 2005). The relative humidity of sapropel is associated with organic matter, and its value increases with the content of organic matter.

Useful characteristics of organically rich lake sediments are their adhesive properties and water repellence (Balčiūnas et al. 2016; Obuka et al. 2016). The adhesive capacity of sapropel is due to its components of animal and vegetable residues. Green algae shells consist mostly of cellulose, which has weak decomposing properties. Organic sapropel, whose organic matter proportion is formed by green algae, is rich in cellulose but poor in minerals, as it consists of ash and low-level humic substances formed mainly by peat meristem. It should be noted that the high content of organic nitrogen, including free amino acids, contribute to the adhesive properties of sapropel (Balčiūnas et al. 2016; Obuka et al. 2016).

The organic adhesion capacity of lake sediments also affects the composition of humic molecule structure and concentration. This contributes to the emergence of strong links between the molecules of the material at the time of creation. The molecules of humic substances remain flexible, and their presence improve the durability and strength of the material. Therefore, with the above mentioned properties, it is expedient to use sapropel as an adhesive in various ecological construction and plaster/finishing materials.

The adhesive properties of sapropel are practicable in the production of composite materials for construction industry, interior design objects with both cold compaction techniques and hot–glue press at elevated temperature and pressure (Zach et al. 2013; Mounika et al. 2012). Thus, a prospective area of application of sapropel includes creation of different composites, for example, sapropel—wood fibre, sapropel—hemp shives and sapropel—wood sanding dust, sapropel—flax fibre and moss In these composite materials sapropel is used as a binder.

Sapropel samples have different origins (lake) and different moisture (%), organic matter (%) and carbonate (%) content. For example, a sapropel sample from Lake Padelis contains 35.57% carbonates and 85.97% moisture, its colour is pale gray-pink and density −1.24 g/cm^3. A sapropel sample from Lake Pilvelis, in turn, is dark greenish-brown, with homogeneous, jelly-like structure and a density of 1.10 g/cm^3. A sapropel sample from Lake Veveru has a high moisture level −97.66%, low density −1.08 g/cm^3, and its organic matter content reaches 86.25%.

The thermal properties of composite materials (sapropel—wood sanding dust and 'Aerosil'; wood sanding dust; wood fibre; hemp shives) was analysed. (Obuka et al. 2015).The silica product Aerosil (colloidal silicon dioxide), is a thickening agent, used as a filler for composite materials. It creates a smooth mixture, which is often used in combination with other fillers. Particular attention was paid to shrinkage cracks that affected the quality of material for further tests. In this case, gypsum was used to fill the cracks (sapropel—wood sanding dust and 'Aerosil'), so as to make the material fit for the thermal conductivity test. The measurement results also indicate a higher thermal conductivity of the material, that is, 0.080 W/m*K. Such a high result can be explained by the fact that not all of the cracks were sufficiently filled. In addition, gypsum may have influenced the result, as its thermal conductivity is around 0.18 W/m*K. The thermal conductivity test of sapropel—filler composite was carried out with variable types of sapropel and fillers. The results obtained are shown in Table 1.

Table 1 The thermal conductivity of studied materials

Material: binder-filler	Density, kg/m^3	Thermal conductivity, W/m*K
Sample 1,3—hemp shives	191	0.063
Sample 1,2—hemp shives	200	0.059
Sample 3—wood fibre	153	0.055
Sample 3—wood fibre	202	0.060
Sample 3—wood sanding dust	214	0.061
Sample 3—wood sanding dust—'Aerosil'	376	0.080

Fig. 1 Sample 1,3—hemp shives (authors photo)

According to the results obtained, the composite made of sapropel sample 3—wood fibre (density 153 kg/m^3) has the superior results (Fig. 1). Visually, this material differs from the composite of sapropel sample 3—wood fibre (Fig. 2) that has a density of 202 kg/m^3, because it has a denser structure and comparatively better resistance to deformation. In any case, the results indicate that these composites have similar characteristics and thus have a similar potential of use. The thermal conductivity of air-dried composites is relatively low because of the organic origin of the raw materials, detailed and mixed cellural structure and homogeneous fibres with interconnected and open pores.

Since the composites of sapropel—hemp shives, due to different sizes of particles, have a heterogenous structure with cavities and uneven composition with weaker inclusions, they deform more quickly. However, the composite (density 376 kg/m^3) made of wood sanding dust and 'Aerosil' with sapropel sample 3 binder and the composite made of sapropel sample 3 and wood sanding dust (Fig. 3) with density 214 kg/m^3 both formed shrinkage cracks during drying, indicating the inferiority of technology. The structure of the composite is made of densely grouped wood sanding dust particles mixed with a sapropel binder.

In a study of sapropel—sawdust and peat—sawdust composite materials (Obuka et al. 2013) the obtained thermal conductivity measurement results were 0.067 W/m*K and 0.060 W/m*K. The study considered the freezing cycles and moisture of tested materials. It shows that the thermal conductivity coefficient of the sapropel—sawdust composite becomes lower after freezing cycles. This is explained by the fact that the refrigeration process causes moisture loss and drying of the plate. This causes composite pores to fill with air, which is known to be the best heat insulation material. It is explained in literature that humidity in the

Fig. 2 Sample 3—wood fibre (authors photo)

Fig. 3 Sample 3—wood sanding dust (authors photo)

composite reaches the wood fibre saturation point, and so the mechanical properties of the material do not deteriorate. If a composite material is resistant to freezing cycles, such material is usable in North European conditions. The thermal conductivity coefficient for the sapropel—sawdust composite material is 0.050–0.060 W/m*K and for the peat—sawdust composite material—0.055–0.064 W/m*K. The obtained results are similar to composite materials made of sapropel—wood sanding dust and 'Aerosil'; wood sanding dust; wood fibre; hemp shives (Fig. 4).

In Binici et al. (2014) study of the insulation materials made from sunflower stems, textiles and agricultural by-products as fillers, epoxy resin was used as a binder for better fibre strength and binding efficiency. An average thermal conductivity coefficient of 0.1642 W/m*K was obtained. However, the thermal conductivity coefficient of a sample with less epoxy resin added and made only from sunflower stems and sunflower stem fibres reached 0.0728 W/m*K

Fig. 4 Sample 1,2—hemp shives (authors photo)

(Binici et al. 2014). In a study of bamboo fibre and polyester composite, Mounika et al. (2012) conclude that the thermal conductivity coefficient decreases with an increase in the proportion of fibre. The thermal conductivity coefficient ranged from 0.185 to 0.196 W/m*K (Binici et al. 2014). Literature on materials based on natural fibres from renewable raw material sources (flax, hemp, wood, bamboo, sheep wool) shows very good sound and thermal insulation properties. This is due to the low density of the materials and natural character of input fibres ("airy", lightweight material).

In a study of lime-hemp concrete, a variety of binders were compared, including metakaolin obtained by burning kaolin clay (40% by mass) at 800 °C and dolomitic lime (60% by mass) produced by Saulkalne Ltd. Eight different types of hemp shives were used as fillers. The obtained results show that the lime-hemp concrete material has a density of 312–337 kg/m^3, and its thermal conductivity is more diverse and ranges from 0.0718 to 0.0778 W/m*K (Sinka et al. 2015).

In a study of the use of agricultural waste in sustainable construction materials, Madurwar et al. compared measurements of insulation materials of different origins. The results ranged from 0.046 to 0.056 W/m*K for rice husks and from 0.118 to 0.240 W/m*K for oil palm leaves. The authors concluded that there are significant similarities between the corn cob and extruded polystyrene (XPS) materials in terms of microstructure and chemical composition. Materials made from rice husk and coconut coir reached the best results (Shea et al. 2012).

The results obtainted for compressive deformation (Fig. 5) show that the composites with a filler of birchwood sanding dust and 'Aerosil' have good results. The compressive deformation results vary from 0.724 MPa for the sapropel sample 3—wood sanding dust composite to 0.674 MPa for sapropel sample 3—wood

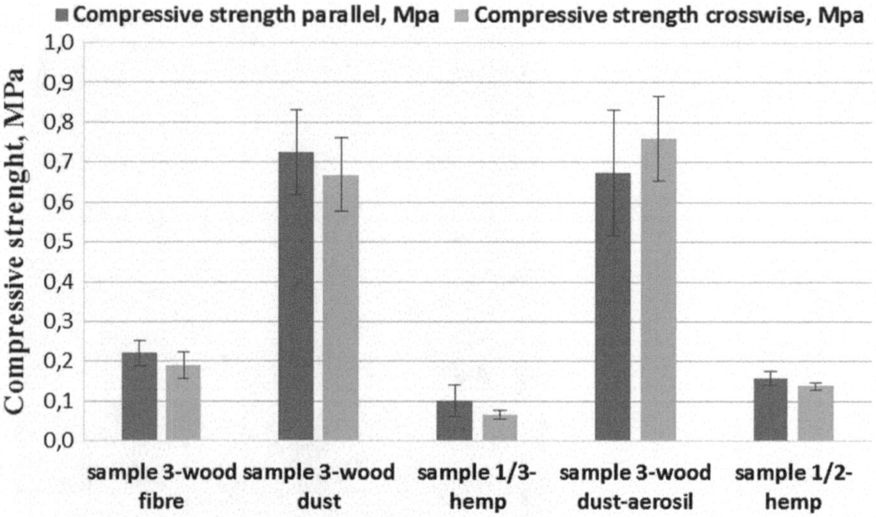

Fig. 5 Compressive strength of different composites, parallel and crosswise

sanding dust and 'Aerosil' as a filler, while the compressive strength crosswise shows 0.669 and 0.760 MPa respectively.

In compressive deformation, the composites made from hemp shives and wood fibre as a filler obtain a rate of 0.221 MPa for sapropel sample 3—wood fibre and hemp shives materials 0.101 (Sample 1,3—hemp shives) and 0.159 MPa (Sample 1,2—hemp shives), while the compressive strength crosswise measurement shows 0.191, 0.066 and 0.138 MPa respectively. These results are relatively lower due to lower intensity of filler and binder bonding. The mixing of binder with filler is more complex, because it is more difficult for the binder to enter the filler's structure, as the particles of the former are larger.

This was observed with the wood fibre composite and sample 3 as the binder. However, birch wood sanding dust and 'Aerosil' binder, consisting of dust particles, mixes with filler evenly, entering the structure of the filler. The structure of the material is smoother, increasing the mechanical resistance of the composite. The sample preparation techniques affect the compressive strength results: the preceding crushing of the material disrupts its structure. The use of various fillers and binders affects the physical and mechanical properties of composite materials; so, the hidrophysical and mechanical properties of composites can be changed by changing the types of fillers and binders.

Compared to the compressive strength results, the flexural deformation strength results (Fig. 6) show that the material strength is relatively lower. This can be seen from the results: for example, the samples made from hemp shives show lower results. The reason could be the granulometric composition, as there are many large shives that create voids and uneven composition with weaker inclusions, resulting in a faster deformation of the samples. In the process of making the wood

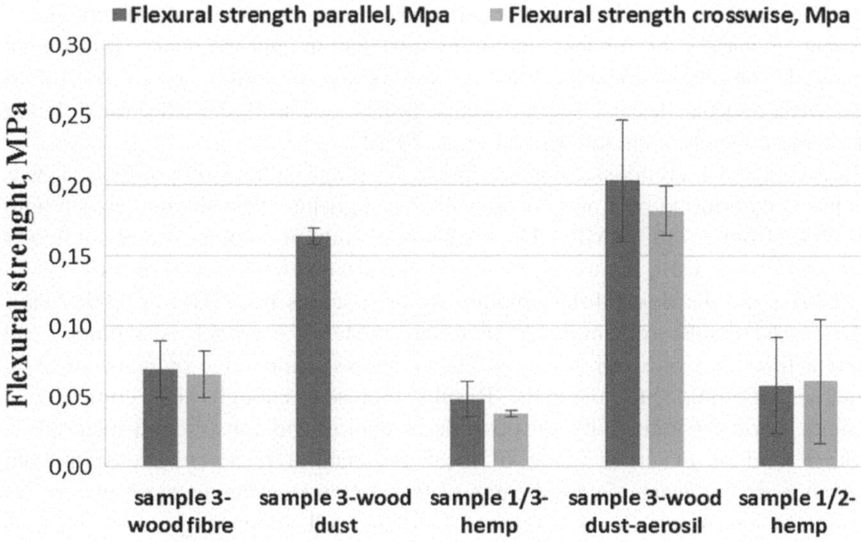

Fig. 6 Flexural strengths of different composites, parallel and crosswise

fibre—sapropel composite, the mixing of the binder with the filler is more complex, because it is more difficult for the binder to enter the structure of the filler forming tangles and air gaps. Voids remain that reduce the mechanical strength.

In flexural deformation strength, the composites made from hemp shives and wood sanding dust as a filler obtain a rate of 0.069 MPa for sapropel sample 3—wood fibre and 0.164 MPa (sapropel sample 3—wood sanding dust) and 0.203 MPa (sapropel sample 3—wood sanding dust—'Aerosil'), while the compressive strength crosswise shows 0.066 MPa for sapropel sample 3—wood fibre and 0.182 MPa for sapropel sample 3—wood sanding dust—'Aerosil' respectively. These results are relatively lower due to lower intensity of the filler and binder bonding. The mixing of the binder with the filler is more complex, because it is more difficult for the binder to enter the structure of the filler, as the particles of the former are larger and longer when using the composite of sample 1,2 sapropel. The reason for this could be the higher adhesive capacity and lower moisture content of sapropel.

In a study of the composite material of sapropel and peat sawdust, similar composites were created, and the results show that the filler and binder types and the preparation technology options change the compressive and flexural deformation strengths of the composite. According to the indicated results, the average compressive resistance was 0.03 MPa, and the compressive resistance of the peat—sawdust composite (activated peat binder) also was 0.3 MPa.

Comparing the results of this work with those of the research of sapropel—concrete, the materials made of wood fibre and hemp shives have lower rates. During preparation, 1% NaOH solution was added. The results obtained in the studies showed the compressive strength of 0.55 MPa for absolutely dry composite materials and 0.56 MPa for air-dried composite materials. In a study of lime-hemp concrete, a variety of binders, including metakaolin and dolomitic lime, were compared. The obtained results show the compressive strength range from 0.140 to 0.337 MPa and the flexural strength range from 0.021 to 0.059 MPa for different lime-hemp concrete materials (Sinka et al. 2015).

In a study of a composite material made from agricultural by-products, it was concluded that durian peel and coconut-fiber composites yield the highest strength indicators: from 2.9 to 36 MPa. The results of a study of a composite of sunflower stems and epoxy resin, in turn, show the compressive strength test scores of 0.283–0.312 MPa and the flexural deformation strength scores from 0.06 to 0.09 MPa.

Discussed results of composite materials made of sapropel as a binder and different filler materials (Obuka et al. 2016), shows competitive outcome of basic properties of composite materials. Results represent enough high compressive strength, thermal conductivity for possible insulation and construction materials.

Such kind of materials made of local and natural resources are significant research topic to discuss. Strengthening of research in sustainable, environmentally friendly materials and local resources field as well supporting innovations, is important to continue and implement in industry throught bioeconomy instruments. As mentioned before from a historical perspective, natural and local materials are the only ones available for humans and it is needful to find best practice for creating innovative, high added value products using local knowledge. In respect of decreasing the dependence on fossil resources (including fossil fuels) and creating new jobs, local arrangements of entrepeneurship in depleting remote rural areas.

2 Conclusions

From natural materials and local resources, such as sapropel, and industrial by-products, such as birch wood sanding dust and fibre and hemp shives, sawdust it is possible to develop environmentally friendly composite materials for construction for various needs of utilisation. The particle granulometric composition, surface area and other characteristics of the filler have an effect on the binding with sapropel as a binder. Composite materials are characterised by a relatively high mechanical strength, shape-holding ability and easily amenable texture imprint. One of the most important properties of these materials in today's application scope is the ability to sequester CO_2, as both the binder and the filler are bio-based. The mechanical and thermal properties of bio-based materials were evaluated and compared to similar materials. The mechanical and thermal properties of sapropel-based composites were similar to those of synthetic as well as mineral materials, suggesting that sapropel composites could have similar use in the construction industry: as a self-bearing wall thermal insulation material that works together with the structural timber frame. As the sapropel-based materials have high organic content, they are vulnerable to biodegradation; therefore, antimicrobial additives are significant to add.

References

Balčiūnas G, Žvironaite J, Vejelis S, Jagniatinskis A, Gaidučis S (2016) Ecological, thermal and acoustical insulating composite from hemp shives and sapropel binder. Ind Crops Prod J 91:286–294

Binici H, Eken E, Dolaz M, Aksogan O, Kara M (2014) An environmentally friendly thermal insulation material from sunflower stalk, textile waste and stubble fibres. Constr Build Mater 51:24–33

European Commission (2010). Europe 2020: a strategy for smart, sustainable and inclusive growth. http://ec.europa.eu/eu2020/pdf/COMPLET%20EN%20BARROSO%20%20%20007%20-%20Europe%202020%20-%20EN%20version.pdf

European Commission (2012). Commission adopts its strategy for a sustainable bioeconomy to ensure smart green growth in Europe. http://ec.europa.eu/research/bioeconomy/pdf/bioeconomycommunicationstrategy_b5_brochure_web.pdf

Lehmann J, Joseph S (eds) (2012). Biochar for environmental management: science and technology. Routledge, London

Leonova GA, Bobrov VA, Lazareva EA, Bogush AA, Krinovogov SK (2011) Biogenesis contribution of minor elements to organic matter of recent lacustrine sapropels (Lake Kirek as example). Lithol Min Resour 46(2):99–114

Liužinas R, Jankevičius K, Šalkauskas M (2005) Improvement of lake sapropel quality: a new method. Geogrfijos Metraštis 3(82):44–55

Mounika M, Ramaniah K, Ratna Prasad VA, Mohana Rao K, Hema Chandra H (2012) Thermal conductivity characterization of bamboo fiber reinforced polyester composite. J. Mater. Environ. Sci. 3(6):1109–1116

Obuka V, Korjakins A, Brencis R, Preikšs I, Purmalis O, Stankeviča, Ķļaviņš M (2013) Sapropeļa – kūdras, sapropeļa kokskaidu siltumizolācijas plāksnes un to īpašības. Rīgas Tehniskās universitātes zinātniskie raksti, Materiālzinātne un lietišķā ķīmija 29:127–136

Obuka, V., Šinka, M., Kļaviņš, M., Stankeviča, K., Korjakins, A. (2015). Sapropel as a binder: properties and application possibilities for composite materials. In: 2nd international conference on innovative materials, structures and technologies IOP Publishing IOP Conference Series: Materials Science and Engineering, 96:1–10

Obuka V, Veitmans K, Vincēviča-Gaile Z, Stankeviča K, Kļaviņš M (2016) Sapropel as an adhesive: assessment of essential properties. Res Rural Dev 2:77–82

Oliver P (2006) Built to meet needs: cultural issues in vernacular architecture. Architectural Press, Elsevier, Oxord

Priefer C, Jorissen J, Fror O (2017) Pathways to shape the bioeconomy. Resources 6:1–23. https://doi.org/10.3390/resources6010010

Sassi P (2006) Strategies for sustainable architecture. Taylor & Francis Ltd, London

Schmid O, Padel S, Levidov L (2012) The Bio-economy concept and knowledge base in a public goods and farmer perspective. Bio-based Appl Econ 1(1):47–63

Shah B (2009) Textbook on pharmacognosy and phytochemistry. Elsevier India, Delhi

Shea A, Lawrence M, Walker P (2012) Hygrothermal performance of an experimental hemp–lime building. Constr Build Mater 36:270–275

Sinka, M., Radina, L., Sahmenko, G., Korjakins, A., Bajare, D. (2015). Enhancement of lime-hemp concrete properties using different manufacturing technologies. In: First international conference on bio-based building materials, pp 1–8

Stankeviča K, Kļaviņš M (2013) Sapropelis un tā izmantošanas iespējas (Sapropel and its application possibilities). Sci J Riga Tech Univ Mater Sci Appl Chem 29:113–131 (in Latvian)

Stankevica K, Vincevica-Gaile Z, Klavins M (2016) Freshwater sapropel (gyttja): its description, properties and opportunities of use in contemporary agriculture. Agron Res 14(3):929–947

Stankeviča K, Kļaviņš M, Rutiņa L (2012) Accumulation of metals in sapropel. Mater Sci Appl Chem 26:99–105

UN Sustainable Development Goals (2015) http://www.un.org/sustainabledevelopment/sustainable-development-goals/

Urge-Vorsatz D, Danny Harvey LD, Mirasgedis S, Levine MD (2007) Mitigating CO_2 emissions from energy use in the worlds buidings. Build Res Inf 35(2):129–137

Zach J, Hroudová J, Brožovský ZK, Gailius A (2013) Development of thermal insulating materials on natural base for thermal insulation systems. Procedia Eng 57:1288–1294

Tropicalizing Sustainable Bioeconomy: Initial Lessons from Ecuador

Daniel V. Ortega-Pacheco, Ariel Silva, Alfredo López, Ramón Espinel, Diego Inclán and María J. Mendoza-Jiménez

Abstract Sustainable bioeconomy is being revised in tropical megabiodiverse developing countries. Given competing economic interests and development inequities, biodiversity may require becoming a strategic and central resource in national economies to ensure political feasibility of bioeconomic models. This paper attempts to address the need to document alternative approaches to transition to sustainable bioeconomy in the context of extractive economies in tropical and megabiodiverse developing countries. Using a case study approach, it reviews the Ecuadorian experience to developing a bio-industry value chain as an institutional arrangement that can enable a more efficient and integrated use of biological resources towards a sustainable and resilient economy, while addressing structural development and biodiversity protection challenges. Knowledge generated from this research can assist policymakers working on optimal design of instruments aiming at unraveling the full potential for biodiversity as a key resource in development strategies of tropical megabiodiverse countries.

D. V. Ortega-Pacheco (✉) · R. Espinel · M. J. Mendoza-Jiménez
Escuela Superior Politecnica del Litoral (ESPOL), Guayaquil, Ecuador
e-mail: daviorte@espol.edu.ec

R. Espinel
e-mail: respinel@espol.edu.ec

M. J. Mendoza-Jiménez
e-mail: mjmendoz@espol.edu.ec

A. Silva
United Nations Development Program—BioFin Initiative, Quito, Ecuador
e-mail: aosventas@gmail.com

A. López · D. Inclán
National Biodiversity Institute, Quito, Ecuador
e-mail: alfredo.lopez@ambiente.gob.ec

D. Inclán
Universidad Central, Quito, Ecuador
e-mail: diego.inclan@ambiente.gob.ec

W. Leal Filho et al. (eds.), *Towards a Sustainable Bioeconomy: Principles, Challenges and Perspectives*, World Sustainability Series,
https://doi.org/10.1007/978-3-319-73028-8_11

Keywords Bioeconomy · Megabiodiversity · Sustainable transition
Bioindustry · Ecuador

1 Introduction

Bioeconomy is receiving increasing attention as an alternative to a global transition to sustainable development. Countries across the globe more and more confront development decisions that require arrangements to align socio-economic dynamics in harmony with nature while addressing poverty challenges. These decisions typically require consideration of, among others, bio-physical, political and economic factors. Bioeconomy needs to address those challenges in order to become a sustainable economic model (Sillanpää and Ncibi 2017). The following chapter explores the tropicalization of bioeconomy for unraveling the biodiversity potential and catalyzing a transition to sustainability in a tropical megadiverse country—Ecuador.

Bioeconomy is defined as the industrial transition to the sustainable use of aquatic and terrestrial biological resources in intermediate and final products for economic, environmental, social and national security benefits (Golden and Handfield 2014). It incorporates a set of economic activities related to the research, development, production and use of biological products and processes (OECD 2009). Regardless of different definitions, there is evidence of its global importance. Unlike the declining trend in investments in sectors such as oil, investments in bio-industry sectors show a sustained growth. For instance, US and European investments in bio-technology since the late 1990s have grown by about 20% (Hill 2000). In Europe, the bioeconomy sectors are worth 2 trillion euros in annual turnover and account for more than 22 million jobs (about 9% of the workforce) (EU 2012). At the beginning of 2015, public-private investments reached more than 2.1 billion euros (Piotrowski et al. 2016). Improvements and innovative adaptations of the bioeconomic model—by economies in different development stages—seem to be significant drivers of its sustainability and globalization, which is of course strategic by nature.

Tropical megadiverse developing countries face increasing pressures to biodiversity conservation during on-going development efforts. While countries like Ecuador have made significant efforts to combat poverty during the last decade, recent research highlights the increasing rate of biodiversity loss and the need for ambitious policy approaches to address it (WWF 2016). Megadiverse countries with sufficiently large domestic markets to develop cost effective manufacturing capacities at different stages of the supply chain may be more interested in bioeconomy. However, in many tropical developing countries several non-technological and economic factors may stand in the way of bioindustry: insufficient technical knowledge and absorption capacity to produce innovative technologies locally, insufficient purchasing power to acquire innovative products, and insufficient market size to justify local production units (Jha 2008). However, a more important consideration is perhaps the existence of preferences toward bio-products: a real domestic demand. This would be essential to sustain the economy until it redeems its export potential, starting with markets sharing similar demand structures, following Linder's hypothesis.

Ecuador is perceived to follow a process of endogenous socioeconomic transformation, driven by public policy (Cypher and Alfaro 2016). Empirical evidence has revealed a tendency towards an increasing imbalance between exports and imports in physical terms for Ecuador (Vallejo 2006), as well as other Latin American countries (Giljum and Eisenmenger 2004). Ecuador's physical deficit, when exported tons exceed imported tons, was the highest and did not decrease between 1980–2000, when the export sector was mainly based on oil and biomass (Russi et al. 2008). In response to this and other threats, the country has created an innovative policy arrangement aimed at promoting biodiversity to achieve a long-term economic performance that diverts from the middle-income trap and oil-dependence.

This chapter uses an institutional economics framework to study a institutional arrangement for bioeconomy in Ecuador. It follows previous research arguing that development models must move away from narrow concerns with macroeconomic imbalances and poverty-alleviation safety nets to confront head on the structural inequalities, while addressing sustainability (Rival, Muradian and Larrea 2015). It also builds on literature underpinning the critical importance of fostering specific uses for biodiversity as means to ensure its conservation and address rural inequities (Espinel 2009; Golinelli et al. 2015).

To organize our analysis, we use the *Situation, Structure* and *Performance* (SSP) framework articulated by Schmid in 1978 (Schmid 1987). We first describe key macroeconomic elements of the Ecuadorian economy, the Situation. Next, we describe key attributes of the National Strategy for Biodiversity, the Structure, followed by predictions concerning the operation of those structures with emphasis on rural development, the Performance. These predictions will be evaluated using evidence from field research, previous experience, and other data. We then go on to discuss in more depth the expected relative performance of these structures in light of alternatives changes, including strengths and weaknesses. The chapter concludes by identifying key lessons for policy makers and possible avenues for future research.

2 Tropicalizing Bioeconomy

Tropical megadiverse countries have many singularities from legislative, historical and geographical development gaps requiring innovative models of bioeconomy or its adaptation ("tropicalization"). In the case of oil-based economies, the model's complexity increases and requires political feasibility through the material delivery of benefits to equate and weigh-off conflicting interests detracting from the competitive advantages that biodiversity-based knowledge and bio-industry development offer. Identified challenges to tropicalize a bioeconomy are presented in Fig. 1.

Any configuration of bioeconomy should account for complex interdependencies and trade-offs between the relative participation of the sector in GDP, labor migration, deforestation, biodiversity loss, industrial and services sectors' development. Tropical megadiverse countries emphasizing biodiversity in a transition to sustainable patterns of production have the potential to materialize economic benefits through employment, wealth creation and higher level of self-sufficiency for farm and non-farm rural communities (Golden and Handfield 2014), which can in turn have a positive effect in mobilizing societal support in the initial stages of implementation. One option for the bioeconomy to play a critical role in development of tropical megadiverse countries is to address rural sector challenges in the context of implementing the United Nations 2030 Agenda for Sustainable Development. Furthermore, existing literature highlights the role of the agricultural sector in providing a source of informal social protection in the form of food security and the conservation effect on biodiversity of small-farming (see Espinel 2009). Thus, transition strategies to bioeconomy can also be understood as an indirect mechanism to incentivize the sustainable provision of these public goods.

Ecuador, particularly, faces challenges in converging with developed nations, due to its inability to complete the productive transition from low value-added sectors (commodities and natural resource and labor-intensive manufactured goods) to high value-added sectors (technology-intensive manufactured goods) (Domínguez and Caria 2016). In a well-developed strategy, Ecuador may be able to offer initial lessons to the world's sustainable bioeconomy policy approaches. While developed countries' efforts to achieve a transition to bioeconomy heavily rely on bioenergy and biomass (Sillanpää and Ncibi 2017), the Ecuadorian strategy

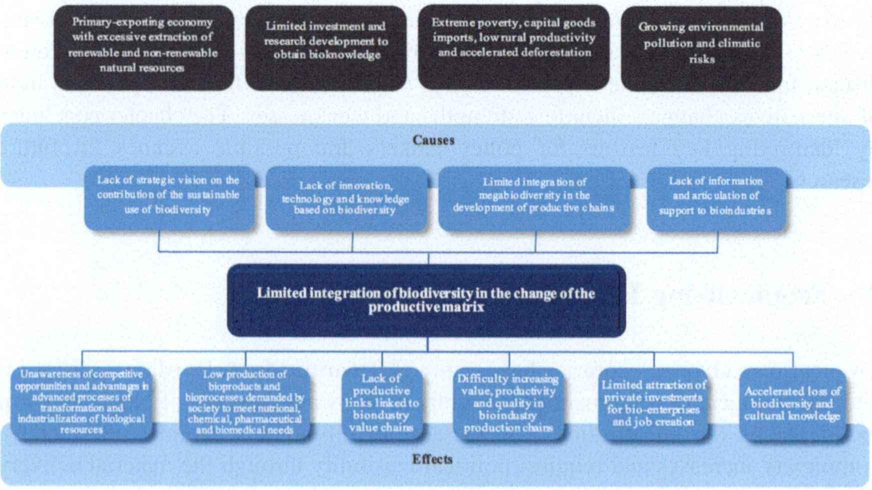

Fig. 1 Summary of identified challenges to tropicalized bioeconomy

focuses on biodiversity as a key building block of a sustainable economy, by means of—not limited to—knowledge-intensive products. This should add to the sustainability of the bioeconomy model proposed in Ecuador. From the recognition of nature's rights in its constitution, to the promotion of the concept of net avoided emissions at the United Nations Framework Convention with the Yasuní-ITT Initiative, Ecuador's innovative approach to bio-industry and bioeconomy is a demonstration of its restless efforts to achieve sustainability and poverty eradication.

Some authors argue for a reassessment of the role that biodiversity should play in the development plan and the national strategies to eradicate poverty and change the productive matrix in a transition from a primary extractive economy to a tertiary one based on knowledge (see Walsh 2010). The expected results of a strategy for bioeconomy may be observable in the long run, yet the industry is starting to acknowledge the need for further development of bio-industry value chains. Initial lessons from Ecuador can offer some insights into combating biodiversity loss and putting into value its central role in transition to other megadiverse tropical countries facing similar economic trade-offs, historical inequities and conflicting interests.

3 Method

Institutions structure incentives, shape people's beliefs and preferences, and introduce predictability to human interaction (Schmid 2004). As such, human institutions (i.e., ways of organizing activities) may contribute to biodiversity loss (Barbier et al. 1994). Institutions and biodiversity protection may thus be evaluated and compared in contexts that consider institutional performance and relationships (Wells 1998). Focusing on institutions can help policymakers evaluate alternative arrangements for achieving sustainable development and poverty eradication.

3.1 Situation, Structure and Performance Analytical Framework

The Situation, Structure and Performance (SSP) framework (Schmid 1987) allows for an analytical institutional impact assessment. By observing "variables" in their context (i.e., the situation), which are taken as given, interdependencies arising from characteristics of the "good" and actors involved in "transactions" are identified. These transactions, the interplay of individuals and their contexts including ever-present costs, are the unit of observation. Institutional impact analysis of an established situation (e.g., the sustainable use of biodiversity to foster economic development) allows for comparing institutional alternatives (i.e., structures) that may impact interdependencies related to the good(s) and individuals (Boerrke 2001). That is, by holding the "situation" constant, alternative "structures" may be evaluated to predict relative "performance" in terms of substantive measures (e.g., economic outcomes and distributions). Evaluating presumed or actual performance allows the articulation of conclusions and/or insights concerning the alternative structures' likelihood of achieving targeted social goals.

Situation

Despite significant progress in reducing inequality, particularly since 2000, Ecuador shows evidence of multidimensional inequalities, including socioeconomic, ecological, and power asymmetries, and on their historical and transnational character (Braig et al. 2015). We now present the macroeconomic variables that influence the "transaction" under study, i.e., the provision of bio-based goods and services in Ecuador, enabled by the sustainable use of biodiversity. Namely, we will look at last decade's performance of the most relevant figures under two subtopics: Growth and inequality (GDP and Gini Index) and Structural changes in the economy (agricultural sector as a percentage of GDP, consumption patterns, agro-trade balance). Given its bio-based nature, we look at the agricultural sector as a proxy to assess the sustainability of prevailing bio-economic sectors. Before moving on to the Structure, we will summarize future perspectives that authors have elaborated, based on the aforementioned figures, regarding socio-ecological transitions in Ecuador.

GDP growth and inequality

On the one hand, Ecuador's 2007–2014 period showed strong growth (4.3%)—compared to 1990–2001 (2.4%)—trade surplus (9%), and moderate inflation (4.8%) (BCE 2015). The favorable terms of trade for Ecuadorian commodities, such as cacao and oil, as well as large investments in public infrastructure strongly added to growth. At the international level, commodity prices remained high, expanding mining and export agriculture activities. Both terms of trade and financial inflows pressed for an appreciation of the real exchange rate of around 10.3% between 2007 and 2012 and 8% in 2012–2016 (BCE 2012, 2016).

The strong economic growth observed since 2007—with the exception of 2009—has morphed into an economic downturn. The most recent forecast by the International Monetary Fund shows negative growth from 2016 until 2020 and a cumulative fall in GDP of almost 7% (IMF 2016). Between 2009 and 2016, private external debt increased by USD 2.1 billion and public debt, by USD 18.3 billion (BCE 2016). The 7.8-magnitude earthquake that hit Ecuador in April 2016 represented an even greater challenge to public finances than low oil prices or plummeting export revenues due to unfavorable exchange rates. Among several other measures, the Solidarity Bill increased the Valued-Added Tax by two percentage points in order to collect additional resources for reconstruction. As a result, tax collection decreased by 8% between 2015 and 2016. Furthermore, between the first quarter of 2016 and the same period of 2015 (t/t-4), real GDP's variation was −4% (BCE 2016).

Regarding inequality, the Gini coefficient lowered from 0.57 to 0.45 between 2000 and 2016 (BCE 2016). Inequality between groups has also decreased. The ratio of household per capita income between urban and rural households fell from 2.38 in 2005 to 1.69 in 2014 (SIISE 2016). A decomposition of the Gini fall over 2002–2012 in Ecuador suggests another explanation: a drop in the skilled-unskilled wage ratio (Cornia 2014). This reflects a 9% reduction in overall poverty between 2001 and 2014 (i.e., from 54.9 to 33.6%) (Cypher and Alfaro 2016). Some authors, however, have argued that once the commodities boom ended—worldwide oil prices reduced from USD 109,45 in 2012 to USD 26,5 in 2016—government policies aimed at reducing inequality turned out to be unsustainable and inequality may start to rise again (see Gachet et al. 2016). That is, reductions in inequality in Ecuador may have been subject to temporal characteristics of resource booms.

Structural changes in the economy

The main drivers of Ecuador's growth have not resulted in productivity gains nor in an increase of the technological content in agricultural and industrial products. Much to the contrary, the boom has been central in the domestic service sector. The net trade deficit of manufactured or semi-manufactured goods reflects this process. During the last decade, increased imports of capital goods were absorbed by different sectors of society (i.e., education, health, agriculture, communications, energy, construction, etc.), including the industrial subsystem, and—in support of the expansion of public and private research—development and implementation institutions.

Changes in the sectoral composition of production are, among others, obvious signs of structural transformation. They are generally associated with changes in demand, foreign trade and the use of factors. Regarding changes in consumption patterns, the decreasing representation of the agricultural sector in the Ecuadorian economy can be observed in the evolution of food consumption in relation to GDP. Despite household consumption being the main contributor to GDP growth between 1966 and 1994, food consumption accounted for almost half of total consumption in the 60 s but its representation reduced to 32% by 1994 (Marconi and Samaniego 1995). This transformation has involved processes of industrialization, rising tertiary sectors (services) and technological intensity, inter-sectoral and regional labor displacements and changes in income distribution. In fact, while per capita income nominally increased by 8.8 times, the share of the agricultural sector in the Ecuadorian economy decreased from 26.5% in 1966 to 11.5% in 1981 (Marconi 1985) and to 8% by 2016 (BCE 2016). This evidence of structural change has been accompanied by a trade surplus in the agricultural sector. It exported a yearly average of USD 9 billion of agricultural goods between 2012 and 2014 and imported just USD 1.3 billion. Given that demand is less than supply capacity, Ecuador may not be fully utilizing its available resources, which could be transformed into bio-products.

Ecuador's export sectors are maturing and should see organic growth in volume terms over the coming five years (Duff and Padilla 2015). In addition, based on a computable general equilibrium model, Wong and Kulmer (2011) find that the European Union-Ecuador trade agreement, which came into effect in January 2017, will likely rise real wages for skilled and unskilled workers in the rural sector. However, this may also be accompanied by falling employment among unskilled workers in urban areas. Jobs for rural unskilled wage work account for around 15% of total employment, thereby potentially reducing poverty.

During the last decades, a wide range of public agencies have implemented efforts to enhance productive activities in the rural sector (i.e., outside and within forests) through a mix of incentive-based and command-and-control conservation and productive programs. In addition to marginally promoting bio-trade of non-forest products within protected zones, the Ministry of Environment introduced Socio Bosque, a conditional payment to align landowner's incentives to protect private forests areas. On the other hand, the Ministry of Agriculture has implemented large initiatives aiming at increasing productivity in rural production. Furthermore, the strategic sectors (hydroelectric, oil and mining) have influenced the rural sector through policies promoting the change of the energy and productive matrix. The aggregated effect of these policies is an increased relative and absolute participation of the rural sector in the Ecuadorian GDP, significantly altering the tradeoffs between the contribution of the primary and industrial productive activities to GDP, labor migration, deforestation and biodiversity (Falconí and Larrea 2004).

The evidence above indicates that structural change in Ecuador has been accompanied by social, economic, political and environmental changes. Policies have directly and indirectly induced the expansion of the agricultural, hydroelectric, oil and mining frontier at the expense of forests and biodiversity.

Socio-ecological transitions in Ecuador

Falconí and Vallejo (2012) argue that key determinants that promote socio-ecological transitions in Ecuador, as well as in other Andean countries, are economic efficiency, income redistribution and physical sustainability, for extractive economies: (i) induce environmental pressures and deepen inequalities, and (ii) imply economic growth prospects limited by the ecosystem's carrying capacity.

The need to promote public policies that guarantee the environmental, social and economic sustainability of the rural sector creates an opportunity for a pattern of specialization that considers biodiversity as a strategic resource for Ecuador's sustainable industrial development. Additionally, low prices of commodities generate a crisis of alternatives in the rural sector, but also an opportunity for production schemes based on biodiversity that reduce the vulnerability to external shocks. Consequently, it can be argued that Ecuador has sufficient conditions to include bio-industry within its industrial sector. In fact, the National Strategy for Biodiversity (NaBiS) 2015–2030 has been designed to pave the way for the industrialization of biodiversity based on bio-knowledge.

Structure

We now consider the characteristics of the NaBiS 2015–2030, an institutional arrangement designed to protect biodiversity to catalyze a sustainable transition of the Ecuadorian economy (MAE 2016a). This transition is not only formally aligned to the Aichi Targets agreed upon by the Convention on Biological Diversity, but more importantly, to the country's own development approach, and seeks to overcome structural inequities and relations of dependence generated by the current primary-extractivist and export-dependent model through bio-knowledge accumulation (SENPLADES 2013). According to a recent study, its full and effective implementation requires an investment equivalent to a 0.38% of Ecuador's GDP (MAE 2016b).

On implementation, it should be noted that NaBiS 2015–2030 and its Plan of Action 2016–2021 have been built with an integrative vision of planning at different government levels (see Fig. 2) (MAE 2017). On the one hand, their targets are closely connected with the priorities established in the National Development Plan, the guidelines under the environmental sustainability axis of the National Territorial Strategy, and the strategic value-chains prioritized in the National Strategy for Changing the Productive Matrix; on the other hand, with the local demands collected in the last participatory exercise to update Zonal Planning Agendas. The challenges now lie in ensuring that decentralized governments internalize the proposals obtained from participatory processes in their local planning, and, given the cross-sectoral nature of biodiversity, in articulating with all the agendas of inter-sectoral coordination as part of a multi-level governance structure.

We explore NaBiS in the context of the National Strategy for Changing the Productive Matrix (Vicepresidencia del Ecuador 2015), which implies the strengthening of an articulated institutional arrangement between public, private

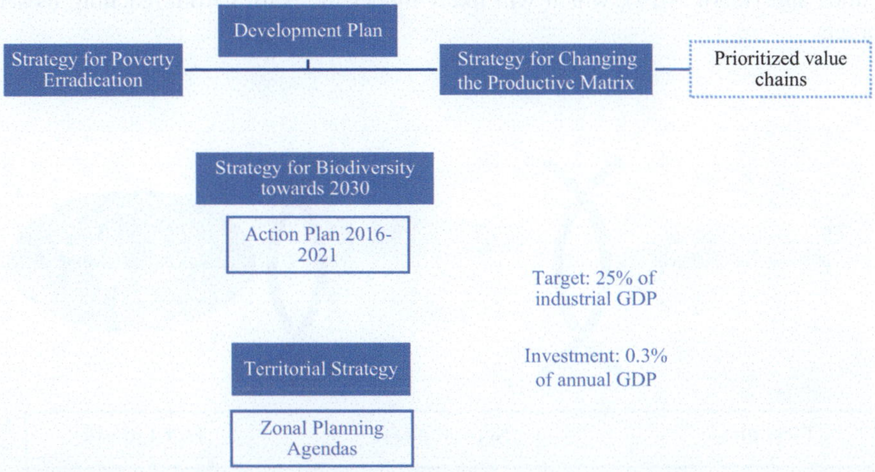

Fig. 2 Ecuador's bioeconomy institutional arrangement

and community actors that take advantage of territorial comparative advantages and—fundamentally—the economic decisions in the scientific and technological knowledge; all within a framework of environmental sustainability and respect for the rights of nature. Considering this arrangement is essential to construct a new economy based on biodiversity as its main competitive advantage, in line with the Constitutional provision that sets biodiversity as key strategic resource.

The National Strategy for Changing the Productive Matrix involves three main phases for the transformation of specialization patterns, wealth generation and income distribution (see Fig. 3). Its objective is that the relative shares of knowledge-based goods and services—including bio-knowledge—and tourism services have a greater weight in GDP than those generated by the primary sector. For this matter, the strategy has prioritized nine sectors and thirteen chains to diversify production, generate added-value and increase the exportable supply based on inputs from ecosystems and biodiversity. Nascent industries will benefit over a four-year horizon from a public investment of approximately USD 47 billion.

The most salient NaBiS attributes setting forth the development of an Ecuadorian bioeconomy are presented in Table 1. In addition to some discussed earlier, we have included attributes impacting different scopes of action.

Performance

We present three predictions regarding possible outcomes to the ruling development approach in Ecuador, interpreted as the aftermath of the NaBiS in joint performance with the institutional arrangements that informed it. The first two scenarios describe outcomes provided that no additional efforts are placed in delegating biodiversity the central role for development, the basic idea underpinning a bioeconomy model. Recent research points to critical issues arising from the implementation of Ecuador's development strategy (see Cypher and Alfaro 2016; Childs and Hearn 2017), which will likely hinder the NaBiS from reaching its set targets.

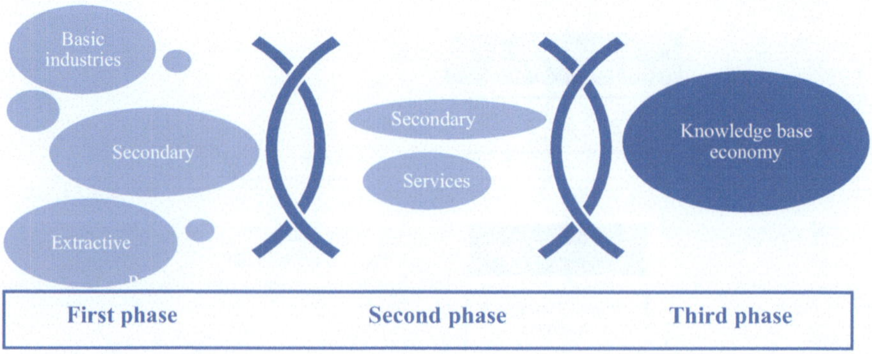

Fig. 3 Transition process underlying the National Strategy for Changing the Productive Matrix

Table 1 Identified NaBiS 2015–2030 attributes fostering bioeconomy

Scope	Description
Institutional sector	
Governance	– Includes mechanisms of **local participation** and their link with sectoral and national authorities and national planning systems – Promotes **access to ICTs** and improvement of information systems (e.g., System of Environmental and Economic Accounts) – Requires consideration of internal (scale, specialization, technological practices, etc.) and external economies (raw materials suppliers, labor force, agglomeration, etc.), thus a more **detailed planning** and **better-quality information** generation
Fiscal policy	– Promotes **coordination** of environmental public investment with selective actions for productive purposes, as well as channelling resources from the international cooperation – Aims at **mobilizing financial resources** from taxes, loans, equity investments and securities to promote rural productivity and R&D
Regulation	– Enables a conducive regulatory environment to explore **border regulation** and works on policies to **avoid merchandise flows** – **Coordinates** environmental and agricultural **incentives** with those set in the Organic Code of Production, Trade and Investment, Organic Law on Internal Tax Regime and Organic Law on Incentives for Public-Private Partnerships and Foreign Investment
Economic sector	
Real GDP	– Seeks to a) **maximize** the benefits of **exports** based on internationally competitive inputs, and b) **maximize domestic** input **production**, establishing reference prices and strategic storage stocks to achieve food sovereignty – Generates new **incentives to increase** the **productivity** and **added value** of the agricultural and livestock sector – Introduces **incentives** related to the **demand**, that foment the processes of certification or through the systems of public purchases
R&D	– Promotes basic and applied **research** to innovate and **diversify solutions and products**, improving efficiency, the implementation of best practices and incorporation of state-of-the-art technology from the use of biodiverse inputs
Local development	– Promotes the development of **physical communication** (terrestrial, fluvial, air) and availability of **communication systems** – Promotes the **anchoring of human capital in these territories**, by boosting the national bio-industry – **Anchors biodiversity to territory** through increasing local demands (i.e., gourmet food, handicrafts, bio enterprises, etc.), promoting the domestication of species for on-farm production and sustainable use in native forests
Regional development	– Plan **terrestrial connectivit**y between cities, populations and Amazonian settlements to improve territorial competitiveness, maximizing the external economies for companies and minimizing risks of deforestation – Creates a **cluster of biodiversity** to promote the development of bio-enterprises to foster regional development

(continued)

Table 1 (continued)

Scope	Description
Private sector/ entrepreneurs	– Supports programs for **small and medium size initiatives**, e.g., an incubator and trading facilities – Supports the development of **replicas of successful business models**, based on the use of biodiversity – Promotes the bio-industry with technologies that **optimize protein production** for animal/human consumption purposes – Promotes **strategic alliances** with dominant companies to integrate them into the input-product matrix to biodiversity

Status quo

First, the expected performance of the NaBiS is not likely to achieve ambitious outcomes as the current transition is set to start heavy industry without having established the bridge with the prevailing primary sector. Similar experiences in Asian countries have initially fostered great capacities to export simple and basic consumer goods, before using their already high national industrial capacities to jump into the heavy industry (Amsden 2001). In addition, basic industry shows high idle capacity worldwide, thus a likely oversupply in the market. Second, Ecuador's prioritized strategic sectors have high capital-intensive production functions and will probably incur high expected socio-environmental effects. On the one hand, these sectors' development involves imports of capital-intensive inputs, such as heavy construction, specialized machinery or creating a new endogenous machine-tool manufacturing industry. On the other hand, there is a gap in terms of investments needed for R&D and the availability of highly-trained human capacities associated to its development.

The implied challenges to rural development are evident if we consider the limited opportunities for economic stimulation (i.e., job creation, wage stabilization, investment) that this scenario offers in these locations. Just as critical, we consider the danger to biodiversity that this capital-intensive development approach might represent, for it is Ecuador's main asset and the main provider under a tropicalized bioeconomy model.

Biodiversity as a priority

If, however, efforts were allocated in prioritizing the use of biodiversity, by means of increased support and promotion of R&D projects and infrastructure that generate and accumulate bioknowledge, the scenario is much more promising. As explained by Golinelli et al. (2015), the Ecuadorian bioeconomy policy has considered alternative types of research to encourage early open collaboration oriented to knowledge generation, as well as commercial research oriented to income generation through an active exchange of knowledge, technology and materials in advanced stages. While the first is expected to reward research and discourage monopolization, the latter will likely generate exportable solutions to international markets and include not only physical resources but also genetic, genomic and metagenomic information, models, methods and protocols. These opportunities may be further promoted by an increased attractiveness for international investors searching for projects or settings related to biodiversity. Current institutional arrangement (i.e., Environmental and Science and Technology Codes) is key to continue promoting bio-prospection for applied research projects, as well as throughout the strategic stages during R&D. It also incorporates opportunities for a collective license model addressing the shortcomings of the patent model (Safrin 2004; Kloppenburg 2014). Furthermore, total factor productivity might rise if micro, small, and medium-sized enterprises increasingly participate in an environment favorable to innovation rather than one paralyzed by monopolies, oligopolies, or private oligopsonies (Vivarelli 2014).

The expected consequent increase of employment and labor wages (agricultural and non-agricultural) in the rural sector might be the tip of the iceberg regarding the tangible benefits of a knowledge-based, conservation-oriented and sustainable development approach. The establishment of bio-industrial clusters strategically located near rural areas and along geographical borders not only tackles the concerns regarding lack of jobs and income instability in the sector, but will likely revitalize local consumption (endogenous economic impulse) and even call for foreign actors' attention (private or sovereign) to invest. Such strategy is not unknown to the region. Large national firms in Brazil opted for a transition through technical change, as a result of multinationals' behavior (Cassiolato and Schmitz 2002).

4 Discussion

Like any other development strategy, bioeconomy is a long term prospect. The aforementioned potential effects of strengthening bioeconomy (increased productivity, improved job generation and increased wages) should be interpreted with caution, for the bioindustry's productivity might lower at the time of policy implementation until the system's dynamics, supported by distribution infrastructure investments, produce tangible results. When such sustainable transition occurs, bioeconomy has the potential for affecting regional trade patterns (i.e., Andean region) and transforming Ecuador into a consumer of raw materials from biodiversity to add value, substitute oil-based imports and ultimately export surplus bio-products. Import substitution is especially relevant in Ecuador due to its limited monetary policy (dollarization) and vulnerability to dollars fleeing the economy.

By boosting bioeconomy, the probable increase of consumption levels within individuals in the rural sector may happen as a result of bio-products' reasonable prices. More importantly, however, sustainable bioeconomy assumes that, in the long-term, aggregated consumption may not change but its composition will be significantly altered by changing productive systems and increasing consciousness on consumption. In this sense, public procurement of bio-products can play a critical role to accelerate the transition by linking primary and secondary bio-industry through increasing demand of basic commercial bio-products.

By diverting consumption towards the bio-industry sectors and providing job stability in the rural sector, an increase of R&D investments may impact the aggregate demand, leading to increased output, despite having no effect on the individual's overall consumption level. We also note that these investments can not only reduce rural-urban migration, but also reduce the deflationary pressure in the non-rural labor market in the same way that bio-industry shows labor absorption capacity and a potential positive effect on aggregate wages in the rural sector.

Experiences in globalization of local industrial R&D of niche-based firms from developing countries indicate that high-technological competences can help local actors of bio-industry value chains become part of global R&D networks. Offshoring R&D and industrial activities to megabiodiverse developing countries of nascent stage companies in developed countries should become a regular phenomenon. This could additionally provide opportunities for local and foreign affiliates to work synergistically along de supply chain.

Our analysis also highlights the critical importance of added-value in bio-industry value chains. Considering that the share of food in expenditure falls as income rises, bioindustry should render highly added-value products to ensure those historically excluded groups the access to wealth creation productive factors (i.e., bio-knowledge and entrepreneur opportunities). In fact, we believe that bioeconomy meets sustainable development as rural sector productivity increases in tropical megabiodiverse countries, because as biodiversity assumes its central role in the economy, it contributes to conservation by satisfying poverty needs (e.g., food security) and promoting workers' transition into sustainable practices (e.g., small farming) and industries.

It is beyond the scope of this chapter to work out the aggregate implications of these several competing effects without more assumptions and a fully-specified structural model. Future research should aim to provide empirical evidence of macroeconomics effects of the Ecuadorian bioeconomy as a useful case study to derive further lessons. It is nonetheless useful to consider that bioindustry has potential to absorb the rural sector surplus labor in a demand-constrained economy.

There is a large literature suggesting that exposure to markets and new technologies affects social relations and can erode traditional forms of social protection (Scott 1985; Polanyi 1944; Marx 1847). Li (2009) notes the tension between inequality in access to land for the poor and the promotion of modernization in agricultural production. These works call for policy consideration to ensure equitable access to new wealth creation and productive factors during development of bio-industry value chains, and an overall inclusive and rights-based transition to bioeconomy.

Despite its alignment to the Strategy for Changing the Productive Matrix, further emphasis should be given to biodiversity as a key central resource in initial stages, especially to overcome current challenges (Fig. 1). We argue that bioeconomy offers greater cost-benefit advantages as the investments required will make use of increasing primary sector productivity and a blooming development of high added-value chains with less foreseeable negative social and environmental externalities relative to basic and intermediate industries. In this context, bioeconomy may offer a sustainable option for a socioecological transition. In fact, given the current economic situation, there can be expected significant larger gains in wealth generation, employment and inequalities reduction from accelerating development phases by promoting simple and basic consumer goods derive from bio-industry value chains.

5 Conclusions

Our analysis indicates that sustainable bioeconomy is being revised in Ecuador. Due to complex social and economic interdependencies, the development of the bioindustry requires tropicalizing foreign models and technologies as well as policies that support and encourage the development of value chains. The Ecuadorian NaBiS is an institutional arrangement that provides initial lessons to tropical megadiverse countries, for it has a promising role to enable a more efficient and integrated use of biological resources towards a sustainable and resilient economy, while addressing structural development and biodiversity protection challenges.

The economic potential for bioeconomy is observed to, with the appropriate investment in distribution infrastructure and increase in rural productivity, affect fossil fuel-based imports and reduce inequalities in the rural sector. To take advantage of the potential of Ecuador's bio-industry, there is a clear need to strengthen up the central role of biodiversity in the change of the productive matrix. This can be achieved by developing value chains with a structured and defined strategic vision—nourished by R&D efforts—that coordinates suppliers, producers and users in a way that is consistent with the institutional framework.

Finally, this analysis observes that the development of bioindustry in Ecuador should occur similarly to that of other intermediate and final industries included in the National Strategy to Changing the Productive Matrix. Bioindustry is likely to induce growth for the local industry adding further value into current primary production activities. The creation of a domestic market in this sector occurs more slowly, because, unlike international markets, the initial phase of a bioindustry is characterized by strong public leverage and national consumption. For this to occur, a communication strategy is recommended to increase awareness about the bioindustry's importance within a context of change of the production matrix and its long-term feasibility. Given the current strong focus on developing basic industries as a first step to start the economic transition, it seems relevant to reassess biodiversity and its role in industry impact (i.e., employment, balance of payments, public and private financial inflows). Future research should aim to develop a structural model to explore the potential for bioeconomy in a dynamic and multi-temporal context.

References

Amsden A (2001) The rise of "The Rest". Oxford University Press, Oxford

Barbier EB, Burgess JC, Folke C (1994) Paradise lost? The ecological economics of biodiversity. Earthscan, London

BCE (2012) Información Estadística Mensual No. 1930. Banco Central del Ecuador, Quito

BCE (2015) Información Estadística Mensual No. 1966. Banco Central del Ecuador, Quito

BCE (2016) Información Estadística Mensual No. 1978. Banco Central del Ecuador, Quito

Boerrke PJ (2001) Putting the 'political' back into political economy. In: Biddle JE, Davis JB, Medema SG (eds) Economics broadly considered. Routledge, London and New York

Braig M, Costa S, Göbel B (2015) Social inequalities and global interdependences in Latin America: a provisional appraisal. Revista Mexicana de Ciencias Políticas Y Sociales 60 (1):209–236

Cassiolato JE, Schmitz H (2002) Hi-tech for industrial development: lessons from the Brazilian experience in electronics and automation. Routledge, London

Childs J, Hearn J (2017) "New" nations: resource-based development imaginaries in Ghana and Ecuador. Third World Q 38(4):844–861

Cornia GA (2014) Falling inequality in Latin America: policy changes and lessons. Oxford University Press, Oxford

Cypher JM, Alfaro Y (2016) Triángulo del neo-desarrollismo en el Ecuador. Problemas del Desarrollo 47(185):163–186

Domínguez R, Caria S (2016) Ecuador in the middle income trap. Problemas Del Desarrollo 47 (187):89–112

Duff A, Padilla A (2015) Latin America: agricultural perspectives. Retrieved from https://economics.rabobank.com/publications/2015/september/latin-america-agricultural-perspectives/

Espinel R (2009) Economía política de la biodiversidad: conocimientos ancestrales y derechos de propiedad. In: Martín Mayoral F (ed) Deuda externa y economía ecológica: dos visiones críticas. FLACSO Ecuador, Quito, p 13

EU (2012) Innovating for sustainable growth: a bioeconomy for Europe. Communication from the Commission to the European Parliament, the Council, the European Economic and Social Committee and the Committee of the Regions. European Commission COM(2012)—Brussels

Falconí F, Larrea C (2004) Impactos ambientales de las políticas de liberalización externa y los flujos de capital: el caso de Ecuador. In: Falconí F, Hercowitz M, Muradian R (eds) Globalización y desarrollo en América Latina. FLACSO, Quito

Falconí F, Vallejo MC (2012) Transiciones socioecológicas en la región andina. Revista Iberoamericana de Economía Ecológica 18:53–71

Gachet I, Grijalva DF, Ponce P, Rodríguez D (2016) Vertical and horizontal inequality in Ecuador: the lack of sustainability (WIDER Working Paper No. 2016/106). p 37

Giljum S, Eisenmenger N (2004) North-south trade and the distribution of environmental goods and burdens: a biophysical perspective. The J Environ Dev 13(1):73–100

Golden JS, Handfield R (2014) The emergent industrial bioeconomy. Ind Biotechnol 10(6):371–375

Golinelli S, Vega-Villa K, Villa-Romero JF (2015) Biodiversidad: ciencia ciudadana, saberes originarios y biodiversidad aplicada en la economía social del conocimiento. In: Vila-Viñas D, Barandiaran XE (eds) Buen Conocer/ FLOK Society: Modelos sostenibles y políticas públicas para una economía social del conocimiento común y abierto en Ecuador. IAEN-CIESPAL, Quito, Ecuador, pp 345–396 Digital

Hill AB (2000) International markets for genetic resources: opportunities for Colombia. Science, technology and innovation. Discussion Paper No. 2, Center for International Development, Harvard University, Cambridge, MA

IMF (2016) Ecuador: purchase under the rapid financing instrument (Tech. Rep. International Monetary Fund—IMF Country Report No. 16/288), p 38

Jha V (2008) Environmental priorities and trade policy for environmental goods: a reality check. ICTSD Issue Paper, No 7
Kloppenburg J (2014) Re-purposing the master's tools: the open source seed initiative and the struggle for seed sovereignty. The J Peasant Stud 41(6):1225–1246
Li TM (2009) Exit from agriculture: a step forward or a step backward for the rural poor? The J Peasant Stud 36(3):629–636
MAE (2016a) Estrategia Nacional de Biodiversidad 2015–2030. Ecuador, Quito
MAE (2016b) Estrategia Nacional de Biodiversidad 2030. Ambiente 2035- Aportes Al Debate de La Política Ambiental, 5
MAE (2017) Plan de Acción 2015–2021 para implementar la Estrategia Nacional de Biodiversidad. Quito, Ecuador
Marconi S (1985) Tendencias estructurales del sector agrícola ecuatoriano. Cuestiones Económicas, 11, Banco Central del Ecuador
Marconi S, Samaniego P (1995) Las fuentes del crecimiento económico: una perspectiva a partir de la demanda. Notas Técnicas, 19, Banco Central del Ecuador
Marx K (1847) The poverty of philosophy: answer to the "philosophy of poverty" by M. Proudhon. Progress Publishers, Moscow; (reprinted 1955)
OECD (2009) The bioeconomy to 2030: designing a policy agenda. Organization for Economic Co-operation and Development (OECD), Paris
Piotrowski S, Carus M, Carrez D (2016) European bioeconomy in figures. Brussels
Polanyi K (1944) The great transformation: the political and economic origins of our time. Beacon Press, Boston, MA
Rival L, Muradian R, Larrea C (2015) New trends confronting old structures or old threats frustrating new hopes? ECLAC's compacts for equality. Dev Change 46(4):961–978
Russi D, Gonzalez-Martinez AC, Silva-Macher JC, Giljum S, Martínez-Alier J, Vallejo MC (2008) Material flows in Latin America. J Ind Ecol 12(5–6):704–720
Safrin S (2004) Hyperownership in a time of biotechnological promise: the international conflict to control the building blocks of life. The Am J Inter Law 98(4):641
Schmid AA (1987) Property, power, and public choice: an inquiry into law and economics. Praeger Publishers, New York
Schmid A (2004) Methods. In: Conflict and cooperation. Blackwell Publishing Ltd., Oxford, pp 138–162
Scott JC (1985) Weapons of the weak. Yale University Press, New Haven, CT
Secretaría Nacional de Planificación y Desarrollo (SENPLADES) (2013) Plan Nacional de Desarrollo del Buen Vivir 2013–2017. SENPLADES, Quito
SIISE (2016) Distribución del Ingreso: Coeficiente de Gini. Available at http://www.siise.gob.ec/siiseweb/siiseweb.html?sistema=1#
Sillanpää M, Ncibi C (2017) A sustainable bioeconomy: the green industrial revolution. Springer, Berlin
Vallejo MC (2006) Estructura biofísica de la economía ecuatoriana: un estudio de los flujos directos de materiales. Revibec: Revista Iberoamericana de Economía Ecológica 4:55–72
Vicepresidencia del Ecuador (2015) Estrategia Nacional para el Cambio de la Matriz Productiva. Quito
Vivarelli M (2014) Innovation, employment and skills in advanced and developing countries: a survey of economic literature. J Econ Issues 48(1):123–154
Walsh C (2010) Development as Buen Vivir: institutional arrangements and (de)colonial entanglements. Development 53(1):15–21
Wells MP (1998) Institutions and incentives for biodiversity conservation. Biodivers Conserv 7 (6):815–835
Wong S, Kulmer V (2011) Poverty impacts of trade integration with the European Union: lessons for Ecuador. In: Durán Lima J, LaFleur M, Pellandra A (eds) Trade, poverty and complementary policies in Latin America. CEPAL, Santiago, pp 35–64
WWF (2016) Living planet report 2016. Gland, Switzerland

Economic Assessment of Tourism Based on Shark-Seeing and Diving as a More Profitable Activity Than Commercial Fishing

Yessica Miriam Plata Zepeda, Patricia Ramírez Romero and Fabíola S. Sosa-Rodríguez

Abstract Even though sharks are top predators, there is little information available aimed to understand their habitats, movements and dynamics (Hoyos-Padilla et al. 2014). Today, there are 1236 species of *chondrichthus* in the world including sharks, rays and chimeras. However, the population and variety of sharks are declining due, mainly, to overfishing (Klimley 2015). In Mexico, there are 111 species of sharks, which represent 8.98% of the world's diversity (Del Moral et al. 2016; CONABIO 2016). Sharks bioaccumulate mercury (Hg), which is a toxic compound for humans and other species because of their place in the trophic chain. Human health can be affected by the consumption of sharks' meat contaminated with this metal (Escobar Sánchez 2011). The effect of mercury in human bodies ranges from tremors to neurotoxic and teratogenic effects (Raimann et al. 2014). The observation of sharks in the wild has increased—it generates economic benefits that go beyond the price paid for its enjoyment, since it promotes the conservation of the species. For instance, in 2013, these economic benefits were estimated at 314 USD millions by year (Cisneros et al. 2013). In Baja California Sur, Mexico, fishing represents 1.28% of the state's GDP and shark fishing only 0.03%. However, the tourism sector generates 16.61% of the GDP, mainly explained by the gray whale sighting. This activity generates 4.12 USD

The original version of this chapter was revised: Corresponding author's corrections have been incorporated. The erratum to this chapter is available at https://doi.org/10.1007/978-3-319-73028-8_30

Y. M. Plata Zepeda
Master Program in Energy and Environment, Metropolitan Autonomous University-Iztapalapa, Iztapalapa, Mexico
e-mail: yessica.plata.zepeda@gmail.com

P. Ramírez Romero
Department of Hidrobiology, Metropolitan Autonomous University-Iztapalapa, Iztapalapa, Mexico
e-mail: patt@xanum.uam.mx

F. S. Sosa-Rodríguez (✉)
Department of Economics, Metropolitan Autonomous University-Azcapotzalco, Azcapotzalco, Mexico
e-mail: fssosa@gmail.com

W. Leal Filho et al. (eds.), *Towards a Sustainable Bioeconomy: Principles, Challenges and Perspectives*, World Sustainability Series,
https://doi.org/10.1007/978-3-319-73028-8_12

millions by year, which represents 0.06% of the state's GDP (GBCS 2015a, b). If more whales or sharks sighting sites were developed, the revenue of this activity to local communities would increase. The purpose of this research was to determine whether ecotourism with sharks is an activity that represents greater economic benefits to the local population than shark fishing for human consumption since the ingestion of shark meat potentially contaminated with mercury may endanger people's health. Some main findings include that tourists are willing to pay for sighting and swimming with sharks more than 212 USD compared to 53 USD paid for whale watching. Therefore, ecotourism with sharks is a more profitable activity to the local population compared to shark fishing, in addition to promoting the conservation of several species.

Keywords Economic assessment · Shark-seeing · Commercial fishing
Contingent valuation method · Market price · Baja California Sur

1 Introduction

Ecosystem services are defined as the benefits obtained from ecosystems that are essential for social, economic, and human survival. MEA (2005) identify four categories of ecosystem services, namely provision, regulation, support and cultural services. This research analyzed two ecosystem services: the first one is related to food supply due to fishing, and the second is recreation as a result of shark-seeing and diving—an alternative activity that favors conservation.

Ecosystem-services assessment has increased during the last decade to guide decision-makers in the definition of conservation strategies. The results of these investigations have created greater awareness in society in regard to the value of nature and the services it provides to humans. Although there are several methods for economic assessment, they use mainly depend on the ecosystem service analyzed, its context and the information available. The methods used in the research to assess the ecosystem services previously mentioned include contingent valuation, market prices and cost-benefit analysis.

The method of contingent valuation is intended to determine the willingness to pay of individuals in order to receive a good or service. To this end, interviews need to be conducted to key actors, so respondents can express their preferences in terms of economic amounts they are willing to pay to have access to environmental goods or services. With this method, respondents can also express their willingness to receive some type of compensation in case the benefits from those environmental goods or services get reduced due to a reduction in their quality (Montenegro 2007; Medina et al. 2012).

The method based on market prices is used to determine the income that anglers obtain from the commercialization of various species of fish, including shark meat. This method was utilized because shark meat is traded in different distribution centers for sea products; therefore, prices per kilogram were easily identified (Van Damme et al. 2011). It is worth mentioning that sometimes consumers do not know that they are consuming shark meat because it is sold as other species such as cod or dried fish meat.

Finally, the cost-benefit analysis allows us to assess costs and benefits resulting from decision making or projects in order to determine if an activity is economically, socially and environmentally profitable. In contrast, if the costs outweigh the benefits, the activity is not considered profitable and should not be executed (Serra et al. 2012).

Sharks are top predators of marine ecosystems, so they regulate the populations of other organisms (Navia 2013). Despite this, there is not enough information regarding their movements and dynamics; this information will allow to understand the characteristics of their habitants and to identify their location in order to develop strategies for their conservation (Hoyos-Padilla et al. 2014). Due to the great interest in studying these organisms, new technologies have been generated for their observation and monitoring; examples of these technologies include satellite tags and underwater vehicles (Skomal et al. 2015). As a result of the use of these devices, new information of diverse species of sharks was known. For instance, it was identified that the hammerhead shark of the genus *Sphyrna* has annual migration patterns. Another example is related to the horizontal and vertical movements of many species of sharks made at night in order to feed themselves (Hoyos-Padilla et al. 2014). It is during migrations that some sharks can reach places that lack protection strategies, putting entire populations at risk (Klimley 2015). To avoid this, it is vital to create marine reserves where shark fishing is prohibited, as well as to innovate in conservation strategies based on shark-seeing and diving, which would also generate higher revenues to the local population.

To date, 1236 species of sharks are known worldwide. In Mexico 111 species were identified. This figure represents 8.98% of the total shark world's diversity. In the Mexican Pacific Ocean, 62 species have been identified while in the Mexican Atlantic Ocean 69 species were registered (Del Moral et al. 2016). However, the number and variety of sharks are declining mainly due to overfishing and incidental catching since they could be trapped in gillnets and longlines. This situation has endangered the survival of many species. In consequence, the International Convention on Trade in Endangered Species (CITES) defined that shark-meet trade is subject to strict international control (Klimley 2015). In addition, there is no adequate control of shark fishing because economic resources are limited; however, this activity must be regulated to support this species survival.

The management of shark fishery was formulated by the National Fisheries Institute (INAPESCA) in 1993. This institution recommended the suspension of new permits that allow shark fishing. In Mexico, 39 species of sharks have commercial value, of which 12 are more abundant and belong to the families *Alopiidae, Carcharhinidae, Squatinidae, Sphyrnidae and Triakidae* (DOF 2007a).

Medium-height and high-altitude fisheries have contributed to 60% of the national shark production; nevertheless, 50% of the production of commercial species is composed of undeveloped organisms because of over-exploitation of some species (DOF 2007b). Another reason to limit the consumption of sharks is that their meat contains mercury (Hg) (INECC 2009), which is a toxic element that bioaccumulates in organisms and biomagnifies through the trophic chain, with a higher concentration of this metal in top predators such as sharks (Escobar Sánchez 2011). The main source of mercury exposure to the population is the consumption of certain types of fish; children, the elderly and women in reproductive age are the most vulnerable to the effects of this metal. Although mercury can have a natural origin, the biggest contribution in the oceans is anthropogenic. Some sources of mercury that pollute oceans include emissions from incinerators, waste treatment plants, and the electrochemical and power generation industry, among others (Llop et al. 2013). The effect of this pollutant range from tremors, limb numbness, sensory disturbances, to neurotoxic and teratogenic effects (Raimann et al. 2014), but depending on its concentration in the blood and main organs can also cause death (Health Canada 2007).

It is estimated that 90% of mercury present in organisms is found in an organic form as methylmercury (INECC 2010). While mercury binds to proteins, it is bioaccumulated in the fish muscle tissue as methylmercury (Health Canada 2007). In Mexico, the norm NOM-242-SSA1-2009 regulates the concentration of mercury allowed for fresh, chilled, frozen and processed fish products to avoid health risks. The concentration of methylmercury allowed in fish for human consumption based on the aforementioned norm, range from 0 to 1 mg/kg in fresh fish, and from 0 to 0.5 mg/kg for dry fish (DOF 2011). Nevertheless, there are studies that have found concentrations above the norm for fresh meat from diverse species of shark, which represents a risk for human health. As mentioned, frequently shark meat is sold as other species of fish such as cod. For instance, in Mazatlán (Sinaloa) the concentration of methylmercury in shark meat was 2.1 mg/kg, while in Ciudad del Carmen (Campeche) was 6.020 mg/kg, and in Coatzacalcos (Veracruz) was 8.303 mg/kg (INECC 2009). There are a number of studies that have measured the concentration of methylmercury in shark meat in several sites around Mexico (Guzman 2009; INECC 2009, 2010). Table 1 describes the location where samples were collected and the concentration of methylmercury in shark meat in mg/kg. In most of the sites where these samples were collected, with the exception of Isla Magdalena and Ensenada (both in Baja California), and Chachalacas (Veracruz), the methylmercury concentration exceeded the norm. It should be noted that in the case of Coatzacoalcos (Veracruz) and Ciudad del Carmen (Campeche), this concentration far exceeded the norm (1 mg/kg). Human consumption of shark meat from these sites is dangerous. Since shark meat can be considered as a biomarker, the presence of high concentration of methylmercury in shark tissue provide evidence of important levels of pollution in the area, where samples were taken as a result of hydrocarbons exploitation and petrochemicals processing. Map 1 shows the sites where samples were taken (Fig. 1).

The observation of wildlife has increased in Mexico and all over the world. This activity not only promotes the conservation of species but also generates economic benefits for local populations related to the ecotourism sector. Wildlife sightings

Table 1 Methylmercury concentration in shark meat

N.	Species	Sample	Site	Methylmercury Concentration (mg/kg)	References
1	Mako shark	Muscle	Isla Magdalena, Baja California Sur	0.491	Velez (2009)
2	Hammerhead shark	Muscle	Mazatlán, Sinaloa	2.101	INECC (2009)
3	Shark[a]	Muscle	Ciudad del Carmen, Campeche	6.020	INECC (2009)
4	Shark[a]	Muscle	Coatzacoalcos, Veracruz	8.303	INECC (2009)
5	Shark[a]	Muscle	Tampico, Tamaulipas	2.109	INECC (2010)
6	Shark[a]	Muscle	Coatzacoalcos, Veracruz	1.100	INECC (2010)
7	Shark[a]	Muscle	Puerto Progreso, Yucatán	3.725	INECC (2010)
8	Shark[a]	Muscle	Ensenada, Baja California	0.844	INECC (2010)
9	Shark[a]	Muscle	Mazatlán, Sinaloa	3.301	INECC (2010)
10	Shark[a]	Muscle	Puerto Madero, Chiapas	1.500	INECC (2010)
11	Shark[a]	Fin	Las Barrancas, Baja California Sur	0.048	INECC (2010)
12	Blue shark	Muscle	Punta Belcher, Baja California Sur	1.03	Escobar (2011)
13	Hammerhead shark	Fin	Chachalacas, Veracruz	0.1057	Barrera (2013)

[a]Unidentified species

Fig. 1 Location of the sites where samples were collected. *Source* Based on Velez (2009), INECC (2009), Escobar (2011) and Barrera (2013)

Table 2 Sites for shark-seeing and diving worldwide

N	Site	Annual Income (Million US dollars)	References
1	Africa	14,465	Cisneros et al. (2013)
2	America	171,246	Cisneros et al. (2013)
3	Asia	30,539	Cisneros et al. (2013)
4	Europe	28,315	Cisneros et al. (2013)
5	Oceania	69,785	Cisneros et al. (2013)
6	Palau	18	Vianna et al. (2010)
7	Fiji	42	Vianna et al. (2011)
8	Bahamas	109,4	Haas et al. (2017)
9	Indonesia and the Philippines	152,341	De Brauwer et al. (2017)

have direct and indirect use values, as well as non-use values due to the intrinsic characteristics of nature (intrinsic values) and the relevance of preserving it for future generations (legacy values), since they can continue enjoying this experience (Cardenas 2006). Thus, ecotourism is an alternative way of recreation that allows tourists to know and enjoy natural areas and their ecosystem services in a responsible manner. This type of tourism also helps to create awareness among people of the problems that can affect ecosystems and the great amount of benefits that nature provides to humans (García and Diaz 2007).

Cisneros et al. (2013) conducted a study to estimate the economic benefits provided by shark-seeing and found that in 2011 this activity generated 314 USD millions. The authors also calculated that in 20 years, this amount could reach 780 USD millions per year. This provides evidence of the economic contribution that this activity could have in Mexico if shark-seeing and diving are promoted in those areas, where these species can be found. Currently, shark tourism is carried out at 70 sites worldwide, mainly in 45 countries. Among the most popular places are Bahamas, Fiji, Costa Rica, South Africa, Indonesia, Philippines, Mexico and USA (Hawaii).

Other studies on shark-seeing and diving in specific sites estimate that the annual income related to this activity ranges from $18 to $152,341 million USD (Vianna et al. 2010, 2011; Cisneros et al. 2013; Haas et al. 2017; De Brauwer et al. 2017) (Table 2); therefore, this activity could have a relevant contribution to the economy of Mexico if this strategic sector were promoted as a priority for sustainable economic development (Fig. 2).

In Mexico, there are several sites were shark-seeing and diving are conducted; for example, the white shark (*Carcharodon carcharias*) can be found in Isla Guadalupe (Baja California Norte); the tiger shark (*Galeocerdo cuvier*) in Cabo Pulmo (Baja California Sur); the whale shark (*Rhincodon Typus*) in Isla Mujeres, YumBalam, Isla Contoy and Holbox (Quintana Roo), as well as in Bahia de Los Angeles (Baja California Norte) and La Paz (Baja California Sur); the bull shark (*Carcharinus leucas*) in Playa del Carmen (Quintana Roo), and the nurse shark (*Ginglymostoma cirratum*) in Cancun and the Riviera Maya (Quintana Roo). The

Fig. 2 Map of the sites for shark-seeing and diving worldwide. *Source* Based on Cisneros et al. (2013), Vianna et al. (2010), Vianna et al. (2011), Haas et al. (2017) and De Brauwer et al. (2017)

Fig. 3 Map of the locations where shark-seeing and diving take place in Mexico. *Source* Based on Velez (2009), INECC (2009), Escobar (2011) and Barrera (2013)

sites of sighting and diving with sharks in Mexico are described in Table 2, and the location of these sites is shown on Fig. 3 (Table 3).

In 2014, 20.35 thousand tons of fish were caught in Mexico; this volume represented 300.02 USD millions. For the same year, the National Gross Domestic Product (GDP) was 15.21 USD trillion, and the state of Baja California Sur, where

Table 3 Sites where shark-seeing and diving take place in Mexico

ID	Site	Species
1	Isla mujeres	Whale shark (Rhincodon typus)
2	Holbox	
3	Bahía de los Angeles	
4	La Paz	
5	YumBalam	
6	Isla Contoy	
7	Isla Guadalupe	White shark (Carcharodon carcharias)
8	Playa del Carmen	Bull shark (Carcharinus leucas)
9	Cabo Pulmo	Tiger shark (Galeocerdo cuvier)
10	Cancún	Nurse shark (Ginglymostoma cirratum)
11	Riviera Maya	
12	Magdalena Bay (Port San Carlos)	Hammerhead Shark (Sphyrna zygaena) Mako Shark (Isurus oxyrinchus) Silky Shark (Carcharinus falciformis) Blue Shark (Prionace glauca)

Source CONABIO (2016)

the study site is located, contributed to the total national GDP in 0.7%. Although Baja California Sur is not one of the most productive states in the country such as Mexico City, Monterrey and Guadalajara, it concentrates much of the terrestrial and marine biodiversity. The fishing sector for Baja California Sur generates 1.28% of the total state GDP, including artisanal and industrial fishing; this summed up 3.32 thousand tons of fish extracted per year (CONAPESCA 2014). From the total, shark fishing accounts for only 0.03% of the total state GDP. Although shark fishing contributes very little to the state's wealth generation, this activity has caused some species to be in danger of extinction. For this reason, it is essential to identify alternative strategies for the conservation of sharks that also represent a way of living for local populations such as ecotourism. If these populations do not guarantee their livelihoods, there will be no successful strategy for the conservation of any endangered species.

The tourism sector generates 16.61% of the total state GDP, being the gray-whale watching and diving two of the activities that contribute the most to the economic output of the sector. Nevertheless, this activity is only carried out from January to March every year. In 2014, revenues that resulted from gray-whale watching in three sites (San Ignacio Lagoon, Ojo de Liebre Lagoon and Banderas Bay) amounted to 5.52 USD millions; figure that represents 0.06% of the total state GDP (GBCS 2015a, b). Therefore, if more sighting sites were included not only for the gray whale but also for different shark species that can be found in Baja California Sur, such as the hammerhead shark, the shortfin mako shark and the blue shark, the income of tourism would increase significantly for the state.

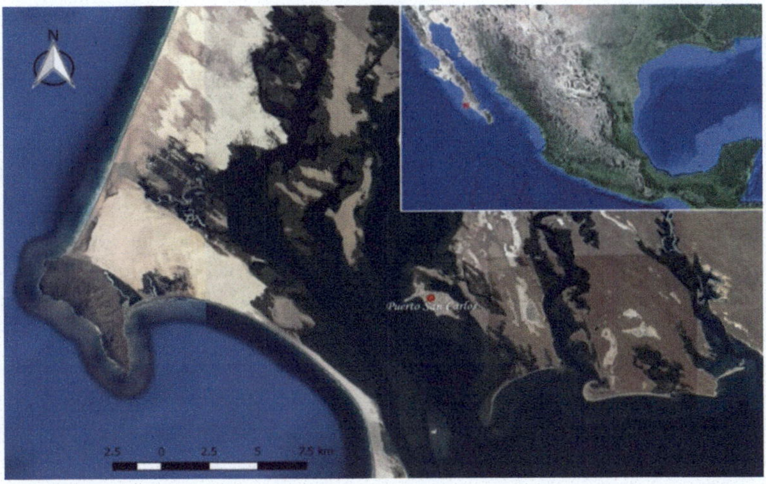

Fig. 4 Port San Carlos in Baja California Sur

The study area where this research was carried out is the Port San Carlos, in the municipality of Comundu, Baja California Sur. It is located in the maritime region called North Pacific and has an approximate of 6.5 thousand inhabitants (GBCS 2015a, b). Port San Carlos is a fishing port, and the species of sharks that have a commercial value in this area include: the blue shark (*Prionace glauca*), the silky shark (*Carcharinus falciformis*), the mako shark (*Isurus oxyrinchus*) and the hammerhead shark (*Sphyrna zigaena*) (Ojeda Ruiz de la Peña and Ramírez Rodríguez 2012). During winter, this port becomes a touristic destination due to the arrival of a large number of gray whales. Based on our results, shark conservation is more profitable for the state and municipality than fishing; since greater economic benefits could be obtained from shark-seeing and diving. The following section of the manuscript describe the methodology used to estimate the economic benefits that could be obtain from shark-seeing and diving compared to commercial fishing. Figure 4 shows the location of the study area.

2 Methodology

Ecosystem services can have different types of values and be classified in two main categories: use values and non-use values. While use values are based on the actual use of ecosystem goods and services, non-use values are related to intangible benefits provided by ecosystem services. Use values include direct and indirect uses. For example, trees provide direct use values, since their wood can be used to build furniture. Indirect use values can be exemplified by the ecological functions provided by ecosystems such as climate regulation, pollination, and photosynthesis.

Option value refers to the benefits that people receive if they have the option of enjoying some ecosystems service in the future, although they may not currently use them (Kolstad 2000; Field and Field 2002). For example, a person may hope to visit Greenland in order to see the northern lights sometime in the future, and thus would be willing to pay to preserve the area.

Non-use values refer to the value defined by people for knowing that future generations will have the option to enjoy the benefits of ecosystem services (bequest value) or the value that people define for knowing that something exist (existence value) (Kolstad 2000; Field and Field 2002). For example, a person is willing to pay to protect Greenland because he or she values the fact that future generations could enjoy it or the fact that it exists even though he or she never expects to go there. Since a person may benefit in several ways from the same ecosystem service, the total economic value is the sum of the benefits that result from both use and non-use values provided by an ecosystem service. The following sections explain the methods selected to assess use and non-use values from shark-seeing and diving compared to shark fishing for human consumption.

2.1 Economic Valuation Methods

The economic benefits generated by fishing and ecotourism were analyzed by using different economic valuation methods. Fishing and the sale of shark meat for human consumption were analyzed by using the method of the market prices. In the case of the assessment of the benefits provided by ecotourism with sharks, the contingent valuation method was utilized. To define the more profitable activity, comparing ecotourism with fishing, a cost-benefit analysis was conducted. Next, those methods are explained.

a. **Contingent valuation**

This method was used to assess the tourists' willingness to pay to carry out activities such as shark-seeing or diving. This method was based on an auction market simulation, being the starting price similar to the price payed by tourists for gray-whale watching. The ranges were defined as the following: (1) less than 53 USD; (2) between 53 and 106 USD; (3) between 106 and 159 USD, and (4) more than 159 USD. Tourist willingness to pay was identified by interviews with people who might be interested in carrying out shark-seeing or diving. The total number of interviews conducted in the study area accounted for 121. This method was chosen because one of the objectives of this research was to identify the potential willingness to pay of consumers. In addition, this method reduced the problem of value underestimation, since it considers both use values (direct and indirect) and non-use values (existence and bequest).

b. **Market prices**

It was used to determine the income resulting from the sale of shark meat, including the benefits received by anglers and intermediate and final sellers; therefore, it was identified the price per kilogram of shark meat. The total revenues were subtracted from the costs incurred by anglers to obtain shark meat (i.e. boat maintenance, gasoline, salaries), as well as the costs related to the distribution and marketing for the commercialization of this type of meat. This allowed us to know the net benefits of carrying out this activity. Interviews were conducted with shark fishermen, and intermediate and final sellers. To this end, four visits were made to two markets in the study area, where shark meat is sold: Bravo Market and Francisco I. Madero Market. This method was chosen due to the importance of estimating the net benefits of selling shark meat for human consumption.

c. **Cost-benefit analysis**

The cost-benefit analysis was used to determine the more profitable activity comparing whether the conservation of sharks through sustainable tourism represents higher benefits than fishing for human consumption. Bellow, we describe the criteria used to select key actors to interview.

2.2 *Identification of Key Stakeholders*

Key stakeholders were identified based on their potential to be involved in shark-seeing and diving (e.g. anglers, intermediate and final sellers), as well as based on their interest in ecotourism with sharks (e.g. whale watchers). These key actors were interviewed in order to generate the information required for estimating the benefits of shark-seeing and diving compared to shark fishing. Among the actors interviewed, there are tourists interested in the activity, service providers offering sustainable activities with sharks, NGOs focused on marine biodiversity protection, shark fishers, intermediate and final sellers, and hotels and restaurants owners in the area. A questionnaire was designed to interview each sector; actors interviewed were selected randomly.

a. **Questionnaires design**

Semi-structured questionnaires were designed. Each questionnaire contains a) common questions for all sectors and b) specific questions for each sector, based on their activities. In total, 121 persons were interviewed: 50 anglers, 39 fish vendors, 26 tourists, 34 service providers, 27 NGOs, 26 hotels, and 30 restaurants. The interviews were conducted in person during two shark-seeing and diving seasons: one in November 2016, and another in April 2017. The questionnaire included the following topics: (1) socioeconomic information, (2) income, (3) restrictions to carry out their activities (e.g. programs, permits), (4) Expenses (e.g. expenses to maintain boats, equipment, gasoline, salaries), and (5) interest in shark-seeing and diving. The questions asked in particular to each sector were the following:

a. 1 For fishermen: it was relevant to know the main species of shark exploited, those with a relevant commercial value, the distribution of shark meat, and prices per kilogram.
a. 2 For fish sellers: questions were focused on how shark meat is sold, species with commercial value, buyer preferences, main distribution points, and prices per kilogram.
a. 3 For tourists: it was important to identify their interest in carrying out shark-seeing and diving, and their willingness to pay to enjoy this experience.
a. 4 For services providers: questions were oriented to determine their opinion on their interest in offering and promoting ecotourism with sharks.

3 Results

Results of the interviews are analyzed in this section. It was identified that having access to education influences that people understand ecotourism with sharks as an alternative activity that can promote conservation without affecting the incomes of local population. In general, there is great interest in shark-seeing and diving by a large number of local people, as well as national and international tourists. Indeed, people are willing to pay a larger amount of money for doing this activity compared to gray-whale watching. Unfortunately, there is a denial of many anglers to participate in ecotourism for cultural reasons since they consider that being an angler is part of their cultural identity and a family tradition.

a. **Education as a variable that influence shark-seeing and diving interest**

Figure 5 shows the level of education of the interviewees by sector. The educational level that prevails in anglers and sellers of shark is secondary education. This is a relevant factor that influences their way of life and limits their capacities to carry out other economic activities for diversifying their income. Children in the study area are taught by their families to be anglers, so most of them have no

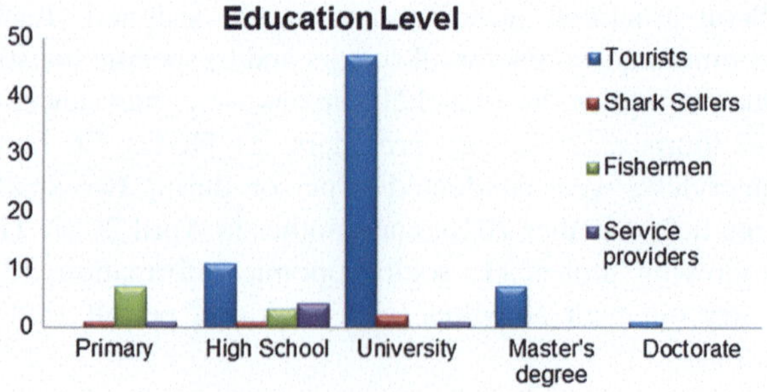

Fig. 5 Educational levels of key sectors

interest in studying or be involved in other economic activities. In the case of tourists, most have studied at the university, while the education level of service providers is high school. Low education levels among anglers and service providers have an influence in their abilities and knowledge to identify different species of sharks and the risks entailed for endangered species (De Ferranti 2002).

b. **Interest in the activity**

Among the tourists interviewed, there are people who enjoy carrying out activities such as sport fishing, whale-watching, diving, and hiking, among others. Some people interviewed in the study area (Port San Carlos, B.C.S) purchase a package known as "Water Safari" offered by the NGO "Pelagiclife". This package included the option of shark-seeing and diving. In general, tourists come from countries such as Mexico, France and the United States; most of the tourists interviewed came from Mexico City.

Figure 6 shows the great interest of tourists in shark-seeing and diving: 89% mentioned to be interested in sighting and swimming with sharks, and only 11% of respondents were not interested in this activity for fear of having an accident. Respondents interested in ecotourism with sharks represent a potential increase in tourism with sharks. These activities will help the protection of these species, in addition to generating economic benefits to local population that could be higher than those received by shark fishing. Shark fishing in the municipality of Comondú, BCS, where Port San Carlos and Port Adolfo Lopez Mateos are located, accounted for 41.08 USD millions. This figure represents 1.4% of the total GDP for the municipality (GBCS 2015a, b).

People interested in the ecotourism with sharks mentioned that they need more information about the different species of sharks that they could see, the places where the sighting and diving could be carried out, the best season to conduct the activity, potential risks of the activity, special requirements, and promotions. Figure 7a details the income level of tourists interviewed: 33% of the respondents earn less than 266 USD monthly; 29% between 267 and 534 USD; 14% between 535 and 802 USD; 1% between 803 and 1070 USD; finally, 23% receive income higher than 1071 USD per month. Figure 7b shows the willingness to pay for shark-seeing and diving: 46% is willing to pay more than 212 USD to conduct this activity; only 1% between 159 and 212 USD; 7% between 106 and 159 USD; 16% between 53 and 106 USD; 19% less than 53 USD; and 11% is not interested in doing the activity because they are afraid of being involved in an accident, so they would not pay anything. The fact that 44% of the interviewees are willing to pay

Fig. 6 Interest in shark free diving

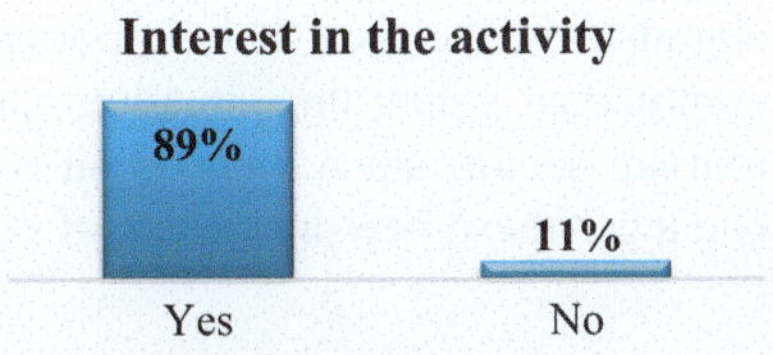

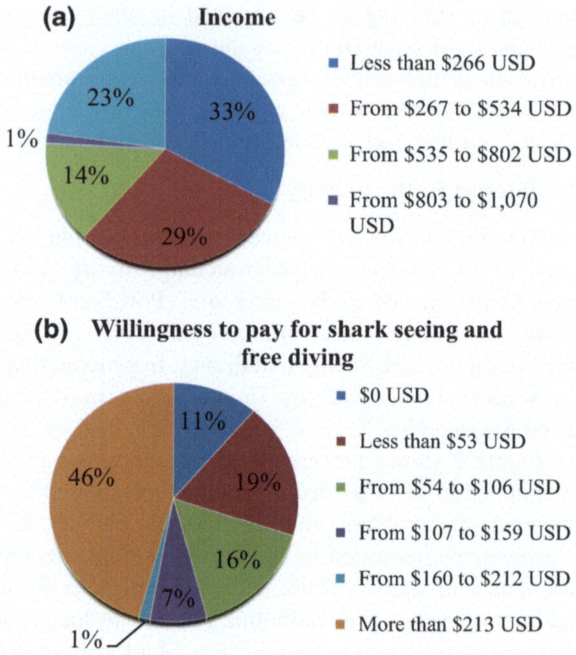

Fig. 7 **a** Interviewee's income and **b** Willingness to pay of tourists for shark-seeing and diving

less than 334 USD is explained by the restriction of their income. Despite this, only 20% of the total would be willing to pay less than the amount charged for whale watching, which can be considered a close substitute of shark-seeing and diving. In consequence, there is evidence that this activity has great economic potential, given that although most of the people interested in taking part of it have income of less than 265 USD per month, most of them are willing to pay more than 212 USD for enjoying this experience. This situation implies that they are willing to save the required amount of money for carrying out this activity. Indeed, people with higher income are interested in experiencing these activities with sharks more than once.

Figure 8 shows main species that tourists are interested in for sighting and diving. The most charismatic species are: (1) the white shark (*Carcharodon carcharias*), (2) the whale shark (*Rhincodon typus*), (3) the smooth hammerhead shark (*Sphyrna zigaena*), (4) the shortfin mako shark (*Isurus oxyrinchus*), and (5) the blue shark (*Prionace glauca*). In the Port San Carlos there are not white sharks or whale sharks, so these species would be discarded from the offer of shark-seeing and diving. Instead, tourists could carry out ecotourism with shark-seeing of smooth hammerhead shark, shortfin mako shark, and blue shark; species that are the most abundant in the area. It is important to emphasize that these species also have a commercial value; this situation has created conflicts between the fishing and the tourism sector, since there is an over-exploitation of sharks in the area that can endanger these new initiatives of ecotourism with sharks.

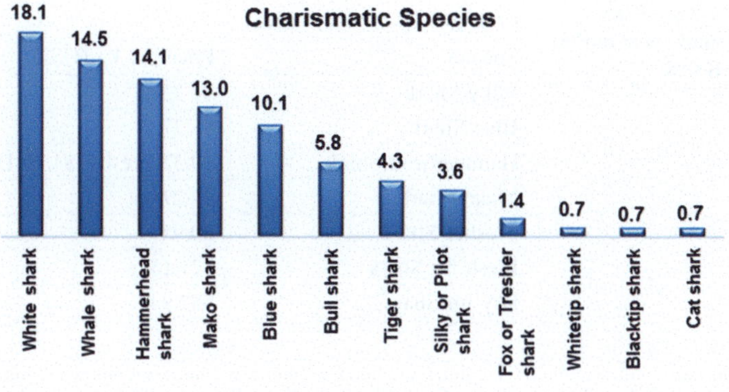

Fig. 8 Species of interest for sighting and diving

Fig. 9 Species of shark with commercial value

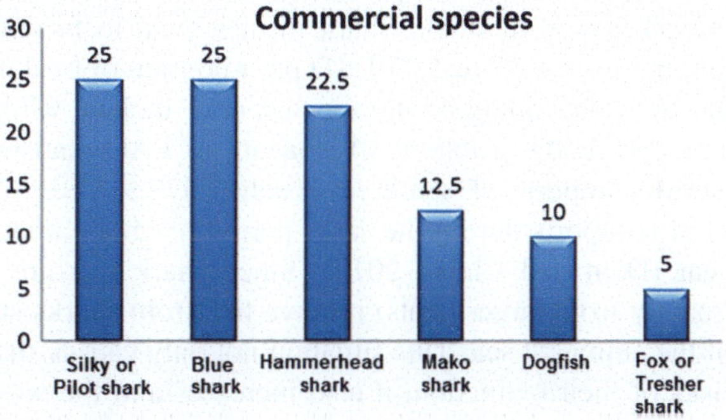

c. **Exploited species and their commercialization**

Figure 9 details the species of sharks that are caught in Port San Carlos and sold in different markets. These varieties include the hammerhead shark, the mako shark, the blue shark and the silky shark. The meat of these species is sold at national level, being commercialized in Ensenada (Baja California Norte), Guadalajara and Mexico City. In interviews conducted, some shark sellers only know that the meat they are selling is dogfish *(Squalus acanthias),* although they could be selling other species of sharks.

Sharks caught measure less than 1.5 m and weigh less than 5 kg (Dent and Clarke 2015). Based on their measures, shark sellers called them cazon, meaning that they could be "juveniles" of different species. This situation provides evidence of the over-exploitation of different shark species (DOF 2007b), many of which are protected in the CITES list. Some of these endangered species such as the smooth hammerhead (*Sphyrna zigaena*) and the silky shark (*Carcharinus falciformis*) have

Table 4 Prices paid per kilogram of shark meat in Port San Carlos, B.C.S

Price per kg of shark meat	
Species	Price per kg (USD)
Silky Shark	\$1.32
Blue Shark	\$1.27
Hammerhead Shark	\$0.37 Fresh—\$4.76 Dry
Mako Shark	\$1.27
Tresher Shark	\$0.95
Fresh fin Shark	\$21.18
Dry fin Shark	\$52.94

also a commercial value in Port San Carlos; they are caught, even though they should be protected under this norm (UNEP-WCMC 2014).

The process that anglers follow is described below. They arrive to the beach, eviscerate sharks and separate the fins for their individual sale and the rest of the animal is sold in stock. Shark meat is paid to fishers at an excessively low price ranging from 0.37 to 1.27 USD per kilogram of fresh meat, and up to 4.76 USD for dry meat depending on the species; nevertheless, while the fin is sold between 1.17 and 2.94 USD (Table 4). According to FAO statistics, between 2000 and 2011, world's imports of shark fin reaches 377.9 USD million per year, while shark meat's imports during the same period reach an amount of 239.9 USD million per year (Dent and Clarke 2015). Since shark fins are priced higher, some anglers (mainly from larger boats) remove fins from sharks still alive and abandon the rest of the animal at sea. This situation not only causes sharks over-exploitation and the risks of their extinction, it also increases and the waste of rest of the meat of this animal that can also be used for human or animal consumption.

Based on the interviews conducted, it was determined that there is a number of anglers that caught sharks in Port San Carlos who do not have a fishing license; indeed, some anglers use illegal documents to fish sharks. In addition to this problem, which explains the over-exploitation of diverse shark species, many anglers who fish for other animals are able to extract sharks sporadically, and those interested in swordfish fishing, illegally extract sharks to obtain shark fins. The lack of regulation has increased the number of extracted organisms and how the meat is processed. Another problem is related to the increase in competition among fishermen for shark fishing among traditional anglers and larger vessels. It is observed that there is no proper control or management of shark debris, since anglers eviscerate sharks at the edge of their camps; the remains of some of these organisms are left on the beach and, later, washed away by the seawater (Fig. 10). There is not a control on who can fish, or the amount of sharks that are allowed to be fished, or on watching that the forbidden species to be exploited do not be fished, since they are at risk of extinction. Vessels from other places such as Mazatlán, Sinaloa, and United States also fish in Port San Carlos illegally, according to the information provided by the interviewees.

Fig. 10 **a** Bird devouring shark remains. **b** Shark heads on the beach. Photo by Y. M. Plata Zepeda

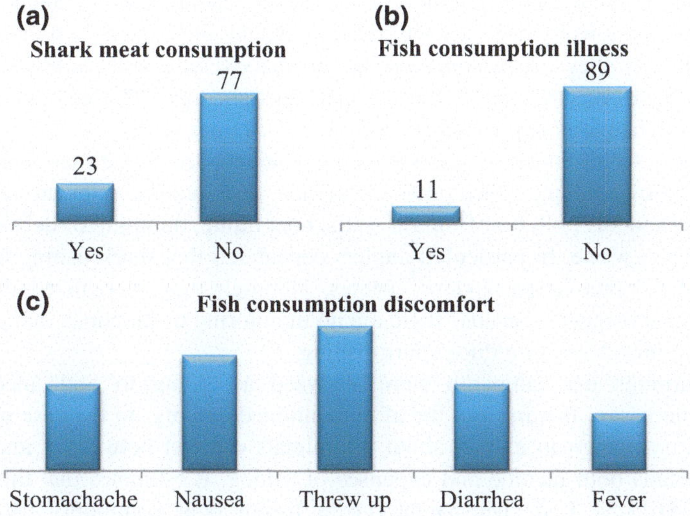

Fig. 11 **a** Shark meat consumption. **b** Illness caused by fish consumption. **c** Symptoms from consumption of polluted fish meat

Figure 11a describes the consumption of shark meat according to interviews with key actors in Port San Carlos. Few people responded that they consume shark meat, or if they do it, it has a very low frequency. Nevertheless, shark meat is mainly sold dry, although there is also a market for fresh shark meat. Anglers recognized that shark meat is sent to Mexico City where it is sold as other species such as cod. It has been found that in some places where fish is sold, people are not aware that they are consuming shark meat, since they think that they are buying other fish species (Bornatowski et al. 2013).

Figure 11b analyzes potential diseases caused by fish consumption according to respondents. Most of the interviewees have not got sick due to fish consumption, which could be shark meat. However, some mentioned that they got sick and present the following symptoms: vomit, nausea, stomachache, diarrhea and even fever (Fig. 11c). If shark meat is contaminated with methylmercury, its consumption can cause neurotoxic effects such as tremors, limb numbness, sensory disturbances, and neurons' degeneration, among others. Samples of shark (fin and muscle) taken in Port San Carlos are being analyzed to identify the presence and concentration of mercury in order to determine if their consumption could potentially represent a health risk.

Of the total anglers interviewed, 78% would be willing to stop fishing and engage in shark tourism if they receive the appropriate training to be successful in carrying out this activity, the remaining 22% mentioned that they were not interested in changing their way of life due to insufficient promotion of shark tourism in the area. This group believes that the demand would not be sufficient to provide enough income to all anglers. Nevertheless, tourism benefits other sectors such as transportation, hotels, and restaurants. It is also expected that the maintenance costs of boats will be lower since they will need to change the engines' oil every two or three weeks, if they are to be engaged in shark-seeing and diving instead of changing them every week, as they have to do right now. The amount spent per engine's oil change is approximately of 44.45 USD per week.

According to all analyzed interviews, respondents are aware that sharks are at risk due to overfishing. They also mentioned that another relevant factor that endangers sharks is pollution from oil spills, boat traffic, dumping of debris into the sea, and water waste. In particular, anglers considered that shark-fishing ban could be helpful for recovering diverse species, although they recognized that many anglers do not respect it because there are no monitoring or sanctions that create the appropriate incentives to reduce shark fishing.

The earnings per campaign were analyzed according to both income and expenses necessary to carry out the aforementioned activity. In the case of fishing, one single campaign to extract about 40 animals can last between 1 and 3 days. Table 5 details both income and expenses of fishing. It estimates that this activity generates $611.54 USD per panga (i.e. a fishing boat commonly used in the developing world). The main economic activity in the study area is fishing. There are three shark-fishing camps: "Punta Arena", "San Lázaro", and "Magdalena". In each fishing camp, from 5 and 8 pangas work formally, except in "Magdalena". From 5 and 8 pangas go out fishing in this camp during the low season, whereas in the high season (August, September and October) this figure reaches from 20 to 25 pangas. Consequently, more than 58 pangas could be fishing around 2320 animals in the high season. This situation shows evidence of the over-exploitation of fishery resources and lack of effective policies for shark species' conservation. In the case of tourism with sharks, an activity carried out by NGOs working in Port San Carlos, the campaign lasts 3 days, reaching a profit that ranges from $1,548.93 to $2,756.30 USD. Therefore, ecotourism with sharks is a more profitable activity with greater social, environmental and economic benefits (Table 6; Fig. 12).

Table 5 Revenues per fishing season (1–3 days)

Concept	Amount US dollars (USD)
Fishing	2.29
Total kg	34.31
Total income	686.16
Gasoline	55.75
Oil change	9.15
Equipment	2.86
Engine repair	6.86
Total expenses	70.50
Total revenues	611.54

Table 6 Revenues per ecotourism campaign (3 days)

Concept	Amount US dollars (USD)	
Participants	4	8
Total income	2,744.63	5,489.26
Boat (3 days)	1,115.01	1,115.01
Transportation and gasoline	514.62	1,029.24
Hotel	291.62	583.24
Breakfast	102.92	205.84
Dinner	171.54	343.08
Total expenses	1,195.71	3,191.87
Total revenues	1,548.93	2,297.39

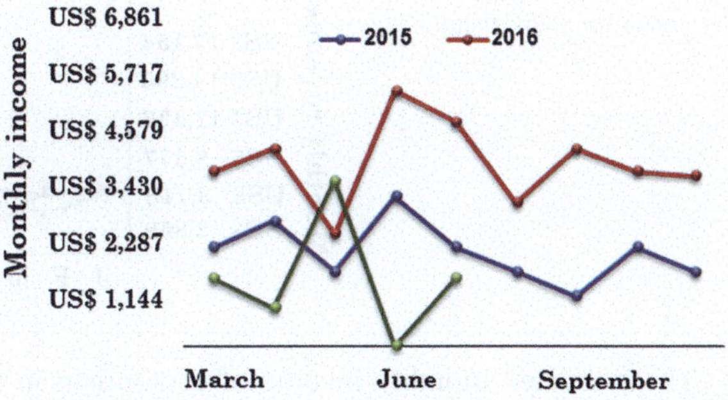

Fig. 12 Monthly income for shark sighting in Port San Carlos

Sport fishing in Port San Carlos is carried out throughout the year, while whale-watching takes place from January to March. Tourists who take part on these activities may also be interested in shark-seeing or in swimming with sharks—activities that could foster the conservation of these species. Other touristic attractions in the area include mangroves, sand dunes (Fig. 13c), and wild animals observation—mainly birds, sea lions (Fig. 13b), whales (Fig. 13a), and swordfish.

Fig. 13 Ecotourism offer in Port San Carlos **a** Orcas, **b** sea lions, **c** mangrove area and sand dunes

Table 7 Annual ecotourism revenues in Port San Carlos

Concept	Total 2015	Total 2016
Participants	67	118
Total income per participant	42,141.53	80,966.63
Boat (3 days)	10,378.14	20,584.74
Transportation	5,031.82	9,148.77
Gasoline	628.98	1,143.60
Hotel	3,679.52	7,101.73
Breakfast and dinner	4,281.62	6,340.10
Total expenses	24,000.09	45,554.02
Total revenues	18,141.44	35,412.61

Fig. 14 Monthly revenues per panga for shark fishing

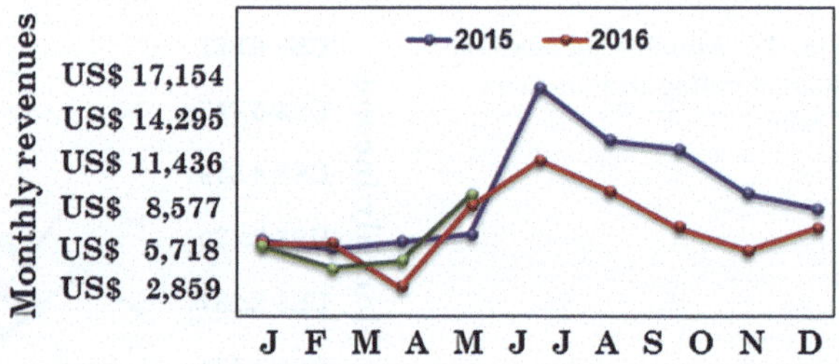

The estimated monthly income of ecotourism in Port San Carlos has doubled from 2015 to 2016, because the number of tourists interested in shark-seeing and diving has also increased. Table 7 shows the income received by the different sectors related to tourism (including hotels, restaurants, and transportation from the airport of La Paz to Porto San Carlos). We want to emphasize that shark-seeing supports the local economic development of many actors.

On the other hand, the estimated monthly income related to shark fishing in Port San Carlos has decreased from 2015 to 2017 in 50,465.77 USD as a result of the decline in fishery stock (Fig. 14). Besides, expenses related to the boats'

Table 8 Annual revenues from shark fishing in Port San Carlos

Concept	Total 2015	Total 2016	Total 2017
Fisheries (number of organisms)	3,588	2,796	1,080
Total (kg)	71,760	55,920	21,600
Total income	82,064.48	63,949.91	24,701.68
Miles traveled	14.70	13.89	6.86
Gasoline	10,622.30	10,806.99	5,146.18
Oil change	1,646.78	1,646.78	731.90
Equipment	514.62	514.62	228.72
Engine repair	437.43	463.16	217.28
Total expenses	231,220	234,900	110,600
Total revenues	68,843.36	50,518.37	18,377.59

Fig. 15 Sharks caught by a panga

maintenance are high since pangas require their oil to be changed once per week, as well as motor repairs to be performed every two years (Table 8) (Fig. 15).

In order to adapt fishing boats so they are suitable for conducting touristic activities, it is necessary to fulfill several requirements in order to have access to a safety certificate and nautical permit. The cost of these adjustments and permits is estimated in $2,406.70 USD (Table 9); nevertheless, this amount exceeds the economic budget of fishermen to perform other activities such as ecotourism with sharks.

Comparing the expenses and profits from the aforementioned activities, fishermen can obtain higher revenues in ecotourism than in fishing. In the case of tourism in Port San Carlos, sport fishing is carried out throughout the year; whale-watching is conducted from January to March. People who perform these activities may be interested in shark-seeing and diving, thus contributing to sharks' conservation.

Table 9 Expenses for fishing boats adjustments to conduct tourism

Concept	Expenses US dollars (USD)
Shadow for the boat	514.62
Life jackets (14)	228.72
Seats	200.13
Carpet for the boat	257.31
BHF Radio	400.26
First aid kit	14.29
GPS	285.90
Anchor	57.18
Hand flares	34.31
Tools	17.15
Oars	45.74
Flashlight	11.44
Security certificate	8.01
Nautical permit (tourism)	102.92
Insurance (captain and boat)	228.72
Total	2406.70

4 Conclusions

The low education level of anglers has a strong influence on their lives; since they are children, their families teach them how to fish. They need to help in these activities to support the family income. In consequence, some of them consider fishing as part of their traditional way of life; they do not have interest in keeping on studying or carrying out a different economic activity. Some of them are interested in knowing more about the resources they exploit, as well as the endangered species. The tourists' education level also influences strongly their interest in the environment and their curiosity to know more about marine charismatic species including sharks, whales and other species.

Our research identified that anglers receive a quite low payment per kilogram of shark meat (1.32 USD per kg) compared to the price of shark fin, which is almost 20 times higher (21.18 USD). This situation has created the wrong incentives for encouraging illegal exploitation of sharks. Not all anglers fish shark illegally; however, there are large boats that indiscriminately catch sharks for their fins, and waste the rest of the animal, which is left in the beach or thrown in the sea. This is evidence of the lack of regulation of the vessels that fish in Port San Carlos. In addition, not all of them are local; many vessels come from other places such as Sinaloa and the United States of America. In general, fishing practices, specificities of the species caught, and the volume of animals extracted are not regulated nor monitored. In consequence, some anglers that fish sharks illegally catch them during banned fishing seasons. Finally, there is a mismanagement of waste by some anglers, which increases pollution in the area.

To improve not only shark conservation, but other species too, it is essential to increase monitoring and improve regulation in the sector, as well as to provide information and create awareness about the species at risk, the importance of the banned fishing seasons, the differences among species, and potential alternatives of living, such as ecotourism with charismatic species (e.g. sharks species).

Results provide evidence that there is a growing interest in shark-seeing and diving. The willingness to pay for the activity is high compared to the amount paid for other ecotourism activities such as whale watching. Therefore, tourism with sharks has a high potential to become the main economic activity in Port San Carlos, apart from representing an innovative way of promoting the conservation of shark species. Ecotourism with sharks could generate great benefits for local populations compared to the income received by fishing. Moreover, the consumption of shark meat that may contain high levels of methylmercury could endanger people′s health. If ecotourism with sharks is promoted in Port San Carlos, it is important that local populations get involved, and appropriate knowledge and abilities be built, for otherwise there will be few beneficiaries of changing economic activities from fishing to ecotourism.

Port San Carlos has a great potential for developing different activities based on ecotourism. Gray whales are in the area from January to March, but different species (e.g., mako shark, blue shark, hammerhead shark) that cannot be seen in other places are there all the year around. Besides, there are mangroves in the area as well as dunes, and a large number of birds—all of which could awaken the interest of tourists. However, there is a need for improving infrastructure and encouraging local groups to organize cooperatives in order to engage in ecotourism. It is equally important to have proper regulation and monitoring tools in order to create the right conditions for a sustainable development in the area.

Profits per campaign in both activities are similar; however, fishing can endanger shark′s survival since current fishing practices are not sustainable. On the other hand, profits from tourism with sharks are higher than fishing; promoting tourism can foster local economic development in addition to increasing the conservation of sharks. However, potential financial funding must be found in order to switch from one activity to the other. Fishermen do not have enough economic resources to obtain permits and to adapt their boats.

References

Barrera García AM (2013) Elementos traza e indicadores de estrés oxidativo en el Tiburón Azul (Prionace glauca) de la costa occidental de Baja California Sur, México [Trace elements and indicators of oxidative stress in the Blue Shark (Prionace glauca) of the western coast of Baja California Sur, Mexico]. Doctorate Dissertation in Science. Center of Biological Investigations of the Northwest, Baja California Sur, p 98

Bornatowski H, Braga RR, Vitule JRS (2013) Shark mislabeling threatens biodiversity. Science 340(6135):923

Cárdenas Torres AN (2006) Valoración económica de la actividad recreativa con tiburón ballena y su relación con la calidad del hábitat en Bahía de los Ángeles, Baja California [Economic valuation of recreational activities with whale shark and its relation with the quality of life of the inhabitants of Bay of the Angels, Baja California]. Master Dissertation. Autonomous University of Baja California, Ensenada, p 66

Cisneros MAM, Barnes MM, AlAbdulrazzak D, Navarro HE, Rashid US (2013) Global economic value of shark ecotourism: implications for conservation. Oryx. https://doi.org/10.1017/S0030605312001718

Comisión Nacional de Acuacultura y Pesca [National Commission of Aquaculture and Fisheries] (CONAPESCA) (2014) Base de datos de producción anuario 2014 [Production database. Yearbook 2014]. CONAPESCA, Mexico

Comisión Nacional para el Conocimiento y Uso de la Biodiversidad [Commission for the Knowledge and Use of Biodiversity] (CONABIO) (2016) Listado de especies de tiburones presentes en México [List of shark species present in Mexico], CONABIO, Mexico.

De Brauwer M, Harvey E, McIlwain J, Hobbs J, Jompa J, Burton M (2017) The economic contribution of the muck dive industry to tourism in Southeast Asia. Marine Policy 83:92–99

De Ferranti DM (2002) De los recursos naturales a la economía del conocimiento: comercio y calidad del empleo [From natural resources to the knowledge economy: trade and quality of employment]. World Bank, Washington, DC

Del Moral-Flores LF, Morrone JJ, Alcocer J, Ponce de León GP (2016) Diversidad y afinidades biogeográficas de los tiburones, rayas y quimeras (Chondrichthyes: Elasmobranchii, Holocephali) de México [Diversity and biogeographical affinities of sharks, rays and chimeras (Chondrichthyes: Elasmobranchii, Holocephali) of Mexico]. Int J Trop Biol 64(4):1469–1486

Dent F, Clarke S (2015) State of the global market for shark products. FAO Fisheries and Aquaculture Technical Paper No. 590. FAO, Rome, p 187

Deutsche Gesellschaft für Internationale Zusammenarbeit (GIZ) (2012) Integración de los servicios ecosistémicos en la planificación del desarrollo: Un enfoque sistemático en pasos para profesionales basado en TEEB [Integrating Ecosystem Services into Development Planning: A Systematic Approach for Professionals Based on TEEB]. Department of Environment and Climate Change, GIZ, Berlin, Germany

Diario Oficial de la Federación [Official Journal of the Federation] (DOF) (2007a) Norma Oficial Mexicana [Mexican Official Standard]. NOM-064-Pesc-2006 sobre Sistemas, Métodos y Técnicas de Captura Prohibidos en la Pesca en Aguas de Jurisdicción Federal de los Estados Unidos Mexicanos [NOM-064-Pesc-2006 on Systems, Methods and Capture Techniques Prohibited in Fishing in Mexican Waters]. Available online in: http://www.ordenjuridico.gob.mx/Publicaciones/CDs2011/CDPesca/pdf/PNM04.pdf

Diario Oficial de la Federación [Official Journal of the Federation] (DOF) (2007b) Norma Oficial Mexicana [Mexican Official Standard]. NOM-029-PESC-2006 Pesca responsable de tiburones y rayas. Especificaciones para su aprovechamiento [NOM-029-PESC-2006 for responsible fishing for sharks and rays. Specifications for use]. Available online in: http://www.diariooficial.gvamundial.com.mx/historico/2007/Febrero/sagarpya14_1.pdf

Diario Oficial de la Federación [Official Journal of the Federation] (DOF) (2011a) Norma Oficial Mexicana [Mexican Official Standard]. NOM-059-SEMARNAT-2010 Protección ambiental-Especies nativas de México de flora y fauna silvestres. Categorías de riesgo y especificaciones para su inclusión, exclusión o cambio. Lista de especies en riesgo [NOM-059-SEMARNAT-2010 for environmental Protection for Mexican Native Species of Wild Flora and Fauna. Risk Categories and Specifications for Inclusion, Exclusion or Change. List of Species at Risk]. Available online in: http://dof.gob.mx/nota_detalle_popup.php?codigo=5173091

Diario Oficial de la Federación [Official Journal of the Federation] (DOF) (2011b) Norma Oficial Mexicana [Mexican Official Standard]. NOM-242-SSA1-2009 Productos y servicios. Productos de la pesca frescos, refrigerados, congelados y procesados. Especificaciones sanitarias y métodos de prueba [NOM-242-SSA1-2009 for products and services. Fresh fishery products, chilled, frozen and processed. Health specifications and test methods]. Available online in: http://www.dof.gob.mx/nota_detalle.php?codigo=5177531&fecha=10/02/2011

Escobar Sánchez O (2011) Bioacumulación y Biomagnificación de mercurio y selenio en peces pelágicos mayores de la Costa Occidental de Baja California Sur, México. National Polytechnic Institute. Interdisciplinary Center of Marine Sciences. Available online in: http://www.repositoriodigital.ipn.mx/handle/123456789/16358

Field BC, Field MK (2002) Chapter 5. The economic analysis of environmental quality. Environmental economics. McGraw Hill, Madrid, pp 93–118

García JR, y Díaz MM (2007) Los indicadores de sostenibilidad en el turismo [The sustainability indicators in tourism]. RESTMA 6 (2007):27–62

Gobierno del Estado de Baja California Sur [Government of the State of Baja California Sur] (GBCSa) (2015a) Baja California Sur: Información estratégica [Baja California Sur: Strategic Information]. Secretariat of Promotion and Economic Development, Directorate of Information Technology and Statistics. GBCS, La Paz. p 52

Gobierno del Estado de Baja California Sur [Government of the State of Baja California Sur] (GBCSb) (2015b) Comondú: Información Estratégica [Comondú: Strategic Information]. Secretary of Economic Development, Environment and Natural Resources Directorate of Informatics and Statistics. GBCS, La Paz. p 31

Guzmán Mar JL (2009) Análisis de especies de Selenio y Mercurio en muestras de pescado [Analysis of Selenium and Mercury species in fish samples]. UANL, Monterrey. Available online in: http://respyn.uanl.mx/especiales/2010/ee-01-2010/documentos/trabajos_libres/TL06.pdf

Haas RA, Fedler T, Brooks EJ (2017) The contemporary economic value of elasmobranchs in The Bahamas: Reaping the rewards of 25 years of stewardship and conservation. Biological Conservation 207

Health Canada (2007) Human Health Risk Assessment of Mercury in Fish and Health Benefits of Fish Consumption. Health Canada, Ottawa. ISBN: 978-0-662-47023-6

Hoyos-Padilla EM, Ketchum JT, Klimley AP, Galván-Magaña F (2014) Ontogenetic migration of a female scalloped hammerhead shark Sphyrna lewini in the Gulf of California. Anim Biotelemetry (2):17

Instituto Nacional de Acuacultura y Pesca [National Institute of Aquaculture and Fisheries] (INAPESCA) (2016) El aprovechamiento del tiburón en México [The use of sharks in Mexico]. INAPESCA, Mexico. Available online in: http://www.gob.mx/inapesca/articulos/el-tiburon

Instituto Nacional de Ecología y Cambio Climático [National Institute of Ecology and Climate Change] (INECC) (2009) Selección y muestreo de peces de alto consumo humano en México con objeto de determinar el contenido de mercurio (Hg) [Selection and sampling of fish for human consumption in Mexico in order to determine mercury (Hg) concentration]. INECC, Mexico. p 121

Instituto Nacional de Ecología y Cambio Climático [National Institute of Ecology and Climate Change] (INECC) (2010) Selección y muestreo de peces de alto consumo humano en México con objeto de determinar el contenido de mercurio (Hg) [Selection and sampling of fish for human consumption in Mexico in order to determine mercury (Hg) concentration]. INECC, Mexico. 64 pp

Klimley AP (2015) Shark Trails of the Eastern Pacific. American Scientist the magazine of Sigma Xi, The Scientific Research Society. Vol. 103

Kolstad C (2000) Chapter 4. Efficiency and markets. Environmental economics. Oxford University Press, New York, pp 58–63, 72–74

Llop S, Ibarlucea J, Sunyer J, Ballester F (2013) Estado actual sobre la exposición alimentaria al mercurio durante el embarazo y la infancia, y recomendaciones en salud pública [Current status on dietary exposure to mercury during pregnancy and childhood, and recommendations on public health]. Gaceta Sanitaria, 27(3):273–278. Available online in: http://doi.org/10.1016/j.gaceta.2012.09.002

Mazaba Lara JJ (2015) Determinación de Plomo (Pb) y Mercurio (Hg) en aletas e hígados de Tiburón Martillo (Sphyrna lewini), provenientes de Chachalacas, Veracruz, México

[Determination of Lead (Pb) and Mercury (Hg) in fins and livers of Hammerhead Shark (Sphyrna lewini) from Chachalacas, Veracruz, Mexico]. Thesis Dissertation. Faculty of Biology. University of Veeracruz, Xalapa

Medina C, Aravena C, Vásquez F (2012) Valoración económica de la conservación de tiburones en la Reserva Marina de Galápagos [Economic valuation of shark conservation in the Galapagos Marine Reserve]. Latin American and Caribbean Environmental Economics Program (LACEEP)

Millennium Ecosystem Assessment (MEA) (2005) Ecosystems and human well-being: current state & trends assessment. Island Press, Washington, DC

Montenegro R (2007) Valoración Económica de los Recursos Turísticos y Pesqueros del Parque Nacional Coiba [Economic Assessment of the Tourist and Fishing Resources of the Coiba National Park]. Conservation Strategy Fund, Panama

Navia AF (2013) Función ecológica de tiburones y rayas en un ecosistema costero tropical del Pacífico colombiano [Ecological function of sharks and rays in a tropical coastal ecosystem of the Colombian Pacific]. PhD Thesis Dissertation in Marine Sciences, National Polytechnic Institute, La Paz, p 190

Ojeda Ruiz de la Peña, MA, Ramírez Rodríguez M (2012) Interacciones de pesquerías ribereñas en Bahía Magdalena-Almejas, Baja California Sur [Interactions of riparian fisheries in Magdalena-Almejas Bay, Baja California Sur: Region and society]. Region Society 24 (53):189–204

Raimann X, Rodríguez OL, Chávez P, Torrejón C (2014) Mercurio en pescados y su importancia en la salud [Mercury in fish and its importance in health]. Rev Med Chile 142:1174–1180

Serra VJ, Malky A, Reid J (2012) Costos y beneficios del proyecto hidroeléctrico del río Inambari: Conservación estratégica [Costs and benefits of the Inambari river hydroelectric project: Strategic conservatio]. Available online in: http://conservation-strategy.org/sites/default/files/field-file/ACB_INAMBARI_FINAL_abril2012_0.pdf

Skomal GB, Hoyos-Padilla EM, Kukulya A, Stokey R (2015) Subsurface observations of white shark Carcharodon carcharias predatory behaviour using an autonomous underwater vehicle. J Fish Biol 87:1293–1312

UNEP-WCMC (Comps.) (2014) Checklist of CITES species. CITES Secretariat, Geneva, Switzerland, and UNEP-WCMC, Cambridge, United Kingdom, p 1268

Van Damme PA, Carvajal-Vallejos FM, Molina Carpio J (2011) Los peces y delfines de la Amazonía boliviana: hábitats, potencialidades y amenazas [Fish and dolphins in the Bolivian Amazon: habitats, potentialities and threats]. Edit. INIA, Cochabamba, Bolivia, p 490

Velez Álvarez M (2009) Indicadores de estrés oxidativo relacionados con la presencia de elementos traza (Plomo, Cadmio, Mercurio y Arsénico), en diferentes tejidos de Tiburón Mako (Isurus oxyrinchus) [Indicators of oxidative stress related to the presence of trace elements (Lead, Cadmium, Mercury and Arsenic), in different tissues of Mako Shark (Isurus oxyrinchus)]. Master Dissertation. Center of Biological Investigations of the Northwest, Baja California Sur, p 86

Vianna GMS, Meekan MG, Pannell D, Marsh S, Meeuwig J (2010) Wanted Dead or Alive? The relative value of reef sharks as a fishery and an ecotourism asset in Palau. Australian Institute of Marine Science and University of Western Australia, Perth

Vianna GMS, JJ Meeuwig, D Pannell, H Sykes y MG Meekan (2011) The socio-economic value of the shark-diving industry in Fiji. Australian Institute of Marine Science. University of Western Australia, Perth, p 26

Part IV
Bioeconomy: Advances on Agriculture, Biofarming and Food Production

Potential for Sustainable Urban Food Production in a Medium Scale City in Germany

Kay Plat, Andreas Meyer, Petra Schneider and Kai Perret

Abstract In Germany, the percentage of sustainable food production like organic farming increased in the last decades, but is generally still low in comparison to conventional farming. Organic farming provides an approach to increase the sustainability potential of the food supply in the frame of bioeconomy. The assessment of the potential of sustainable urban food supply was in the focus of the present investigation, having as investigation scale the city of Magdeburg in Saxony-Anhalt, Germany. Scope of the investigation was the feasibility assessment of implementation options as well as the perception of the consumers. Following systems were investigated for Magdeburg: urban farming, vertical farming and aquaponics. In terms of the use of urban spaces were considered roof farming, land recycling, as well as the refurbishment of former farms and greenhouses. The feasibility analysis was supported by an option analysis regarding the potential for the use of renewable energies. The results show that even the consumers are willing to pay for organic food, but there are too few sustainable, organic and local urban food products on the market yet. The needed energy for modern urban farming projects is still high in climate areas like the north of Germany. New technologies and the assessment of renewable energy source potential for urban food production is a site specific decision, which might ensure the support of the operational cost of the urban food production systems.

K. Plat · A. Meyer · P. Schneider (✉)
Hochschule Magdeburg-Stendal (FH), University of Applied Sciences
Magdeburg-Stendal, Breitscheidstraße 2, 39114 Magdeburg, Germany
e-mail: petra.schneider@hs-magdeburg.de

K. Plat
e-mail: kay.plat@gmx.de

A. Meyer
e-mail: andreas.meyer@hs-magdeburg.de

K. Perret
Zoologischer Garten Magdeburg gGmbH, Zooallee 1, 39124 Magdeburg, Germany
e-mail: perret.zoomd@icloud.com

W. Leal Filho et al. (eds.), *Towards a Sustainable Bioeconomy: Principles, Challenges and Perspectives*, World Sustainability Series,
https://doi.org/10.1007/978-3-319-73028-8_13

keywords Sustainable urban food production · Urban agriculture
Urban gardening

1 Introduction

Food is a basic necessity of every human being. For this reason, food production and supply is of great importance for the energy supply of every human being. The direct proximity to the consumer and thus direct marketing of the cultivated food within a city has made urban agriculture an integral part of many cities for generations (Gehrke 2012). Urban agriculture has been operating as long as people and cities exist. In the beginnings of urban agriculture food was usually cultivated by the inhabitants of the town in small gardens. Gradually the anchoring methods continued to develop.

There is no clear definition for the current concept of "urban farming". However, some authors have had different understandings to conceptually approach this term. Often, in the literature, there are different approaches for the term "urban farming", it is widely understood to mean any form of agricultural activity within cities and their peripheral areas (Schulz et al. 2013). These include all forms, from food production to animal husbandry to the marketing of non-food products. The size also ranges from shrine gardens to large commercial areas (Schulz et al. 2013). From scientific point of view, Lohrberg (2010) distinguishes between the "actor-oriented approach" and the "space-oriented approach" (Lohrberg 2010). The actor-oriented approach is mainly related to the benefit of the actor (gardener). This approach has mostly personal or non-profit motives and is an important part of many households, especially in developing countries. The actor-oriented approach is referred to as urban gardening and serves above all to improve the quality of life of the individual actor or community (Lohrberg 2010). The space-oriented approach considers the available areas of a city in which commercial agriculture can be operated. These include all economic agricultural uses of urban and peri-urban areas. These uses have both social, economic and ecological motivations (sustainability) and should have a positive impact on the entire urban development (Lohrberg 2010).

In Table 1 according to Lohrberg and Timpe (2011) both forms (urban agriculture and urban gardening) are compared and characteristic properties are presented.

Table 1 Differentiation of urban agriculture and urban gardening according to Lohrberg and Timpe (2011)

Urban agriculture	Urban gardening
• Market-oriented • Professional • Specialized • Adaptable • Land based • Low media effect • Sustainable	• Subsistence-oriented (self-use) • Civically • Quality of life-oriented territoriality • Neighbourhood or neighbourhood reference • Large public awareness (media)

According to the definition of Lohrberg and Timpe (2011), urban farming stands above urban agriculture and urban gardening. It combines all agricultural forms of urban development. To ensure adequate food security in the future, cities with integrated sustainable agriculture could be the solution to ensure the food security of the growing world population (FAO 2009). Germany is currently not affected as much by a food shortage as comparatively developing countries. Nevertheless, sustainable processes and production methods are also indispensable in highly developed countries such as Germany in order to continue to produce food in the future. The cornerstone of sustainability according to the Brundtland Commission (World Commission on Environment and Development 1987) illustrates how the framework conditions for sustainability are and what factors play a role in making a measure sustainable. The Federal Ministry for Economic Cooperation and Development (BMZ) gives information on the definition and sustainability of agricultural production. The criteria for sustainable agriculture are listed below (Ministry for Economic Cooperation and Development 2015):

- Focuses on methods and procedures that improve soil productivity while minimizing the adverse impact on the climate, soil, water, air and biodiversity as well as human health.
- Aims to use as little as possible non-renewable and petroleum-based equipment and replace it with renewable resources.
- Focuses on the local population with its needs, knowledge, skills and socio-cultural values and institutional structures.
- Ensures that the basic needs of food and agricultural raw materials of today and future generations are met in a qualitative and quantitative manner.
- Ensures long-term employment, satisfactory income and dignified and equitable living and working conditions for all people engaged in agricultural value chains.
- Reduces the vulnerability of the agricultural sector to unfavourable natural (e.g. climatic) and socio-economic (e.g. high price fluctuations) conditions and other risks.
- Encourages sustainable institutions in rural areas to promote the participation of all stakeholders and the balance of interests.

Agriculture in the city often differs significantly from conventional agriculture in rural areas. This difference has many origins. The motivation of the individual actors in urban agriculture is different. Today's urban agriculture is mainly due to its social, ecological and economic benefits, which is related to the aforementioned sustainability. A city offers many advantages with its many differences to rural areas. Existing infrastructures, the different climate and short supply chains within a city are just some of the many advantages of urban agriculture (Keuter et al. 2014). The World Meteorological Organization (WMO) defines the city climate as a "local climate changed over the surrounding area." One also speaks of microclimate in a city (Stone 2016). Usually, cities with their climate show a warmer and milder climate than in the surrounding countryside. This is mainly due to the high degree of sealed surfaces and the degree of development (radiant heat and wind reduction) (Stone 2016). In this work, the principles of sustainable agriculture are used as a basis for the further development of sustainable urban agriculture.

2 Overview on Strategies for Sustainable Urban Food Production

2.1 General Overview

The forms and options of urban agriculture are very diverse. In this work, a distinction is made between cultivation strategies and the types of used spaces and/or land. In terms of spaces, respectively lands, the following types are distinguished:

- Use of fallow and brownfield land for food production
- Rooftop farming
- Vertical land use (vertical farming).

The spaces can be used as listed above and managed with the following types of cultivation:

- Organic cultivation of vegetables in soil
- Aquaponics
  - Aquaculture
  - Hydroponics.

2.2 Types of Used Spaces and/or Land

2.2.1 Use of fallow and brownfield land for food production

With increasing building development and sealing of surfaces, there is also a shortage of space in German cities. This consumption is one of the most severe environmental problems of our time (Kaelberer et al. 2005). In the course of the sustainability strategy of the Federal Government of Germany, land consumption in Germany shall be reduced to 30 hectares per day by 2030 (Ministry for Economic Cooperation and Development 2015). Fallow areas might be urban lands which were once used and are no longer in use or which are too small for a building purpose, while brownfields represent former industrial sites, where the industrial use is finished and might have left behind soil contamination but also large spaces eventually with ruins. Inner-urban fallow lands might especially be used for the food cultivation of vegetables and herbs. With the help of the fallow land use for food production, an area can be recycled in principle and new land consumption can

Table 2 Advantages and disadvantages of reuse of brownfield sites according to Kaelberer et al. (2005)

Advantages	Disadvantages
• Sufficient location (centrality, public transport, urban environment, proximity to existing business partners) • Higher value stability or growth, especially in regions with population decline • Saved costs of deployment through the use of existing infrastructure, • Encouraged ecological compensatory measures • Marketing advantage and longer binding time due to building stock with a special flair • Possibility of direct marketing	• Higher costs and delays due to inefficient organization as well as insufficient co-operation between the actors involved • Potential soil risks, e.g. with regard to contaminated sites, munition, etc. in the ground • Costly or restrictive conditions, e.g. preservation of existing buildings • Contour-producing support structures that encourage new greenfield greening • Potential marketing problems due to negative imagery as a "breach area"

be prevented. This land recycling offers many advantages, as well as some disadvantages, which are listed below (Kaelberer et al. 2005) according to the German National Environmental Agency (Table 2).

The greatest challenges for a planned agricultural measure on a fallow land are the legacy of the area. The usage strategy depends on the former use of the fallow land. It might be necessary a waste site investigation and sanitation. The use of land to urban agriculture is of particular interest to planners and cities because of their ecological and aesthetic qualities (Schulz et al. 2013). The larger open space connections resulting from the meadows and fields cultivated in the city ensure unsealed soils and support cold air paths (Lohrberg 2010).

2.2.2 Rooftop farming

There are roofing concepts and agricultural roof surfaces around the world. In Germany, approximately 10 million square meters of roof area are planted every year (Demling and Eppel 2014). In addition to classical roofing, which is generally not used for food production, there are more and more cities in the world that are used primarily for food production. Similar to the use of fallow land, agricultural roofs also offer many advantages for a city such as the improvement of the working and living environment, an improvement of the surrounding climate and noise reduction by good sound absorption (Mann 2013). The increasing use of space can also be countered by the use of roof surfaces.

The commercial cultivation of vegetables on roof surfaces is often done in greenhouses, which can be operated with a variety of cultivation systems (e.g. aquaponics, hydroponics, organic cultivation in soil etc.). As such, the Gotham Greens roof tile in New York is one of the largest commercially used roofing plants worldwide (about 1500 m^2). In principle, every urban agricultural concept depends primarily on the respective location and a detailed site analysis must be carried out in advance. When it comes to the use of the roof for vegetable cultivation, a large number of requirements on the location and the building have to be taken into account in Germany.

The Leibniz Centre for Agricultural Landscape Research (ZALF) provides a recommendation for action and a location analysis in the guide for roof greenhouses "There's something growing on the roof—rooftop greenhouses". The ZALF criteria consider the roof structure, the building, the environment and the macro level as an indispensable location criterion for the planning of roof greenhouses (Freisinger et al. 2013).

2.2.3 Vertical farming

The term "vertical farming" comes from English and means "vertical agriculture". There are different constructions. As a general rule, each type of construction, which is built vertically in height, is designated. In most cases, vertical-farming is building-related and forward-looking, an innovative technology to optimally exploit a space or the surface within a building (Klanten et al. 2011). This can be inside the building as well as at the facade. In buildings there are often shapes that are similar to high-rack bearings. On each level, food is grown with artificial lighting in various systems. The decisive advantage of vertical-farming is, above all, the higher space utilization efficiency of a room. The cultivation area for, for example, vegetables can be expanded a lot.

In case of vertical farming outside of buildings, the façades are more green than farmed. The focus here is less on the production of food, but on the improvement of the city climate (the air purifying effect of the plants) and other advantages for the environment and life quality. The potential for cities is also very large in such a case. Vertical farming buildings might be, for example, unused industrial halls or other unused buildings. Also the vertical construction of shelves in greenhouses is a variant of vertical farming. The shelf load distribution in the building and the statics of the shelves must be taken into account in the case of racking systems. In addition, a modular design of the system should be attempted. With a modular design, one is flexible and the system can be extended or changed as desired (while respecting the room limits).

2.3 Cultivation Strategies

2.3.1 Organic urban vegetable cultivation/Organic cultivation of vegetables in soil

In order to define ecological urban vegetable cultivation, it is necessary to consider in advance what ecological agriculture in general means in Germany and the legal framework conditions. According to the BMZ, the term "organic agriculture" is explained as follows: "Organic farming excludes the use of synthetic plant protection products and mineral fertilizers and tries to work with natural methods and closed plant cycles. There are several different associations and certifications, but it can also be ecologically produced without certification." (Ministry for Economic Cooperation and Development 2015).

In order to be able to label food in the European Union with an ecolabel (or organic label), the criteria of the EC Eco-Regulation must be complied with. These requirements are very comprehensive and complex. The main criterion of the EC eco-regulation (2007) is the cultivation of plant products in soil: "The supply of plants with nutrients is to be carried out mainly through the ecosystem of the soil" (European Parliament 2007). An objective of the EC Eco-Regulation is to "maintain and promote soil and natural fertility of soil, soil stability and biodiversity of soil to prevent and control soil compaction and erosion and to provide plants Nutrients mainly via the ecosystem of the soil "(European Parliament 2007).

In this investigation organic vegetable cultivation is understood as the cultivation of plants in the soil ecosystem. Here, again, the term soil has to be defined before. The European Commission (2002) defines soil "… as the top layer of the earth's crust. It consists of mineral particles, organic matter, water, air and living organisms. The soil forms the interface between the Earth (Geosphere), Air (Atmosphere) and Water (Hydrosphere)" (European Commission 2002). Apart from the EC Eco-Regulation and the associated "eco-label", there are other associations and organizations which provide a "bio-seal" (e.g. Demeter). The requirements for the organic cultivation conditions must always correspond to at least those of the EC Eco-Regulation. Having in view the definition of sustainable urban agriculture, it can be concluded that sustainable agriculture can, but does not have to, comply with the definitions of the legal framework for organic farming. In other respects, ecologically produced products cannot be sustainable in some cases if, for example, the products have long supply and cooling chains in order to reach the customer.

2.3.2 Aquaponics

In 2015, the European Parliament launched a publication entitled "Ten technologies which could change our lives: possible consequences and political effects". Among these ten technologies, aquaponics is also described as a future-changing and inevitable technology (Woensel and Archer 2015). The term aquaponics is made up of the words aquaculture (fish farming/fattening) and hydroponics (earthless plant breeding on water basis). Aquaponics combines the cultivation of plants and fishes and can be regarded as a closed cycle. The advantages of aquaculture and that of hydroponics can be combined and above all the disadvantages are eliminated to a large extent (Rakocy et al. 2004). The cycle consists of the three main components, fish, bacteria and plants (see Fig. 1).

Fig. 1 Cycle of aquaponics

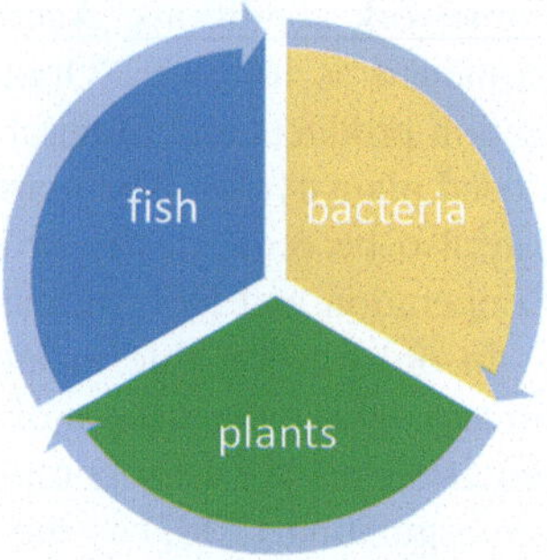

The excrements of the fish are converted by the bacteria into nutrients, which the plants need to grow. The plants clean the water with the absorption of the nutrients (in a high quantity harmful to fish), which then comes clean again to the fish. The circuit closes (Rakocy et al. 2004, 2006; Bernstein, 2013; Wilson 2013). This basic principle of aquaponics is understood in theory worldwide. The application ranges from small home systems to very large commercial installations. Due to the different applications and objectives, the systems differ in their design partly clearly from each other.

2.3.3 Aquaculture

The term "aquaculture" means a controlled breeding of all aquatic organisms, such as fish, molluscs, cretaceous fish and others, in natural and artificial waters and containers. These cultivated organisms are usually owned by the aquaculture operator. In contrast, in the open sea is done fishing Hubold and Klepper (2013). In order to conserve natural fish stocks, it is imperative that the share of fish produced in aquaculture continue to rise and that of the wild catches decrease. The breeding or the fattening of fish (aquaculture) is now spread around the world and is used in various forms. The main strategies are a) aquaculture in buildings, b) in ponds, and c) in the sea. Some of them do not consider a closed cycle, and usually consume enormous fresh water and energy.

Aquaculture in pond farming is practiced almost anywhere in the world (in fresh water). It can be cultivated extensively (e.g. natural ponds with carp) or intensively (e.g. Pangasius fattening with net cages in tropical countries). In the case of intensive aquaculture in pond farms, negative impacts on the environment can usually not be excluded (pollution of soil and water). Aquaculture in the sea is usually realized with the aid of net cages near the coast. Here too, the environmental consequences are usually large. Due to the massive feed rate in e.g. some Salmon farms on the Norwegian coast eutrophicate the coastal sections very much, which has a detrimental effect on the local environment.

2.3.4 Hydroponics

Hydroponics means plant cultivation without soil (or without the soil ecosystem) and with the aid of water. The required nutrients of the crop plants are transported to the roots of the plant by a nutrient solution (nutrients enriched with nutrients) and taken up there. In general, there are many different forms of plants in hydroponics to cultivate. The most important and most frequently used forms are the following according to (LetsGrow 2016; Rokocy et al. 2004).

- NFT (Nutrient-flow-technique)
- Raft-Deep-Water-Culture (DWC)
- Low tide flood systems with substrate beds.

In contrast to conventional plant cultivation in soil, a controlled cultivation in the hydroponics offers some advantages, as there are (LetsGrow 2016; Texier 2013):

- Water savings
- Precise dosage of fertilization
- Reduction of nutrients
- Reduced use of pesticides (preferred use of pest control agents)
- Faster growth of plants
- Monitored and closed environment (e.g., greenhouse).

2.3.5 Further Types of Urban Farming

In addition to the abovementioned land use forms and forms of cultivation for urban farming, further urban forms are described or named briefly. However, these forms are not considered in the variant assessment.

Intercultural and community gardens: are mostly run for social reasons. The main focus is on the community of the individual gardeners, the desire for self-sufficiency and the exchange of information. As a rule, such gardens are not operated with the aim of profitability, but are seen as leisure activities.

Ornamental plants and flowers: refers to the cultivation and marketing of ornamental plants and flowers by nurseries, where the production site often is also a sales location. The cultivation methods are different depending on the nursery. Cultivation is usually carried out in an enclosed site in the open air or in the greenhouse in the city or town. Ornamental plants and flowers generally fulfil aesthetic purposes and are not pre-grown as foodstuffs. For this reason, there is no further elaboration on this topic.

Small animals, bees, algae, mushrooms etc: In today's urban agriculture, small animals, barnacles, algae and mushrooms are increasingly being cultivated and bred. Since these forms are not considered further in the further investigation.

2.4 Scope of the Present Investigation

Scope of the present investigation was the assessment of the potential for sustainable urban food production in Magdeburg, Germany. The investigation approach comprised interviews on the consumer behaviour, material flow analysis of the various forms and options of sustainable urban agriculture, as well as a feasibility and SWOT analysis.

3 Overview on the Situation in Magdeburg

Magdeburg is the capital of Saxony-Anhalt. The city on the Elbe is one of the three upper centres of Saxony-Anhalt and has about 241,000 inhabitants (as of 2016). The city covers approximately 202 km^2 and comprises 129,500 households. Magdeburg is located at the intersection of the river Elbe, Elbe-Havel and Mittelland canal has an important inland port and is an industrial and trading centre. Of economic importance are machine and plant construction, health care, environmental technologies and recycling, logistics as well as the production of chemical products, iron and steel products, paper and textiles.

The Magdeburg area is characterized by intensive agriculture. A total of 7606 hectares of agricultural land are used in the Magdeburg region (Federal Statistical Office Saxony-Anhalt 2017). The Federal Statistical Office (2017) states vegetables were grown in Saxony-Anhalt on an area of 3701 hectares (data as of 2012). The agricultural (vegetables) activities are characterised by the fact that few farms manage large areas (about 29 hectares per operation on average). These are about 124 vegetable farms. Compared to the other federal states, Saxony-Anhalt has the largest vegetable farms (Statistical Office Germany 2013), e.g. approximately 0.5% of agricultural land is used for vegetable production (Ministry of Agriculture and Food Saxony-Anhalt (MULE) 2017). According to Behr and Niehues (2009) the annual lettuce consumption in 2008 amounted to approx. 380 kg per 100 households in Germany. Applied to Magdeburg, this would compare to an approximate consumption of 492,100 kg of lettuce per year. If it is assumed that approximately 10% of the lettuce demand are covered by organic lettuce, an annual consumption of 49,210 kg of organic lettuce can be concluded for Magdeburg. With an average fish consumption in Saxony-Anhalt of 5.6 kg fish per year, the market demand for fish can be concluded with 1350 tonnes per year in Magdeburg.

In Magdeburg are 13 locations with activities on urban organic gardening (as of 2017). One example is the communal garden "Jardin de Rayon" (see Fig. 2). The garden was established on a former fallow land in the middle of the city.

Fig. 2 Development of inner-urban fallow land in Magdeburg: the intercultural and community garden "Jardin de Rayon" (Leipziger Str.)

4 Investigation Strategy

The methodology used for the assessment of the potential for sustainable urban food production in Magdeburg consisted of a background analysis (with a literature review), complemented with the collection of empirical data from interviews of the public. Further, the background analysis was supported with a SWOT analysis. Further was prepared a material flow analysis with the scope of the comparison of the options of sustainable urban food production methodologies in Magdeburg. Within the scope of this investigation, a survey was conducted to derive a tendency for the potential sales of urban food products in Magdeburg. The title of the survey was: "Urban Farming Potential in Magdeburg". The survey was created and implemented online with the help of www.umfrageonline.com. The survey was active from 15.02.2017 to 31.03.2017 (six weeks). A total of 347 people participated.

In the first part of the survey, general demographic background to respondents were determined, like place of residence, the age, the sex and the current professional status. The statement about the place of residence is decisive, since only answers from people who live within 100 km Magdeburg are taken into account. This serves to determine the most exact purchasing and consumption behaviour of Magdeburg citizens. In the second part of the survey, the consumer behaviour of Magdeburg citizens was questioned with regard to fresh vegetables, fresh herbs and fish products. It was asked where the products are purchased/purchased, how often they are purchased/purchased, how much customers are satisfied with the range of products in Magdeburg. The final part of the survey was devoted to the potential implementation of urban agriculture in Magdeburg and the willingness to buy and accept urban food products.

From the potential forms of sustainable urban agriculture described in Chap. "Service-Based Bioeconomy—Multilevel Perspective to Assess the Evolving Bioeconomy with a Service Lens", there are a total of six possible variants which are considered in the investigation, as shown in Fig. 3. The varieties

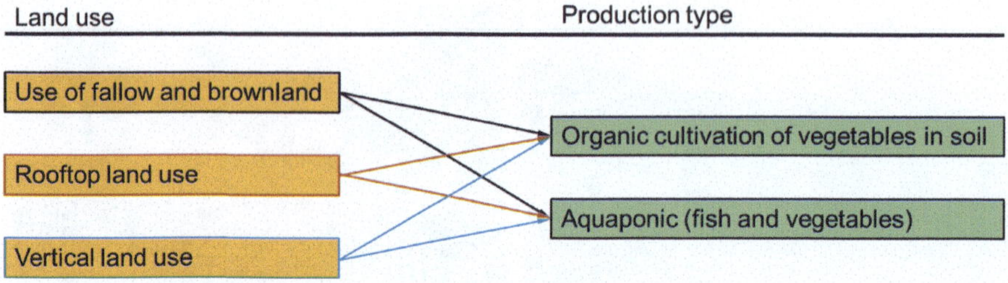

Fig. 3 considered variants for the sustainable urban farming option analysis in Magdeburg

of fallow land use in combination with aquaponics, fallow land use in combination with organic cultivation of the soil, use of the roof in combination with aquaponics, use of the roof in combination with organic cultivation of vegetables in soil, vertical farming use in combination with aquaponics and vertical farming use in combination with organic vegetable cultivation in soil. Beside the technical evaluation of the variants, was prepared a SWOT (Strength, Weakness, Opportunities, Threats) analysis, which is an instrument for positioning and strategy development (David 1993; Helms and Nixon 2010). The existing strengths, weaknesses, opportunities and risks are compared. The goal is a view of the situation on the basis of strategic decisions that are taken by Strengths—*obtained or expanded*, Weaknesses—*to be reduced*, Opportunities, and Threats—*to be eliminated*. SWOT is an abbreviation for Strengths, Weaknesses, Opportunities and Threats.

The resulting variants are summarised as follows and are illustrated in Fig. 4:

- Variant I—Use of fallow land and brownfields with aquaponics
- Variant II—Use of fallow land and brownfields with organic farming
- Variant III—Rooftop with aquaponics
- Variant IV—Rooftop with organic farming
- Variant V—Vertical farming with aquaponics
- Variant VI—Vertical organic farming.

The general scope of investigation is defined by the claimed area of 2400 m^2. The resulting considered six variants (in the following numbered I to VI) require an area of 2400 m^2. In addition, an optimal location for the variant consideration is assumed. Factors that ensure a fundamental decision for the location for the variants are:

- Low shading through adjacent buildings
- No pollution of the land and the floor
- Good infrastructure
- Location: Magdeburg (climatic conditions).

For the aquaponics variants (I, III and V), was considered the fish tilapias (*Oreochromis Niloticus*) and for the hydroculture lettuces *(Lactuca sativa var. Capitata)*. The fish are kept at a stocking density of 40 kg/m^3. The stocking density indicates how many kilograms of fish that are grown in one cubic meter of water.

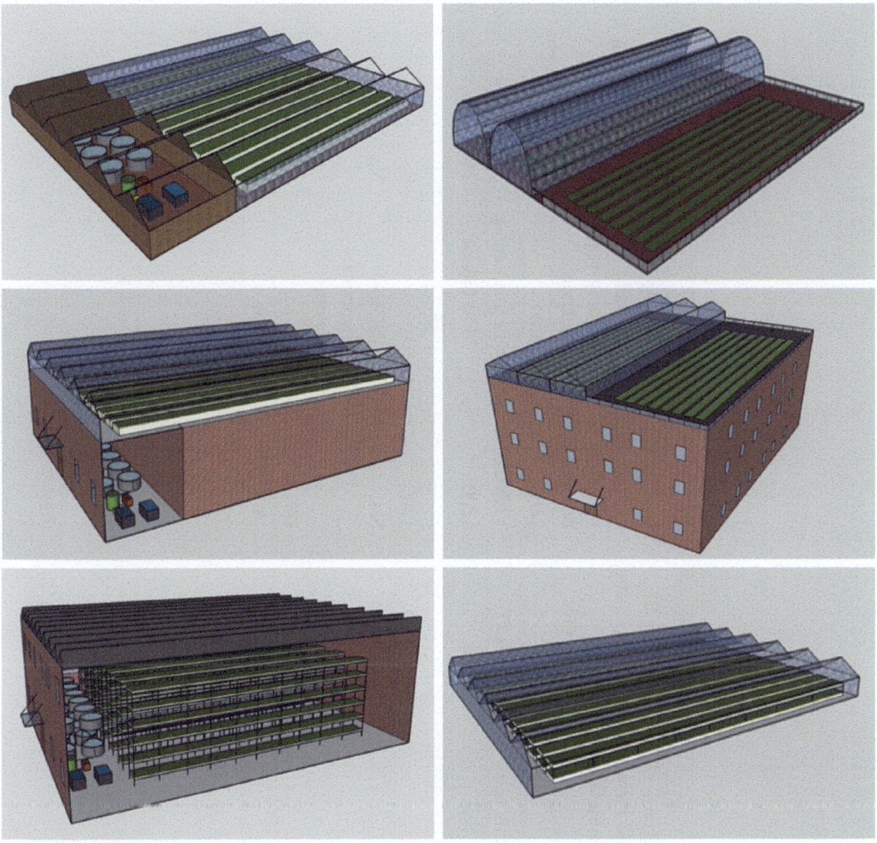

Fig. 4 Illustration of the considered variants for the sustainable urban farming option analysis in Magdeburg

The feed rate is 1–5% of the total fish mass per day. This ensures a continuous growth of fish until slaughter, which is carried out after approximately six months. A fish weighs between 400 and 600 g of total weight during slaughter. In the illustrated aquaponics systems, a ratio of 1: 1.5 (1 m^3 of water volume fish tank to 1.5 m^3 nutrient solution for plants) is striven for. The fish grow in fish tanks with different capacities (depending on the variant). In addition, a drum filter for the separation of the solid particles (sewage sludge) and a biofilter (nitrification and denitrification) are connected to the aquaponic system.

The water cycles for aquaculture and those of the hydroculture are decoupled systems. The process water, which is in the water cycle for the plants, does not return to the fish, but circulates continuously in the hydroculture system. The sewage of the fish reaches the water cycle for the plants after filtering and denitrification. It is treated in a separate basin and the water parameters adjusted to the needs of the plants (pH value, additional trace elements, etc.). An UV filter also

Table 3 Technical planning data

Variant criteria	Variant I, use of fallow and brownland with aquaponic	Variant II, use of fallow and brownland with organic farming	Variant III, rooftop using with aquaponic	Variant IV, rooftop using with organic farming	Variant V, vertical farming with aquaponic	Variant VI, vertical farming with organic farming
Area [m^2]	2400	2400	2400	2400	2400	2400
Greenhouse	VENLO	Foil greenhouse	VENLO	VENLO	No greenhouse	VENLO
Greenhouse used area [m^2]	1520	1200	2400	1200	2400 indoor area	2400
Outdoor area [m^2]	0	1200	0	1200	0	0
Building use (indoor)	Yes for aquaponics system	No	Yes for aquaponics system	No	Yes for aquaponics system	No
Temperature minimum [°C]	20	5	20	10	20	10
Plant cultivation area [m^2]	1064	1680	1540	1800	5760	3520
Fish tank size (total) [m^3]	320	0	504	0	625	0

disinfects the water. The amount of water added to the cycle of hydroculture is about 5–10% of the total water volume of aquaculture per day. Since the water does not return into the fish water cycle, the water loss must be compensated by fresh water (e.g. from rainwater).

Hydroculture is planted in Deep-Water-Culture (DWC). The height of the bees is 0.3 m, which also corresponds to the water level in the planting basin. The beds have a width of 2 m and vary depending on the variant in your length. The lettuce plants are pre-grown prior to hydroculture, for example, in coconut substrate before they can be used in the hydroculture. The lettuce cuttings are then inserted into openings of a styrodur plate after pre-breeding. The plant roots are then directly in the water and the lettuce heads are held by the styrodur plates on the surface. A plant density of 20 lettuce heads per square meter is considered. The lettuce heads in hydroculture have a harvest weight of approximately 300 grams and can be harvested in a harvest interval of 4 weeks.

For the organic farming variants (II, IV and VI), lettuce (*Lactuca sativa var. Capitata*) is used as planting culture. The plant density is between 8-11 plants per square meter. The ecosystem soil is used as a plant substrate. The composition varies depending on location. A compost allowance of 2 kg/m^2 is calculated per harvest interval. The harvest interval differs depending on the variant and lies between 6-12 weeks for organic lettuce cultivation. The water requirement is assumed to be about 640 L/m^2 and harvest interval. Depending on the variant, this results in an additional water requirement (apart from rainfall) of approx. 140–640 L per crop interval and square meter. Table 3 shows the planning details of each variant.

5 Results

5.1 Consumer Behaviour

Because of the criterion that only respondents within a radius of 100 km of Magdeburg can be taken into account, the number of participants was 322 (n = 322). 81.7% of the respondents live directly in Magdeburg, 10.7% within a radius of 50 km and 7.7% within a maximum of 100 km. Gender distribution is relatively balanced (52% male, 47% female, 1% no response). The largest share of respondents (43.3%) was in the age between 26 and 35 years. The majority of respondents to the survey were employees with a percentage of almost 50%. 35% of respondents were students.

The reported frequency of purchase of fresh lettuce and herbs was 70% of respondents at least 0–1 kg per week. 18% of them informed to buy more than 1 kg per week. Approximately 18% of the respondents consume at least 1 kg fresh leaf vegetables and herbs monthly. The main sources of supply are supermarkets and discounter with approximately 83%. Nearly 96% of the interviewees said they buy up to 1 kg of fresh vegetables every week. Approximately 8% of the respondents consume at least 1 kg of vegetables per day and 46% of them per week. The main portion of fresh vegetables is purchased from supermarkets. Approximately 1% of the respondents stated that they cultivate the majority of the vegetables themselves, and about 3% obtain their vegetables from a weekly market and organic food store. Generally, the majority of respondents is satisfied with the supply of fresh vegetables in Magdeburg. It is, however, to be stated that the criteria of the regionality of the products was answered as unsatisfactory by 33.5% of the participants. The supply of climate-friendly vegetables in Magdeburg was also stated to not be satisfactory for approximately 27% of the respondents. 87% of the interviewees pay attention to freshness and quality when purchasing fresh vegetables. In the third place (74%), the respondents are buying vegetables that are of regional origin. For 68% of the respondents is important the price and for 67% climate friendly production.

In addition to the consumption behaviour of vegetables, the survey also examined the consumption behaviour of fish in Magdeburg. To the question about the frequency of purchasing fresh fish 41% of the interviewees responded "never or less than monthly". Only about 7% of the interviewees buy fresh fish every week. Nearly 24% buy their fish in frozen form and 6% only as a processed product (e.g. smoked salmon). 16% buy up to 1 kg of fish per month. Due to the small consumption level of fish, only 185 interviewees responded to the question "Where do you buy/buy the bulk of your fish", informing that almost 80% buy in the supermarket and discounter. The percentage of fish bought in the market is 1.6%, 2.7% directly at the producer, 5.4% are fishing themselves, and 5% refer to their fish without mentioning other sources. Many participants did not give any information about their satisfaction on the fish availability. In addition, those interviewees who responded are not very satisfied with the supply of fresh fish in Magdeburg. The interviewees, who are not satisfied with the supply of locally produced fish in Magdeburg, sum up to 25%. According to the responses, 17% of the respondents would like more fish from climate friendly breeding. 84% regard freshness and quality as the main criterion when buying fish. 75% of the respondents appreciate a seal of sustainability, 67% consider most important the price, 63% the climate-friendly breed, 62% the fish's regionality, and for 53% of the respondents it is important that the fish comes from aquaculture.

67.5% of the interviewees would buy more vegetables grown on roof surfaces, 63.4% of the participants prefer a fallow land use for vegetable cultivation, and 42.3% vertical farming. 35.5% prefer rural food products. 30.9% consider aquaponics products to be attractive. The results of the survey indicate a potential for urban fish and lettuce. Consumers want more regionality and a large proportion of respondents would support urban agriculture with the purchase of products. In addition, there is no commercially operated urban farming company in Magdeburg at this time (2017).

5.2 *Results of the Option Analysis*

In order to achieve comparable results, the functional unit of the material flow analysis (MFA) was set to 1 kg of lettuce. In the case of a material flow analysis, a determination of the examination framework (system limits) is mandatory. Furthermore, the system limits are defined and listed, which is considered within the substance flow analysis and what is not. In the MFA was looked at:

- Feed requirements of the fish or according to the variant the fertilizer application [Kg]
- Energy supplies (thermal and electrical energy in [kWh])
- Water requirements [m^3], waste production [Kg], and land consumption [m^2].

It was not considered:

- Pre-value-chains of energy generation, water supply, and fodder production
- Individual biochemical intermediates within the system
- Nutrient composition of the waste
- Manufacture of plant seeds, breeding of fish
- Cost
- Processing, sale and consumption
- Production of the system components: greenhouse, building, substrate and medium (e.g. stone wool), fertilizer, fish tanks, plant beds, pots, beneficiaries, etc.

The individual biochemical intermediates within the sub-processes of the systems (nitrogen, phosphates, etc.) were not considered because of the system delineation. The general material flows for lettuce and fish production are summarised in the Figs. 5 and 6. All calculations are based on the functional unit 1 kg of lettuce, and

The specific consumption of the urban organic agriculture variants to produce 1 kg of lettuce differ, in some cases considerably. It becomes clear that the size of the cultivation area also reduces the specific substances consumption for the production. In addition, it must be noted that the required thermal energy depends considerably on the installed premises (e.g. greenhouse or greenhouse and isolated building). Variant V (vertical aquaponics in a building) has the best heat balance, but requires an artificial lighting due to the lack of solar radiation in the building.

Variant V has the lowest specific consumption for the other parameters. It has also to be stated that the specific water consumption of the ecological variants (II, IV and VI) is very high. This is due in particular to the high degree of evaporation and infiltration of the water. Also in terms of the land consumption, the organic farming options have significantly worse results than the aquaponics variants. The input and the output material flows of the production process of 1 kg lettuce are shown in Figs. 7 and 8.

As can be seen in Fig. 9, the specific outputs of the individual variants do not differ significantly. The smallest specific waste is produced by variant VI.

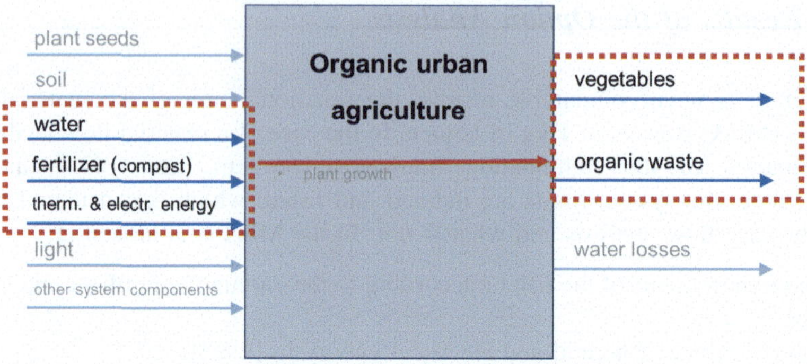

Fig. 5 Qualitative material flow of the urban organic agriculture system

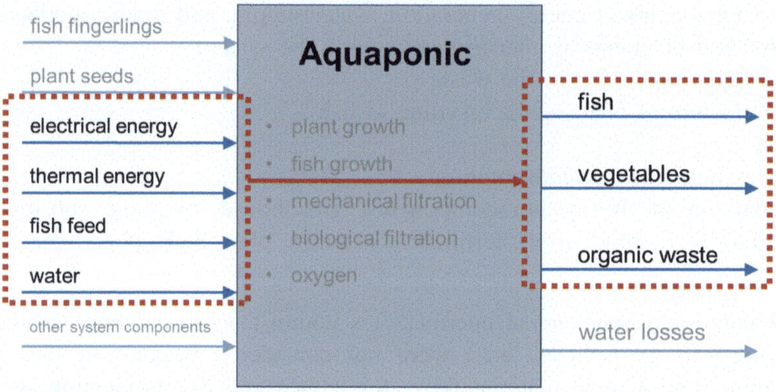

Fig. 6 Qualitative material flow of the aquaponic system

This is mainly due to the vertically used area. Variant V also has a relatively low specific waste output. The organic farming options do not have fish products as output. For the aquaponic options, the unutilized feed and the mortality of the fish are included in the balance of the resulting waste (Tables 4 and 5). This is the reason why the specific wastes are slightly higher in variants I, III and V than in variants II, IV and VI. Both options, urban organic farming and aquaponics, are assessed through a SWOT analysis. A final evaluation of the results is made in Table 6.

On the basis of the market analysis, the assessment according to sustainability criteria, the material flow analyses, the carbon footprint and the SWOT analysis, a rating matrix was developed (see Table 6). This evaluation matrix with the defined evaluation criteria is used to determine the variant with the highest potential for implementation in Magdeburg. The following criteria have been taken into account in the assessment: negative effects on the soil ecosystem, water use, nutrient requirements (compost or feed), fertilizer use, pesticide, herbicide and fungicide

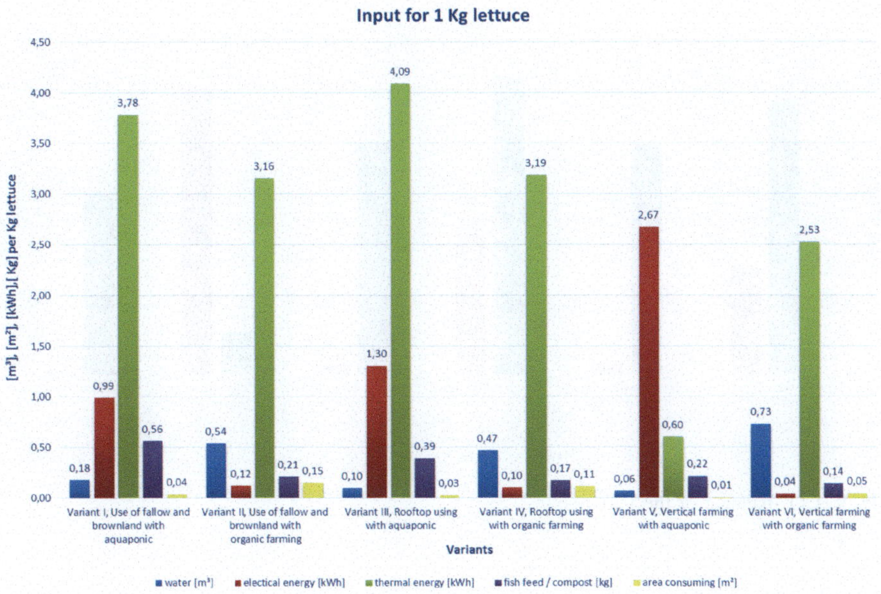

Fig. 7 Input comparison of the variants

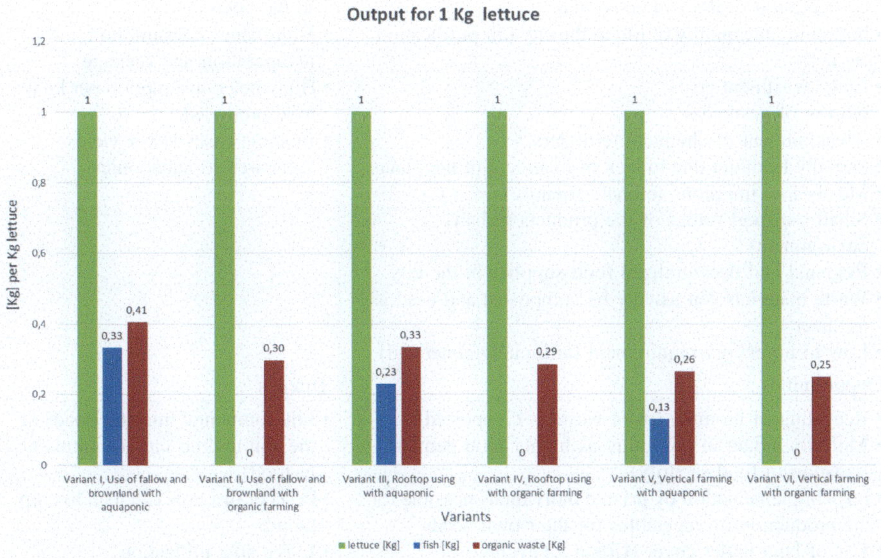

Fig. 8 Output comparison of the variants

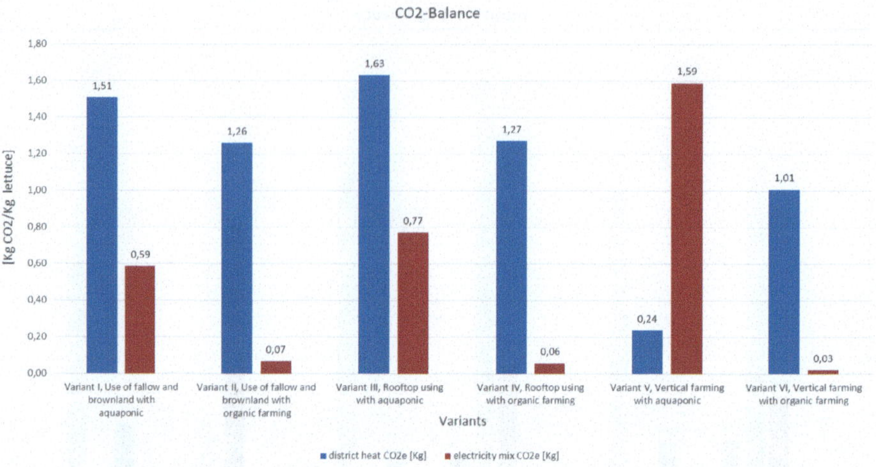

Fig. 9 CO_2-balance of the variants

Table 4 SWOT analysis of organic vegetable cultivation in soil

Strengths	Weaknesses
• Creation of green spaces in the city promotion and conservation of the soil ecosystem • Improving the quality of life in the city (air purification, etc.) • Low investment costs • No use of pesticides • Abandonment of chemical fertilizers • Partially healthier due to lack of treatment of the plants • Mostly low imissions (climate friendly) • Small transport routes of the products (urban environment) • Regional and decentralized food supplies in the city • Waste materials can usually be composted and used as fertilizer • Low to no energy requirement (artificially generated)	• Exposed to weather conditions in the open air • High water consumption (evaporation and seepage) • High area consumption per kg of food produced • In some cases lower yields compared to aquaponics
Opportunities	**Threads**
• Recycling of biomass (plant waste is composted) • More important in the future as higher food demand (creation of food security) • Easy implementation by private individuals possible for the production of vegetables for their own needs • Use of land under cover (Urban Upcycling) • Possibility of promoting environmental education in the city	• Environmental impacts (loads of the soil and air can accumulate in food) • Pests or diseases can lead to crop failure • Large area utilization

Table 5 SWOT analysis for the aquaponics system

Strengths	Weaknesses
• Closed material flow • Reduced consumption of materials (water, energy, etc.) • High yields • No use of pesticides through abandonment of chemical fertilizers • Space saving • Low imissions (climate friendly) • Low impact on air, water and soil because closed system (delimitation through, for example, greenhouse) • Small transport routes of the products (urban environment) • Fresh fish and fresh vegetables (regional and decentralized food supply in the city) • Time-efficient (plants grow faster in hydroculture) • very sustainable (through energy supply with renewable energies)	• Mostly high investment costs • In case of non-self-sufficient energy supply, high energy costs • In the case of incorrect implementation, a form of mass factory farming • No organic certificate for vegetables and fish according to European Union law • Fish feed mostly still consists to a large extent of fish meal or - oils from fish from wild • In large scale high labour costs
Opportunities	**Threads**
• In the future even more important, since higher food demand (creation of food security) • Possibility of automation • Decentralized energy supply through renewable energies possible • Utilization of biomass (plant waste, sewage sludge, fish excrements, fish residues after, for example, filleting) • Development of a label or certificate for organic production in aquaponics • Use of buildings (Urban Upcycling) • Use of waste heat from large heat producers (such as power plants, waste incineration plants, etc.) • Possibility of promoting environmental education in the city • Alternative feed for the fish (e.g. insects, water lenses, etc.)	• By cultivating non-urban areas, loss of habitat for native flora and fauna. • The cultivation of carnivorous fish usually encourages the overfishing of wild catches (often 60% of captive fish) • Excessive stress levels and associated disease susceptibility (medication use) • Close to nature fish breeding difficult and challenging

use, waste generated, negative effects on water, the nutritional value of the products, the market potential for Magdeburg and the land consumption. A value was determined for each criterion. These range from 1 (very low) to 3 (medium) to 5 (very high). The evaluation is based on the previous analyses and evaluations of the individual variants.

Table 6 Evaluation matrix of variants

Evaluation criterion	Variant					
	Variant I, use of fallow and brownland with aquaponic	Variant II, use of fallow and brownland with organic farming	Variant lll, rooftop using with aquaponic	Variant IV, rooftop using with organic farming	Variant V, vertical farming with aquaponic	Variant VI, vertical farming with organic farming
Negative impact on the soil ecosystem	2	1	2	1	2	2
Water consumption	2	4	1	4	1	5
Nutrient requirements (fish feed/ compost)	4	3	4	3	3	2
Fertilizer use	1	1	1	1	1	1
Pesticide, herbicide and fungicide use	1	1	1	1	1	1
Electrical energy consumption	3	1	3	1	4	1
Thermal energy consumption	4	2	4	2	1	3
CO_2 emissions (greenhouse effect)	4	1	5	1	3	1
Emerging waste	2	3	2	3	2	2
Negative impact on water bodies	1	2	1	2	1	2
Nutritional value of products 1 very high and 5 very low)	1	4	1	4	1	4

(continued)

Table 6 (continued)

Evaluation criterion	Variant					
	Variant I, use of fallow and brownland with aquaponic	Variant II, use of fallow and brownland with organic farming	Variant lll, rooftop using with aquaponic	Variant IV, rooftop using with organic farming	Variant V, vertical farming with aquaponic	Variant VI, vertical farming with organic farming
Market potential Magdeburg (1 very hiqh and 5 very low)	2	1	2	1	2	1
Land use	2	5	2	4	1	2
Total	29	29	28	28	23	27

1 very low; *2* rather low; *3* medium; *4* rather high; *5* very high

For each variant a total sum was determined. In analogy to the ecological footprint a small total sum is the optimum. The smallest total of the evaluation was determined for variant V: vertical area use in combination with aquaponics.

5.3 Design Sample for Sustainable Urban Food Supply

A pilot application is foreseen at the Zoo in Magdeburg, to be used as "Farm in the Zoo". In the project is planned to expand the existing site with an aquarium and a river landscape. Beside this investment there will be still enough space of of 3000–6000 m^2, depending on the detailed design, to host the urban farming space, which considers a rooftop farming with an aquaponics system. For the new construction of an aquaponics system, two planning variants were developed, each depending on the available space. The first variant is to build the facility evenly, the other option is to build the hydroculture on the roof of the new building. In both cases, aquaculture is to be integrated into the future building. The process engineering operation is the same in both possibilities, but not the heat demand. An aquaponic system with a total area of 2556 m^2 is planned for both cases, out of which 2256 m^2 are foreseen for the hydroponics system containing 1430 m^2 for plant cultivation. The remaining 300 m^2 of space are allocated to the aquaponics system. Depending on the season, the hydroponic part of the plant is operated as a warm greenhouse (April–September, 20 °C) or cold greenhouse (October–March, 15 °C). In addition to the 1440 m^2 of the plant area, the system includes 12 fish tanks each with 18.5 m^3 water volume and a stocking density of 50 kg fish per basin (Fig. 10).

The pre-designed option was assessed through a CO_2-balance, taking into account several options for the energy supply of the system with renewable energies. When looking at CO_2 emissions, it becomes obvious that the emission of the

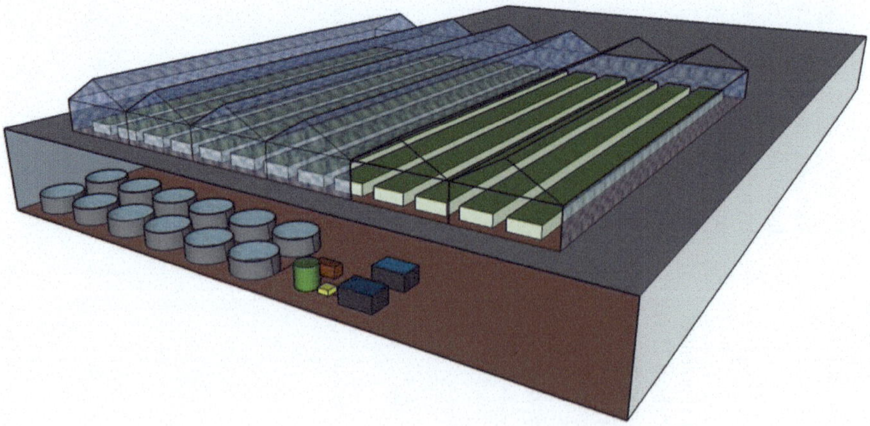

Fig. 10 design sample for sustainable urban food supply at Zoo Magdeburg

power supply is determined by the operation of the process technology, since it has the highest energy requirement. In relation to the process technology, solar thermal energy or geothermal energy needs about 1.7–3.3% of its demand. CO_2 emissions are correspondingly the same. When natural gas is used for the supply of electricity, the CO_2 release is 32,667 kg for process technology (33,790 kg for a process with geothermal energy). As comparison, 80,239 kg are released for the process technology with energy supply through brown coal (82,999 kg for a process with geothermal energy), resulting in the case of brown coal in 145.6% more CO_2 than with natural gas.

In the case of heat supply, the differences are caused by the design of the planning variants, since the hot water requirement is the same in both cases. For example, when installing rooftop cultivation 7 m above ground, 14.5% less heat is required in comparison to an installation on ground level resulting from more intensive global radiation which reaches the building. In the case of ground level installation, the shadow volume caused from adjacent buildings is significant larger. With heating through household waste, 21,743 kg of CO_2 are generated in the design variant with ground level planning in comparison to 18,585 kg CO_2 at 7 m level. In contrast to this, a heating with brown coal releases 54,089 kg CO_2 in the case of ground planning (46,233 kg in the design option at 7 m above ground). This results in 149% more CO_2 released with brown coal as energy source in comparison to household waste incineration. This means, the consumption of primary energy sources for the power generation reaches its maximum value for brown coal utilization. The operation power for the process technology needs 41,583 kg brown coal (43,013 kg for geothermal engineering). By comparison, for natural gas use, 7,321.01 kg (7,572.81 kg for geothermal engineering) must be provided. Thus the mass consumption of brown coal is 468% above that of natural gas.

In terms of renewable energy supply for sustainable urban farming systems, it is to be summarised that not all forms energy supply and storage are realizable for any plant planning size and system. The significant factors for an autonomous energy supply are the following four important characteristics which must be taken into account during planning: the meteorological conditions at the site, the energy requirements especially of the aquaponics system, the urban conditions, and the potentially applicable renewable energy systems. The determining condition for the sustainable urban farming systems is the fact, if a fish production through aquaponics is included as this needs significantly more energy. Further, the urban conditions play a significant role, especially for the performance of the solar and photovoltaic collectors. Due to this fact, the design indicates an advantage for a separation of hydroponics for a roof surface and aquaculture on a level ground.

6 Conclusions

The scope of this work was to conduct a variant assessment for urban agriculture in order to evaluate the option with the highest implementation potential in Magdeburg. The study of the variants has shown that urban agriculture has a high potential in Magdeburg, based on the results of the survey and the option analysis. The consumers in Magdeburg are willing to buy urban agriculture products.

If all variants are considered, variant V (vertical farming with aquaponics) has the highest implementation potential for Magdeburg. A cost estimate has shown that the costs of the labour for a non-automated system are very high compared to other costs. If the Vertical Farm would be automated, the labor costs would decrease. It has been shown that the valuation of the variables is mainly based on the planned planning variables of the system. If individual parameters such as, the location or the types of greenhouses, etc. are modified, the entire result of the specific consumption of a variant modifies accordingly. The main planning parameters are the location, the size and the type of the system which is used (e.g. aquaponics or organic culture in soil). These criteria have an enormous impact on the required electrical and thermal energy, water demand and other input and output variables.

The location has been the same in the considered variants and has been prioritized from an optimal location (little shading and good infrastructure). The analysis of the energy requirement showed that a selection of the materials used (greenhouse glazing and building insulation) had a significant impact on the energy requirement. Good and modern insulation materials with a low heat transfer coefficient have the properties to reduce the heat energy requirement by 70%. The electrical energy supply depends on the system used and the process technology involved. With the help of the study on self-sufficient energy supply from renewable energies, it has been shown that large urban farming projects such as "Farm in the Zoo" can be supplied only partly through renewable energies. The energy demand for the analyzed variant is too big in a climate like Magdeburg.

Due to the multiple influences, the preferred variant for urban agriculture is always to be derived from the location and the project size. Each project must be considered and analyzed. The success of each urban agriculture project also depends on the respective demand of the consumers. This work can be used as a basis for further planning in urban agriculture.

References

Behr H-C, Niehues R (2009) Market and sales. Land Building Research—Special Issue 330 (2009) 69 (in German)

Bernstein S (2013) Aquaponic gardening: a step-by-step guide to raising vegetables and fish together

David F (1993) Strategic management, 4th edn. Macmillan Publishing Company, New York, NY

Demling F, Eppel J (2014) *Vegetables on the roof.* Gemüse auf dem Dach, Bayerische Landesanstalt für Weinbau und Gartenbau, Bavarian Wine And Gardening Agency, In German
European Commission (2002) Towards a strategy for soil protection—COM 179
European Parliament (2007) Organic production and labelling of organic products. Council Regulation (EC) No 834/2007 of 28 June 2007 and repealing Regulation (EEC) No 2092/91
Federal Statistical Office Saxony-Anhalt (2017) Land survey and actual use of agricultural land in hectares Magdeburg, interactive database. www.statistik.sachsen-anhalt.de, 18 Apr 2017
Food and Agriculture Organization of the United Nations (FAO) (2009) Food for the cities
Freisinger UB, Specht K, Sawicka M, Busse M, Siebert R, Werner A, Thomaier S, Henckel D, Galda A, Dierich A, Wurbs S, Grosse-Heitmeyer J, Schön S, Walk H (2013) There's something growing on the roof—Rooftop greenhouses—Idea planning implementation. Leibniz Centre for Agricultural Landscape Research (ZALF), Muncheberg
Gehrke J (2012) Urban gardening—How the gardens return to the city, NABU Federal Association. In German: *Urban Gardening – Wie die Gärten in die Stadt zurückkehren,* NABU-Bundesverband
Helms MM, Nixon J (2010) Exploring SWOT analysis—Where are we now? A review of academic research from the last decade. Journal of Strategy and Management 3(3):215–251
Hubold G., Klepper R. (2013). The importance of fishing and aquaculture for the global food security. Thuenen Institute for market analysis (in German)
Kaelberer A, Klever, S, Lepke, T (2005) The future lies on brownfield sites—reactivation of urban land reserves, ICSS of the Umweltbundesamt (ed.), In German: *Die Zukunft liegt auf Brachflächen – Reaktivierung urbaner Flächenreserven*, ICSS des Umweltbundesamt (Hrsg.)
Keuter V et al (2014) Building-Based Food Production—Where we stand today and what will be possible in the future. In Farming and Fraunhofer Institute for Environmental Safety and Energy Technology UMSICHT, Symposium: "Urban Horticulture—Production returns to the city"(in German)
Klanten R, Ehmann S, Bolhöfer K (2011) My Green City – Back to nature with attitude and style. Gestalten Verlag GmbH & Co, KG
Letsgrow (2016). Workshop on aquaponics and its components, 05 Nov 2016
Lohrberg F, Timpe A (2011) *Urban agriculture—new forms of primary production in the city.* Specialist magazine "Planerin" 5/2 (in German)
Lohrberg F (2010) Urban agricultural landscapes. In: Published in Valentin (ed) The return of the landscape (in German)
Mann G (2013) Gründach – Urban Farming. Optigrün international AG, Journal Dachbaumagazin 1–2
Ministry of Agriculture and Food Saxony-Anhalt (MULE) (2017) Plant production and economic importance (in German). Online available: https://mule.sachsen-anhalt.de/stichworte-a-z/landwirtschaft/pflanzenproduktion/gemueseanbau/gemueseanbauvolkswirtschaftlichebedeutung/?q=schule, 18 May 2017
Ministry for Economic Cooperation and Development (BMZ) (2015). What is sustainable agriculture? German Association for International Cooperation (GIZ) GmbH (in German)
Rakocy JE, Masser MP, Losordo TM (2006) Recirculating aqaculture tank production systems: Aquaponics—Integrating fish and plant culture, Southern Regional Aquaculture Center (SRAC)
Rakocy JE, Bailey DS, Shultz C, Thoman ES (2004) Update on tilapia and vegetable production in the UVI aquaponic system. University of the Virgin Islands—Agricultural Experiment Station
Schulz K, Weith TH, Bokelmann W, Petzke (2013) Urban agriculture and "green production" as part of sustainable land management, Discussion paper No. 6, Leibniz Center for Agricultural Landscape Research (ZALF) e.V., Institute of Socioeconomy (in German)
Statistical Office Germany (2013) New surveys on vegetable growing in Germany. Online available: https://www.destatis.de/DE/ZahlenFakten/Wirtschaftsbereiche/LandForstwirtschaftFischerei/ObstGemueseGartenbau/Methoden/Downloads/GemueseanbauDeutschland_82013.pdf?__blob=publicationFile, 19 May 2017 (in German)

Stone C (2016) The urban farmer—Growing food for profit on leased and borrowed land. New Society Publishers, Canada

Texier W (2013) Hydroponics for everybody. Online available: http://www.mamaeditions.net/pdf/9782845940871_intro.pdf, 02 Oct 2016

World Commission on Environment and Development (1987) Our Common Future, Oxford. Oxford University Press, New York

Wilson L (2013) Aquaponic system design parameters: Fish to plant ratios, Australia

Woensel L, Archer G (2015) Ten technologies which could change our lives. European Parliament, Brussels

Building Bioeconomy in Agriculture: Harnessing Soil Microbes for Sustaining Ecosystem Services

Raj Rengalakshmi, Manjula M., V. R. Prabavathy, S. Jegan and B. Selvamukilan

Abstract Agriculture in Indiahas been heavily dependent on the use of inorganic fertilizers and pesticides for the past 4 to 5 decades. Such practices has had negative environmental consequences in terms of reduced ability of the agro-ecosystems in regulating (maintenance of hydrological services, water quality, carbon sequestration) and supporting (soil microbial diversity, soil structure and fertility, nutrient cycling, biological pest control, pollination) Ecosystem Services (ES). Bioeconomy in agriculture, which stresses on supply of food and agricultural outputs through extraction and use of high quality inputs from renewable resources, gains significance in this context. Use of renewable factors of production reduces agriculture-induced environmental impacts, reduces the yield gaps and strengthens agro-ecosystems ability to regulate and support ES. Organic farming approach that promotes bioeconomy in agriculture is highly dependent on diverse soil microflora consisting of beneficial microbes. They facilitate ecological services such as nutrient cycling, disease control, drought tolerance, degradation of organic matter, water lifting etc. Biofertilizers and bio-pesticides are the biological products necessary to augment the soil microflora, in this way they are linked to building bioeconomy in agriculture. The renewable use of bio resources called bio-inputs henceforth primarily offers a mean to reduce the dependence on chemical inputs

R. Rengalakshmi (✉) · Manjula M. · V. R. Prabavathy · S. Jegan · B. Selvamukilan
M. S. Swaminathan Research Foundation – MSSRF, III Cross,
Taramani Institutional Area, Chennai, India
e-mail: rengalakshmi@mssrf.res.in

Manjula M.
e-mail: manjula@mssrf.res.in

V. R. Prabavathy
e-mail: prabavathyvr@mssrf.res.in

S. Jegan
e-mail: jegan.sekar@mssrf.res.in

B. Selvamukilan
e-mail: selvamukilan@mssrf.res.in

W. Leal Filho et al. (eds.), *Towards a Sustainable Bioeconomy: Principles, Challenges and Perspectives*, World Sustainability Series,
https://doi.org/10.1007/978-3-319-73028-8_14

and sustain the provision of valuable ES. These bio-inputs play an integral role in maintaining soil quality nutrient fixation, mobilization and solubilisation processes, induces abiotic and biotic stress tolerance and manages pest and diseases through maintaining a healthy prey-predator balance in environment. Microbial based bio products are an important input for organic and sustainable agriculture production systems as it works on the same principles as bio-economy and could be considered as an important pathway for promoting bio-economy in agriculture. The recent progress in research and development of microbial consortium and microbiome approaches adds value to the use of bio-inputs in agriculture. This paper is an attempt to detail the processes through which the principles of bioeconomy in the soil ecosystem could be effectively harnessed to achieve productivity in perpetuity by the use of renewable microbial bio resources.

Keywords Soil ecosystems · Ecosystem services · Bioeconomy
Sustainable production and bio-inputs

1 Introduction

Sustainable management of agricultural production systems calls for a paradigm shift in production processes and the concept of bioeconomy which advocates the use of renewable and biological resources for food production with reduced damage to environment assumes great significance. In this context, use of biological inputs viz., biofertilizers and biopesticides play an important role in augmenting bioeconomy in agriculture sector. Business-as-usual approaches in increasing agricultural output through fossil fuel based inputs will only result in exacerbating the existing disservices from agriculture. Moreover, countries committed to UN-SDGs are bound by their commitment to promotion of sustainable agriculture envisioned in Goal 2 of the UN SDG's on achieving zero hunger (UNDP 2010). Given this, the global food demand has to be met through innovative and sustainable solutions that optimize resource use, minimize ecosystem disservices and build robust and resilient agricultural production systems.

The term 'bioeconomy' has been defined differently by many authors, and the definition is evolving continuously and undergoing improvements with relevance for application across novel issues transcending different sectors in the economy. The European Commission defines the bioeconomy as, "the production of renewable biological resources and the conversion of these resources and waste streams into value added products, such as food, feed, bio-based products and bioenergy" (European Commission 2012). Whereas OECD defines it as economic outputs derived from biotechnologies (OECD 2009). In the Indian context also largely the concept is linked to biotechnology industries. According to Leitao (2016) it has three important basic elements; application of advanced developments in gene and cell science to evolve innovative products, utilization of renewable biomass and bioprocesses for sustainable production and the incorporation of biotechnological knowledge across sectors for new products. Here the term 'biomass' is defined as materials of biological origin like 'whole or parts of plants, trees, algae, marine organisms, microorganisms and animals' (CEN 2011). It clearly shows the need for paradigm shift from fossil fuel economy based approach to a bio-product based bio-economy approach of sustainable agriculture to improve the soil biological properties and thereby ecological services for sustainable production and conservation of environment.

1.1 Bio-economy policies in India

The concept of Bio-economy is gaining significance at the country level, Government of India launched the National Mission on Bioeconomy in 2016 as per the National Biotechnology Development Strategy 2015–2020 (GoI, MoST, PIB, 2015). The mission is said to be unique to Southeast Asia and with the launch of this mission, India joins the club of the select countries who are promoting bio-resources for sustainable development. The Indian bioeconomy industry is estimated at USD 7 billion in 2015 (BIRAC 2016) and projected to reach USD 11.6 billion by 2017 and USD100 billion by the year 2025. The industry has been growing at a compound annual growth rate of 20% for over a decade and needs to grow at 30% in the next decade to meet the target of USD100 billion (BIRAC 2016; Burril Media 2014). To achieve this, the Department of Biotechnology (DBT) established the Biotechnology Industry Research Assistance Council (BIRAC) for developing biotechnology based products and propelling the sector to reach the ambitious target of USD100 billion by 2025.

Also, Bioeconomy is one of the key sectors identified and included in the Government of India's flagship 'Make in India' initiative launched in 2014. Towards this, DBT has established a 'Make in India' facilitation centre at BIRAC (BIRAC 2016). Bio-Agri which comprises primarily of hybrid seeds, GM crops and biofertilisers and bio-pesticides is the third largest segment, after bio-pharmaceuticals and bio services, and accounts for 15% of the value of the Indian bioeconomy industry. Bio-agri segment is expected to grow to USD 5.4 billion by 2025 from the current 0.9 billion (BIRAC, 2016). Government of India lays a lot of emphasis on the growth of the bio-agriculture sector, and had allocated 22% of the total fund allocation in the biotechnology industry to promotion of research and development in bio-agriculture in the 12th Five Year Plan. Of the different components in the bio-agri sector, the role of biofertilzers and bio-pesticides are crucial in restoring and building the soil Ecosystem Services and there by contribute to sustainable Bioeconomy in Agriculture.

1.2 Current status of Agroecosystems and its Services

The agro-ecosystems are primarily evolved for its provisioning services mainly for the supply of food, fibre, and fuel. Apart from this, it provides valuable Ecosystem Services (ES) to maintain the balance and interactions among different components in the environment. The potential of agriculture in providing such ecosystem services is gaining research attention globally (MEA 2005; Dale and Polasky 2007; Power 2010). Climate regulation, water purification, managing surface water flow, maintaining ground water level and assimilation and breakdown of waste are some of the regulating services offered by agro-ecosystems (Swinton et al. 2006). The ability of the agro-ecosystems to provide valuable ES are determined by the nature of production systems, management practices and technique of production adopted (Björklund et al. 1999). Global food production was reported to have increased 2.5 times over the 40 year period ranging from 1960 to 2000 (MEA 2005) and its further estimated that an additional 70% increase in food production is needed to feed the growing population of 9 million by 2050 (UN 2013). Increase in agricultural production has traditionally been triggered by the promotion of input intensive agricultural systems.

The major contributor to this exponential increase in crop output reportedly came from increase in yield achieved through a spurt in irrigation and vast increase in use of chemical fertilisers and pesticides (Tilman et al. 2001). As globally, and in India too, the breakthrough in crop productivity achieved through the green revolution was the result of promotion of intensive agriculture practices, which laid emphasis on the use of chemical fertilisers and pesticides as catalytic technologies, aimed at increasing output. The chemical fertilizer consumption in India has been increasing sharply over the three decades (1970–2000) as witnessed from the data that the total nitrogen, phosphorus and potassium fertilizers consumption has increased from 2 million to 18 million tonnes and the per-hectare consumption has also risen from 11 to 95 kg in the similar period (FAO 2005). The quantum jump in the use of chemical fertilisers has had its negative environmental impacts in terms of loss of soil microbial diversity, decline in flow of soil based ecosystem services and loss of soil fertility. Further, the continuous use of chemical inputs damaged the environment and reduced its capacity to provide a number of ES (Costanza et al. 1997; Krebs et al. 1999; Tilman et al. 2001).

Promotion of input intensive agricultural production systems has had its impacts in terms of provision of ES from agriculture landscapes especially from soil and irreversible damage to the environment (Matson et al. 1997). Region specific studies conducted across the globe also establish the strong link between chemical input intensive agricultural production systems and the loss of ecosystem services. Mäder et al. (2002) reported that the rate of decrease of soil organic carbon was 85% under conventional farming when compared to its original levels, while it was significantly lesser under organic system due to increased soil microbial processes.

A study conducted by Li et al. (2009) in China on the impact of shift from traditional farming practices to modern intensive agriculture on ES, reported huge differences in ES value per unit area under different land use practices, with modern intensive agriculture practice reporting the lowest values.

A comparative assessment of soil microbial load and diversity on a low input intensive cropping system (Paddy) and a high intensive cropping system (banana) in Kerala, India, reports low levels of soil microbial activity in banana compared to paddy (MSSRF 2017 unpublished work). Postma-Blaauw et al. (2012) reported that the heavy use of chemical inputs along with intensification practices reduces the quality and biodiversity of the soil ecosystem. As a consequence, the soil loses its resilience, sustainability and productivity in long-run. Indira Devi (2007), using dose response model and cost of illness, quantitatively estimated the physical impact of pesticide applied in paddy fields in Kuttanad region in Kerala. The study found that application of chemical fertilisers and pesticides caused health damage to the extent of Rs. 38/day (US $0.86) per individual. A study carried out by Nellemann et al. (2009) forewarns that "up to 25% of the world's food production may become lost due to environmental breakdown by 2050 unless action is taken". A key focus area for this action should be conserving soil microbial diversity and augmenting the natural ability of soils to provide valuable ecosystem services.

Soil Based Ecosystem Services and Agriculture production: The provisional, regulating, supporting and cultural categories of ES provided by soil are broadly sub-categorised into six areas: (i) buffering and moderation of the hydrological services; (ii) physical support to plants; (iii) retention and delivery of nutrients to plants; (iv) disposal of wastes and organic matter: (v) renewal of soil fertility; and (vi) regulation of major elements cycle (Daily et al. 1997). Of the four main categories of ES, the one associated with supporting services (nutrient cycling, decomposition, improving soil structure and fertility, etc.) is highly critical to agriculture production. Specifically, it gives basic ES including nutrient cycling, conversion of organic materials and toxic compounds, control of pests and diseases and water regulation (Doran and Zeiss 2000).

The ES provided by soil are essentially based on the biotic community in the soil, hence the application of chemical inputs to soil are critical in maintaining soil quality. In this context, at the system level, the microbiome (microorganisms in a particular environment) plays an essential role in soil processes (Barrios 2007) and thus the composition, quantity and quality of microbial communities in soil assumes importance for its sustainable productivity (van der Heijden et al. 2008). Studies have also proven the role of soil microbial biomass in the nutrient recycling process; as the soil biology plays an inevitable role in soil chemistry and structure, maintaining the well balanced microbiological activity in the soil is essential for its ecosystem services (OFRF 2017).

1.3 Pathways for Nourishing Soil based Ecosystems Services in Building Bioeconomy

Bio-economy in the context of agriculture refers to the sustainable production and use of biological resources for achieving productivity in perpetuity. Bio-economy is based on knowledge and innovation in biosciences and biotechnology in synergy with other technologies. This section elaborates two potential agriculture production systems that have the potential to use ecologically conducive, renewable and environmentally safe bioeconomy based products and practices to spur and improve the soil ecosystem services:

I. **Organic Farming Practices**

The increasing concerns about intensive agriculture and its detrimental effects on environment have led to the development of sustainable agricultural practices such as organic farming (FiBL 2000; FAO 2002; IFOAM 2005) or agroecological farming, zero budget farming, natural farming, low external input sustainable agriculture etc. Of which organic farming is widely adopted as a tool or approach to restore the ES and increase the farm productivity. Bio-economy based products, namely bio-fertilisers and bio-pesticides (bio-inoculants) are indispensable for promotion of organic farming that advocates sustainable resource conserving agricultural practices, ensures food and nutrition safety and adds to soil biodiversity (Megali et al. 2013; Patil et al. 2013; Venkateswarlu 2008).

Gomiero et al. (2008; 2011) reports that low-input farming systems such as organic farming with reduced use of chemical inputs aim to lessen the negative impacts, improves production and maintains ES. Organic agriculture is defined as "a holistic production management (whose) primary goal is to optimize the health and productivity of interdependent communities of soil, life, plants, animals and people" (UNCTAD 2006). Therefore, it aims to utilise and maintain ES by improving the natural environment, increased water retention, reduced soil erosion and increased agro-biodiversity (UN 2008). At present, organic farming is gaining importance at the global and national level. Currently 31 million ha worldwide is under organic agriculture at a market value of US $26.8 billion, and growing at a rate of 20% per year (Willer and Yussefi 2006). In case of India, the area under organic farming is expected to reach 2 million ha in 2020 from its current level of 7 million ha (APEDA 2017).

In organic farming systems, priority is given to the use of on-farm and locally available renewable inputs and resources by adopting diversification and integration principles. This aims to foster self-regulating ecological interactions and processes for a better ES. The organic agriculture promotes important soil ecosystem based ES such as building up of soil organic matter and soil biological processes, enhancing the bioavailability of nutrient, improving the soil aeration and bulk density which are directly related to water holding and infiltration properties and providing soil borne pest and disease control measures. In the recent past, a study conducted by Ramesh et al. (2010) in five states in India clearly showed the differences in soil quality (bulk density, organic carbon, soil microbial activities, availability of major and micro nutrients to crops) between organic and conventional farms which use chemical inputs. For example, there was an increase of 29.7% soil organic carbon in organic farms (1.22%) compared to 0.90% in conventional farms. Similarly dehydrogenase, alkaline phosphatise and microbial biomass carbon were higher in organic soils by 52.3, 28.4 and 34.4% respectively when compared to soils in conventional farms which used chemical inputs. The fact was reiterated by the study of Melero et al. (2006) in Spain in which soil microbial biomass was considerably high under organic farming than under conventional farming due to greater supply of available carbon. Araújo et al. (2008) also reported similar findings in Brazil with "acerola" orchard (*Malphigia glaba*).

The agriculture practices in organic farming, mainly the addition of organic matter through compost, crop residues, amendments etc. promotes the changes in the soil microbial biomass. Along with this, the application of bio-inoculants—biofertilizers or biocontrol agents to augment the beneficial microflora is gaining attention (Truu et al. 2008). Hence, the enhanced soil biological activities under organic systems are exceptionally vital for the increase of soil fertility and upholding the environmental sustainability.

II. Use of Bio-inoculants in Sustainable Agriculture

Another interesting and direct pathway of introducing bioeconomy in agriculture is through the direct use of products of bio-inoculants. There has been great interest in the use of microbial products in agriculture to enhance productivity, improving crop health under adverse environmental conditions, controlling pest and disease with reduced pesticide, environmental impact and health issues (Deepak et al. 2014). Thus harnessing the 'phytomicrobiome' especially the rhizomicrobiome has proven to be the most effective approach to improve farm productivity and food quality in a sustainable way, which also promotes positive environmental and social outcomes (Jegan et al. 2016; Singh and Trivedi 2017). The current market for the agriculture biological is predicted to be $2.9 billion, and currently biopesticides account for over $2.0 billion dollars, which constitutes 5% of $44 billion of chemical fertilizers and pesticides (Reed and Green 2013).

Currently, a number of microbial products are available in the market and number of new products to be launched by leading chemical companies. Indigo has launched a new cotton seed coated with microbes reported to yield 10% higher under saline conditions, and similarly Bioag Alliance has announced the launch of a number of new microbial products. Increasing investment by public, private and start-up companies have announced an exponential growth in this sector (Singh 2017). Intensive research focus for the discovery of novel microbial products, as biopesticides and biofertilizers, has currently led to the availability of over 149 registered microbial strains for agricultural products (Copping 2009). Recently, adoption of rhizosphere microbial communities as an alternative for chemical fertilizers has evolved as an innovative method in sustainable agriculture and bio-safety programmes (Adesemoye et al. 2008, 2009; Adesemoye and Kloepper

2009; Bhardwaj et al. 2014). Also, Gopal et al. (2013) tested the concept of tailor-made core microbiome transfer therapy in maize, atriplex, sugarbeet by harnessing the technological advancements in molecular biology to manage plant diseases for different crops. On the same line, Mendes et al. (2011) tested the soils that control the *Rhizoctonia solani* by mixing with disease conducive soil in sugarbeet by mixing at a ratio of 1:9 which showed the successful control of infection. The metagenomic molecular analysis showed 17 bacterial communities belong to three groups as core microbiome in disease suppression. Similarly Rosenzweig et al. (2012a, b) identified core microbiome for potato common scab disease and Kyselkova et al. (2009) for tobacco black root rot.

Of the two pathways, the second one is having greater potential in improving the soil based ES by its use in both organic and sustainable agriculture production systems.

1.4 Significance of Bioinoculants in Augmenting Soil Based Ecosystem Services

Agriculturally important microflora interacts with soil and plants and provides beneficial services like plant growth effects, disease control by building the resistance capacity against the pathogens, and provides drought tolerance capacity to plants through different physiological mechanisms. The role of bio-inoculants in improving soil fertility, soil quality, improving soil microbial diversity, increasing crop productivity and improving crop adaptability to environmental stress has been well established (Paul and Nair 2008; Singh et al. 2011; Bhardwaj et al. 2014; Sekar and Prabavathy 2014; Viswanath et al. 2015). Bio-pesticides play an important part in Integrated Pest Management (IPM) strategies in crop production. They are effective in small quantities and affect only the target pest in contrast to the broad spectrum conventional chemical pesticides (Kachawa 2017; Kawalekar 2013). Thus, bio-pesticides conserve the beneficial organisms and protect biodiversity, maintaining optimum pest-predator balance. The active and inert ingredients in a biopesticide formulation are considered safe and the carrier media is an easily biodegradable material (Soesanto 2012).

Bio-pesticides by their inherent ability to target specific pests and property of biodegradability do not have any negative externality on the environment unlike the chemical pesticides which pollutes the soil, water and air through deposition of chemical residues. The role of N_2-fixing bacteria that support Nitrogen (N) plant nutrition in leguminous crops is important in soil biological properties. Approximately 80% of nitrogen fixation, a process where atmospheric nitrogen gas (N_2) is transformed into ammonia, is carried out by diazotrophs, which can achieve >150 kg N ha^{-1} yr^{-1} in unmanaged ecosystem, which is higher than the recommended mineral N application, which is around 100 kg N ha^{-1} $yr^{-1;}$ but, in some parts of the world it can reach >500 kg N ha^{-1} yr^{-1} (Vitousek et al. 2009). The N_2

fixing Rhizobium associated with legumes can provide up to 90% of the N requirements of the host through atmospheric N_2 fixation (Franche et al. 2009). Free living N fixers like cyanobacteria, *Pseudomonas*, *Azospirillum*, and *Azotobacter* fix significant amounts of nitrogen (0–60 kg N ha^{-1} $year^{-1}$) (Kahindi et al. 1997; Burgmann et al. 2004; Rodrigues et al. 2008). In case of Phosphorus (P), though the P content of soil is relatively high, the plant available form is relatively low, with about 1 mg/kg soil. Nearly 70–90% of the P added to soil as phosphate fertilizer is rapidly converted into insoluble form and is not available.

The Phosphate Solubilising Bacteria (PSB) play an important role in the P cycling in soils by the production of (i) organic acid such as gluconic acid, 2-Ketogluconic acid, oxalic acid, citric acid, malonic acid, succinic acid, glycolic acid, and lactic acid (ii) lowering the soil pH (iii) chelation of the cations bound to P (iv) and release of the P from the phosphates and the phytates by producing the corresponding hydrolases (Siddiqui 2006; Hayat et al. 2010). The strains of *Bacillus*, *Pseudomonas*, and *Rhizobium* and Mycorrhizal fungi are the most powerful phosphate-solubilizers (Avis et al. 2008; Rodriguez and Fraga 1999; Tilak et al. 2005; Bagyaraj et al. 2015). Strains of genera *Burkholderia*, *Agrobacterium*, *Micrococcus*, *Aerobacter*, *Flavobacterium*, and *Erwinia* have also been shown to convert the insoluble inorganic phosphate in the soil to a soluble plant-usable form (Rodriguez and Fraga 1999; Hayat et al. 2010; Tilak et al. 2005).

Similarly, some specific species of the microbial communities in the rhizosphere contribute to plant growth by mobilizing nutrients and making them available (Lynch and Whipps 1990), increasing root health through competition with root pathogens (Weller et al. 2002) or enhancing nutrient uptake (Smith and Read 1997). The use of microbial products has been shown to improve crop diversity, productivity, acquisition of P, Iron and Zinc from soil and enhances grain nutrient quality (Briat et al. 2014; Dragicevic et al. 2015). White and Broadley (2009) advocates microbe mediated approach in tackling malnutrition through enhancing the nutritional quality of food. A two-fold increase in Iron content in rice due to bioinoculation of plant growth promoting bacterial strains—*Psudomonas putida*, *P. fluorescens*, and *Azospirillum lipoferum* was reported (Sharma et al. 2013). Tariq et al. (2007) demonstrated the efficiency of a commercial mixed PGPR consortium (containing *Pseudomonas* sp. and other strains of PGPR) acting as Zinc solubilizer and increasing Zinc concentration up to 157% in rice.

With reference to plant growth promotion and biocontrol activity, the soil microbiome especially the root rhizobiome greatly influences plant health, soil fertility (Philippot et al. 2013; Sekar and Prabavathy 2014; Jegan et al. 2016) and agriculture productivity, and is often referred to as the plant's second genome (Berendsen et al. 2012; Chaparro and Sabot 2012). The structure of the rhizosphere microbial community is influenced by soil type and the amount of root exudates (Uren 2001; Bais et al. 2006; Moe 2013). In the natural rhizosphere system there are numerous beneficial groups with multi-beneficial plant growth and promoting traits and biocontrol of pest and pathogens (Raupach and Kloepper 1998; Picard and Bosco 2008; Ryan et al. 2008; Hartmann et al. 2009; Sekar and Prabavathy 2014; Viswanath et al. 2015; Krishnan et al. 2016; Raju et al. 2016).

Combination of potential PGPRs strains with multi- beneficial plant growth and promoting traits will provide a green solution to upcoming problems like, drought, salinity, increasing temperature, pest and phytopathogenic infections in the agricultural system for global food safety and security. (Miransari 2011). Combined application of two or more biocontrol strains consistently the biocontrol efficiency due to multiple modes of action over a wider range of environmental conditions (Larkin and Fravel 1998; Sekar et al. 2016).

Varying combinations of multiple and compatible biocontrol agents have been reported for improved disease control (Paulitz et al. 1990; Raupach and Kloepper 1998; Shanmugam et al. 2002). Inoculation of more than one PGPR or PGPR together with AM fungi is more beneficial in promoting plant growth compared to inoculation with either one of them (Anuroopa and Bagyaraj 2017; 2014; Bagyaraj et al. 2015). Tilak et al. (2005) reported increased nodulation and plant growth when PGPR is co-inoculated with Rhizobium in pigeon pea. Association of AM fungi in legumes alleviates P stress, which also enhances nitrogen fixation by the plant leading to improved plant growth, yield and increased protein content (Feng et al. 2002; Garag and Manchanda 2008). Nadeem et al. (2010) reported the use of multi-strain microbial consortia showed efficient performance, survival and competence of the inoculum in natural environment and field conditions.

Also, Methylotrophic bacteria are known to play a significant role in the biogeochemical cycle in soil ecosystems, ultimately fortifying plants and sustaining agriculture (Kumar et al. 2016). They utilize the greenhouse gases and therefore minimize global warming, in addition to actively participating in carbon cycling in the soil (Iguchi et al. 2015; Kolb and Stacheter 2013). Methylotrophs are associated with the plant rhizosphere or phyllosphere or both and show no pathogenic activity. In seed coating or a seed inoculum, methylotrophs can be used to enhance seed germination (Meena et al. 2012. Methylotrophs regulate carbon cycling and therefore may reduce global warming. In the carbon cycling, methylotrophs plays a key between greenhouse gases like CO_2 and methane (Iguchi et al. 2015).

Considering the fact that in general, 60–90% of the total fertilizer applied is lost and only 30–50% of applied N fertilizers and 10–45% of P fertilizers are utilized by crops, the use of microbial products in agriculture is the immediate alternate to promote ES. But the promotion of microbial products in organic agriculture has limitations in terms of application, production, compatibility, registration, etc. (Chen 2006). Compared to conventional agrochemicals which are broad spectrum products with wide application, the PGPRs on the other hand are highly targeted and results in variable quality and efficacy in the field condition, which is a complex environment with multiple players acting simultaneously (Timmusk et al. 2017). Thus, the varied agro environmental conditions posses a great challenge on the efficacy of the microbial products (Cummings 2009; Owen et al. 2015). The performance of the bioinoculants in the field may vary, as its survival is dependent on the soil, compatibility with the crop on which it is inoculated, and the interaction ability with indigenous microflora in soil, and environmental factors (Martínez-Viveros et al. 2010).

To harness the benefit of potential microbial products the following factors need to be addressed (i) the selectivity of the strain should be crop specific and ecological niche based (Dey et al. 2004; Choudhary 2011). (ii) A sufficiently long shelf life of the inoculants is key for assuring the efficacy of the biofertilizer, and is a major challenge for any kind of formulated product (Bashan et al. 2014). (iii) Creating awareness on the use of bioinoculants among the farming community is crucial to determine the usage and demand (Bhattacharjee and Dey 2014). (iv) The regulatory process involved in the registration of the products needs to be strengthened. Thus, the assurance of the efficacy of a microbial product in a particular soil with a specific crop is a complex task which should focus on facilitating reproducible results in field application and sustainable food production under changing climate.

2 Concluding Remarks

To sum up, there are several scientifically sound and innovative ways in which the potential of bioeconomy could be harnessed for creating an environmentally safe, ecologically sound and economically viable agricultural production system. Bioeconomy can be the answer to conserve and optimally maintain ecosystem service flow from agro-ecosystems. However, the full potential of any scientific and technological advancement to provide societal benefit will remain a dream if it is not backed and supported by sound policies that are designed to effectively link science and society. The success of the green revolution to usher in agricultural productivity that transformed India from 'being a ship to mouth country' to one that is 'self reliant in food production' is testimony to this. On this line, the GOI's recently introduced strategy of National Mission on Bioeconomy is trying to promote the bioproducts relevant to bio-agriculture sector, of which bioinputs assumes greater significance in the current context of agro-ecosystems.

The bio-based products like bio-fertilisers and bio-pesticides when used to increase agricultural productivity can create a stable and sustainable agro-ecosystem that is capable of meeting the goals of Bioeconomy such as food security of the burgeoning population, reducing dependency on the non-renewable resources for food production, sustainable natural resource management without degrading the environment and its services. The advanced molecular tools will be highly helpful to understand the microbial diversity in terms of structure and functions, to develop a location specific core microbiome for the major diseases and microbial consortia for biofertilizers which will be a novel and innovative products in the market for sustainable production and ensuring environmental health. In short, bio-economy is a means to improve productivity and efficiency of agriculture products, while at the same time it ensures conservation and sustenance of valuable ecosystem services from agriculture landscapes and thus reduces its ecological footprints.

References

Adesemoye AO, Kloepper JW (2009) Plant-microbes interactions in enhanced fertilizer-use efficiency. Appl Microbiol Biotechnol 85:1–12

Adesemoye AO, Torbert HA, Kloepper JW (2008) Enhanced plant nutrient use efficiency with PGPR and AMF in an integrated nutrient management system. Can J Microbiol 54:876–886

Adesemoye AO, Torbert HA, Kloepper JW (2009) Plant growth-promoting rhizobacteria allow reduced application rates of chemical fertilizers. Microb Ecol 58:921–929

Agricultural and Processed Food Products Export Development Authority (APEDA) (2017) National Programme for Organic Production, Present status in India. Retrieved on 20th May 2017 from www.apeda.com/organic.htm

Anuroopa N, Bagyaraj DJ (2017) Inoculation with selected microbial consortia enhanced the growth and yield of *Withania somnifera* under polyhouse conditions. Imperial J Interdisc Res 3(2):127–133

Araújo ASF, Santos VB, Monteiro RTR (2008) Responses of soil microbial biomass and activity for practices of organic and conventional farming systems in Piauí state, Brazil. Eur J Soil Biol 44:225–230

Avis T, Gravel V, Antoun H, Tweddell R (2008) Multifaceted beneficial effects of rhizosphere microorganisms on plant health and productivity. Soil Biol Biochem 40(7):1733–1740

Bagyaraj DJ, Sharma MP, Maiti D (2015) Phosphorus nutrition of crops through arbuscular mycorrhizal fungi. Curr Sci 108:1288–1293

Bais HP, Weir TL, Perry LG, Gilroy S, Vivanco JM (2006) The role of root exudates in rhizosphere interactions with plants and other organisms. Annu Rev Plant Biol 57:233–266

Barrios E (2007) Soil biota, ecosystem services and land productivity. Ecol Econ 64:269–285

Bashan Y, de-Bashan LE, Prabhu SR, Hernandez JP (2014) Advances in plant growth-promoting bacterialinoculant technology: formulations and practical perspectives (1998–2013). Plant Soil 378:1–33

Berendsen RL, Pieterse CM, Bakker PA (2012) The rhizosphere microbiome and plant health. Trends Plant Sci 17(8):478–486

Bhattacharjee R, Dey U (2014) Biofertilizer, a way towards organic agriculture: a review. Afr J Microbiol Res 8:2332–2343

Bhardwaj D, Ansari MW, Sahoo RK, Tuteja N (2014) Biofertilizers function as key player in sustainable agriculture by improving soil fertility, plant tolerance and crop productivity. Microb Cell Fact 13:66

Biotechnology Industry Research Assistance Council (BIRAC) (2016) Make in India for bio-tech-the way forward, biotechnology industry research assistance council, department of biotechnology, Government of India, New Delhi: BIRAC

Björklund Johanna, Limburg Karin E, Rydberg Torbjörn (1999) Impact of production intensity on the ability of the agricultural landscape to generate ecosystem services: an example from Sweden. Ecol Econ 29(2):269–291

Briat JF, Dubos C, Gaymard F (2014) Iron nutrition, biomass production, and plant product quality. Trends Plant Sci 20:33–40

Burgmann H, Widmer F, Von Sigler W, Zeyer J (2004) New molecular screening tools for analysis of free-living diazotrophs in soil. Appl Environ Microbiol 70:240–247

Burril Media (2014) Accelerating growth: forging india's bioeconomy, biotechnology industry organization and association of biotechnology led enterprises. Retrieved on 2 May 2017 from https://www.bio.org/sites/default/files/files/Burrill_AcceleratingGrowth_India-6-9-final.pdf

CEN (2011) Biobased products—overview of standards (CEN/TR 16208:2011). European Committee for Standardisation (CEN)

Chaparro C, Sabotin F (2012) Methods and software in NGS for TE analysis. In: Clifton NJ (eds) Methods in molecular biology vol 859. Springer, pp 105–14

Chen YP, Rekha PD, Arun AB, Shen FT, Lai WA, Young CC (2006) Phosphate solubilizing bacteria from subtropical soil and their tricalcium phosphate solubilizing abilities. Appl Soil Ecol 34:33–41

Choudhary DK (2011) Plant growth-promotion (PGP) activities and molecular characterization of rhizobacterial strains isolated from soybean (Glycine max L. Merril) plants against charcoal rot pathogen, Macrophomina phaseolina. Biotechnol Lett 33(11):2287–2295

Copping LG (2009) Manual of biocontrol agents, 4th edn. British Crop Protection Council, Alton, p 1350

Costanza R, d'Arge R, De Groot R, Farber S, Grasso M et al (1997) The value of the world's ecosystem services and natural capital. Nature 387:253–260

Cummings SP (2009) The application of plant growth promoting rhizobacteria (PGPR) in low input and organic cultivation of graminaceous crops; potential and problems. Environ Biotechnol 5:43–50

Daily GC, Matson PA, Vitousek PM (1997) Ecosystem services supplied by soil. In: Daily GC (ed) Nature services: societal dependence on natural ecosystems. Island Press, Washington DC, pp 113–132

Dale H Virginia, Polasky Stephen (2007) Measures of the effects of agricultural practices on ecosystem services. Ecol Econ 64:286–296

Deepak B, Wahid AM, Sahoo RK, Tuteja N (2014) Biofertilizers function as key player in sustainable agriculture by improving soil fertility, plant tolerance and crop productivity. Microb Cell Fact 13:66

Dey R, Pal K, Bhatt D, Chauhan S (2004) Growth promotion and yield enhancement of peanut (Arachis hypogaea L.) by application of plant growth-promoting rhizobacteria. Microbiol Res 159:371–394

Doran JW, Zeiss MR (2000) Soil health and sustainability: managing the biotic component of soil quality. Appl Soil Ecol 15:3–11

Dragicevic V, Oljaca S, Stojiljkovic M, Simic M, Dolijanovic Z, Kravic N (2015) Effect of the maize–soybean intercropping system on the potential bioavailability of magnesium, iron and zinc. Crop and Pasture Science 66:1118–1127

European Commission (EC) (2012) EU bioeconomy strategy and action plan- innovating for sustainable growth. Retrieved 14th Sept 2016, from http://ec.europa.eu/research/bioeconomy/pdf/bioeconomycommunicationstrategy_b5_brochure_web.pdf

FAO (2002) Organic agriculture, environment and food security. In: Nadia ELS, Caroline H (eds) Environment and natural resources series 4. Retrieved on 28th May 2017 from http://www.fao.org/DOCREP/005/Y4137E/Y4137E00.htm

FAO (2005) The fertilizer use by crops in India, Rome. Retrieved 10th May 2017 from http://www.fao.org/docrep/009/a0257e/A0257E00.htm#TOC. United Nations Food and Agriculture Organization, Rome

Feng G, Zhang FS, Xl Li, Tian CY, Tang C, Rengel Z (2002) Improved tolerance of maize plants to salt stress by arbuscular mycorrhiza is related to higher accumulation of soluble sugars in roots. Mycorrhiza 12:185–190

FiBL (2000) Organic farming enhances soil fertility ad biodiversity: results from a 21 year old field trial. Research Institute of Organic Farming (FiBL), Frick, Switzerland

Franche C, Lindstrom K, Elmerich C (2009) Nitrogen fixing bacteria associated with leguminous and non-leguminous plants. Plant Soil 321:35–59

Garg N, Manchanda G (2008) Effect of arbuscular mycorrhizal inoculation on salt-induced nodule senescence in *Cajanus cajan* (pigeonpea). J Plant Growth Regul 27(2):115

Gomiero T, Paoletti MG, Pimentel D (2008) Energy and environmental issues in organic and conventional agriculture. Crit Rev Plant Sci 27:239–254

Gomiero T, Pimentel D, Paoletti MG (2011) Is there a need for a more sustainable agriculture? Crit Rev Plant Sci 30(1–2):7–19

Gopal M, Gupta A, Thomas GV (2013) Bespoke microbiome therapy to manage plant diseases. Front Microbiol 5:15

Hartmann A, Schmid M, van Tuinen D, Berg G (2009) Plant-driven selection of microbes. Plant Soil 321:235–257

Hayat R, Safdar Ali S, Amara U, Khalid R, Ahmed I (2010) Soil beneficial bacteria and their role in plant growth promotion: a review. Ann Microbiol 60:579–598

IFOAM (2005) The IFOAM norms for organic production and processing. International Federatoin of Organic Agriculture Movements, Bonn

Iguchi H, Yurimoto H, Sakai Y (2015) Interactions of methylotrophs with plants and other heterotrophic bacteria. Microorganisms 3:137–151

Indira Devi P (2007) Pesticide use in the rice bowl of Kerala: health costs and policy options. SANDEE Working Paper No.20-07, South Asian Network for Development and Environmental Economics, Nepal

Jegan S, Baskaran V, Ganga V, Kathiravan R, Prabavathy VR (2016) Bioinoculants a tool for the alleviation of plant stress and soil fertility. In: Bagyaraj DJ, Jamaluddin (eds) Microbes for plant stress management and restoration of degraded land. New India Publishing Agency, New Delhi, India, pp 25–53

Kachhawa D (2017) Microorganisms as bio-pesticides. J Entomol Zool Stud 5(3):468–473

Kahindi J, Woomer P, George T, de Souza Moreira F, Karanja N, Giller K (1997) Agricultural intensification, soil biodiversity and ecosystem function in the tropics: the role of nitrogen-fixing bacteria. Appl Soil Ecol 6:55–76

Kawalekkar JS (2013) Role of biofertilisers and bio-pesticides for sustainable agriculture. J Bio Innov 2(3):73–78

Kolb S, Stacheter A (2013) Prerequisites for amplicon pyrosequencing of microbial methanol utilizers in the environment. Front Microbiol 4:268

Krebs JR, Wilson JD, Bradbury RB, Siriwardena GM (1999) The second silent spring? Nature 400:611–612

Krishnan R, Menon RR, Tanaka N, Busse HJ, Krishnamurthi S, Rameshkumar N (2016) Arthrobacter pokkalii sp nov, a novel plant associated Actinobacterium with plant beneficial properties, isolated from saline tolerant Pokkali rice, Kerala, India. PLoS ONE 11:e0150322

Kumar M, Tomar RS, Lade, Paul D (2016) Methylotrophic bacteria in sustainable agriculture. World J Microbiol Biotechnol 32:120

Kyselkova M, Kopecky J, Frapolli M, Defago G, Sagova-Mareckova M, Grundmann GL et al (2009) Comparison of rhizobacterial community composition in soil suppressive or conducive to tobacco black root rot disease. ISME J 3:1127–1138

Larkin RP, Fravel DR (1998) Efficacy of various fungal and bacterial biocontrol organisms for control of Fusarium wilt of tomato. Plant Dis 82:1022–1028

Laurent Philippot, Spor. Ayme′, He′nault Catherine, Bru David, Bizouard Florian, Jones Christopher M, Sarr. Amadou, Maron Pierre-Alain (2013) Loss in microbial diversity affects nitrogen cycling in soil. The ISME J 7:1609–1619

Leitão A (2016) Bioeconomy: the challenge in the management of natural resources in the 21st century. Open J Soc Sci 4:26–42. Retrieved on 10th May 2017 from http://dx.doi.org/10.4236/jss.2016.411002

Li J, Zhang D, Yang X, Tong Y (2009) Effects of land use changes on values of ecosystem functions on coastal plain of South Hanghzhou Bay Bank, China. Afr J Agric Res 4(5): 542–547

Lynch JM, Whipps JM (1990) Substrate flow in the rhizosphere. Plant Soil 129:1–10

Mäder P, FlieBbach A, Dubois D, Gunst L, Fried P, Niggli U (2002) Soil fertility and biodiversity in organic farming. Science 296:1694–1697

Malusà E, Vassilev N (2014) A contribution to set a legal framework for biofertilizers. Appl Microbiol Biotechnol 98:6599–6607

Manjula M, Gopi G (2016): In search of ecological rationale for 'The Kerala Conservation of Paddy Land and Wetland Act'. SANDEE Final Technical Report 64, South Asian Network for Development and Environmental Economics, Nepal

Martínez-Viveros O, Jorquera M, Crowley D, Gajardo G, Mora M (2010) Mechanisms and practical considerations involved in plant growth promotion by rhizobacteria. J Soil Sci Plant Nutr 10:293–319

Matson PA, Parton WJ, Power AG, Swift MJ (1997) Agricultural intensification and ecosystem properties. Science 277(5325):504–509

MEA (Millennium Ecosystem Assessment) (2005) Ecosystems and human well-being: a framework for assessment. Island Press, Washington DC

Meena KK, Kumar M, Kalyuzhnaya MG, Yandigeri MS, Singh DP, Saxena AK, Arora DK (2012) Epiphytic pink-pigmented methylotrophic bacteria enhance germination and seedling growth of wheat (*Triticum aestivum*) by producing phytohormone. Antonie Van Leeuwenhoek 101:777–786

Megali L, Glauser G, Rasmann S (2013) Fertilization with beneficial microorganisms decreases tomato defenses against insect pests. Agron Sustain Dev. https://doi.org/10.1007/s13593-013-0187-0

Melero S, Ruiz Porras JC, Herencia JF, Madejón E (2006) Chemical and biochemical properties in a silty loam soil under conventional and organic management. Soil Tillage Res 90:162–170

Mendes R, Kruijt K, de Bruijn I, Dekkers E, van der Voort M, Schneider JHM et al (2011) Deciphering the rhizosphere microbiome for disease-suppressive bacteria. Science 332:1097–1100

Miransari M (2011) Interactions between arbuscular mycorrhizal fungi and soil bacteria. Appl Microbiol Biotechnol 89:917–930

Moe LA (2013) Amino acids in the rhizosphere: from plants to microbes. Am J Bot 100:1692–1705

Nadeem SM, Zahair ZA, Naveed M, Asghar HN, Asghar (2010) Rhizobacteria capable of producing ACC-deaminase may mitigate salt stress in wheat. Soil Sci Soc Am J 74:533–542

Nellemann C, MacDevette M, Manders T, Eickhout B, Svihus B, Prins AG, Kaltenborn BP (eds) (2009) The environmental food crisis—The environment's role in averting future food crises. A UNEP rapid response assessment. United Nations Environment Programme, GRID—Arendal. Retrieved on 24th April 2017 from http://old.unep-wcmc.org/medialibrary/2010/09/07/51d38855/FoodCrisis.pdf, ISBN: 978-82-7701-054-0

OECD (2009) The Bioeconomy to 2030 designing a policy agenda: main findings and policy conclusions. Retrieved on 20th May 2017 from https://www.oecd.org/futures/long-termtechnologicalsocietalchallenges/42837897.pdf

Organic Farming Research Foundation (OFRP) (2017) Soil microbial interactions and organic farming. Retrieved 25th May 2017 from http://ofrf.org/sites/ofrf.org/files/staff/OFRF.Soil_.brochure.4.16.v4.Web_.pdf

Owen D, Williams AP, Griffi th GW, Withers PJA (2015) Use of commercial bio-inoculants to increase agricultural production through improved phosphorus acquisition. Appl Soil Ecol 86:41–54

Patil P, Ghag P, Patil S (2013) Use of bio-fertilizers and organic inputs—as LISA technology by farmers of Sangamner. Int J Adv Res Technol 2:28–33

Paul D, Nair S (2008) Stress adaptations in a plant growth promoting rhizobacterium with increasing salinity in the coastal agricultural soils. J Basic Microbiol 48(5):378–384

Pauliyz TC, Ahmad TS, Baker R (1990) Integration of *Pythium nunn* and *Trichoderma harzianum* isolate T-95 for the biological control of *Pseudomonas cepacia* or *P. fluorescence* to seed. Plant Dis 75:987–992

Picard C, Bosco M (2008) Genotypic and phenotypic diversity in populations of plant-probiotic Pseudomonas spp. colonizing roots. Naturwissenschaften 95:1–16

Postma-Blaauw M, de Goede R, Bloem J, Faber J, Brussaard L (2012) Agriculture intensification and de-intensification differently affect taxonomic diversity of predatory mites, earthworms, enchytraeids, nematodes and bacteria. Appl Soil Ecol 57:39–49

Power G Alison (2010) Ecosystem services and agriculture: tradeoffs and synergies. Philos Trans R Soc B 365:2959–2971

Raju K, Sekar J, Vaiyapuri Ramalingam P (2016) *Salinicola rhizosphaerae* sp. nov., isolated from the rhizosphere of the mangrove *Avicennia marina* L. Int J Syst Evol Microbiol 66:1074–1079
Ramesh P, Panwar NR, Singh AB, Ramana S, Yadav SK, Shrivastava R, Subba Rao A (2010) Status of organic farming in India. Curr Sci 98(9):1190–1194
Raupach GS, Kloepper JW (1998) Mixtures of plant growth-promoting rhizobacteria enhance biological control of multiple cucumber pathogens. Phytopathology 88:1158–1164
Reed and Green (2013) How microbes can help feed the world report on American Academy of Microbiology Colloquium. Retrieved on 4th May 2017 from http://academy.asm.org/images/stories/documents/Feed the world.pdf
Rodríguez H, Fraga R (1999) Phosphate solubilizing bacteria and their role in plant growth promotion. Biotechnol Adv 17 (4–5):319–339
Rodrigues EP, Rodrigues CS, de Oliveira ALM, Baldani VL, Teixeira da Silva JA (2008) *Azospirillum amazonense* inoculation: effects on growth, yield and N2 fixation of rice (*Oryza sativa* L.). Plant Soil 302:249–261
Rosenzweig N, Tiedje JM, Quensen JF, Meng QX, Hao JJJ (2012a) Microbial communities associated with potato common scab-suppressive soil determined by pyrosequencing analyses. Plant Dis 96:718–725
Rosenzweig C, Jones JW, Hatfield JL, Mutter CZ, Adiku SGK, Ahmad A, Beletse Y, Gangwar B, Guntuku D, Kihara J, Masikati P, Paramasivan P, Rao KPC, Zubair L (2012b) The agricultural model inter comparison and improvement project (AgMIP): integrated regional assessment projects. In: Hillel D, Rosenzweig C, (eds) Handbook of climate change and agroecosystems: global and regional aspects and implications. ICP Series on climate change impacts, adaptation, and mitigation, vol 2. Imperial College Press, pp 263–280
Ryan RP, Germaine K, Franks A, Ryan DJ, Dowling DN (2008) Bacterial endophytes: recent developments and applications. FEMS Microbiol Lett 278:1–9
Sekar J, Prabavathy VR (2014) Novel Phl-producing genotypes of finger millet rhizosphere associated pseudomonads and assessment of their functional and genetic diversity. FEMS Microbiol Ecol 89:32–46
Sekar J, Rengalakshmi R, Prabavathy VR (2016) Microbial consortial products for sustainable agriculture: commercialization and regulatory issues in india. In: Singh HB (eds.) Agriculturally important microorganisms. https://doi.org/10.1007/978-981-10-2576-1_7
Shanmugam V, Senthil N, Raguchander T, Ramanathan A, Samiyappan R (2002) Interaction of *Pseudomonas fluorescens* with *Rhizobium* for their effect on the management of peanut root rot. Phytoparasitica 30:169–176
Sharma SB, Sayyed RZ, Trivedi MH, Gobi TA (2013) Phosphate solubilizing microbes: sustainable approach for managing phosphorus deficiency in agricultural soils. Springer Plus 2:587
Siddiqui ZA (2006) PGPR: Prospective biocontrol agents of plant pathogens. In: Siddiqui ZA (ed) PGPR: biocontrol and biofertilization. Springer, Netherlands, pp 111–142
Singh BK (2017) Creating new business, economic growth and regional prosperity through microbiome-based products in the agriculture industry. Microb Biotechnol 10:224–227
Singh BL, Trivedi P (2017) Microbiome and the future for food and Nutrient security. Microb Biotechnol 10(1):50–53
Singh JS, Pandey VC, Singh DP (2011) Efficient soil microorganisms: a new dimension for sustainable agriculture and environmental development. Agric Ecosyst Environ 140:339–353
Smith SE, Read DJ (1997) Mycorrhizal symbiosis, 2nd edn. Academic Press, London
Soesanto L (2012) Biopesticides- the impact to the environment. In: The 5th international seminar of Indonesian society for microbiology (ISISM), vol 2, Indonesia
Swinton SM, Lupi F, Robertson GP, Lundis DA (2006) Ecosystem services from agriculture: looking beyond the usual suspects. Amer J Agr Econ 88(5):1160–1166
Tariq M, Hameed S, Malik KA, Hafeez FY (2007) Plant root associated bacteria for zinc mobilization in rice. Pak J Bot 39:245–253
Tilak K, Ranganayaki N, Pal K, De R et al (2005) Diversity of plant growth and soil health-supporting bacteria. Curr Sci 89(1):136–150

Tilman D, Farigione J, Wolff B, D'Antonio C, Dobson et al (2001) Forecasting agriculturally driven global environmental change. Science 292:281–284

Timmusk S, Behers L, Muthoni J, Muraya M, Aronsson AC (2017) Perspectives and challenges of microbial application for crop improvement. Front Plant Sci. https://doi.org/10.3389/fpls.2017.00049

Truu M, Truu J, Ivosk M (2008) Soil microbiological and biochemical properties for assessing the effect of agricultural management practices in Estonian cultivated soils. Eur J Soil Biol 44:231–237

UN (2008) Organic agriculture and food security in Africa. In: United Nations conference on trade and development—United Nations environment programme, capacity building task force on trade, environment and development. Retrieved on 16 May 2017 from http://www.unep-unctad.org/cbtf/publications/UNCTAD_DITC_TED_2007_15.pdf

UN (2013) World population prospects: the 2012 revision. UN Department of Economic and Social Affairs, New York

UNCTAD (2006) Trade and environment review. Environmental requirements and market access for developing countries: Developing pro-active approaches and strategies. United Nations Publication, NY. 275

UNDP (2010) Goal 2- Zero Hunger. Sustainable development goals. Retrieved on 4th April 2017 from http://www.undp.org/content/undp/en/home/sustainable-development-goals/goal-2-zero-hunger.html

Uren NC (2001) Types, amounts, and possible functions of compounds released into the rhizosphere by soil-grown plants. In: Pinton R, Varanini Z, Nannipieri P (eds.) The rhizosphere—biochemistry and organic substances at the soil-plant interface. Marcel Dekker, Inc, New York, pp 19–40

van der Heijden MGA, Bardgett RD, Van Straalen NM (2008) The unseen majority: soil microbes as drivers of plant diversity and productivity in terrestrial ecosystems. Ecol Lett 1:296–310

Venkatashwarlu B (2008) Role of bio-fertilizers in organic farming: organic farming in rain fed agriculture. Central institute for Dry Land Agriculture, Hyderabad, pp 85–95

Viswanath G, Jegan S, Baskaran V, Kathiravan R, Prabavathy VR (2015) Diversity and N-acyl-homoserine lactone production by Gammaproteobacteria associated with *Avicennia marina* rhizosphere of South Indian mangroves. Syst Appl Microbiol 38:340–345

Vitousek PM, Naylor R, Crews T, David MB, Drinkwater LE et al (2009) Agriculture. Nutrient imbalances in agricultural development. Science 324:1519–1520

Weller DM, Raaijmakers JM, Gardener BB, Thomashow LS (2002) Microbial populations responsible for specific soil suppressiveness to plant pathogens. Annu Rev Phytopathol 40:309–348

White PJ, Broadley MR (2009) Biofortification of crops with seven mineral elements often lacking in human diets—iron, zinc, copper, calcium, magnesium, selenium and iodine. New Phytol 49–84

Willer H, Yuseffi M (eds) (2006) The World of organic agriculture: statistics and emerging trends. IFOAM, Bonn

Revitalisation of Agricultural Biomass for an Industrial Bio-economy, Case Studies on South Africa and Netherlands Bio-economy

Marie Blanche Ting and Jim Philp

Abstract Re-industrialisation of agricultural biomass holds much promise in sectors which have reached maturity. Advancement towards functionalization of biomass for the production multiple products as part of an industrial bio-economy is gaining traction. This chapter discusses the potential of revitalizing, South Africa's mature sugar industry using green chemistry, for potential production of platform chemicals. Platform chemicals are precursors for making diverse products, which includes chemicals, materials, polymers and fuels. In an effort to build an industrial bio-economy, South Africa has made these intentions clear through its National Bio-economy strategy published in 2014. However, to do so requires, a fundamental enabling support in knowledge infrastructure, intra-regional linkages and industry, human capital development particularly interdisciplinary skills in chemistry and engineering. Such ambitions do not have to start anew, as noted in this chapter, Netherlands could provide a key partner. Netherlands has ample science cluster and biotechnology parks, strong knowledge generation in biocatalysis, bioprocessing, and has access to biorefinery facilities. We are argued that collaboration of both countries could provide a catalyst effect in realizing a truly functional industrial bio-economy within its economy and neighbouring regions.

Keywords Industrial bio-economy · Platform chemicals · Biomass
Commodity chemicals · Agriculture · Green chemistry

M. B. Ting (✉)
Science Policy Research Unit (SPRU), University of Sussex, Brighton, UK
e-mail: M.B.Ting@sussex.ac.uk

J. Philp
OECD, 2 Rue André-Pascal, 75775 Paris, France
e-mail: james.philp@oecd.org

W. Leal Filho et al. (eds.), *Towards a Sustainable Bioeconomy: Principles, Challenges and Perspectives*, World Sustainability Series,
https://doi.org/10.1007/978-3-319-73028-8_15

1 Introduction

The contribution of agriculture into Gross Domestic Product (GDP) varies significantly across African countries. For the most part the average is 15%, but the trend is in a decline owing to low productivity and limited value addition to agricultural commodities (Moyo et al. 2015:37). This paper argues that, an important opportunity arises if agricultural biomass could effectively be harnessed for the production of materials, chemicals and energy. Thus, in addition to the traditional route of food security, expanding the application of agriculture residues towards industrial feedstock could provide a revolution in this sector. We focus on the potential complementary needs outlined in South Africa's industrial Bio-economy strategy (DST 2014a, b) with the Netherlands advanced biotechnology skills. The increasing trends towards renewable bio-resources for industry applications will spur the development of new integrated bio-refinery concepts for the co-production of food and non-food products (feed, chemicals, materials and energy). Mainstream thinking usually adopts views that economies transition from primary (resource), secondary (manufacturing) to tertiary (service) sectors. As agriculture is deemed to be primary and secondary sectors, there is resistance in utilising it for economic transformation, as it can be considered backward transitions. However, our focal argument in this paper is that socio-technological innovation on agricultural biomass can go beyond the traditional notion of utilising agriculture for food security and consumption. We contend that an industrial bio-economy are means towards revitalisation in all three sections of the economy and beyond (knowledge base economy). This is because it can encourage entrepreneurships, create new markets, intra-regional trade, re-thinking infrastructure developments, enhance and leverage local knowledge for increase agricultural output, and regenerate agricultural based industries (Juma 2011).

South Africa and Netherlands were chosen as case studies because the former has recognised the needs for an industrial bio-economy for the production of chemicals, materials and fuels in contributing towards a greener economy. Whilst, the latter, has realised advances towards biomass applications, but lacks sufficient feedstock. Therefore a mutual alignment between the two countries provides meaningful collaboration for a sustainable industrial bio-economy. This chapter recognises the underlying role of science, technology and innovation, as crucial enabler in achieving the outlined preamble. Key policy recommendations arising from this paper includes the following: knowledge infrastructure, promotion of linkages (intra-regional and industry), policy and strategy alignment and human capital development (with specific focus on green chemistry). This chapter is organised according to the following sections. First, a brief overview of South Africa national innovation system, the need for industry participation to enhance the innovation gap. Then a discussion on South Africa's mature sugar industry which are great sources of biomass functionalisation. The potential application of the biomass to the chemical industry are discussed with specific references to platform chemicals. The next section discussed the Netherlands Bio-economy, with the beneficial role of an industrial bio-economy to its chemicals and energy sector. Further discussion focuses on the expansive public research in industrial bio-economy, and extensive public-private partnerships. Overall, the two countries have ambitions for a fully functional industrial bio-economy. There is remarkable potential implications not only for their own economies but the rippling impact it has on neighbouring regions (Africa and Europe).

2 Brief Overview of South Africa's National System of Innovation (NSI)

South Africa has a fairly well established National System of Innovation (NSI) for science and technology. Historically the foundations for the NSI was started under the White Paper on Science and Technology in 1998, which was followed by the National Research and Technology Foresight in 1999 and the National Research and Development Strategy (NRDS) in 2002. Throughout this time, science and technology was recognised to play a crucial role in contributing to improved competitiveness, with sustainable economic growth and inclusive development for the country's socio-economic goals.

However, there were some failures in achieving commercialisation of scientific research, in what is commonly known as the 'innovation chasm'. Furthermore, the Gross Expenditure on Research and Development (GERD) as a percentage of GDP has been on a decline since 2007/2008 (Fig. 1). The South African government has set out a target of increasing the GERD to 1.5% by 2019 that would require fundamental changes in the economic conditions and government approach towards research and development (R&D).

Key to the increase in GERD is the proportion of business on R&D. The business proportion usually accounts for around 50% of GERD but this has been on a consistent decline since 2008/2009 (Fig. 2, DST 2014b). The business sector R&D expenditure is important because it part of the NSI which extends knowledge generation through to new products and process development.

Thus, an important initiative would be to enhance industry participation to bridge the innovation gap. Relevant to this chapter is the discussion on industrial bio-economy contribution to the country socio-economic goals as earlier outlined. The industrial bio-economy focuses on two areas, these are; industry and sustainable environmental management. The former involves bio-based chemicals, biomaterials and bio-energy. The latter involves water and waste as a means of providing environmental sustainability for the industrial bio-economy (DST 2014a). The South African Bio-economy Strategy was approved by Cabinet and officially launched on January 2014. The Strategy provides a roadmap on how to develop South Africa's natural biological resources into commercial products in the fields of health, agriculture and industry. The Bio-economy Strategy placed a much stronger emphasis on inclusive development with impacts on socio-economic needs.

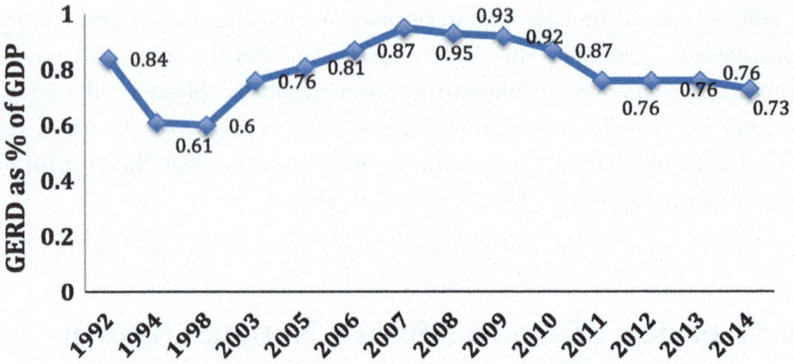

Fig. 1 Gross Expenditure on Research and Development (GERD) as a percentage on of GDP (DST 2014b)

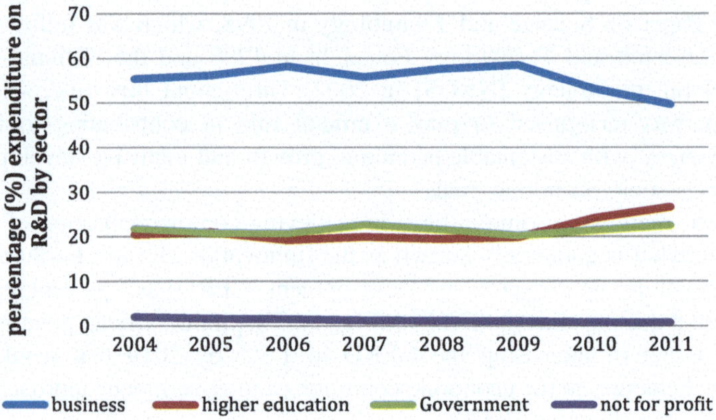

Fig. 2 Percentage of expenditure on R&D by sector (DST 2014b)

3 Revitalisation of Mature Industries

If the industrial bio-economy were to be realised, special attention should go to utilisation of biomass from mature industries, which require revitalisation in South Africa. Mature industries include sugar cane, textiles, pulp and paper that can greatly benefit from an industrial bio-economy through technological innovation. For South Africa, the sugar industry is an important source of biomass that can greatly benefit from an industrial bio-economy process. The South African sugar industry is highly beneficial to the economy, because it contributes to foreign exchange revenues, employment and serves linkages to major suppliers (DAFF 2014). It has been reported that the sugar industry employs around 79,000 jobs, and indirectly support industries that includes fertilizers, fuel, chemical, transport, food and services which employ another 350,000 jobs. This means considering the social multiplier effects of both direct and indirect jobs, it is approximated that 1 million people and more than 2% of South Africa's population is dependent on this industry (DAFF 2011). Furthermore the average sugar cane biomass produce each year is around 20 million tonnes per year (Fig. 3).

However, the industry has suffered due to major drought conditions, and also a decline of exports due to reduction in demand for sugar. In line with global trends, biomass functionalisation is moving away from single products (such as sugar cane for sugar) and more towards multiple products to chemicals, materials and fuels. Revitalisation would require these industries to shift in meeting global demand for products other than sugar, and more towards biomass beneficiation. These include: biofuels, or polymerised ethanol into plastics that give rise to new markets and needs.

According to Filho et al. (2011), production of chemicals from sugar cane can either be through fermentation or the alcohol chemistry route (Fig. 4). It is well known that sugar cane can be processed through a C4 biochemical synthesis route (de Jong and Jungmeier 2015; Filho et al. 2011). The sugar cane biomass has been referred to as "green crude" for the multitude products that can be derived either through fermentation or alcohol chemistry (Filho et al. 2011). Through the fermentation route, the classic example is the production of ethanol via sugar cane biomass. Moreover, as advances in metabolic engineering, metabolomics, and proteomics have progressed; it is now possible to produce Poly hydroxy butyrate (PHB) a bioplastic (Filho et al. 2011). Sugar cane has also been shown to produce amino acids, which can be used in the food industry as an additive, and for animal nutrition (Filho et al. 2011).

On the waste side, there are numerous useful products that can be produced. The torrefaction process of sugarcane bagasse at 150 °C has been shown to produce high quality syngas (Daniyanto et al. 2015). Furthermore, once the syngas is obtained, this can further be used to produce bio-methanol, which is chemically identical to conventional methanol. Methanol is an important platform chemical, because it is a precursor for the production of anti-knocking agents, blended with fuels or as a solvent and anti-freeze (Irena 2013).

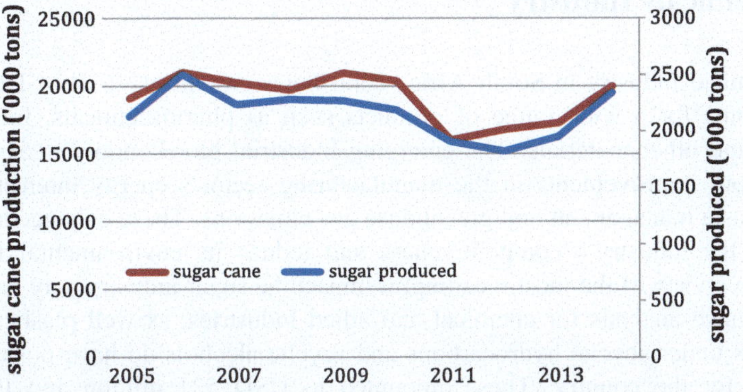

Fig. 3 A comparison between the volumes of sugar cane and sugar produced in South Africa for 2014 (DAFF 2014)

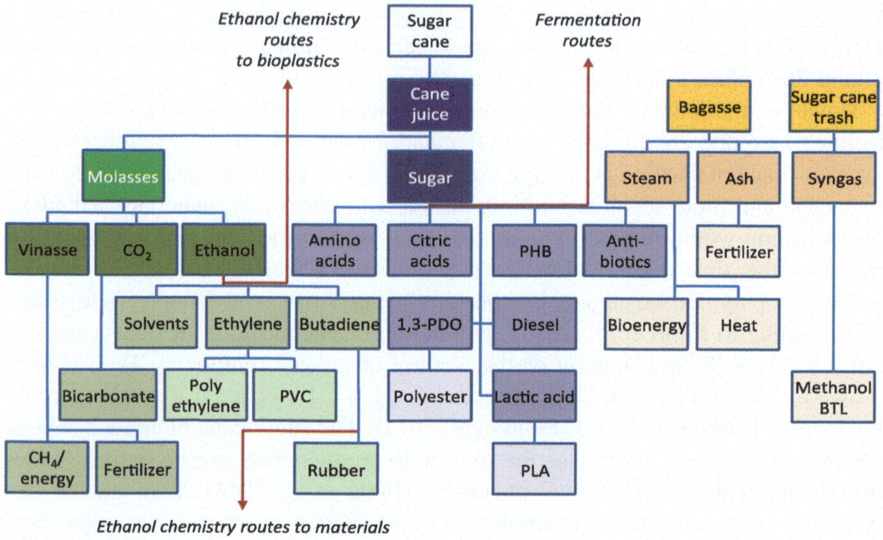

Fig. 4 A schematic illustration of the multitude of products that can be derived from sugar cane. Adapted from Filho et al. (2011)

Thus, South Africa's mature industries have great opportunities to respond to a change in market demand particularly favouring bio-based feedstock. The industrial bio-economy can form a key contribution in bridging the gap between research and commercialisation because it is premised on industrial trends and demand. Taking into consideration the immense potential of converting the sugar industry biomass into multiple products. The most appropriate application that should be considered is in the chemicals industry.

4 Chemicals Industry

The chemical industry in South Africa contributes around 5% to its GDP. It provides inputs for a wide range of products such as pharmaceuticals, food, fuels, plastics and other materials. The emerging industrial bio-economy is expected to bring about improvements in the manufacturing sector's energy intensity, water usage, waste management and greenhouse gas emissions. These enhancements will improve the industry's competitiveness and reduce its environmental footprint. A brief overview of the most meaningful impact the sugar industry may have, is on trade balance analysis for chemical and allied industries, as well plastics. A few chemicals under acyclic hydrocarbons and acyclic alcohols do have positive trade balances for the country. These amounted to USD 521 million and USD 168 million respectively for 2014. However, enzymes, insecticides and fertilizers

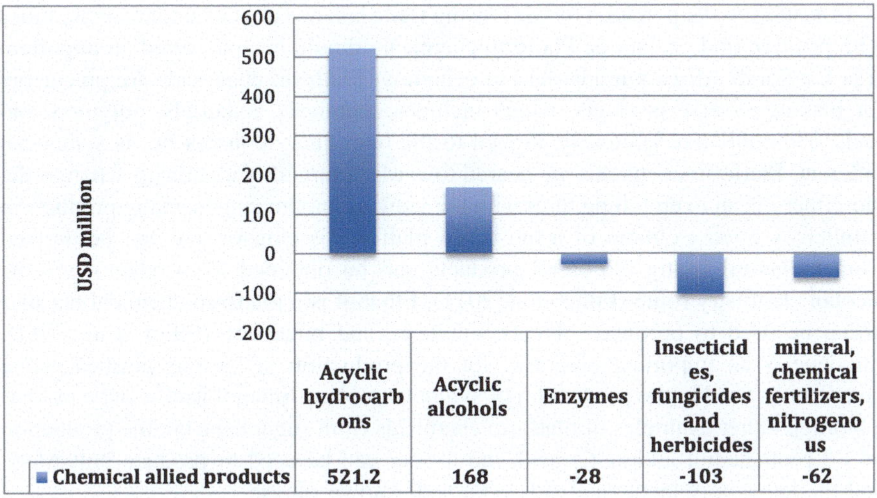

Fig. 5 A trade balance analysis of selected products under South Africa's chemicals and allied industries for 2014 (Harvard Economic Atlas 2014)

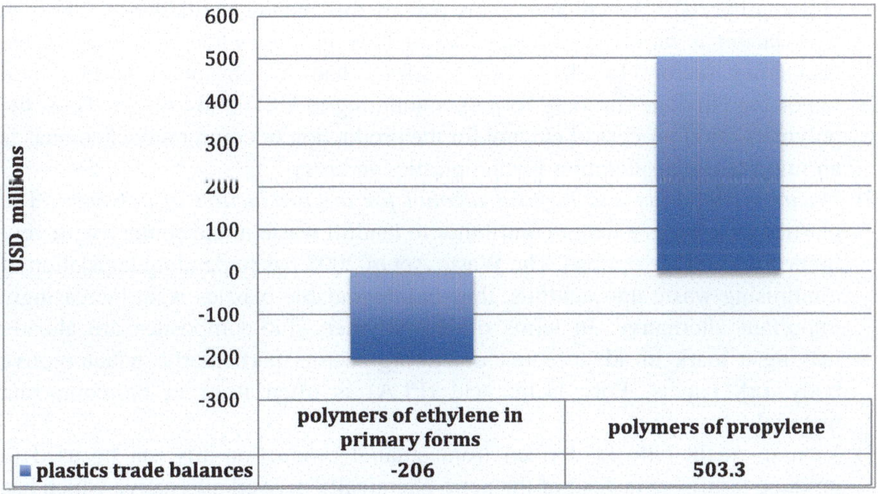

Fig. 6 A trade balance analysis of selected chemicals under South Africa's plastics industries for 2014 (Harvard Economic Atlas 2014)

derived from nitrogenous compounds are on negative trade balances, totalling USD 193 million (Harvard economic atlas 2014) (Fig. 5).

In terms of the plastics industry, chemicals that pertain to ethylene and propylene provide negative and positive trade balances respectively (Fig. 6).

Considering the potential of minimising trade balance deficits or enhancing those with positive trade balances. Platform chemicals should be considered an important area for South Africa's industrial bio-economy. Platform chemicals are precursors for making diverse products, which includes chemicals, materials, polymers and fuels. This initiative is closely aligned to the biorefinery concept but is somewhat different. Platform chemicals are geared towards commodity chemicals whereas the biorefinery is an overarching term with the view of maximising as many products in parallel. A good example of a bio-based platform is ethanol that can be derived from molasses. Many bio-based products can be obtained from ethanol via the alcohol chemistry route (Filho et al. 2011). Ethanol is a platform chemical because it can be used to produce solvents, ethylene, and butadiene (Filho et al. 2011). Ethylene is an important precursor for the production of various plastics, while butadiene is used in the production of synthetic rubber. Ethanol itself can be used as a biofuel. Other examples of platform chemicals from sugar cane are the production of 1.3-propanediol and lactic acid, the former can be used to produce polyesters, and the latter poly lactic acid (PLAs) a well-known plastic.

Based on export and import trade analysis of the South African market and the relevance to the bio-economy, there are three platform chemicals for consideration:

(1) *Ethanol*. South Africa has a biofuel strategy mandate to implement 2% of biofuels by 2015, bioethanol forms part of this mandate. Ethylene is the raw material for a variety of products particularly plastics where it is used in the packaging sector. In 2014, South Africa had a substantial trade deficit importing ethylene (primary forms), amounting to USD 206 million. Thus, the polymerisation process of ethanol for the production of commodity chemicals is an important consideration for the plastics industry.

(2) *Dicarboxylic acids and furfural alcohol for the production of polymers*. The plastics industry is a major contributor to landfill waste in the country with very limited end-of-life options. The Waste Act of 2008 has impending legislation of minimising waste into landfills, thus bio-degradable plastics is an increasingly important alternative. In terms of bio-polymers, bio-composites are already making a mark in advance manufacturing sectors particularly in automotive bodywork panels. Poly lactic acid (PLA) is often used in bio-composite materials.

(3) Vinasse- main effluent derived from ethanol production-this can be used to produce fertilizers instead of the petro chemically derived alternative which has caused an increase in anthropogenic inputs into the nitrogen cycle. Considering the trade deficit of importing nitrogenous-based fertilizers, bio-based fertilizers are an important alternative for the agriculture sector.

There is clearly a case to initiate a bio-based economy in South Africa considering the immense potential of the sugar industry biomass, as well the complementary application to the country's chemicals and plastic industry. However, given that it is a nascent industry in the country there are challenges that concern the right skills base, enabling environment for research and development, and industrial translation to scalable productions. Where there maybe complementarity in the case of the Netherlands bio-economy. The discussion now turns to the relevance of the Netherlands bio-economy.

5 The Netherlands and the Bio-economy

The Netherlands represents a very interesting test case for the bio-economy, with strikingly different conditions from South Africa, and yet with common problems. It has a highly advanced and diversified economy, with a pronounced focus on agriculture. It is one of the most densely populated countries in the world, and yet is the world's second-largest exporter of food and agricultural products, after the United States. But in common with many advanced economies, especially in Europe, the major limiting factor for a domestic bio-economy is the availability of domestic biomass.

Without large volumes of domestic fossil fuels other than gas, it is another of the advanced economies that relies on imports. The Netherlands is one of the most carbon-intensive nations in Europe. Around 90% of energy generation is fossil-based. Although currently the fifth-largest exporter of natural gas, it is predicted that by 2025 the Netherlands could be a net importer of gas (Deloitte 2015). The Netherlands has lagged behind other European Union (EU) countries in growing the renewable share of its overall energy portfolio. And yet the Netherlands is also highly visible in industrial biotechnology, in both research and industry, and also in green chemistry. Part of the reason is to maintain its competitive position in the chemicals sector as little can be done about energy and labour costs.

Currently, around 10% of the land surface in the Netherlands is covered by forest. However, Dutch forests are rather fragmented, with typically small forest patches. Production forests are of limited economic importance as far as wood production is concerned. Economic results have in fact been shown to be negative (Van der Maaten-Theunissen and Schuck 2013). Therefore importation of biomass has been deployed to meet emissions reduction targets. The country serves as an important trade hub for Western Europe, bolstered by the region's largest ports, especially the port of Rotterdam, which remains the busiest in Europe. As a result, access to raw materials is simplified for the Netherlands and the port of Rotterdam is also one of the most important petrochemicals sites in the world.

5.1 Bio-economy in the Dutch Government

The Dutch government had a dedicated directorate for bio-economy (the Directorate Bio-based Economy of the Ministry of Economic Affairs, Agriculture, and Innovation). This has evolved into the Unit of Green Growth and Bio-based

Economy at the Ministry of Economic Affairs. Other ministries have interests in bio-economy e.g. the Sustainability Directorate, in the Ministry of Infrastructure and the Environment; and research, skills and education aspects in the Ministry of Education, Culture and Science. There have been inputs in past from the Ministry of Foreign Affairs.

5.2 *The Chemicals Industry is Vital to the Netherlands*

The Netherlands is a country at the heart of the very large and successful European chemicals industry. In 2013, sales of chemicals accounted for over 7% of EU manufacturing output (Oxford Economics 2014). The centre of European chemicals production is the triangle of Belgium, the Netherlands and North West Germany (the interconnected Antwerp-Rotterdam-Rhine-Ruhr Area, ARRRA). Chemistry is central to the economies of these nations. For example, there are close to half a million industrial chemists in Germany (OECD, unpublished data). Around 80% of the 25 top chemical companies in the world maintain significant operations in the Netherlands. The port of Rotterdam is one of the strongest refining and chemical clusters in the world. The chemicals sector is vital to the Dutch economy. The Netherlands is the third largest chemical producer in Europe after Germany and France. It provides work for 64,000 people, distributed among more than 400 companies. It is the second-largest business sector in the Netherlands after the food, beverages and tobacco industry. The sector (including the rubber and plastics industry) contributed 3% to the Dutch GDP in 2012. The export of chemical products amounted to 18% of all goods exports, and 20% of all exported goods produced in the Netherlands. The chemical industry made a positive contribution of EUR 27 billion to the balance of trade in 2012, which was 64% of all goods (VNCI 2012).

While the population of the Netherlands represents a mere 0.2% of the global total, the chemical industry is responsible for a full 2% of global output. Such is the importance of chemistry to the Netherlands that the government selected the industry as one of nine national growth sectors, with the aim of putting into place a tailored industrial policy. By 2030 the chemicals industry is expected to have begun to reduce its dependence on hydrocarbons by increasing the use of secondary (waste) and bio-based feedstock. Hydrocarbons will still form the predominant feedstock, but bio-based alternatives will represent as much as 15–20% of inputs (Deloitte-VNCI 2012).

However, the chemicals sector in many OECD nations is threatened by the rise of Asia, particularly China, and the Middle East in volumes and competitiveness in chemicals. Industrial biotechnology and green chemistry are seen as ways to keep the chemicals sector competitive, and this is clearly highlighted by BE-Basic (Bio-based Ecologically Balanced Sustainable Industrial Chemistry).

5.3 *Broader Climate Change Ambitions*

The Dutch government supports the EU goal of GHG emissions reduction of 40% by 2030, and regards 40% as the minimum. Regarding transport, the government wishes to cap 2030 emissions at a level 17% lower than those of 1990. By 2035 all new passenger cars will have to run CO_2 free. In energy, future renewables may be dominated by onshore and offshore wind energy. However, there remain uncertainties over the long term costs and benefits. The government wishes to build new nuclear power stations, but decisions are in limbo following Fukushima.

5.4 *Energy Security*

The main sources of biomass for electricity generation are wood pellets and residues from agriculture and industry.[1] The importance of wood pellets for large scale power generation is increasing dramatically, such that many countries have become net importers, (Table 1), as an earlier evaluation predicted (Banse et al. 2008). Wood pellets are imported to the EU mainly from Canada, the United States and Russia, which together have a total share of over 80% of pellet imports to the EU.

These patterns create different policy requirements for different countries. For example, the Netherlands mainly co-fires wood pellets in large coal power stations for electricity production although it has limited capacity to increase its production of wood pellets, so relies heavily on imports. Comparing Germany and the Netherlands on a *per capita* basis, by these 2010 figures Germany produced roughly double the quantity of wood pellets per person compared to the Netherlands, but the Netherlands consumed much more than double per person compared to Germany. Key factors to achieve the national targets of the Netherlands have been identified as enhanced technological development and the import of sustainable biomass resources (Hoefnagels et al. 2013).

6 Bio-economy Policies

The term bio-based economy is relatively well known in the Netherlands and is used to address that part of the economy that is active in producing bio-based materials and products and bioenergy, with interest in the biomass needed for that, as biomass availability is a major issue in the Netherlands. In the Netherlands, the Ministry of Economic Affairs, Agriculture and Innovation has decided that the bio-based economy is one of the emerging economic pillars to be supported. The development of the national strategy was the result of an on-going interaction

[1]http://epp.eurostat.ec.europa.eu/portal/page/portal/eurostat/home/.

Table 1 Wood pellet production and consumption patterns in various countries (thousand tonnes)

Country	Production		Consumption	
	2005	2010	2005	2010
Austria	440	850	305	660
Denmark	187	180	820	1600
Finland	190	253	55	213
Germany	240	1200	200	1845
Italy	240	750	290	1450
The Netherlands	110	120	487	913
Sweden	1100	1645	1480	2200
EU	2628	9260	3835	11,400

Adapted from Scarlat et al. (2013)

Note One source has predicted that Europe will be importing 80 million tonnes of solid biomass per annum by 2020 (Cocchi et al. 2011)

between business, society and science, stimulated by policy makers. In April 2012, the Cabinet presented a mid- and long-term vision and strategy for the bio-based economy (Dutch Cabinet 2012) via the *Framework Memorandum Bio-based Economy*. It looked at the entire bio-based economy chain, from sustainable biomass supply to bio-based applications.

The *Innovation Contract Bio-based Economy* is a joint agenda developed by industry and research organisations. It contains six work packages, each covering the entire innovation chain (from more basic research to valorisation). These work packages are: bio-based materials; bioenergy and biochemistry; biorefinery; biomass cultivation and production; re-winning and re-use: water, nutrients and soil, and; economics, policy and sustainability. In total, more than 100 companies will participate in the projects and have committed more than EUR 200 million (Innovatiecontract Bio-based Economy 2011–2016 2012).

Energy Agreement for Sustainable Growth from the Netherlands government of Rutte/Asscher set the country on a course to achieving a completely sustainable energy supply system by 2050. Among other objectives of the *Energy Agreement for Sustainable Growth* is a rapid expansion in energy generated from renewable sources (from 4.4% in 2014 to 14% in 2020 and 16% in 2023). This aligned the Dutch trajectory with broader EU climate change mitigation targets. Published in 2013 by the Social and Economic Council of the Netherlands (SER 2013), it specified that requirements on carbon debt should be implemented in 2015. It also set out sustainability requirements for biomass used for co-firing in power stations.

The SDE-plus (*Stimuleringsregeling Duurzam Energieproductie*) is an enabling policy towards the 14% goal of 2020. It is a production-based subsidy that bridges the revenue gap for power producers between fossil fuel-based energy and a renewable alternative, including biomass co-firing. It compensates producers for the higher cost of a renewable option for a fixed number of years, depending on the technology used.

As part of the Netherlands green growth policy, the *Green Deal Programme* is a business support policy that aims to involve the private sector in transition to a greener economy. The programme was first launched in October 2011 by the Netherlands Ministry of Economic Affairs.[2] Green Deals cover nine themes: energy, bio-based economy, mobility, water, food, biodiversity, resources, construction and climate. The main aim is to remove barriers in order to help sustainable initiatives get off the ground and to accelerate this process where possible. *Groene Groei: voor een sterke, duurzame economie*[3] (2013) jointly addresses economic growth and environmental improvement, and continued support for the bio-economy.[4]

Similar challenges faced by South Africa in increasing private sector involvement in research and development. The Dutch were also struggling to derive benefit for its SME (small medium enterprises) from the research in public knowledge institutes. The SMEs were having difficulty in sourcing funding for relatively high-risk innovative projects. Furthermore, there was also incoherent alignment between government policies and private sector needs To address these challenges, the Dutch Ministry of Economic Affairs and the Ministry of Education, Culture and Science worked together with other Ministries to ensure that in the run-up to Horizon 2020[5] the research themes and other topics, such as SME participation, were formulated in a way that benefits Dutch interests. Netherlands Enterprise Agency provides a stimulus and support for Dutch participants in Horizon 2020.

The *Top Sectors* policy is aligned with Horizon 2020 and it aims to strengthen the position of nine economic sectors in which the Netherlands has a leading position internationally (Government of the Netherlands 2014). The *Top Sectors* are characterised by a strong market and export position, a solid knowledge base, close collaboration between entrepreneurs and knowledge institutes and the potential to provide input in solving societal challenges. South Africa should learn from this, because it is based on policies that match industry need with high societal impact. The *Top Sectors* approach was planned where investments are targeted in nine leading sectors of the economy (Table 2). It sets out an integrated policy for each sector. The key basic components of this integrated approach are: research and innovation, human capital, regulatory framework and the international dimension.

Arguable, all of the nine sectors have a link to bio-economy, but there is a direct and demonstrable link in at least six of the nine sectors. *Top Sectors* is effectively responding to the societal challenges set out in Horizon 2020. One of these four challenges explicitly mentions bio-economy. The four challenges are:

---

[2]http://iepd.iipnetwork.org/policy/green-deal.

[3]Green Growth: a strong, sustainable economy.

[4]https://www.rijksoverheid.nl/onderwerpen/duurzame-economie/documenten/kamerstukken/2013/03/28/kamerbrief-groene-groei-voor-een-sterke-duurzame-economie.

[5]https://ec.europa.eu/programmes/horizon2020/.

Table 2 The Top 9 Sectors chosen by the Dutch government as their areas of important focus, relative to their GNP

Top 9 sectors	Activities	% GNP
Agriculture and food	Agro-food sector: various (animal and vegetable) food chains, food valley	4.4
Life sciences and health	Vaccines, biomedical materials, Health Valley, bioscience park Ledien	3.7%
Horticulture and seed stock	Plant breeding, vegetables, fruit, trees, flowers, bulbs, green ports	1.4
Logistics	International supply chains, service logistics, freight, main ports in Rotterdam	3.4%
Energy	Energy valley, international energy market (Gas roundabout)	3.4%
Chemical	Petrochemicals, fine chemicals, maintenance valley	2.2%
Water	Water and delta technology, water purification	0.4%
High tech systems and materials	High tech materials, nanotechnology, aircraft, steel	6.7%
Creative industry	Fashion, design and media	1.6%

1. Health, demographic change and well-being;
2. Food security, sustainable agriculture and forestry, marine, maritime and inland water research and the bio-economy;
3. Secure, clean and efficient energy, and;
4. Smart, green and integrated transport.

The following sections will now discuss the broad research and development activities that pertains to industrial bio-economy in the Netherlands. The main point is to outline the significant developments and strength in biorefinery, green chemistry, and biocatalysis with partnerships from industries.

7 Education and Training Are Key to Producing the Bio-economy Workforce

The Netherlands has a long history in microbiology, fermentation and bioprocess. A demand for 10,000 bio-based experts is expected in the next ten years (Langeveld et al. 2016). Many universities and schools of applied sciences are responding by developing dedicated courses, BSc and MSc programmes. Bioprocessing is a key component of future bio-economy objectives, and the country has particular depth in bioprocess education, with a nucleus at Delft. Examples are:

- A Masters programme in *Biochemical Engineering*, in which students learn how to design bioprocesses and understand the underlying cellular, molecular and physical principles for the engineering aspects of bioprocesses;
- A two-year Post-MSc programme *Designer in Bioprocess Engineering*.[6] This programme develops MSc-holders in (bio-)chemical engineering or related disciplines into multi-disciplinary specialists and prepares for an industrial career. Graduates of the programme receive the title Professional Doctorate in Engineering (PDEng).

The traditional on-campus experience is bound to be revolutionised by the explosion of Massive Open Online Courses (MOOCs), which will enhance, if not partially replace classroom and laboratory work. MOOCs are especially suited to disciplines that are evolving fast and require multi-disciplinary skills, such as synthetic biology and industrial biotechnology (Delebecque and Philp 2015). A specialist MOOC for industrial biotechnology is offered jointly by the Technical University of Delft and the University of Campinas (Box 1).

7.1 Multi-Disciplinary Bio-economy Research and Development

Multi-disciplinarity in R&D is a central feature of bio-economy research as the societal needs that demonstrate the need for a bio-economy do not conveniently fall into single disciplines. For example, there is clearly cross-over between food security and energy security as biomass has dual roles in a bio-economy. Research is being conducted in the Netherlands *Top Sectors Chemicals* policy into process technologies for the application of new raw materials, such as biomass and CO_2. A striking example of this approach is the artificial leaves project (Box 1), which combines energy security and climate change mitigation with the chemicals sector, and crosses several disciplines, including industrial biotechnology and green chemistry.

Box 1. The BioSolar Cell project Artificial leaves contain catalysts, light antennas and other components that are 'copied' from the natural photosynthesis system. Theoretically, these artificial leaves can convert solar energy into chemical energy with a very high efficiency of more than 40%. They are therefore about 100 times more efficient than natural leaves. Using CO_2 as a source of carbon helps to reduce the use of fossil oil and the emission of greenhouse gases. Also, atmospheric CO_2 can be bonded with hydrogen to make methanol. Methanol can then be used as a raw material in

[6] http://www.tnw.tudelft.nl/?id=33083&L=1.

standard chemical processes to make other chemicals and fuels, such as kerosene.

This work is being conducted in the BioSolar Cells Project, a public-private partnership of companies and knowledge institutions led by Wageningen University, with other knowledge institutions from Amsterdam, Delft, Den Bosch, Eindhoven, Groningen, Leiden and Twente. It is supported by the Dutch government (principally the Ministry of Economic Affairs, Agriculture and Innovation). The programme will last five years and has a total budget of EUR 42 million.

Source: Various, and http://www.biosolarcells.nl/en/home.html

Box 2. edX course in Industrial Biotechnology The course is a joint initiative of TU Delft (Netherlands), the international BE-Basic consortium and University of Campinas (Brazil). It provides the insights and tools for the design of sustainable biotechnology processes. The basics of industrial biotechnology are used by students for the design of fermentation processes for the production of fuels, chemicals and foodstuffs.

The TU Delft MOOCs are offered through the online edX platform, where MIT, Harvard and other universities have been making courses available to anyone with an internet connection since 2012. One of the reasons that TU Delft chose to use edX is that it allows the publication of materials with an open licence, making it possible for others to use the materials as well.

Source: adapted from Delebecque and Philp (2015)

7.2 *Public-Private Partnerships and Competitive Clusters*

The cluster concept is far from new, but national programmes based on the cluster model continue to be prominent government strategies for technology development and deployment. The cluster model is widely applied in industrial biotechnology, and the first international clusters are forming. It is particularly appropriate to industrial biotechnology given the wide range of stakeholders from farmers to small and big industry, and the uniquely expensive biorefinery facilities.

7.2.1 BE-Basic

The Biotechnologically Based Ecologically Balanced Sustainable Industrial Consortium (BE-BASIC) coordinated by TU Delft is an international public-private partnership (PPP) that develops industrial bio-based solutions to build a sustainable society. BE-Basic[7] has a research and development budget of more than EUR 120 million. The Ministry of Economic Affairs, Agriculture and Innovation as part of the Economic Structure Enhancement Fund (FES) fund half and the other comes from industry and knowledge institutions. BE-Basic was founded early in 2010, and puts its international focus into practice through strategic partnerships in a selected number of countries: Brazil, Malaysia, the US and Vietnam. The BE-Basic Foundation initiates and stimulates collaborations between academia and industry, between scientists and entrepreneurs and between the Netherlands and abroad. It has plenty of industrial partners, of which DSM and Corbion are the largest. At present there are 40 partners with an international focus whereby TU Delft has set up a joint office in Brazil.

7.2.2 Bio-Based Delta

Based in the Southwest of the Netherlands, Bio-based Delta[8] combines entrepreneurs, knowledge institutes and governments in Zeeland, Zuid-Holland and Brabant to work together towards a bio-based economy. The region has large agricultural, horticulture and chemical sectors, and an advantageous geographic location (along the Antwerp-Rotterdam axis). Bio-based Delta focuses on three pillars: green raw materials; green building blocks, and; making the process industry more sustainable.

7.2.3 Bio Base NWE

Bio Base NWE is a project funded by the Interreg North-West Europe 2014–2020 Programme.[9] Eight partners from five European countries (Belgium, Germany, Ireland, The Netherlands and the United Kingdom) are involved in this three-year project. The overall aim of Bio Base NWE is to support the development of the bio-based economy in NWE by facilitating innovation and business development by small and medium sized enterprises (SMEs) and improving professional training and education for the bio-based economy.

7.2.4 Big-C

BIG-C is an initiative of North Rhine-Westphalia (NRW), Flanders and The Netherlands[10] in sustainable chemicals (this region represents 30% of the entire EU chemical industry). It is a cross-border Smart Specialisation initiative aiming at

[7]www.be-basic.org.

[8]http://biobaseddelta.nl/pagina/english.

[9]http://www.nweurope.eu/5b/.

[10]http://www.fi-sch.be/nl/wp-content/uploads/Version-180414-BIG-C-position-paper.pdf.

transforming Europe's industrial mega cluster in Flanders (Belgium), The Netherlands and the German state of NRW into the global leader of bio-based innovation growth.

7.2.5 Leiden University and Bioscience Park

Leiden University is the oldest in the Netherlands, hosting 32,000 students of which 8000 are Masters Candidates, 9000 Ph.D. candidates and 2000 international students. Education is provided in a research-intensive environment with problem solving taught through research. The Leiden Bioscience Park covers 120 hectares, hosts 100 companies and 5000 employees. These include international companies, including two Japanese companies. The interaction of companies is actively managed. The park hosts five of the top ten biotechnology deals in Europe. A biotechnology training facility is being built where expertise in Good Manufacturing Practice (GMP), and Standard Operating Procedures (SOPs) are being developed, as an international facility, opening in 2015 at a cost of EUR 7.7 million.

7.2.6 List of Industries Playing a Significant Role as Uptakes for the Products Related to the Industrial Bio-economy

The examples below are not meant to be exhaustive. They are chosen to illustrate the range of activities being developed in the Netherlands.

- ***DuPont***

DuPont is located within the Leiden University Bio Science Park. DuPont is a global company that is involved in very broad areas of products, from electronics to materials, through chemicals, to foods. The company's total sales for 2012 were $34.8 billion of which industrial biosciences contributed around $1.2 billion.

- **AkzoNobel *and Royal Cosun***

AkzoNobel, a Dutch multinational paints, coatings and specialty chemicals company headquartered in Amsterdam, and agro-industrial cooperative Royal Cosun, headquartered in Breda, formed a partnership in 2016 to develop novel products from cellulose side streams resulting from sugar beet processing (Il Bioeconomista 2016). This addresses the need for more sustainable raw materials from a variety of industries, such as food and healthcare, as well as the coatings and construction sectors. It highlights both the Royal Cosun focus on the bio-based economy and the AkzoNobel Planet Possible agenda, which includes ongoing efforts to develop and introduce sustainable, bio-based products that contribute to a circular economy.

- **Avantium**

Avantium has developed and commercialised technology for catalytic conversion of plant-based sugars into building blocks for plastics and other applications. Avantium is a chemical technology company with interests in renewable chemistry. It has developed a process in its laboratories in Amsterdam and pilot plant in Geleen for the production of 2,5-furandicarboxylic acid (FDCA). The leading high-volume application is in PEF (polyethylene furanoate), a polyester based on FDCA and MEG (ethylene-glycol). PEF is the 100% bio-based alternative to PET for plastic bottles. In 2016, Avantium and BASF announced that they intend to establish a joint venture to erect a production plant for FDCA based on fructose at Antwerp, Belgium. Currently the technology utilises first generation feedstock (sugar and starch (corn) crops) but Avantium are developing processes for second generation (non-food) crops. This could be an opportunity for South Africa to provide Avantium with sugarcane bagasse for bio catalytic conversions into bioplastics.

- **DSM**

DSM is a global life sciences and materials sciences company headquartered in Heerlen, the Netherlands. DSM is exploring multiple routes using its competencies in both chemistry and biotechnology. This technology is seen as a platform for the development of other high potential bio-based building blocks and/or platform molecules.

- ***Reverdia***

Reverdia is a joint venture between DSM, and Roquette Frères, the global starch and starch-derivatives company of France. In 2012, Reverdia began operations in Cassano Spinola, Italy, as the world's first dedicated, large-scale plant for the production of succinic acid from renewable resources.

- ***POET-DSM***

The POET-DSM Advanced Biofuels joint venture began in 2012. In 2014 it opened a cellulosic ethanol production plant, Project LIBERTY, in Emmetsburg, Iowa, United States. This was one of the first cellulosic ethanol plants in the world, built to convert corn cobs and residues into ethanol.

- ***Biorefinery.nl***

Biorefinery.nl[11] is a joint initiative of Wageningen University and Research Centre (WUR) and the Energy Research Centre of the Netherlands (ECN), Petten, supported by the former SenterNovem. Its role is to inform industry, research institutes, universities, social institutes, and governments about biorefinery research activities, new developments and projects. It also aspires to creating a global biorefinery

[11] http://www.biorefinery.nl/home/.

vision and formulating a roadmap for research and development of biorefinery processes.

Having discussed the multiple industries that has significance for an industrial bio-economy in the Netherlands, which provides scale up opportunities for research and development. The discussion now turns to biomass requirements and sustainability.

8 Biomass Requirements

Given the small size of the country and its high population density, local biomass supply in the Netherlands is limited. Biomass currently is the main source of renewable energy in the Netherlands. Plans show that, by 2020, the amount of biomass used to replace fossil resources will have doubled, compared to 2010 levels (PBL 2014). The total biomass demand for the Netherlands could be 2500 petajoules[12] (PJ) in 2050 (given many variables), whilst it is estimated that the Netherlands, in the long term, may produce or collect no more that around 200 petajoules in biomass. Achieving even this quantity will be a real challenge. Therefore high efficiency is growing and using biomass is necessary. As Dutch production per hectare is already the highest in Europe (Ministry of Economic Affairs 2013), there is limited scope for intensification, and certainly not much scope for land extensification. Therefore, imports will be required.

Moreover, there is strong competition for biomass in different sectors (Fig. 7). For example, the chemical industry in the Netherlands aims to produce 50% of its plastics from biomass by 2050. But chemistry is by far the smaller sector of the top four in demand for biomass. Nevertheless, chemistry creates more jobs and higher value added than energy or biofuels (Piotrowski et al. 2016). Research indicates that bioplastics have lower demands on land for biomass due to much smaller production volumes than fossil-derived fuels and greater land efficiency (Endres and Siebert-Raths 2011). Estimates (Carus and Piotrowski 2009) suggest that the impact of bioplastics on food markets, agricultural prices and land competition in 2008 was about 250 times less than that estimated for biofuels.

8.1 Biomass Sustainability

A future perspective is needed. There will in fact be a worldwide market for biomass due to perhaps the greatest conundrum faced by society. Worldwide demand for energy and resources will increase substantially over the coming

[12]1 petajoule (PJ) is 1×10^{15} J (a million times a billion joules), and 1 exajoule (EJ) is 1×10^{18} J (a billion times a billion joules).

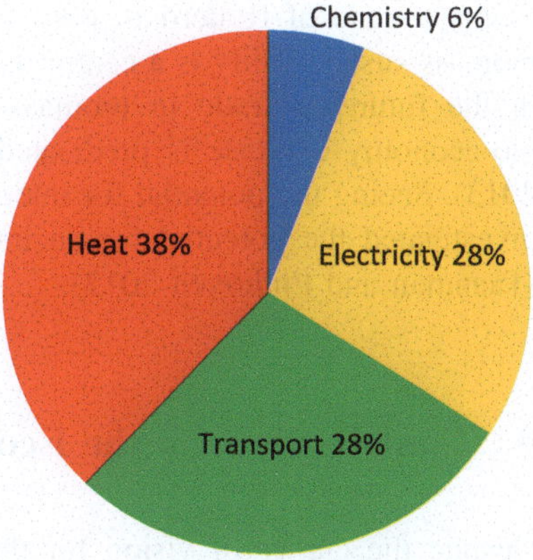

Fig. 7 Requirement for biomass in 2050 for the Netherlands (PBL 2014)

decades. Their increased use will lead to more GHG emissions, while countries have agreed to reduce emissions in order to mitigate climate change. Europe, for example, strives for a low-carbon economy by 2050, with 80–95% lower greenhouse gas (GHG) emissions levels. The best chances for meeting these targets include a substantial contribution from biomass as a source of renewable carbon and energy. The possibilities for the Netherlands to use more biomass in the future not only depend on potential global production levels, but also on the demand in other countries. If biomass becomes profitable and thus an attractive renewable resource for use in the Netherlands, however, it will also be so in other countries, creating a large demand and a whole new market and infrastructure for biomass growth, harvesting and transport.

There is a history of research on biomass sustainability in the Netherlands. The PBL Netherlands Environmental Assessment Agency (PBL) and CE Delft have collaborated to explore the sustainable use of biomass in the bio-based economy. In February 2012, the PBL, in collaboration with CE Delft, published the PBL Note: Sustainability of biomass in a bio-based economy (PBL 2012). A follow-up paper examined cascading use of biomass (CE Delft 2012). Cascading is often mentioned as essential to a sustainable bio-based economy.

8.2 *A Potential International Biomass Dispute Settlement Facility*

Whilst, an industry built on agricultural biomass is beneficial, caution has to be made regarding sustainable biomass use (Bosch et al. 2015). Efforts are being made to harmonize and develop metrics, research on life cycle analysis, integrated methodologies (e.g. biomass stock, inputs of labour and capital) as well as governance framework for biomass sustainability (Bosch et al. 2015). Given the predicted importance of biomass as a feedstock of the future, and the critical role in biomass trade of Rotterdam port, it has been recognised in the Netherlands that biomass sustainability is a central issue in biomass trade and will become more so in the future as trade in biomass expands. In fact, it is clear that the whole bio-economy exercise is predicated upon biomass sustainability (Pavanan et al. 2013). Seeing the potential for international biomass disputes, the Netherlands has investigated the potential for an international biomass dispute settlement facility (Taanman and Enthoven 2012).

9 Contributions to the Economy

Despite the clarity of vision for the future bio-economy clearly evident at government level, implementation has lagged behind in the Netherlands (Fig. 8).

The front-runners in Europe are clearly Germany, Italy and France. Despite the importance of the chemicals sector in Belgium and the Netherlands, turnover and employment figures are relatively low compared to other Western European economies. There could be several reasons for this. First of all, in comparison, Western and Northern European countries generate much higher turnover compared to the employment generated. The countries with the highest ratio between turnover and employment are Ireland, Finland and Belgium. Another could be that the Netherlands, as a knowledge economy, may be creating greater opportunities outside the country. Given some of the joint ventures, this does seem to be the case.

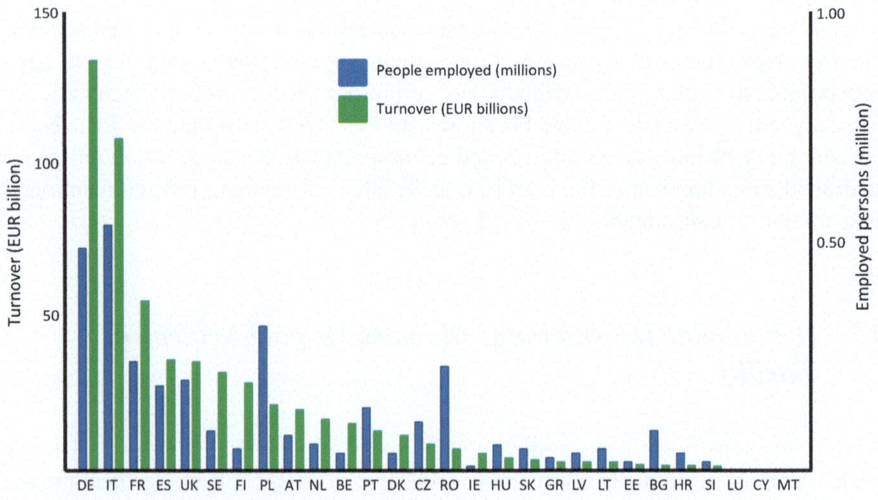

Fig. 8 Turnover and employment in the EU-28 bio-economy, 2013. The figures exclude agriculture, fisheries, food, beverage and tobacco products (Piotrowski et al. 2016)

Another may be the biomass limitation, in which case biomass would be likely to be more expensive in the Netherlands than in biomass-rich countries. This would tend to drive innovation in the Netherlands, but technology and production would tend to be driven elsewhere.

10 Conclusion: South Africa and Netherlands Bio-economy: Combining Each Other's Needs

There are opportunities for both countries to realise each other's ambitions towards an inclusive industrial bio-based economy. South Africa has clearly outlined its intent on implementing its Bio-economy Strategy. There is also tremendous potential in its sugar industry that has plenty of biomass but requires an industrial process overhaul, which should respond to new global demands other than sugar. Moreover, the prospect of revitalising the sugar industry and linking it up to chemicals and plastics are science and technological innovations, which forms the heart of an industrial bio-economy. South Africa does have good research and development programmes in place, and these include Biocatalysis initiative at University of Witswatersrand (Wits) and Council for Scientific Industrial Research (CSIR), Process engineering on the production of fuels and chemicals from plant materials at the University of Stellenbosch. Centre for Bioprocess Engineering (CeBR) at UCT.

In order to realise South Africa's goal for an industrial bio-economy the following would be required:

- Interdisciplinary skills that includes: green chemistry, bioprocess engineering, novel reactor design and configuration for scale up, platform and enabling technologies such as synthetic biology, structural biology and functional genomics, metabolic engineering modelling (steady state and dynamic models), technology transfer and internships with major industry players amongst others.
- Strengthening legislative and policy frameworks: for the production, commercialisation and uptake of bio-based innovation products and services. A good example would be a shift towards bio-plastics that can contribute in minimising waste to landfills.
- Technology brokerage: Support for local production, commercialisation and uptake of bio-based innovation products for food, feed, energy and other purposes.
- Policy engagement: convening high level regional policy roundtables on bio-economy policy strategies.
- Capacity building: training on the global bio-economy policy strategies, opportunities and challenges.
- Industrial participation for large-scale up which is mostly lacking in the country.

The Netherlands is a country that can clearly complement South Africa's shortage in skills and also industrial partners. Since scale-up research is complex, costly and has high levels of risk, it is practical to ensure that big collaborative networks support it. A dedicated facility that has state-of-the-art-facilities, which offers a concentration of a skilled work force, enhances the chances of success for such projects since it enables the bridging of the innovation chasm.

The Netherlands has the following complementary factors that may benefit South Africa if collaboration were to take place between the two countries.

These are the following:

- Ample science cluster and biotechnology parks that enables the association of like-minded enterprises and opens up opportunities for collaboration. Additionally, science parks require a large anchor tenant (DuPont, DSM, Reverdia, Avantium) with good collaborative partnership with a university.
- Strong knowledge generation institutes with long history of excellence in bio-catalysis, bioprocessing, and access to biorefinery facilities. There are also incubation (business and technical/pilot) facilities including physical space and facilities for the development of small, medium and large enterprises.

Where the Netherlands can benefit is the access to biomass, of which South Africa could provide. However, there are tensions that will have to be managed: between the biomass-rich developing countries and the biomass-poor developed countries; between the global North and the global South; and between knowledge-based economies and economies dependent on exporting natural resources. The Netherlands is pioneering in establishing an international biomass dispute facility, which could mitigate a global biomass trade competition. Figure 9 illustrates the situation with international biomass trade: virtually all arrows point to OECD nations, and many of them point to Western Europe.

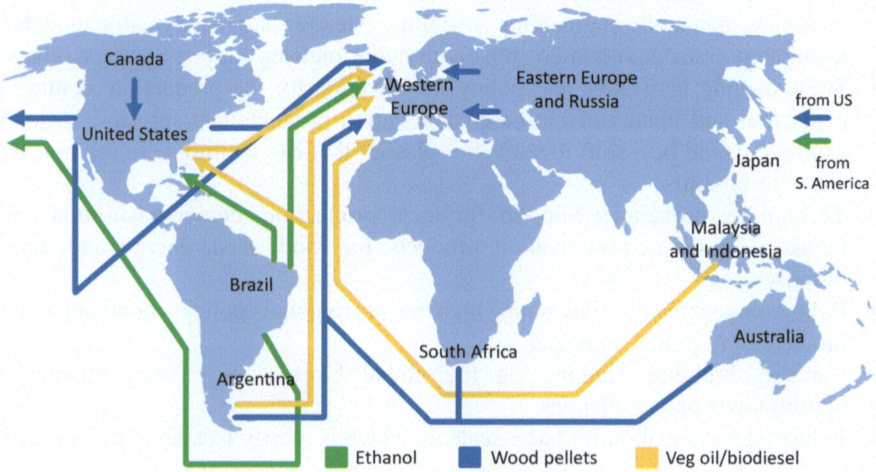

Fig. 9 World biomass trade, 2011. Redrawn from BP-EBI (2014)

Comparing the Netherlands with South Africa actually points to much wider issues at the heart of the future global economy. Monumental events during 2015 have propelled sustainable development and the bio-economy to the highest political visibility they have ever enjoyed. The combination of finite fossil resources and the deliberate actions to drastically reduce GHG emissions make a future bio-economy inevitable. It is no exaggeration to say that this could re-draw the geopolitical power balance of the planet. For example, by 2050 it is likely that there will be much less dependence on the Middle East than now, and by century end, fossil fuel consumption will need to be virtually zero (OECD Policy Brief 2015). Given that biomass can be produced locally, geopolitical patterns can shift to being more distributed. However, other negative aspects will need to be dealt with politically: the potential for increased deforestation as the biomass market opens up; trade barriers, biomass disputes, even conflicts are but a few. Finally, it is one thing to create a bio-economy. It will be quite another to create a *sustainable bio-economy*, and this may well be our last chance to create an enduring economy that operates within the limits of what the planet can provide without causing massive environmental damage. In short, the bio-economy is the way to enable the 'triple bottom line' of economic, social and environmental sustainability, as described in 1994 (The Economist 2009). But this will not happen without massive policy support, regionally, nationally and internationally.

11 Disclaimer

The views expressed are those of the authors and not necessarily those of the OECD or of the governments of its member countries.

References

Banse M, van Meijl H, Tabeau A, Woltjer G (2008) Will EU biofuel policies affect global agricultural markets? Eur Rev Agricu Econ 35:117–141

Bosch R, van de Pol M, Philp J (2015) Define biomass sustainability. Nature 523:526–527

BP-EBI (2014) Biomass in the energy industry. An introduction. BP plc, London, UK

Carus M, Piotrowski S (2009) Land use for bioplastics. Bioplastics Magazine 04/09(4), 46–49

Cocchi M, Nikolaisen L, Junginger M, Goh CS, Heinimö J, Bradley D, Hess R, Jacobson J, Ovard LP, Thrän D, Hennig C, Deutmeyer M, Schouwenberg PP, Marchal D (2011) Global wood pellet industry market and trade study, IEA Bioenergy Task 40. Sustainable International Bioenergy Trade, 190

Daniyanto S, Deendarlianto AB (2015) Torrefaction of Indonesian sugar cane bagasse to improve bio-syngas quality for gasification process. Energy Procedia 68:157–166

de Jong E, Jungmeier G (2015) Biorefinery concepts in comparison to petrochemical refineries. In: Pandey A, Hofer R, Larroche C, Taherzadeh M, Nampporthri KM (eds) Industrial biorefineries and white biotechnology. Elsevier, Amsterdam, Netherlands (Chapter 1), p 3–30

Delebecque C, Philp J (2015) Training for synthetic biology jobs in the new bioeconom. Science. June 02, https://doi.org/10.1126/science.caredit.a1500143
CE Delft (2012). Cascading of biomass. 13 solutions for a sustainable bio-based economy. CE DelftPublication code: 12.2665.52
Deloitte (2015) European energy market reform. Country profile: Netherlands. Deloitte Conseil
Deloitte-VNCI (2012) The chemical industry in the Netherlands: World leading today and in 2030–2050. Deloitte Netherlands
Department of Agriculture and Forestry (Daff) (2011) Profile of the South African sugar market value chain. Pretoria, South Africa
Department of Agriculture and Forestry (Daff) (2014) Profile of the South African sugar market value chain. Pretoria, South Africa
Department of Science and Technology (DST) (2014a) The Bio-economy strategy. Pretoria, South Africa
Department of Science and Technology (DST) (2014b) South African National survey of research and experimental development. Main Analysis report 2011/2012, Pretoria, South Africa
Dutch Cabinet (2012). Hoofdlijnennotitie Bio-based Economy. Kamerstuk 02-04-2012, EL&I
Endres H-J, Siebert-Raths A (2011) Engineering biopolymers: markets, manufacturing, properties and applications. Hanser Publications, Cincinnati, p 674
Filho MV, Araujo C, Bonfa A, Port W (2011) Chemistry based on renewable raw materials: perspectives for a sugar cane-based biorefinery. Enzyme Research, 654596. https://doi.org/10.4061/2011/654596
Government of the Netherlands (2014) Global challenges, Dutch solutions. Ministry of Economic Affairs and Ministry of Education, Culture and Science, The Hague
Harvard Economic atlas of complexity (2014) http://atlas.cid.harvard.edu/explore/tree_map/export/zaf/all/show/2014/ Acccessed 28 Apr 2016
Hoefnagels R, Banse M, Dornburg V, Faaij A (2013) Macro-economic impact of large-scale deployment of biomass resources for energy and materials on a national level—A combined approach for the Netherlands. Energy Policy 59:727–744
Il Bioeconomista (2016). The Dutch bioeconomy takes another step forward thanks to AkzoNobel and Royal Cosun. 07 May 2016
Innovatiecontract Bio-based Economy 2012–2016 (2012) *Groene groei: van biomassa naar business.* www.biobasedeconomy.nl/wp-content/uploads/2012/04/7250-ELI-Innovatierapport-aanpv3.pdf
International Renewable Energy Agency (Irena) (2013) Production of Bio-methanol. Technology brief
Juma C (2011) The new harvest, agricultural innovation in Africa. Oxford University Press
Langeveld JWA, Meesters KPH, Breure MS (2016) The bio-based economy and the bioeconomy in the Netherlands. Biomass Research Report 1601, reference number 59015257. Biomass Research, Wageningen
Moyo MJ, Bah, E-HM, Verdier-Chouchane A (2015) Transforming Africa's agriculture to improve competitiveness. Chapter 2.1 in the African Competitiveness Report, 2015. World Economic Forum (WEF)
Ministry of Economic Affairs (2013). The bio-based economy in the Netherlands. ISSN: 1572-6045
OECD Policy Brief (2015). Three steps to a low-carbon economy. The goal of zero net emissions can be achieved. http://www.oecd.org/policy-briefs/Three-steps-to-a-low-carbon-economy.pdf
Oxford Economics (2014). Evolution of competitiveness in the European chemical industry: historical trends and future prospects. A report for Cefic, October 2014. www.oxfordeconomics.com
Pavanan KC, Bosch RA, Cornelissen R, Philp JC (2013) Biomass sustainability and certification. Trends Biotechnol 31:385–387
PBL (2012) PBL Note. Sustainability of Biomass in a Bio-based Economy. A Quick-scan Analysis of the Biomass Demand of a Bio-based Economy in 2030 Compared to the

Sustainable Supply, PBL Netherlands Environmental Assessment Agency (PBL Publication number 500143001)
PBL (2014) Biomass: wishes and limitations. PBL Netherlands Environmental Assessment Agency, The Hague
Piotrowski, S., M. Carus and D. Carrez (2016). European bioeconomy in figures. Report of the Nova-Institute, commissioned by the Bio-based Industries Consortium
Scarlat N, Dallemand JF, Motola V, Monforti-Ferrario F (2013) Bioenergy production and use in Italy: recent developments, perspectives and potential. Renew Energy 57:448–461
SER (Social and Economic Council of the Netherlands) (2013) Energy agreement for sustainable growth. The Hague, September 2013
Taanman M, Enthoven G (2012) Exploring the opportunity for a biomass dispute settlement facility. The Hague Institute for Global Justice, The Netherlands
The Economist (2009) Triple bottom line. It consists of three Ps: profit, people and planet. 17 Nov 2009
Van der Maaten-Theunissen M, Schuck A (2013) Integration of nature protection in forest policy in the Netherlands. INTEGRATE Country Report, EFICENT-OEF, Freiburg
VNCI (Association of the Dutch Chemical Industry) (2012) Facts and figures about the chemical industry in the Netherlands—2012). https://www.vnci.nl/Content/Files/file/English/vnci-factsfigures-2012-uk.pdf

Mangrove Restoration an Economical Alternative for Generating Incomes

Ángel Sol Sánchez, Gloria Isela Hernández Melchor, Juan Manuel Zaldívar Cruz, Carlos Alberto Zúñiga González and José Luis Santiváñez Galarza

Abstract Mangroves are distributed in the American continent from the southern of the United States to South America and play an important role in the economies of coastal communities that provide a wide diversity of satisfaction ranging from food until protect communities of the wind, hurricanes and storms. However, the presence of pests, diseases, misuse of resources and the fire has been linked to the disappearance of the ecosystem. In 2010, a massive defoliation of black mangrove (*Avicennia germinans* L.) caused by *Caterpillar Anacamptodes sp* was presented, which altered the economic activities practiced in the zone. It problem caused death of this species in an area of 3846 hectares. In 2012 the restoration of 50 hectares began in the ejido Las Coloradas, Cardenas Tabasco to assess the response of the ecosystem and ecological succession. Further growth of mangrove plants, gradual recovery of fishing activity, return of biodiversity, income stabilization and water levels was obtained. Local and national authorities have acknowledged the work done and the process is being transferred to other affected areas. It is expected that the coastal population affected recover their productive activities and keep the restored area. Community participation was 179 people; four workshops were held and given the exchange with other zones similar. The plague disappeared by itself in March 2011 but has been developed insecticides if it presents again. In con-

Á. S. Sánchez · J. M. Z. Cruz
Colegio de Postgraduados, Campus Tabasco, Tabasco, Mexico
e-mail: sol@colpos.mx

J. M. Z. Cruz
e-mail: Zaldivar@colpos.mx

G. I. H. Melchor (✉)
Sociedad Iberoamericana de Bioeconomía y Cambio Climático, Tabasco, Mexico
e-mail: isela7827@gmail.com

C. A. Z. González
Universidad Nacional Autónoma de Nicaragua, Unidad León, León, Nicaragua
e-mail: czuniga@ct.unanleon.edu.ni

J. L. S. Galarza
Centro Agronómico Tropical de Investigación y Enseñanza, La Paz, Bolivia
e-mail: jsantiva@catie.ac.cr

W. Leal Filho et al. (eds.), *Towards a Sustainable Bioeconomy: Principles, Challenges and Perspectives*, World Sustainability Series,
https://doi.org/10.1007/978-3-319-73028-8_16

clusion mangroves it should be considered a priority in the production of food for coastal populations would enhance its ecological restoration as they constitute a niche for feeding, nesting, shelter, perch and reproduction of local and migratory birds as well as being the habitat for many species of mammals, reptiles and invertebrates.

Keywords Mangroves · Restoration · Tabasco · Incomes

1 Introduction

The climate of the planet kept constant changes because of increased emissions of greenhouse gases because of anthropogenic and natural activities. These constant changes force to make decisions that ensure a future with better living conditions of populations and ecosystems, mainly focusing on the actions of adaptation to climate change, which has generated a new vision to address the study of ecosystems from the perspective of structure and function (Magaña 2011; Yáñez-Arancibia 2010).

Increases in atmospheric temperature and sea, reduction and instability in rainfall patterns, rising sea levels, intensifying extreme weather events around the world (droughts, floods, frost, heat stroke), impact on production agriculture and livestock and behavior of ecosystems, which affects the instability of the population and their livelihoods.

Similarly, the effects of climate change impact on mangroves, which represents ecosystems of tropical and subtropical coastal areas, and perturbation generates productive, economic and social instability (Yañez-Arancibia 2010).

In Mexico, the state of Tabasco has a strip of 38,839.52 hectares of mangrove composed of four species *Rhizophora mangle* L. (red mangrove), *Laguncularia racemosa* L. (white mangrove), *Avicennia germinans* L. (black mangrove), and *Conocarpus erectus* L., (botoncillo mangle) in seven municipalities: Huimanguillo, Cardenas, Paraiso, Comalcalco, Centla, Jalpa de Méndez and Nacajuca (Sol et al. 2011). The four species listed in NOM-059-SEMARNAT (2010) as threatened.

In this stretch, there are areas of high vulnerability due to various factors such as coastal erosion, saltwater intrusion, altering the structure of the mangrove by logging and pests, changing land use, increased livestock areas, salinization and siltation, among others. The most critical points are those bordering the Carmen-Pajonal-Machona lagoon complex (Sol et al. 2009).

In addition, during 2010, the area was invaded by an overpopulation of adult and larval inter Anacamptodes sp moth, which devoured 3835.73 hectares of black mangrove approximately (Sol et al. 2012). The main damage was the partial and total defoliation of youth and young adults Mangroves, which affected the continuum of the area deforested suddenly. It increased the solar radiation from 27 to 100%, which brought resulting reduction in volumes fishing scales, oyster and shrimp, and the disappearance by local migration of blue crab (*Cardisoma guanhumi*), species very important in the diet and as a source of income for coastal communities (Sol et al. 2012).

By aerial flights made by the National Forestry Commission it confirmed that 3835.73 hectares were damaged from intermediate to severe level, so it was necessary to reforest to mitigate the impact generated. It was determined that the affected communities were ejido Aquiles Serdan the municipality of Paraiso; ejido Sinaloa, section First and Second, ejido Alacrán, ejido el Matinero, Municipality of Cardenas Tabasco. These communities are located in the border of the Gulf of Mexico and Carmen-Pajonal-Machona lagoon complex. Other communities affected were Las Azucenas, el Golpe, el Mingo and Las Coloradas, Cardenas Tabasco, these north bordering the Carmen-Pajonal-Machona lagoon complex and south to the mainland (Hernández-Melchor et al. 2016).

With this information, in 2011 began the project "Ecological restoration of 50 hectares of black mangrove (*Avicennia germinans* L.) Affected by caterpillars Anacamptodes sp in the Ejido las Coloradas Cardenas, Tabasco, Mexico, with the objectives of reforesting 50 ha of black mangrove, restore hydrological dynamics, monitor the presence of the pest, assess the degree of involvement in human communities, and the response of reforestation as an economic alternative to affected communities (Sol et al. 2002).

2 Background

Mangrove ecosystems develop in tropical and subtropical environments and develop in places subject to flooding and generally under anoxic conditions and saline environments (Lopez and Ezcurra 2002). These form large communities giving rise to a transition zone between aquatic and terrestrial environments, and therefore are home to a high diversity of organisms in aquatic and terrestrial life (Hernández-Melchor et al. 2016).

Such characteristics make them one of the most productive ecosystems in the area, allowing the development of aquatic species in early stages of development such as fish, shrimp and oysters, which begin their life cycle in the mangrove roots. Similarly, these ecosystems reduce the impacts of storms, cyclones and tsunamis; also, they are very important in capturing carbon dioxide. They act as biological filters by removing nutrients and toxins (Calderon et al. 2009).

The beneficial of the mangroves are surprising. Scientific research provides information on estimated mangrove tenure, such as in the Gulf of California, red mangrove cost maintains a fish productivity of about \$ 37,000 ha^{-1} yr^{-1}; forestry revenue is estimated at U.S. \$ 706 ha^{-1} yr^{-1} for Cuba; in the case of estimated volume of firewood, for Honduras it is 11.57 m^3 (Hernández-Melchor et al. 2016).

Similarly, in ecosystem services a worth of 1600 billion per year estimates, carbon sequestration estimates at \$ 1000 per hectare. Gilman et al. (2006) mention that in the relative sea level rise, protection of mangroves is a way to curb coastal erosion, and is less expensive than building dams and control structures erosion.

Of the 170,000 km^2 of mangrove in the world, 48% distributes in five countries: Indonesia (19%), Australia (10%), Brazil (7%), Nigeria (7%), and Mexico (5%) (Hogarth 2001). However, its surface continuously reduces mainly by the oil and gas exploitation, extraction uncontrolled timber and conversion for livestock and aquaculture, in addition to the change of land use for tourism and urban development (Smith et al. 2001; WRM 2002).
In Mexico, the mangrove distributes in 17 states with an estimated 770,057 ha, of which 55% is concentrated in the Yucatan Peninsula (Conabio 2009). However; this figure is unstable because the annual rate of mangrove forestation is 2% for the Pacific slope and 2.8% for the Gulf of Mexico (Zaragoza et al. 2005).

The effects of climate change and anthropogenic threats alter the resilience of mangrove ecosystems (Smith et al. 2001; Uribe and Urrego 2009). In this regard, it has been found that in the island countries overexploitation of the mangroves has resulted in a greater impact of tsunamis, hurricanes, cyclones and storms, which resulted in high economic and social costs.

From a practical point of view, should step up efforts restoration or rehabilitation and incorporate them into strategic environmental planning of the coastal zone, such rehabilitate degraded mangroves, restore hydrology, and reduce deforestation and degradation (Yañez-Arancibia 2010; Broadhead 2011); for such action will have to assess the reasons for the loss of mangroves and work with the natural recovery processes of ecosystems (Lewis 2005; Hernández-Melchor et al. 2016).

Learn about the origins of the problem can be strategic in their management (Uribe and Urrego 2009). In this context, in the period 2001–2009 the coast of Tabasco lost 19 922.9 ha of mangroves due to changing land use, preserving an area of 38 839.52 ha (Sol et al. 2009; Dominguez-Dominguez et al. 2011); said surface is threatened by continued anthropogenic activities aimed at economic development and improved quality of life (Hernández-Melchor et al. 2016).

To counter these mangrove deforestations cannot be studied in isolation, but in a comprehensive way to identify how they are interrelated factors that give rise to deforestation, because as natural ecosystem is influenced by external factors.

Thus, Fig. 1 outlines a categorization of problems in Tabasco induce mangrove deforestation, grouped according to their origin: economic development, marginalization, and natural phenomena (Hernández et al. 2016).

The oil activity on the coast can be considered the most impactful on the mangrove ecosystems, due to for the operation of wells dredging, filling and opening roads are made, with the consequent oil spill that directly impact on regeneration natural site.

The same activities such as agriculture, livestock, forestry, aquaculture or combination affect economic, social and ecological factors for obtaining food, for which a transformation process is performed mangrove mode. For example, the establishment of grasslands in mangrove areas causes compaction and salinization of the soil due to intense trampling of livestock, which feeds mainly on red mango propagules, thereby limiting natural regeneration (Ramirez et al. 2010).

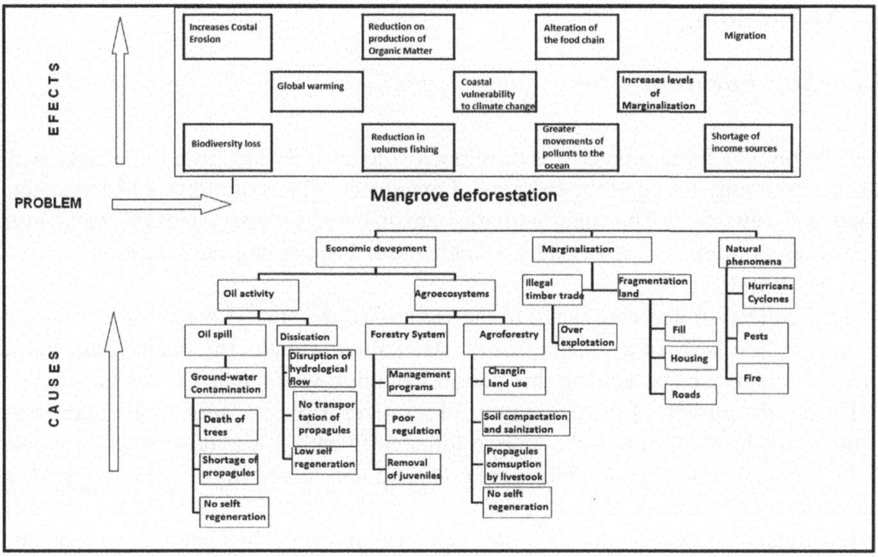

Fig. 1 Causes and effects of mangroves deforestation in Tabasco., Mexico. *Source* taken from Hernandez (2016)

In addition, excessive use of mangrove forest resources and clandestine logging has contributed to overexploitation due to poor forestry regulation and oversight in the area. This is due to various reasons mainly the conditions of marginalization of the communities surrounding the mangrove, which exploit the resources provided by the ecosystem to meet their needs for food and income, which sells mangrove wood shaped struts, coal, screeds, beams and even lumber.

Currently there are areas of high vulnerability due to coastal erosion and intrusion of sea to the mainland, being favored by the continued loss of vegetation caused by land use change. In this case the most critical points are those bordering the Carmen-Pajonal-Machona lagoon complex, where estimate a coastal retreat from −9 to −10 m/year (Hernández et al. 2008).

In addition to this, in 2010 the area was invaded by an overpopulation of adult and inter larval of moth Anacamptodes sp, which devoured little more than 3845 hectares of black mangrove, determining a degree of involvement of level III; where it is necessary to reforest the mangrove in order recover their biological functions (Sol et al. 2012).

For these reasons, this research aimed to produce 50,000 black mangrove plants, and (2) restore 50 ha of mangrove black impacted by caterpillars of Anacamptodes sp in the Ejido las coloradas, Cardenas Tabasco; and evaluate the mangrove importance in the local economy.

3 Methodology

3.1 Site Location

The site on which the project was developed is located 80 km from the county seat. It is a community based in the margins of mangroves. The community lacks potable water and sewerage. The main activities are fishing, shrimp, catch crab and blue crab. The community is reached by a single road. The geographic location indicated in Table 1 and Fig. 2.

Seed collection was conducted in the vicinity of the site. These were germinated in germination beds in a rustic nursery. After three months the plants were transferred to nursery bags. During this period irrigation was applied every two days.

During the growth of plants 5 grams of fertilizer triple 17 were applied per plant in the second and third month. Also, a foliar fertilization was done with grogreen.

In the nursery 191 caterpillars were collected and reared in the laboratory. Also, the presence of diseases was attended.

For transplantation to the field, the field was prepared by removing weeds and trash. Also, trash was removed from the main drain to allow free access of water to the mangroves, and three bridges were built on the main drain to build to ensure water circulation.

The plantation was established in June 2012, at a density of 833 plants per hectare.

After planting, protection monitoring was established to prevent the destruction of plants by people who extract wood and caught crab in the area. Participants received four training courses on production of mangrove, mangrove maintenance,

Table 1 Geographical coordinates of the work area

ID	X	Y	ID	X	Y
1	440501.65	2026788.03	19	438766.53	2025404.75
2	440782.21	2026415.29	21	438662.58	2025580.36
3	440902.32	2026331.94	22	438322.31	2025387.03
4	440975.31	2026187.21	24	438422	2,025,793
5	440779.76	2026001.97	25	438499.68	2026068.12
6	440574.56	2026047.24	26	438341.94	2026467.34
7	440326.76	2025996.62	27	438618.88	2026583.47
8	439380.2	2026151.68	28	438519.51	2026792.49
9	439290.42	2026503.57	30	438841.57	2026789.4
10	439156.93	2026374.84	31	439267.22	2026786.31
11	438964.91	2026336.6	32	439565.32	2026660.44
12	438912.41	2025942.22	33	439722.62	2026634.03
13	439275.43	2025362.14	34	439709.73	2026878.54
14	438895.52	2025325.38	35	439969.43	2026856.71

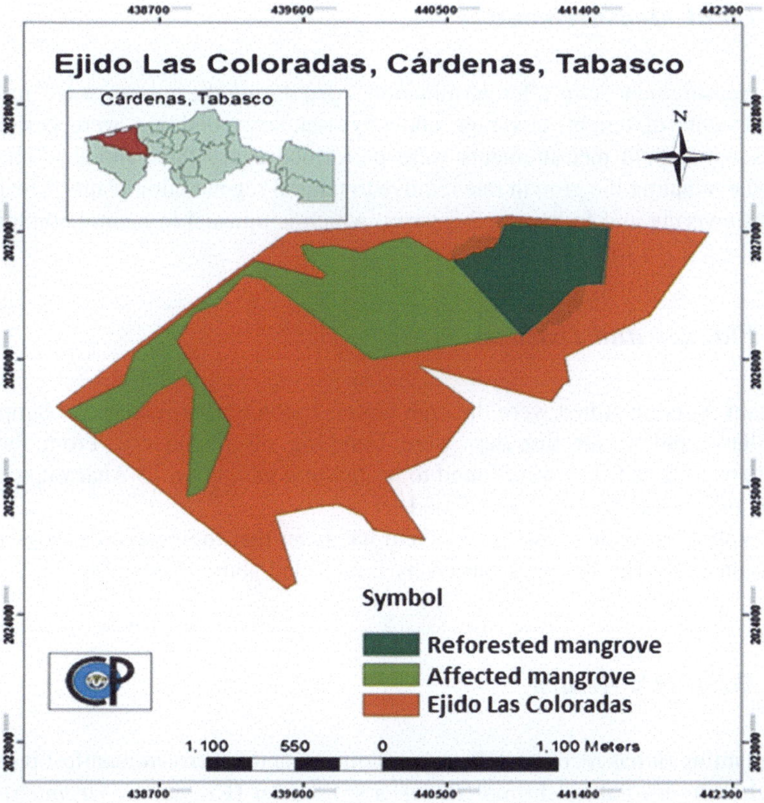

Fig. 2 Location of the study area

soil and water conservation, and development of botanical insecticides. They also visited three states of the Mexican Republic with similar situation to share experiences.

4 Results

4.1 Plant Production

In total 171 kg of seeds from those dragged by the influence of the tides to the coast of the lagoon were collected. Plant production was 45,866, which allowed reforest 50 hectares. Nursery plant growth was homogeneous, with slight variations, the minimum height was 67 cm, and the maximum was 107 cm. Three years after, the site has recovered its total coverage and succession is under development. It is worth mentioning that the work in the area began with 179 people and ended with 39.

4.2 Plant Measurements

Plants measurement starts after germination. Also during the nursery stage, monthly measurements of height, coverage and phytosanitary condition were performed. After sowing, field measurements were performed every three months. The data allowed evaluating the growth rate relative to natural regeneration plants. The plants started flowering and fruiting at 2.8 years, which is unusual in natural conditions.

4.3 Plagues and Diseases

The main pests identified were the crab cutter, Junonia caterpillars, Anacamptodes caterpillars, and as for diseases, were Damping off in nursery. From the 191 caterpilars collected, 119 were found to be *Junonia genoveva*, 13 Anacamptodes sp and the others did not reach adulthood.

In established plantations the problem has been that the people destroys plants with a machete. The flooding caused by tidal killed some plants also.

4.4 Wildlife's Return

After planting, it has recovered the site conditions and fishermen manifest increases in catch volumes flakes, shrimp (*Penaeus* sp), oyster (*Crassostea virginica*), blue crab (*Callinectes sapidus*) and other species.

Fishermen who had a daily income of 70 or 80 pesos a day during the mangrove mortality ecosystem, currently get income on average of 490 pesos in capturing scales, 300 in capturing oyster in shell, 280 in crabs catching and for the case of the blue crab (*Cardisoma guanhumi*) this is variable and depends on the agility to catch them on the run, so even people who catch crab, can have an income of 700 pesos a day's work.

Some people are dedicated to extracting burrowing crabs for sale, and their income is variable from 90 to 140 pesos a day. However, planting is still in the process of stabilization, as the mangrove is too young to provide the resources offered.

5 Discussion

The plant production was considered a success considering that was no germplasm at the site to restore the site with mangrove species had been wiped out by the caterpillar.

Management in germination beds and nurseries was essential for future success. Fertilization with triple 17 and groo green strengthened the plant, which was reflected in a rapid development of Pneumatophores and early flowering. This is very important because in the literature review no makes mentions works on plants mangrove fertilization rather refers to planting bags. This makes the work relevant because under similar conditions can be repeated and accelerate the recovery process in similar ecosystems.

Measurement of plants allowed to compare with other work done with other species of mangrove, though having no previous work on black mangle the comparisons made, are abstract because they are not the same species. However, it was assessed that this species in nursery and planted grows and 2.78 times faster than under natural conditions, this can be attributed to fertilization.

Pests and diseases in general are the main problem in all production systems, in this particular case there was a need to constantly monitor the plants at night because the crab cutter in one night can crosscut up to 36 plants and it reduces the possibility of the success of the nursery.

Initially, it was not clear which species caused the damage, so it was necessary to establish measures to identify pest since in a single night more than 600 plants were cut. After this some crabs were isolated and placed in sections of the nursery to evaluate the damage and mortality of plants. Thus, it was obtained that on average in one night a crab can cut 36 plants 25–30 cm high, which do not regenerate.

Pests and diseases in general are the main problem in all production systems, in this particular case there was a need to constantly monitor the plants at night because the cutter crab can cut until 36 plants each crab in the mean per night and this reduce the possibility of success in the nursery.

6 Conclusion

It is possible to achieve recovery of the mangrove ecosystem whenever aspects are integrated as training of field personnel and continue the restoration process, since the projects are abandoned after concluding the implementation period, are doomed to failure. It is necessary that in projects of this nature, involving the local population.

Promote the restoration of mangrove ecosystem generates ecological, social, and economic benefits in addition to environmental goods and services provided as an ecosystem.

This experience can play an important role in restoring any other altered area.

The reason for the restoration of mangroves is due to the benefits they provide to the local population. For example, the blue crab (*Cardisoma guahumi*) takes four years to reach its commercial size in preserved environments. When the mangrove disappears, the crabs migrate to other sites and the local population suffers from lack of food. The same effect happens with other species such as shrimp, crab and fish.

The wildlife that lives in the mangrove disappears, as are birds, mammals, reptiles and others, so it is necessary to help to restore natural conditions. Likewise, local people do not have to leave their communities when a phenomenon like the present happens.

References

Broadhead JS (2011) Reality check on the potential to generate income from mangroves through carbon credit sales and payments for environmental services. Regional Fisheries Livelihoods Programme for South and Southeast Asia (GCP/RAS/237/SPA) Field Project Document 2011/ REG/2. 37 pp

Calderón C, Aburto O, Ezcurra E (2009) El valor de los manglares. Conabio. Biodiversitas. Número 82, pp 1–6. México

Conabio (2009) Manglares de México: extensión y distribución, p 100. México. www.biodiversidad.gob.mx/.../manglares/.../Manglares_de_MÉxico

Domínguez-Domínguez M, Zavala-Cruz J, Martínez-Zurimendi P (2011) Manejo forestal sustentable de los manglares de Tabasco. Secretaría de Recursos Naturales y Protección Ambiental. Colegio de Postgraduados, p 137 Villahermosa, Tabasco, México

Gilman E, Van Lavieren H, Ellison J, Jungblut V, Wilson L, Areki F, Brighouse G, Bungitak J, Dus E, Henry M, Sauni I Jr, Kilman M, Matthews E, Teariki-Ruatu N, Tukia S, Yuknavage K (2006) Pacific Island Mangroves in a Changing Climate and Rising Sea. In: UNEP Regional Seas Reports and Studies No. 179. United Nations Environment Programme, Regional Seas Programme. Nairobi, Kenya, p 70

Hernández-Melchor GI, Ruíz-Rosado O, Sol-Sánchez Á, Valdez Hernández JI (2016) Cambios de uso del suelo en manglares de la costa de Tabasco. In: Revista Mexicana de Ciencias Agrícolas, núm. 14, febrero-marzo, pp 2757–2767. Instituto Nacional de Investigaciones Forestales, Agrícolas y Pecuarias. Estado de México, México

Hernández SJM, Ortiz PMA, Méndez LAP, Gama CL (2008) Morfodinamica de la línea de costa del estado de Tabasco, México: tendencias desde la segunda mitad del siglo XX hasta el presente. Investigaciones Geograficas, Boletin del Intituto de Geografía, UNAM. Núm. 65. México, pp 7–21

Hogarth PJ (2001). Mangrove Ecosystems. In: Levin SA (Editor(s)-in-Chief) Encyclopedia of biodiversity. Elsevier, New York, pp 853–870

Lewis RR (2005) Ecological engineering for successful management and restoration of mangrove forests. Ecological Engineering 24(4 SI). Florida, USA, pp 403–418

López PJ, Ezcurra E (2002) Los manglares de México: una revisión. Maderas y bosques, vol 8. Número especial 1. Instituto de Ecología A. C. Xalapa, México, pp 27–51

Magaña RVO (2011) Cambio climático: El reto del siglo. Investigación Ambiental 3:63–67, México

Ramírez SAF, Trujillo SO, Zentmyer REH, Martínez RB, Sheseña HIM, Rivas AJ (2010) Identificación y tipificación de áreas potenciales para la restauración de manglares: el caso de los humedales de la cuenca del río Papaloapan Veracruz. México. Pro-natura A.C. Coordinación de proyectos Eco-forestales, p 64

Semarnat (2010) NOM-059-SEMARNAT-2010. http://www.semarnat.gob.mx/leyesynormas/Pages/nom_fauna.aspx

Smith J, Schellinhuber H, Mirza M (2001) Chapter 19. Vulnerability to climate change and reasons for concern: a synthesis. In: McCarthy J, Canziani O, Leary N, Dokken D, White K (eds) Climate Change 2001: Impacts, Adaptation, and Vulnerability. Published for the Intergovernmental Panel on Climate Change. Cambridge University Press, Cambridge, United Kingdom, and New York, NY, USA, p 942

Sol SA, Zenteno RCL, Zamora CLF, Torres RE (2002) Modelo para la Restauración Ecológica en áreas alteradas. Revista de difusión Kuxulcab′ VII(14):48–60

Sol SA, Zamorano CLF, Almeida HY, Hernández MGI, Shirma TED (2009) Estudio Regional Forestal. Unidad de Manejo Forestal Costa de Tabasco, México. CONAFOR-ECODET A. C. Documento Técnico, p 4610

Sol-Sánchez A, Hernández Melchor GI, Sánchez-Gutierrez F (2012) Evaluación de daños y restauración del ecosistema de mangle negro (Avicennia germinans L.) en el ejido las Coloradas, Cárdenas Tabasco, México. Segundo Congreso Mexicano del Ecosistema de Manglar. Cd del Carmen Campeche, pp 153–154

Uribe PJ, Urrego GLE (2009) Gestión ambiental de los ecosistemas de manglar. Aproximación al caso Colombiano. Gestión y Ambiente 12(2):57–71

WRM—World Rainforest Movement (2002) Manglares: Sustento local versus ganancia empresarial. Comité Holandés para la UICN (CH-UICN/TRP), p 69

Yañez-Arancibia A (2010) Los manglares frente al Cambio climático ¿tropicalización global del Golfo de México? In: Yáñez-Arancibia A (ed) Impactos del Cambio Climático sobre la Zona Costera. Instituto de Ecología A. C. (INECOL), Texas Sea Grant Program, Instituto Nacional de Ecología (INE-SEMARNAT), México, pp 91–126

Zaragoza R, Peters E, Vega E (2005) Evaluación de las tasas de pérdida de manglar mediante la comparación de polígonos en 1976 y 2000. In: INE (ed) Evaluación preliminar de las tasas de pérdida de superficie de manglar en México. Dirección General de Investigación para el Ordenamiento Ecológico y Conservación de Ecosistemas. INE-SEMARNAT. México, p 21

Green-Growth Policies and Economic Effects: Lessons Learnt from Organic Farming in the Czech Republic

Radka Redlichová, Věra Bečvářová, Diana Mihaela Pociovălişteanu, Karel Vinohradský and Ivo Zdráhal

Abstract This chapter discusses the findings of research investigating the comparative economic performance of organic farms and conventional agricultural holdings in the Czech Republic. The system of organic farming has become an important component of the ecology-friendly alignment of agriculture in the Czech Republic. This system has extended extremely fast in the past fifteen years and it was stimulated to a large extent by state interference and the increase of subsidy payments, especially after the Czech Republic joined the European Union. A weak point in organic farming in the Czech Republic is its low productivity. In the period 2001 through 2012 organic farming (OrgF) holdings reached 30–40% of the value of agricultural production per ha of that achieved by conventional farming (CoF) holdings. Although OrgF operate with 50% of the inputs per 1 ha of CoF holdings, the overall material and labour demandingness of production is 1.4–1.7 times higher than that of CoF holdings—giving rise to the concept of "ecological paradox". In effect this suggests that, OrgF holdings create a 1.5 times greater adverse ecological footprint per capita food production. Excluding operating subsidies, OrgF holdings reported a loss of 11,000 CZK per ha, which is twice that reported by CoF holdings. Subsidy payments during 2010 through 2012 amounted

R. Redlichová (✉) · V. Bečvářová · K. Vinohradský · I. Zdráhal
Department of Regional and Business Economics, The Faculty of Regional Development and International Studies, Mendel University in Brno, Brno, Czech Republic
e-mail: radka.redlichova@mendelu.cz

V. Bečvářová
e-mail: vera.becvarova@mendelu.cz

K. Vinohradský
e-mail: karel.vinohradsky@mendelu.cz

I. Zdráhal
e-mail: ivo.zdrahal@mendelu.cz

D. M. Pociovălişteanu
Faculty of Economics, "Constantin Brâncuşi" University of Târgu-Jiu, Târgu-Jiu, Romania
e-mail: diana.pociovalisteanu@gmail.com

W. Leal Filho et al. (eds.), *Towards a Sustainable Bioeconomy: Principles, Challenges and Perspectives*, World Sustainability Series,
https://doi.org/10.1007/978-3-319-73028-8_17

to 14,400 CZK/ha for OrgF holdings and 8,500 CZK/ha to CoF holdings. This interference has modified profit making. OrgF holdings, therefore, recorded slightly higher profit per ha and twice the profit per unit of the product when compared to CoF. Subsidies, therefore, represent the main financial source of stability in the financial management of organic farm holdings. Market prices of organic products from OrgF holdings contribute very little to cover higher costs of production. It is mainly caused by the nature of current price transmission in the food commodity chain, which leads to a redistribution of profit away from agricultural producers towards processors and, in particular, retailers. The increase of the contribution of operating subsidies to cost recovery and profit creation together with the development of subsidies per unit of product and per worker (AWU) in OrgF emphasizes the need for further development of the policy strategy for this type of farming. This should include new innovative thinking on improving management techniques, mechanisms for and changes to subsidy structure, particularly in relation to the balance in the market for organic products and income parity. Analyses of the economic aspects of organic agriculture (completed in this study) challenge agrarian policy thinking through highlighting the problem of the economic sustainability of organic farming holdings.

Keywords Organic farming · Efficiency · Profitability · Subsidy payments Agriculture policy · Agribusiness

1 Introduction

Agricultural development in economically developed countries in the twentieth century was characterized by the industrialization of the production process and its comprehensive integration into food chains. This has resulted in food verticals and agribusiness networks increasingly influencing agricultural systems. Current agro-industrial systems are generally characterized by high levels of usage of natural resources and substantial growth in labour productivity. High demands on profitability in agricultural systems are reflected in the variation of production structures of crop and livestock systems, often going beyond the potential offered by the equilibrium status of natural resources. Demands on environmental resources exerted by current agro-industrial systems differ significantly from the demands of systems inherently adapted to natural ecological processes. Current agro-industrial systems are largely associated with the increased use of energy for production mostly from non-renewable sources supported by external inputs of inorganic substances in the nutrition and protection of plants and animals. This increases the risks of damage to the environment.

Recommendations for sustainable agriculture have consistently been suggested, recognizing the risks of modern agricultural practices and the environmental demands of world economic growth. This recognition is particularly well documented in publications such as the Rio Declaration on Environment and

Development and Agenda 21 adopted at the "Earth Summit" in Rio de Janeiro in 1992. This also had direct implication for developing new strategies and techniques of agricultural production and economic development and was clearly reflected in significant changes in agricultural policies.

In the European Union this recognition reflects in the concept of European multifunctional agriculture. This concept orientates the Common Agricultural Policy (CAP) towards promoting balanced food production with agriculture also shaping the landscape, the environment and rural residential areas. The principles of European multifunctional agriculture form the basis for new legislative norms adopted by the EU. It also directs the calculation and distribution of subsidies to farmers who fulfil the requirements set by these principles. The overall perspective of the CAP is also to promote the increased application of ecological considerations in the development of agro-industrial systems in the countries of the European Community.

Environmental demands on agriculture were in recent decades stimulated by the creation and development of various alternative farming practices, amongst them organic farming. Organic farming is presented as a system that aims to fulfil the criteria set for sustainable agriculture through a "specific return" to natural production ways based on crop rotation. Generally, it is characterized by smaller yields and lower intensity of cultivation activity with preference for non-production (environmental and social) functions. Support for organic farming is also included in the CAP through a grant support facility. Other support mechanisms include the legislative classification of organic production and descriptive labelling of organic products. This strategy of the CAP, particularly the orientation of subsidy support, led to an increase in the number of organic farms and in the number of hectares cultivated by them. In 2012 there were already 10 million hectares included in this system. This accounted for 5.5% of the cultivated farm land in the European Union.

In the broader body of literature focusing on the issue of organic farming works addressing the biotechnological and environmental contexts of organic farming generally dominate. Similarly, the focus of informative or promotional journalism is mainly devoted to promoting the environmental aspects of organic production which undoubtedly contributed to shaping the demand for organic products.

In the early stages of the development of organic farming, very little information on its economic aspects were published. Research into the economic aspects of organic farming has only really developed in the past fifteen years. In this context authors like Niggli et al. (2008), Offermann and Nieberg (2000), Sanders (2007), Stefanos et al. (2012), Gassner et al. (2008), Laurence et al. (2013) and Redlichová et al. (2014) have consistently pointed out the need to embrace agro-economic principles in agrarian policy and the management of organic farming systems.

In the same way authors like Hughner et al. (2007), Butler et al. (2008), Kummeling et al. (2008), Richter (2008), Lockie et al. (2006), Goodwin and Harper (2000), Blažková and Chmelíková (2010), Zdráhal and Bečvarová (2013a), Dudová (2014), Hrabalová et al. (2013) and Valeška (2010), have emphasised the need to also focus on market development and prices in the value chain of organic products.

Organing farming has an important role in the rural economy, there is a potential for organic farming to contribute to rural development. (Pugliese 2001).

This study also aims to contribute to deepening the systematic understanding of the economic aspects of organic farming systems in Europe. To achieve this, research results reflecting the economic development of organic farming enterprises in the Czech Republic during the period 2001–2012 is analysed and reported on. This work is also an output of a longer-term research programme at Mendel University in Brno, investigating trends in agribusiness, the formation of markets within the commodity chains and food networks and the socio-economic context of sustainable agriculture.

It summarizes and comments on the findings on changes in the dimensions and extent of use of natural resources, the level of state interventions and the economic efficiency of organic farming enterprises. One of the objectives of this work is also to contribute to the general refining of the objectives of the CAP and the methods of its implementation.

2 Organic Farming in the Czech Republic

During the past twenty years organic farming has become an integral component of the agrarian sector in the Czech Republic. In 2012 there were 3,923 enterprises organically farming an area of 488,000 ha. This amounts to 11.6% of the agricultural land of the Czech Republic (Register of Entrepreneurs).

Most of these organically cultivated farms are situated in the regions classified as Less Favourable Areas (LFA). More than eighty percent of organically farmed agricultural land is covered with permanent grassland, where cattle-rearing systems prevail. The share of organic production in the total production of Czech agriculture is estimated to be between two and four percent. The size structure of these farms is reflected in the below table.

Ways of fostering green economies and green growth have been analysed with respect to political, social and economic aspects (Table 1).

A comparison of the proportion of organically farmed land in the total agricultural land in the EU Member States clearly shows significant differences is offered in Fig. 1.

From Fig. 1 it is clear that these differences in the share of agricultural land under organic production are not only related to the different agro-ecological conditions and the structure of European agriculture, but also to the strategies of individual member states when applying the overall Common Agricultural Policy (CAP) guidelines and its pillars of financing. The reasons for the relatively high proportion of organic farming in the Nordic countries, including Estonia and Latvia, mainly relates to their geographic location and the generally poor conditions for conventional agricultural production compared to the other member states of the EU. The same arguments are relevant to the practicing of agriculture in Austria.

Table 1 Distribution of organic production areas as a percentage of the total area on farms using organic production systems

Farm size (ha)	Area under organic cultivation (%)
<100	16.7
101–500	32.5
>500	50.8

Source FADN CZ (2014); adjusted

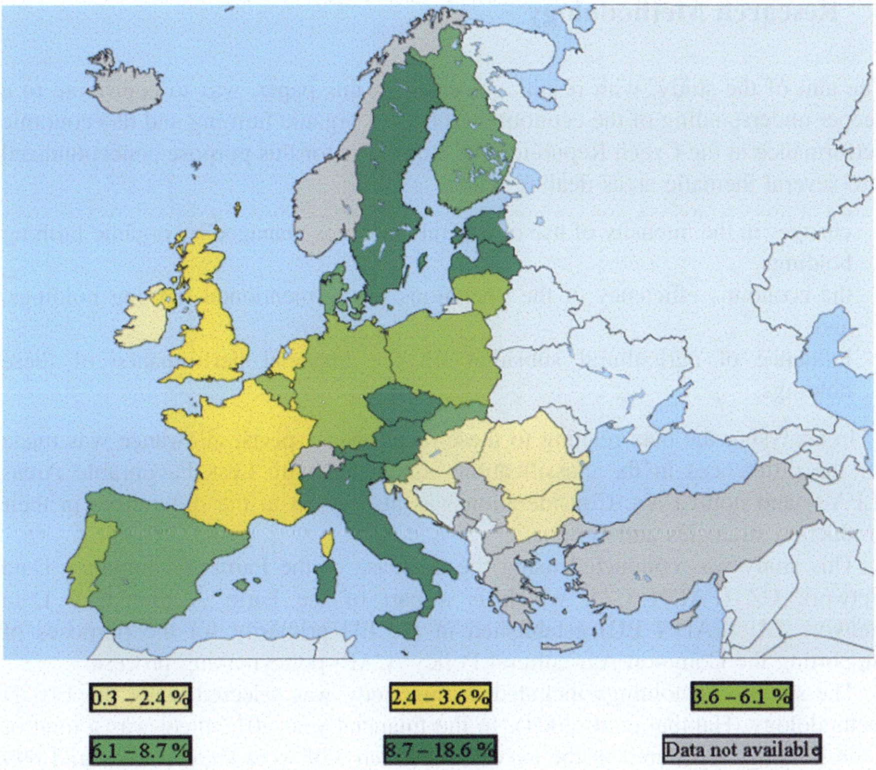

Fig. 1 The distribution of organically farmed land in Europe in 2012. *Source* EUROSTAT (2014)

An assessment of the approach of the structurally much diversified agriculture of Italy is also very complex.

Conversely, in countries that dominate the European agricultural market (like, e.g., France, Germany, Great Britain and the Netherlands), the share of organically farmed land is significantly lower, despite the overall promotion of ecologically and environmentally sensitive farming systems.

In Poland and Lithuania, where the policy focus is stronger on support for the creation of value-adding enterprises downstream in the value chain the share of organically farmed land is also less.

Having said this, the question arises to what extent the current chosen path of support for organic agriculture in the Czech Republic meets the criteria for economic efficiency and sustainability of organic farming. This raises an issue that needs to be considered in further shaping the agricultural policy interventions in supporting environmentally sensitive agricultural systems. This study deals with some of these issues.

3 Research Methodology

The aim of the study, with results presented in this paper, was to contribute to a deeper understanding of the economic aspects of organic farming and its economic performance in the Czech Republic. The study was for this purpose conceptualized into several thematic areas dealing with:

- changes in the intensity of use of natural resources managed by organic farming holdings,
- the economic efficiency of the operations of the mentioned farming holdings, and
- influence of agricultural subsidies on the financial performance of these holdings.

In analysing the data relating to these mentioned aspects, allowance was made for the differences in the classifications associated with Less Favourable Areas (LFA's) and non-LFA's. Consideration was also given to the differences in their production strategies and systems.

This study was conducted using the database of the Farm Accountancy Data Network CZ (FADN CZ), which is a part of the Farm Accountancy Data Network EU (FADN EU) established in the EU primarily for the purposes of supporting the Common Agricultural Policy (CAP) policymaking process.

The sample of holdings included in this study was selected using the FADN methodology (Hanibal et al. 2004). In the financial year 2012 there was a total of 1478 holdings registered in the survey, of which 579 were legal entities and 899 were natural persons. The total area of agricultural land managed by holdings included in the survey amount to 853,219 ha, representing 24.1% of the total utilized agricultural area of the Czech Republic.

Given the objective of the research and the nature of the factual base, the research methodology implies mainly on a comparison between organic holdings (OrgF) and conventional holdings (CoF). The distribution of the number and associated land sizes of agricultural holdings included in the consecutive years are indicated in Table 2.

The data reflects an increase in the total number of holdings included in the study over the recording period. This increase is mainly as a result of an increase in the total number of organic farming holdings in the Czech Republic. The analysis does

Table 2 The distribution and description of organic and conventional holdings of the FADN database included in this study

	Organic holdings		Conventional holdings	
	Number	ha UAA	Number	ha UAA
2001	38	17,549	1,166	805,333
2002	74	25,260	1,351	886,065
2003	48	19,676	1,165	840,174
2004	67	27,207	1,294	821,365
2005	77	28,883	1,285	807,661
2006	75	29,676	1,422	821,090
2007	79	28,829	1,419	847,886
2008	93	28,769	1,482	880,982
2009	100	34,836	1,433	849,557
2010	137	42,758	1,396	823,793
2011	223	64,402	1,315	791,877
2012	229	62,800	1,188	735,417

Source FADN CZ (2014)

not include data of holdings applying organic and conventional regimes simultaneously and/or in-conversion holdings. In the Czech Republic the proportion of land under organic holdings during the period of 2001–2006 represented 5% of the total and by 2012 it was 11%.

The study involved the analysis of a total of 66 variables constructed as follows:

- 12 variables on the legal form, agro-ecological conditions and type of production
- 6 variables on labour force inputs, scope and structure of cultivated land
- 23 variables on capital of holdings, production and financial and economic results
- 25 variables on subsidies and taxes.

Variables reflecting production and financial and economic results are derived from the standard indicators used in the FADN EU database. This construct of indicators is shown in Fig. 2.

The total cost was calculated keeping in mind the methodological restrictions created by the variation in the nature of the legal entities.[1] Remuneration of unpaid labour was derived from the average wage paid during the corresponding year on holdings included in the FADN CZ database. In the text this is indicated using the term “adjusted cost” and the acronym “AdC”.

Indicators used in the study mostly reflect current prices. Quantities reflecting changes in the physical volume of products and/or inputs are calculated from these indicators using changes in the value of agricultural production and changes in

[1]Total cost = sum of (total intermediate consumption + depreciation and costs of external factors (wages + rents + interest)).

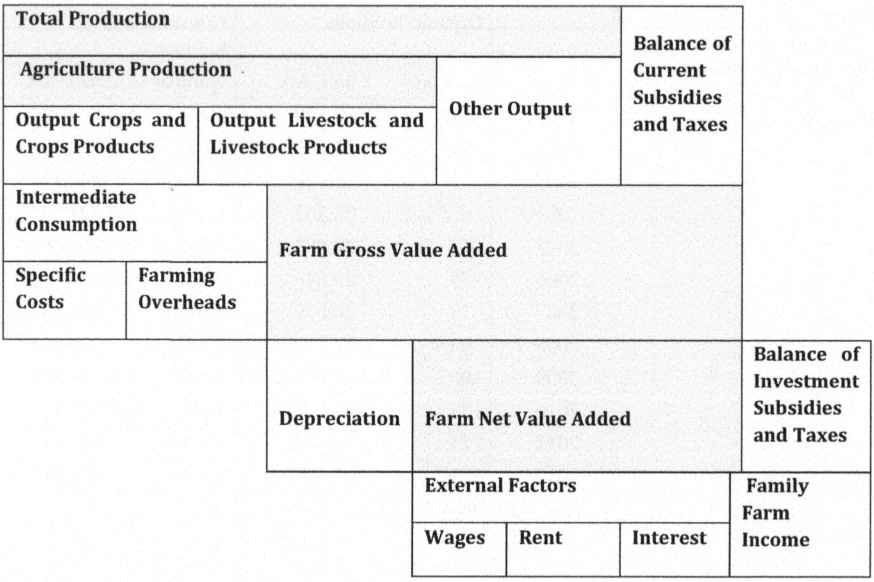

Fig. 2 Conceptual layout of indicators of economic performance used in the FADN EU. *Source* FADN CZ (2014); adjusted

input prices alone. That implies a certain degree of approximation in the final reported figures.

Comparisons in the level of and changes in the economic performance of organic and conventional holdings are based on conceptual values calculated for the trend component and the residual component for of the observed time series 2001–2012.

Graphical illustrations of changes in the values of indicators using the unit area of agricultural land showed slightly concave, slightly convex or almost linear trends. Therefore, the second-order polynomial technique was selected for the modelling of the trend. Figure 3 reflects the illustrations of the trend lines developed with this modelling technique.

4 Results and Discussion

4.1 The Value of Agricultural Production Per Ha in Organic Farming Enterprises in the Czech Republic

An analysis of the economic performance of holdings active in organic farming operations shows significantly lower values of agricultural production per ha, compared to conventional agricultural operations (see Fig. 3). Findings suggest that organic farming holdings (OrgF) in 2012 achieved approximately a quarter of the

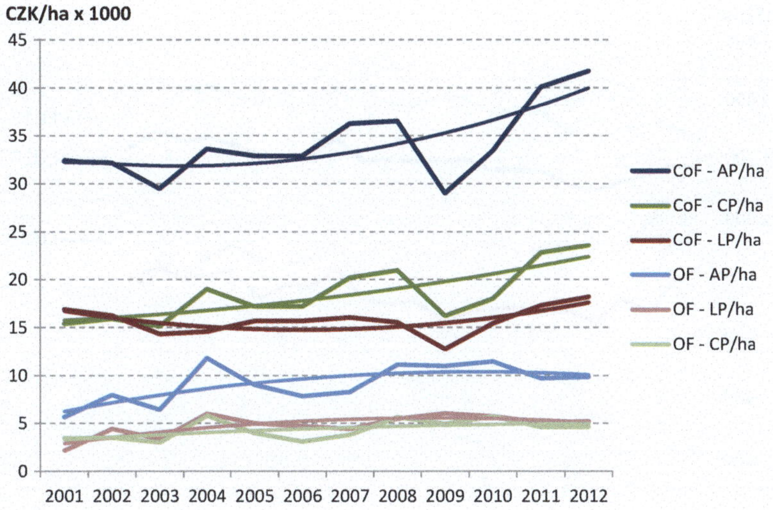

Fig. 3 The value of total agricultural (combined), crop and livestock production per ha at current prices (CZK/ha × 1,000). *Note OF* holdings active in organic sector, *CoF* holdings active in conventional sector, *AP* agriculture production, *CP* crop production, *LP* livestock production. *Source* FADN CZ (2014); own calculations. *Note* Trend lines indicate the results of the second-order polynomial modelling technique

level of agricultural production per hectare (ha) compared to conventional holdings (CoF). Comparisons of the changes in the value of the average annual production shows that this difference persists throughout.

More detailed assessments of the variations in the agricultural producers' price were done. Findings suggested that greater inter-annual variations in the value of agricultural production per ha, as illustrated particularly in the trends of CoF recorded between 2008 and 2010, were influenced mainly by current price changes. This is especially true for the decline recorded in 2009.

To obtain more accurate calculations of the level of and changes in the value of agricultural production associated with different agro-ecological zones, holdings were grouped according to similarity of agro-ecological conditions. These calculations suggested that:

- OrgF holdings within LFA's, on average, realised approximately 30% of the value of production per ha compared to that realised by CoF.
- In more favourable natural environments (non-LFA's) the average value of production per ha realised by OrgF holdings was slightly more than 30% of the value of production per ha realised by CoF holdings.

These calculations further suggest that OrgF holdings in the Czech Republic, considering the total value of production per ha, realize about 30–40% of the value of production of CoF holdings. The value of crop and livestock production per ha,

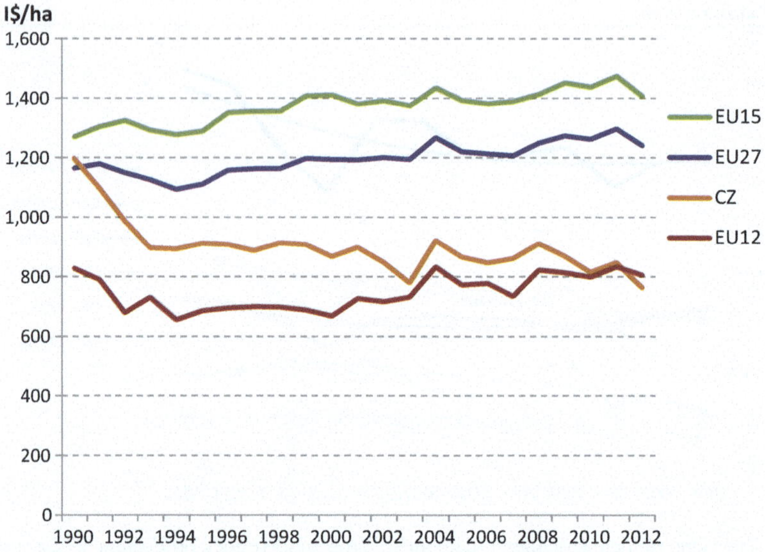

Fig. 4 Changes in average value of total agricultural production per ha in the EU and the Czech Republic. *Note* European Union (EU 27), Old member states (EU 15), New member states (EU 12), *CZ* Czech Republic. *Source* FAOSTAT (2014); own calculations

when calculated independently, follow similar trends and suggests that this difference persists throughout.

When analysing the value of production per ha in organic farming enterprises in the Czech Republic, it is essential to keep in mind the values of agricultural production per ha in other EU countries. Figure 4 illustrates the changes in the average value of agriculture production per ha obtained for the period 1990 through 2013 in the countries of the European Union and in the Czech Republic. Comparisons are calculated using the FAOSTAT database and demonstrates the average changes in the value of agricultural production per ha of UAA.

The value of agriculture production is derived from using the physical volume of production and International Dollar prices (in constant terms; 2004–2006). It therefore is a fairly good indication of the production trends as prices are not based on official exchange rates but on special exchange rates derived from the Geary-Khamis formula.[2]

Figure 4 clearly illustrates that, in comparison to the EU15 countries, the value of agricultural production per ha in the Czech Republic remains significantly lower. This trend remains negative in the last decade, even when compared to the average tendency of EU12 countries.

[2]While official exchange rates are influenced much by sectors (industry, services, finance, etc.) which have nothing to do with the agricultural sector, the Geary-Khamis formula takes into account only the agricultural sector.

Data indicate that the average production in the new member states (EU12) did in fact increase, contributing to the tendencies illustrated in Fig. 4. Having considered the comparative production levels of conventional and organic farming enterprises in the Czech Republic, it stands to reason that an increase in the number of organic holdings could, potentially, reduce the total average value of agricultural production per ha in the Czech Republic.

It is therefore imperative to recognise the risk that a diffusion of organic farming holdings in the Czech Republic holds for the value of agricultural production in the Czech Republic. This risk is of particular importance in favourable natural environments (non-LFA). It will most probably reduce the average value of agricultural production per ha in the Czech Republic.

In countries like France, Germany and Austria, the value of organic agricultural production per ha reaches levels of 40–60% of conventional agriculture (Niggli et al. 2008), while the OrgF holdings realize about 30–40% of the value of production per ha of CoF holdings in the Czech Republic.

It is therefore also important to recognize that the ratio of the value of agricultural production per ha of organic holdings to that of conventional holdings in the Czech Republic is already lower than that for other EU countries. Findings suggest that this scenario exists because of significant differences in production of organic farming holdings in the Czech Republic and the EU countries.

4.2 Value of Agricultural Production, Inputs Per Hectare and Productivity

The volumes of agricultural production per ha is strongly influenced by two possible practices. Firstly, there is the possible quantitative increase of inputs per unit area of production. Secondly there is the increased use of innovations promoting improved labour productivity and material inputs productivity (qualitative). Firstly, it is necessary to acknowledge that increased inputs per ha carries the risk of a higher environmental burden. Secondly, increased productivity of inputs reduces laboriousness (labour per unit of product) and demandingness (material and energy per unit of product). This reasonably accounts for more environmentally friendly effects.

Table 3 shows the results of the analysis of the mentioned quantitative and qualitative factors that influence the volumes of agricultural production for both organic and conventional farms. Results suggest that OrgF use 50–55% of the labour and material inputs per ha, compared to CoF holdings, while realizing about 60–65% of total productivity of these inputs, compared to CoF holdings.

Perspectives on the environmental benefits of organic holdings are strongly influenced by the question whether increased production per ha was the result of quantitative changes (increase of inputs per unit area of production) or qualitative

Table 3 Volumes of agricultural production, input use and productivity of inputs

		Number of farms	AP/ha CZK	AdC/ha CZK	TP/AdC CZK	ha/AWU	TP/AWU CZK
OrgF	2001	38	6,245	10,085	0.72	63.55	464,614
	2012	229	10,089	24,010	0.51	51.81	628,455
	Δ	x	349	1,266	−0.02	−1.07	14,894
	Correl. index	x	0.70	0.98	x	0.74	x
CoF	2001	1,166	32,545	35,774	1.00	24.30	867,704
	2012	1,188	40,001	49,496	0.89	33.44	1,478,984
	Δ	x	678	1,247	−0.01	0.83	55,571
	Correl. index	x	0.71	0.94	x	0.93	x
OrgF/CoF 2012		x	0.25	0.49	0.57	1.55	0.42

Note: *AdC* adjusted costs (incl. remuneration of unpaid labor); *TP* total production; *AP* agricultural production; *AWU* average working unit; *Δ* average change
Source FADN CZ (2014); own calculations

adjustments (better production methods—changed labour and material productivity).

Although OrgF holdings operate with 50% of the inputs per ha of CoF holdings, the overall material and labour demandingness of production is 1.4–1.7 times higher than that of CoF holdings—giving rise to the concept of "ecological paradox". In effect this suggests that, although organic holdings operate on 50% of the input demands per ha of CoF with methodology and technology which is supposed to have lower negative environmental impact, these holdings, however, create a 1.5 times more adverse ecological footprint per capita food production.

4.3 Labour and Energy Demandingness in Organic Farming Holdings in the Czech Republic

Current models of organic farming are generally presumed to increase opportunities for employment in rural areas (Gassner et al. 2008). It is further expected that many of the methodological and technological processes of modern agro-industrial farming systems will be replaced by more labour demanding biological practices.

Findings from this study do not confirm this expectation (assumption) at the aggregate level. Organic farming holdings active in LFA areas achieved around 70–75% of CoF holdings employment per ha. Holdings in non-LFA areas focusing on mixed and field production maintained around 90% of employment per ha of CoF holdings. Analysis of the employment situation at aggregate level suggests that the number of labourers per 1000 ha of UAA for OrgF holdings reached 19.3 AWU

Table 4 Energy consumption and demandingness per ha and per unit value of production

	Number of holdings	Energy consumption in CZK	
		Per ha	Per 1.000 CZK TP
OrgF	229	2,472	204
CoF	1,118	4,939	112
OrgF/CoF	x	0.50	1.82

Note: *TP* total production
Source FADN CZ (2014); own calculations

in 2012, compared to 29.9 AWU for CoF holdings. Similar results for the OrgF were reported by e.g. Laurence et al. (2013).

Significant differences for labour productivity between OrgF and CoF were reported. In 2012 OrgF holdings realised approximately 42% of the labour productivity registered by CoF holdings. This aggregate difference is even more accentuated when comparing between OF in LFA's (difference of 49%) and OrgF holdings in non-LFA's (difference of 33%). Comparisons of the movement in labour productivity between OrgF and CoF holdings maintains this trend.

Consider including a table showing the distribution of labour numbers, labour productivity and energy demandingness factor for each year from 2000 through 2012 for illustration purposes. This is a very important section and this type of information will contribute greatly to increased broader understanding. It will also enhance the understanding and the conceptual positioning of the findings offered following the analyses. Analysis of the energy demandingness of organic farming holdings is indicated in Table 4.

Analysis of the energy consumption in 2012 (as reported in the FADN data base) suggests that OrgF holdings realised around 50% of the energy consumption per ha of UAA compared to CoF holdings. However, energy consumption per unit value of production (demandingness) of OrgF is 1.5–1.8 times higher. Similar findings were reported in studies that compared the energy demandingness of organic and conventional holdings per unit area across Europe (Tuomisto et al. 2012; Laurence et al. 2013; Stefanos et al. 2012).

4.4 Economic Situation of Organic Farming Holdings in the Czech Republic

Further analyses of the data clearly suggest that the reduced level of productivity of the production factors and material inputs of organic holdings lead to reduced levels of economic effectiveness compared to conventional holdings.

Analysis of the data for 2012 suggest that OrgF holdings achieved a lower farm gross value added per ha (gross value added excluding subsidies). OrgF holdings earned approximately 9,000–11,000 CZK per ha UAA and about 330,000–400,000 CZK per worker (AWU) less in comparison to the CoF holdings. These comparisons are indicated in Table 5.

Table 5 Comparisons of the gross value added per hectare and per worker between OF and CoF

		Number of holdings	GVA/ha CZK	GVA/ AWU CZK	After exclusion of subsidies	
					GVA/ha CZK	GVA/ AWU CZK
OrgF	2001	38	4,121	261,890	−805	−51,158
	2012	229	13,846	717,361	−762	−39,479
	Δ	x	884	41,406	−4	1,062
	correl. index	x	0.95	x	x	x
CoF	2001	1,166	11,735	285,161	9771	237,435
	2012	1,188	19,393	648,502	11,046	369,378
	Δ	x	696	33,031	116	−11,995
	correl. index	x	0.81	x	x	x
OF—CoF (2012)		x	−5,547	68,859	−11,808	−398,857

Note: *GVA* gross value added; *AWU* average working unit; *Δ* average change
Source FADN CZ (2014); own calculations

From the findings above, it can reasonably be concluded that the negative values associated with the gross value added of the OrgF holdings suggest that revenues generated from products sold (price of product) were not sufficient to cover the cost of energy, material inputs and services in the production process. It further suggests that the gross value added is not sufficient to cover the depreciation costs of fixed assets and the cost of labour.

Differences in the values of gross value added and profitability for OrgF and CoF respectively are primarily caused by:

- differences in the volume of agricultural production per ha, the productivity of labour and the productivity of material inputs between OrgF and CoF;
- differences in the values of subsidy support for OrgF and CoF holdings; and
- differences in the selling prices of OrgF and CoF produce.

As indicated earlier the lower economic efficiency of the OrgF holdings is caused by lower gross value added (lower value of agricultural production per ha and lower productivity of labour and of material inputs) leading to higher cost per unit of production.

At this point it is important to illustrate the increase in the value of subsidies provided to OrgF during the period 2001 through 2012 (Fig. 5) and demonstrate how subsidy payments conceal the suboptimal gross value added and profitability profile of OrgF holdings.

Figure 5 clearly illustrates the trends in gross value added per ha between OrgF and CoF when subsidy payments are included and excluded from the calculations. From this illustration it is significantly evident that the value of the gross value added per ha of OrgF holdings without subsidies is approximately zero.

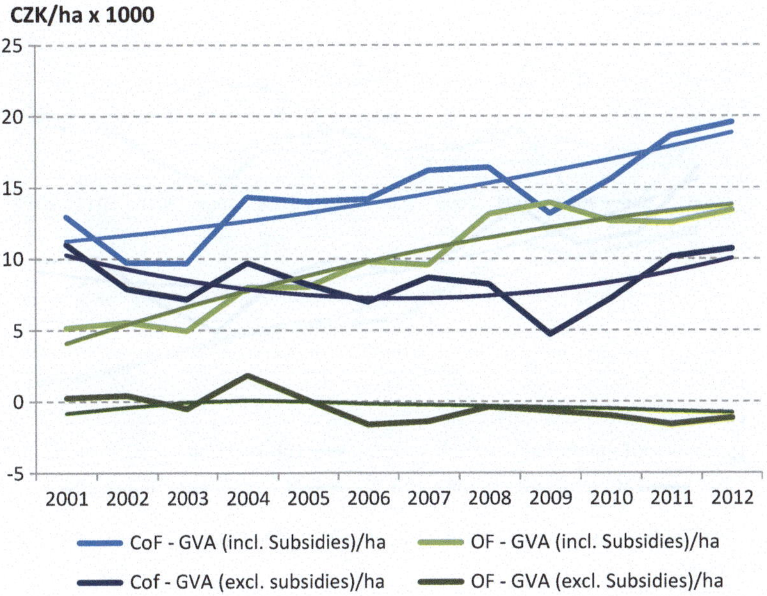

Fig. 5 Value of gross value added per ha for organic and conventional farming holdings including and excluding the value of subsidy payments. *Source* FADN CZ (2014), ČSÚ (2014); own calculations

Analyses of the differences in the levels of and fluctuations in profitability are expressed in terms of profitability per ha of UAA (see Fig. 6 and Table 6) and rate of return (see Table 6).

In reference to calculations shown in Fig. 6, profit is defined as the difference between revenues and adjusted costs. Revenues are defined as the sum of the value of total production and value of operating subsidies (see also the scheme in Fig. 2). Therefore, it does not include the profit from financial operations, extraordinary profit or parts of operating profit/loss from sale of property, creation and accounting of reserves.

Figure 6 and Table 6 illustrate that OrgF holdings realise slightly higher profits per ha and also higher rates of return compared to the CoF holdings. After the exclusion of operational subsidies, OrgF holdings, reported losses per ha are twice that of CoF holdings. Compared to CoF holdings, OrgF holdings also reported losses per unit of production (revenues) five times the magnitude of that of CoF holdings. Subsidies therefore significantly modify the level of gross value added and of profit and profitability.

Excluding operating subsidies, OrgF holdings reported a loss of 11,000 CZK per ha, which is twice that reported by CoF holdings (Table 6). The OrgF holdings recorded the same or slightly higher profit per ha than the CoF holdings. With subsidies the rate of return for both organic and conventional farms is roughly equal —approximately 2% of total capital. In the analysed period, subsidy payments

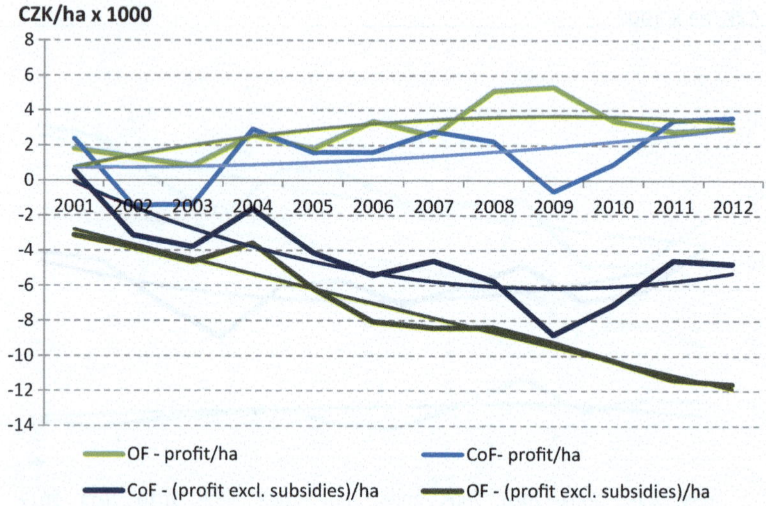

Fig. 6 Variations in profit per ha of UAA of OrgF and CoF holdings. *Source* FADN CZ (2014); own calculations

Table 6 Profit per ha of UAA and rate of return

		Number of holdings	Profit/ha CZK	Rate of return (%)	After exclusion of subsidies	
					Profit/ha CZK	Rate of return (%)
OrgF	2001	38	775	7.14	−4 151	−69.95
	2012	229	3,337	12.20	−11,271	−88.48
	Δ	x	233	0.46	−647	−1.68
CoF	2001	1,166	744	2.04	−950	−2.73
	2012	1,188	2,981	5.68	−5,366	−12.16
	Δ	x	203	0.33	−401	−0.86
OrgF/CoF 2012		x	1.12	2.15	2.10	7.27

Note Revenues = total production + operating subsidies; profit = revenue − total costs adjusted by the remuneration of unpaid labour; rate of return = profit/revenue; rate of return after the exclusion of subsidies = (profit − operational subsidies)/(revenues − operating subsidies); Δ = average growth
Source FADN CZ (2014); own calculations

(as an agricultural policy instrument) clearly modified the profit structure of organic farms.

Above mentioned findings, together with indicators of gross value added and profitability after the exclusion of subsidies, suggest that a significant inefficiency in Czech organic agriculture is its low productivity directly related to the low value of production per ha achieved. Published findings suggest the same situation in the rest of Europe as in organic agriculture in the Czech Republic (Niggli et al. 2008; Offermann and Nieberg 2000; Sanders 2007; Stefanos et al. 2012).

Further analyses also suggest that it is fundamentally important to address the economic sustainability of organic holdings in agricultural policy instrument and in the management of organic agricultural holdings in the Czech Republic. The most apparent solution is considered to be the increase of organic production per ha within the guidelines for good practice set for organic farming in the Czech Republic. The objective should remain to strengthen this eco-functional process of food production.

The selling price of organic products is an important stabilizing factor in the organic production industry. There is, however, a serious shortage of data to assess the impact of selling price of organic products on the financial economic results of the OrgF holdings. Certain perspectives are provided by some sample surveys.

Hrabalová, et al. (2013) conducted a study on the marketing of the organic production of 2,332 organic farmers. More than 65% of the sample of organic producers indicated that a larger portion of their produce is sold on conventional farmers' markets at conventional prices. Up to 91% of organic food sales take place through "mainstream value chains in the Czech Republic", whilst less than 10% is sold through "short value chains" (i.e. direct sales, farmers markets).

Mentioned findings suggest that OrgF holdings in the Czech Republic sell their produce at prices only about 10% higher than those of CoF holdings. A sample survey investigating the prices of organic food sold in retail chains, suggest that the retail prices of organic food are about 1.4 times higher compared to that of conventional foods (Valeška 2010). The higher price levels of organic food when compared to conventional foods were confirmed in a similar study (Živelová and Crhová 2013).

Keeping in mind the lower levels of gross value added per ha (excluding subsidy payments) and subsequent low profitability per ha of OrgF holdings (compared to CoF holdings), it can be inferred that the market price of OrgF produce contributes very little to recover the typically higher costs incurred because of the technical and technological complexities of organic farming systems.

It could therefore be concluded that the basic belief of consumers that they (by paying higher prices for organic products) support organic farmers to recover the higher cost of organic production, whilst also contributing to the preservation of a better environment, corresponds very little to existing realities of price transmissions taking place in organic food chains.

4.5 A Closer Perspective on Subsidy Payments to Organic Agriculture in the Czech Republic

Aforementioned analyses suggest that operating subsidies (included in the profit) have in the Czech Republic substantially increased in the case of OrgF holdings—

Table 7 Operating subsidies for organic and conventional holdings (CZK)

		Average 2001–2003	Average 2004–2006	Average 2007–2009	Average 2010–2012
OrgF	subsidy/ha	5,195	8,530	13,049	14,154
	subsidy/AWU	310,019	408,539	637,859	696,645
	subsidy/AP	0.79	0.96	1.29	1.38
CoF	subsidy/ha	1,976	5758	7,851	8,132
	subsidy/AWU	48,829	153,246	231,869	261,361
	subsidy/AP	0.06	0.17	0.23	0.21

Note: *subsidy* operating subsidy; *AP* agricultural production; *AWU* average working unit
Source FADN CZ (2014); own calculations

particularly after joining the EU in 2004. Table 7 shows subsidy payments per ha in the period 2001 through 2012. From this data is becomes clear that subsidy payments per ha to OrgF holdings were initially approximately 48% higher (2004 when Czech Republic joined the EU) but this discrepancy grew to 74% during the final period from 2010 through 2012. In LFA areas is this difference 10% higher.

Policy measures of state during 2001 through 2012 led to substantial increase in the share of operating subsidies in the income structures of OrgF holdings. This increase in subsidy payments led to increased support for covering farming costs of OrgF holdings. This increasing share of subsidy payments in the total cost recovery and profit making structure is evident from the comparison of OrgF and CoF illustrated in Fig. 7.

During the period 2001 through 2012 there was a considerable increase in subsidy payments to cover the cost of production of OrgF holdings. Subsidy payments, after covering production costs, increased the difference in profit creation between the OrgF holdings and the CoF holdings. In the period between 2010 through 2012, operating subsidies covered 47% of costs in OF holdings and created a 13% rate of return.

This phenomenon stimulated interest in OF farming and significantly contributed to the financial stability of OrgF holdings. There were, however, some negative effects: the interest of OrgF holdings in economic efficiency was reduced and a disproportionate unilateral orientation of OrgF holdings on resources from public budgets was created. This conclusion also reached by other authors (Niggli et al. 2008; Stefanos et al. 2012; Malá 2011).

Findings and data presented in Table 7 and Fig. 7 show that at the end of the analysed period, 2010 through 2012, subsidies covered 47% of OrgF holdings' costs compared to 12% in case of CoF holdings. Subsidies also amounted to 1.38 CZK per 1 CZK of agricultural product in OrgF holdings, while 0.21 CZK per 1 CZK of agricultural product in CoF holdings. Subsidies amounted to 696,000 CZK

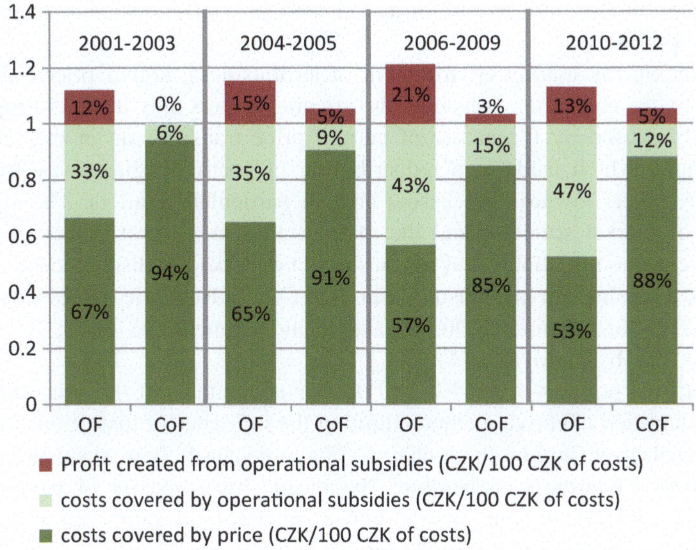

Fig. 7 The differential contributions of product price and subsidy payments to profit making (costs = 100%). *Source* FADN CZ (2014); own calculations

per worker (AWU) in OF holdings compared to 216,000 CZK for CoF holdings. When evaluating this phenomenon, it is necessary to base the assessment of merits and demerits and the amount of subsidies to OrgF holdings on two obligations of OrgF holdings have to fulfil, namely:

- to secure the supply for a specific market segment and satisfy consumer demand for organic products and organic food;
- to deliver public goods contributing to the sustaining of the environment and rural development.

The role and importance of organic agriculture is in this sense also defined by EU legislation and by the Member States of the European Union. The merits of the subsidy payments are also discussed in this context. Support for non-production functions is within the framework of the CAP applied by Member States throughout the European agricultural sector. Greater support for non-production functions by OrgF holdings can be objectively justified by providing greater environmental benefit.

In general, however, it is necessary to keep in mind some significant linkages arising from the state of the current business environment. Technology of organic holdings is based on a lower level of production resulting in an increased cost per unit of product. In a perfectly competitive market, these higher costs should be recovered through higher prices paid by consumers who prefer these products in the belief that they get some benefit from them. The principle of "…the one who wants this product will also have to pay for it..." leaves no room for subsidizing of supply of organic products.

However, in the market environment of agribusiness, higher prices of organic food paid by the consumer, "reaches" the organic farmers only in small proportions. It is mainly caused by the nature of current price transmission in the food commodity chain, which leads to a redistribution of profit margins away from agricultural producers towards processors and, in particular, retailers. The significant influence of market structure and the manifestation of market power in the final phases of commodity chains and agribusiness networks are discussed in a number of studies focusing on agribusiness in the Czech Republic (Bečvárová 2002; Lechanová 2006; Čechura 2006; Blažková and Chmelíková 2010; Zdráhal and Bečvárová 2013b; Dudová 2014).

The antitrust policies of the EU and the Member States of the European Union have not identified effective methods limiting the influence of distortions in product and factor markets within food chains. The impact of these distortions on the income parity of farmers is therefore "balanced" by a transfer of payment from government budgets. Income parity hence becomes an important criterion for determining the amount of subsidies for agricultural producers.

5 Conclusions

The system of organic farming has become an important component of the ecology-friendly alignment of agriculture in the Czech Republic. This system has extended extremely quickly in the past fifteen years. In 2013 there were 4060 organic farming holdings covering 12% of the agricultural land in the Czech Republic. The development of this style of agriculture was stimulated to a large extent by state interference and the increase of subsidy payments, especially after the Czech Republic joined the EU.

A weak point in organic farming in the Czech Republic is its low productivity. In the period 2001 through 2012 OrgF reached 30–40% of the value of agricultural production per ha of that achieved by CoF. The difference in the value of production per ha between OrgF and CoF was persistently increasing. Compared to EU15 countries, OrgF holdings in the Czech Republic achieve a lower value of agricultural production per ha.

The difference in the value of agricultural production per ha between OrgF and CoF holdings is caused mainly by two factors: around 50–60% of the difference is caused by quantitative differences in inputs per ha and around 40–50% by different levels of productivity of these inputs.

Although OrgF operate with 50% of the inputs per ha of CoF holdings, the overall material and labour demandingness of production is 1.4–1.7 times higher than that of CoF holdings—giving rise to the concept of "ecological paradox". In effect this suggests that, although organic holdings operate on 50% of the input demands per ha of CoF with technology which is supposed to have lower negative environmental impact, these holdings, however, create a 1.5 times more adverse ecological footprint per capita food production.

The reduced level of productivity of the production factors and material inputs of organic holdings lead to reduced levels of economic effectiveness compared to conventional holdings. Analysis of the data for 2012 suggest that OrgF holdings achieved a lower farm gross value added per ha (gross value added excluding subsidies). OrgF holdings earned approximately 9,000–11,000 CZK per ha UAA and about 330,000–400,000 CZK per worker (AWU) less in comparison to the CoF holdings.

Excluding operating subsidies, OrgF holdings reported a loss of 11,000 CZK per ha, which is twice that reported by CoF holdings. Subsidy payments during 2010 through 2012 amounted to 14,400 CZK/ha for OrgF holdings and 8,500 CZK/ha to CoF holdings. This interference has modified profit making. OrgF, therefore, recorded slightly higher profit per ha and twice the profit per unit of the product when compared to CoF. With subsidies the rate of return for both organic and conventional farms is roughly equal—approximately two percent of total capital. Hence, for organic farmers subsidies represent the main financial source of stability in their financial management.

Market prices of organic products from OrgF holdings contribute very little to cover higher costs of production. It is estimated that compared to the prices of conventional products they are higher by approximately 10%. On the contrary, consumer prices of organic food are approximately 40–50% higher than prices of conventional food. It is mainly caused by the nature of current price transmission in the food commodity chain, which leads to a redistribution of profit margins away from agricultural producers towards processors and, in particular, retailers.

The increase of the contribution of operating subsidies to cost recovery and profit creation together with the development of subsidies per unit of product and per worker (AWU) in OrgF emphasizes the need for further development of the policy strategy for this type of farming. This should include new thinking on improving management methods, mechanisms for and changes to subsidy structure, particularly in relation to the criterion of balance in the market for organic products and the criterion of income parity.

Analyses of the economic aspects of organic agriculture (completed in this study) challenge agrarian policy thinking through highlighting the problem of the economic sustainability of OrgF holdings. The increase of organic production per ha within the guidelines for good practice set for organic farming in the Czech Republic is considered to be the solution. The objective should remain to strengthen this eco-functional process of food production.

A most likely solution is probably related to the stimulation of Eco-friendly production as a way to increase the productivity of the system. A crucial condition for this strategy concerns the development of markets for organic products and organic food. This should include mechanisms to reduce market distortions within the commodity chains and agribusiness networks.

This study simultaneously supports the argument that the main strategy of eco-friendly alignment of agriculture of the Czech Republic is a transition to a system of integrated agriculture.

Acknowledgements This study was elaborated within the research project: Czech economy in the integration and globalisation process and the agrarian and service sector development under the new conditions of the European integrated market; research branch "The development tendencies of agribusiness, the forming of the segmented markets of commodities' strings and food networks in the integration and globalisation processes and agrarian policy changes" (VZ MSM 6215648904/04).

References

Bečvářová V (2002) The changes of the agribusiness impact on the competitive environment of agricultural enterprises. Agric Econ 48:449–455. ISSN 0139-570X

Blažková I, Chmelíková G (2010) Analýza cenových relací v komoditní vertikále. [CD-ROM]. Brno: In: Sborník příspěvků z mezinárodní vědecké konference Region v rozvoji společnosti 2010, pp 13–17. ISBN 978-80-7375-435-8

Butler G, Nielsen JH, Slots T, Seal C, Eyere MD, Sanderson R, Lleifert C (2008) Fatty acid and fat-soluble antioxidant concentrations in milk from high- and low-input conventional and organic system: seasonal variation. J Sci Food Agric 88:1431–1441.ISSN: 1097-0010

Čechura L (2006) The nature of selected price transmissions in the agri-food chain and their consequences. In: Studies on the agricultural and food sector in Central and Eastern Europe, vol 33, Halle (Saale), IAMO, pp 430–448

Dudová B (2014) Cenová transmise v zemědělsko-potravinářských vertikálách pšenice a hovězího masa na příkladu České republiky. In Hospodářská politika v členských zemích evropské unie, pp 10–20

Gassner B, Freyer B, Leitner H (2008) Labour quality model for organic farming food chains. In: 16th IFOAM Organic World Congress, Modena, Italy, 16–20 June 2008 [online]. http://orgprints.org/view/projects/conference.html

Goodwin BK, Harper DC (2000) Price transmission, threshold behavior, and asymetric adjustment in the U.S. Pork Sector. J Agric Appl Econ 32(3). ISSN 1074-0708

Hanibal J, Dykova E, Foltýn I, Medonos T, Zedníčková I (2004) Uplatnění „Zemědělské účetní datové sítě" (FADN) v České republice. Výzkumná studie č. 78. Praha: Výzkumný ústav zemědělské ekonomiky, 88 pp. ISBN 80-86671-23-2

Hrabalová A, Darmovzalová I, Wollmuthová P (2013) Statistická šetření ekologického zemědělství – Základní statistické údaje 2012, TÚ 4212/2013. Výstup č. 2 [online]. ÚZEI, 52 pp. http://eagri.cz/public/web/file/284641/Zprava_EZ_2012_final.pdf

Hughner RS, McDonach P, Prothero A, Shultz CSI, Stanton J (2007) Who are organic food consumers? A compilation and review of why people purchase organic food. J Consum Behav 6, 94–110. ISSN: 1479-1838

Kummeling I, Thijs C, Huber M, Van De Vijer LP, Snijders BE, Penders J, Stelma F, Van Ree R, Van Der Brandt PA, Dagnelie PC (2008) Consumption of organic foods and risk of atopic disease during the first 2 years of life in the Netherlands. Br J Nutrition 9(3), 598–605. ISSN 0007-1145

Laurence GS, Adrian GW, Bruce DP (2013) The energy efficiency of organic agriculture: a review. In: Renewable agriculture and food systems [online], pp. 1–22, ISSN 1742-1705. http://journals.cambridge.org/download.php?file=%2FRAF%2FS1742170513000471a.pdf&code=f6493a8b372b3be5d563057d88374391

Lechanová I (2006) The process of transmission of supply and demand shocks in Czech Meat Commodity Chain. Agric Econ 52. ISSN 0139-570X

Lockie S, Lyons K, Lawrence G, Halpin D (2006) Going organic. mobilizing networks for environmentally responsible food production. CABI Publishing, Wallingford, 247 pp. ISBN: 9781845931582

Malá Z (2011) Efficiency analysis of Czech organic agriculture. In: E + M Ekonomie a management [online], vol 1, pp 14–28. ISSN: 2336-5604. http://www.ekonomie-management.cz/download/1346061157_405e/2011_01_mala.pdf

Niggli U, Slabe A, Schmid O, Halberg U, Schluter M (2008) Strategic research agenda for organic food and farming. Vision for an organic food and faming research agenda to 2025. IFOAM—EU a FiBL, 46 pp. http://www.tporganics.eu/upload/TPOrganics_VisionResearchAgenda.pdf

Offermann F, Nieberg H (2000) Economic performance of organic farm in Europe. University of Hodenheim, Stuttgart, 68 pp. ISBN: 3-933403-04-9

Pugliese P (2001) Organic farming and sustainable rural development. A multifaceted and promising convergence. Sociologia Ruralis 41(1):112–131

Richter T (2008) Retailing organic food in Europe 2008: latest trends in distribution channels and driving forces. BioFach Congress, Nuernberg, Germany [online]. 21–24 Feb 2008, 19 pp [unpublished]. http://orgprints.org/13189/1/richter-2008-retailing.pdf

Redlichová R, Bečvářová V, Vinohradský K (2014) Vývoj ekologického zemědělství ČR v ekonomických souvislostech. 1. vyd. Brno: Mendelova univerzita v Brně, 91 pp. 1. ISBN: 978-80-7509-173-4

Sanders J (2007) Economic impact of agricultural liberalisation policies of organic farming in Switzerland. Research Institute of Organic Agriculture (FiBL), Frick. ISBN: 978-3-03736-015-6

Stefanos AN, Evangelos P, Savvas Z (2012) Productive efficiency of subsidized organic Alfalfa Farms. J Agric Resour Econ [online] 37(2):280–288. ISSN: 1061-5502. http://ageconsearch.umn.edu/bitstream/134283/2/pp280-288,Nastis.pdf

Tuomisto HL, Hodge ID, Riordan P, MacDonald DW (2012) Does organic farming reduce evironmental impacts? A meta-analysis of European research. J Environ Manage 112:309–320. ISSN: 0301-4797. http://www.sciencedirect.com/science/article/pii/S0301479712004264

Valeška J (2010) Srovnávací test VI výsledky červen 2010. Srovnávací test VI výsledky listopad 2010

Zdráhal I, Bečvářová V (2013a) Rozvoj zemědělství a venkova v evropském modelu agrární politiky - formování strategie v kontextu změn prostředí. 1. vyd. Brno: Mendelova univerzita v Brně,129 pp. ISBN: 978-80-7375-771-7

Zdráhal I, Bečvářová V (2013b) The issues how to express the concentration processes in European agriculture. Acta Universitatis Agriculturae et Silviculturae Mendelianae Brunensis, vol 61, č. 7, pp 2951– 2964. ISSN: 1211-8516

Živělová I, Crhová M (2013) Organic food market in the Czech Republic. Acta Universitatis Agriculturae et Silviculturae Mendelianae Brunensis. 2013. sv. 61, č. 2, pp. 539–546. ISSN 1211-8516

Online Sources [Online During 2014-06 through 2014-10]

ČSÚ (2014). http://www.czso.cz

EUROSTAT (2014). http://ec.europa.eu/eurostat

FADN CZ (2014). https://public.fadn.cz/Cli1login.asp?Key

FAOSTAT (2014). http://faostat.fao.org/site/291/default.aspx

Family Farmers' Cooperative from Ibiúna, São Paulo State, Brazil: An Example of Social Capital as a Driver for Ecological Sustainability Change

Paulo Roberto Borges de Brito

Abstract It has been argued over the years that sustainable agriculture must encompass the "triple bottom line" framework towards a more sustainable bioeconomy. However, much emphasis has been placed on the ecological component ("planet") of the trifecta, neglecting the economical ("profit") and social ("people") dimensions, especially in Sao Paulo state, Brazil. This transition from traditional agriculture to a more sustainable system has excluded family and small farmers, which make up more than 50% of total farmers in Brazil. Social capital literature brings us some insights about how to enable these excluded farmers to more environmentally friendly agricultural practices. Empowerment and participation of local farmers have been proven themselves to be powerful strategies for building the social and economical dimensions of family farmers, equipping them to engage in more ecologically sustainable agricultural practices. This paper shows a practical initiative of a group of family farmers in Ibiuna, Sao Paulo, Brazil, that were empowered and became successful in their activities toward a more sustainable agriculture path, and, ultimately, leading to a more sustainable bioeconomy future in their region.

Keywords Family farming · Sustainable agriculture · Social capital Bioeconomy

P. R. B. de Brito (✉)
College of Business, Colorado State University, Fort Collins, USA
e-mail: Paulo.de_Brito@colostate.edu

W. Leal Filho et al. (eds.), *Towards a Sustainable Bioeconomy: Principles, Challenges and Perspectives*, World Sustainability Series,
https://doi.org/10.1007/978-3-319-73028-8_18

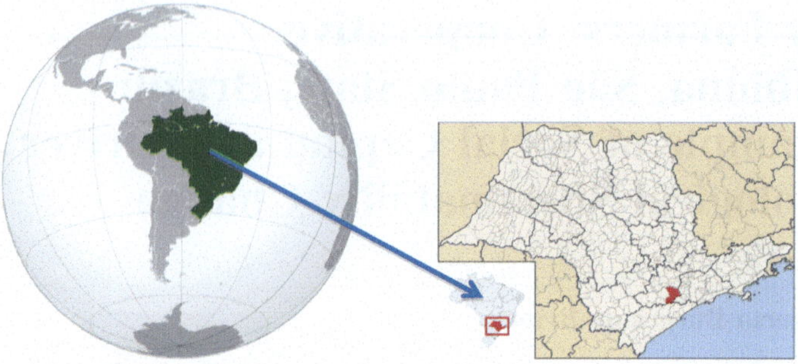

Source http://www.mapsopensource.com/brazil

1 Introduction

Throughout the history of the Brazilian organic movement, the decision for family farmers to begin organic farming was initially influenced by some evidence of poisoning caused by agrochemicals, which, in turn, diminished the demand for conventional products. At the same time, demand for organic products began to rise. It became clear to these farmers that it would be beneficial to organize themselves to successfully enter this new market (Ehlers 1999).

The idea for a new organizational structure came about mainly due to the fact that the family farmer (historically a farm unit of self-production and consumption) values diversity, and collective organization. The family fosters the technical and managerial operations, to the extent that the one who makes the decisions is also the one who puts them into practice. In doing so, the farmer usually achieves greater and wider geographic distribution, as well as develops better methodologies for exploring the environment, as producers are better able to adapt and delineate in more defined and homogeneous ecological units. The family farming production model encourages local development, favoring the planning and collective management of natural resources (Ehlers 1999).

However, despite the favorable trend of integration of family farming in a more environmentally sustainable agriculture, the increasing demand for organic products outside of local markets started to develop trade relations with intermediaries, with the products being sold primarily through supermarkets. In these supermarkets, the distance between consumers and producers and the inability to be certain as to how this organic produce was produced justified the need for production monitoring via certification. In this instance, certification is the procedure whereby an officially recognized, independent, third party assures in writing that a product, process or service meets certain requirements, through the issuance of a certificate (Normative Ruling No 7 1999).

As the demand for certification increases, so does the organic movement's urgency to continuously improve its guidelines aimed at social justice and sustainable development. In this instance, "sustainable" is defined as the use of natural and human resources that meets the needs of the present generations without compromising the ability of future generations to meet their own needs (Organization for Economic Cooperation and Development—OECD 1995). However, the process of crafting a sustainable agricultural methodology, according to Moreira e Carmo (2004) should not only result in a change of inputs, but also strengthen family-based agriculture, profoundly changing the agrarian structure of the country through consistent and coherent government policies for poverty reduction and the empowerment of the poorest citizens, through the revision of the epistemological and methodological assumptions that guide research and development activities. Finally, the authors argue that sustainable rural development should find its beginnings in agro-ecology, which takes sustainability and rural development strategies defined by the participation and "ethno-ecosystem" identity of each location into consideration. This concept relates to the strengthening collective forms of social action that have a transforming (local) endogenous potential. According to Sevilla-Guzman and Molina (1993) and Sevilla-Guzman (2001), the local potential has two dimensions, one is ecological and other is social. The ecological dimension covers the low-input agriculture, the scale of production, energy and agro-ecological base. Its social dimension encompasses the power and social organization of labor, local identity, autonomy, cooperation and other forms of collective social action.

On the other hand, the implementation of an agro-ecology will not easily come to fruition. Such agriculture requires the construction of a new paradigm; one which encompasses both the social and ecological interests of popular social movements as well as the relationship between social and natural sciences. This paradigm must also be understood in the context of socio-environmental problems today; problems which increasingly strive toward the development of truly sustainable solutions. This type of sustainable agriculture mentality, which has been more greatly emphasized in bioeconomic theory in recent years (Schmidt et al. 2012; de Besi and McCormick 2015), has just begun to develop in recent years and is still evolving. (Moreira e Carmo 2004).

In practice, the social dimension of sustainability is still in its infancy. However, as stated previously, this dimension is as important as the ecological component. In São Paulo state, the ecological side of agriculture is much more established than the social dimension. Implicitly, this demonstrates a historical tradition of excluding family farmers in the organic market (Brito 2006). Consequently, what hinders organic family agriculture from flourishing in São Paulo is precisely the lack of emphasis of the social dimension of the sustainable agriculture. In other words, the social organization of labor, local identity, autonomy, cooperation, and other forms of collective social action need to be strengthened in order for organic family agriculture to thrive in São Paulo, and, consequently, to move towards a more sustainable agriculture and bioeconomy.

2 Objetives

The objective of this chapter is to demonstrate that it is possible to fill the social and economic dimensions of the so-called "sustainability gap." This gap can be filled by capacity building and training programs which result in an increase of farmers' social capital, which, it turn, leads to long-term ecological sustainability.

The chapter then uses a family farmer's cooperative *Aprove* as a case study to demonstrate the objective of the study. The chapter then presents a literature review on social capital and an analysis of how *Aprove* producers built their social capital. The chapter then reviews the results of increased social capital and, finally, details conclusions of the findings.

3 Social Capital

As a result of the vast literature in the social capital scientific field, this paper narrowed its focus to authors who focused their study on empowerment, participation, and institutional change. Empowerment and participation are mostly based on the concept of an agent of change as detailed by Amartya Sen (1999) and institutional change based primarily on the research of institutional economists, particularly that of Douglas North (1990).

Empowerment is the process in which learning and knowledge are promoted. Social capital is strengthened through this educational process of learning and knowledge (Abramovay 2003). Empowering producers and enabling them to build their own social capital must be developed from the definition of the active agent of change proposed by 1998 Nobel Prize winner in economics, Amartya Sen (1999).

Sen's concept of the *agent of change* (1999) is defined as someone who acts and brings about change guided by the convictions of his or her own personal values and goals, as opposed to simply accepting the status quo imposed upon them by society. This definition also places the individual in control of his or her actions, offering vast freedom of choice, which, in turn, causes the agent of change to carry out their role thoughtfully and intentionally. The author makes it clear that excluding individuals from the development process is problematic because it distorts information and incentive, causes disutility and higher administrative costs. It is through personifying the role of the agent of change empowers the individual to determine their actions.

In this way, Sen's concept of freedom of choice intersects with Paulo Freire's concept of power (1972). Freire defines power as "the increased awareness and the development of 'critical thinking' of the marginalized and oppressed;" in other words, the power of education. This is the power of "doing" and "being capable," feeling in control of situations. It refers to the recognition of capabilities of such groups to act and play an active role in development initiatives. This also requires overcoming decades of a culture of passive acceptance and strengthening the skills of marginalized groups to get involved as legitimate development actors.

Therefore, based on Sen and Freire, the process of building a social organizational structure should be formed gradually and should be based on norms that value the involvement of each agent (North 1990). The goal should be to seek consensus among the agents, as opposed to a hierarchical confrontation between those who have decision-making power and those who do not (Oakley and Clayton 2003).

Furthermore, according to Sen (1999), exercising the role of an active agent of change depends on each individual's capacity to identify his or her needs along with active participation in group discussions with the goal of influencing decisions, strengthening cooperation, and building norms. This role of an active agent of change is, therefore, dependent on learning and acquisition of new knowledge and skills (North 1990).

Unfortunately, most Brazilian family farmers are not qualified to play the role of active agent of change to the point which they would be able to develop a sustainable agriculture on their own. However, despite this sad reality and the low degree of involvement in social organizational structures among organic family farmers in São Paulo, there does exist one example to the contrary. In the case of the Association of Organic Producers of Verava Subdivision (APROVE) in Ibiúna (São Paulo state), a small group of organic family farmers in search for collective solutions in order to remain in the organic market, has successfully put this agent of change theory into practice through the creation of a socially-organized farmers cooperative, thereby strengthening the social and economic dimension of sustainability.

The case study analyzes the social dimension of sustainability of the *Aprove* producers based on the study of Sen (1999) and North (1990). An individual or a group of individuals will increase their social dimension if they have:

1. Knowledge and learning to identify their needs
2. Participation to influence and have voice to make decisions
3. Social capital: cooperation to craft new institutions.

4 Methodology

In order to measure the three elements of this social dimension identified by the literature review, the following indicators were developed:

Knowledge and learning were measured by cognitive social capital. Cognitive social capital can be defined as capital that promotes cooperation through the sharing of norms, values, etc. This was measured through acquired organic agriculture learning via training, how the knowledge was shared among the families, and how the knowledge was implemented into the production process. These provided a measurement of the farmers' understanding of organic production norms and required documentation for the organic certification process (for more information see Brito 2006).

Participation was measured by group member involvement in the cooperative, and the type of decision-making power he or she held in the group (Brito 2006). Finally, social capital was measured by analyzing the farmers' ability to identify problems, their ability to develop new and clear rules to solve those problems, actions initiated by the group, The use of internal/external communication channels, and enforcement of rules by cooperative's members.

5 A Case Study: The APROVE Producers and Their Social Capital

The history of these farmers begins with a project called "Farm to City" (*Campo Cidade)* in Ibiúna (São Paulo state, Brazil) in the early 90s. The main goal of the project was to bolster local development. Supported by the Catholic Church and the Campo Cidade Foundation, the people who spearheaded the project built a school, which they named Agricultural Family School (*Escola Familia Agricola).* At this point, the project leaders were able to determine the most critical needs of the region. The plan that was created to help them achieve their goals developed eight action items; one of which was to convert the conventional process of production to an organic process.

In mid-1995, a survey conducted by the president of the Organic Agriculture Association (AAO), a non-profit that helps family farmers in organic agriculture through rural extension in São Paulo state, indicated that one of the members of the informal farmer cooperative (the families had not formally organized at that point) had been using an excessive amount of pesticides. This family utilized many pesticide sprays in their cultivation of cassava and other vegetables, which resulted in extensive contamination, and degradation of the soil and water.

The family had historically sold their produce primarily to one of the main wholesale city markets in São Paulo; however, the trading process was complicated and was only successful with the help of an intermediary, the Campo Cidade Foundation. Similarly, because of the farmers' lack of social power in this process, along with the limited market potential inherent in producing only conventional (as opposed to organic) produce, the farmers had limited opportunities for growth in the market. It was when these challenges became apparent that technicians from AAO came to help empower those family farmers through capacity building. Farmers were taken through several educational courses that prepared them for conversion to an organic production system. After the courses, these small producers joined two other relatives and began supplying products to a major organic product trader, greatly strengthening their market power.

Some of the former AAO presidents highlighted that one of the critical components to this family's success was their entrepreneurial spirit and, capacity to organize themselves as a result of their strong family ties. All the family members contributed because they all knew the organic production processes.

Those families all acquired knowledge to change their production process. One of their main issues was contaminated soil. Farmers were able to eliminate pesticides and chemical fertilizer in a short period of time with an understanding of organic production norms and the required documentation based on certification norms and proper forms.

In 2002, an incident occurred which became the catalyst for these family farmers to organize formally into the social cooperative today known as APROVE. After several years of having sold their organic products to a particular trader, the family farmers began to perceive that they were being taken advantage of by the trader. At the time, the organic certification the farmers were given for their produce was owned by the trading organization itself, which had begun to manipulate the requirements of the label in order to gain more favorable pricing. At this point, the leader of the family farmers realized that the farmers had no power over their organic certification seal, and were beholden to whatever requirements the trader wished to impose in order to certify the produce as organic. The farmers again sought the direction of AAO and asked for help to develop a process to self-certify the farmers both individually and collectively.

The farmers developed a cooperative/association for the direct sale of their products in partnership with the local union, today known as APROVE. As of 2006, the association had nine producers and their families with farmers ranging in age from 23 to 57 years old. Of the total, four are landowners of cultivated land, the other five are sharecroppers of owners who are close relatives. The farmers work on average, at least eight hours a day, with at least four people working at any given time. The level of education of families varies from incomplete elementary school to completed high school. Some were attending college at the time of interview.

By that time, those farmers had developed a strong sense of community which had empowered them to change their reality. They knew that in order to succeed in the market they needed to have more group involvement and a more active voice, which included all members of the cooperative. By 2015, active involvement in the group had increased from four member to more than ten during the period of analysis. The decision making process was very top-down before the cooperative (the trade firms setting all the rules). With the implementation of the cooperative, members of the group were able to have a more consensus-based decision-making process using a participatory approach, giving all the members of the group a voice.

The cooperative enabled those farmers to identify their own problems and develop their own rules to solve them. For example, at the inception of the cooperative, farmers had no knowledge of the certification process documentation and lacked methodologies to apply and enforce rules among the group members. By 2015, they were able to define two strategies for documentation, and understand the importance of applying enforcement rules in case of non-compliance.

Aprove farmers were also able to improve the number of actions initiated by the group. In the beginning they only had records of cultivation and crop control. By 2015, they realized the need to have an internal inspector in the group to monitor production and documentation, and a monthly meeting to dicuss the potential risks in the production, storage, and distribution process, as well as develop an assessment and risk plan (Bernardo and Ramos 2016).

Social capital was also increased by internal/external communication rules. In the beginning, communication was limited to telefone calls. By 2015, they have developed written communication channels such as minutes, and they understood the importance having the final meeting result posted and visible on the wall (on a whiteboard) in the warehouse (where the meetings occur) for the members who were absent to remain informed.

Finally, members of the cooperative were able to eventually establish some enforcement mechanisms to ensure all members were conforming to the new organic rules within the group. Instead of relying on external monitoring, the group was able to get a farmer trained to inspect all farms of the cooperative so that they could cut costs in the future. They also developed their own non-compliance rules within the group detailing what would happen if violation of production process occured.

The increased social capital also positively impacted the economic dimension of sustainability. First, gross average monthly income of the group from the organic farming before forming the APROVE cooperative varied between US$132 and US $1,700.00.[1] That number more than doubled in the almost 10 years since the formation of the cooperative.

The income from their collective is divided among all the members equally. Each member also equally shares the administrative expenses associated with running the cooperative. If any work is done outside of the cooperative, it is done so on a voluntary and unpaid basis. These producers have demonstrated their solidarity with each other; helping one another to fulfill commitments, remember tasks, and so on. The sale, freight or delivery of products, input purchases, are activities carried out jointly by the APROVE.

6 Discussion and Conclusions

What we can demonstrate from this case study is that the producers' ability to organize themselves into associations or cooperatives was made possible as a result of the capacity building and training provided by the Campo Cidade Foundation, AAO and the local union. We can conclude that technical and organizational training is a way to empower farmers towards autonomy; to develop themselves and be agents of transformation of their own reality. Without the necessary technical knowledge, this organizational change would not have been possible.

Another important point to highlight is collective organization gave the group greater bargaining power in market channels while simultaneously doubling their

[1]The US dollar on 4/4/2016 was about 3.7 Reais (Brazilian currency).

gross household income over a period of five years. Furthermore, organic farming provides a sustainable solution for families of rural areas with degraded land, where conventional production method is no longer feasible.

The group's ability to cooperate in the various association activities, or, in other words, the social capital of the group was an important driver of this change towards implementation of a bioeconomy in the region. This was largely due to the individual, technical, and organizational capacity of each producer. The large number of connections that the group had with organizations linked to the organic movement, the individual commitment to the principles of organic agriculture, and the existing connections of friendship among group members also contributed to their success.

The example of these producers can serve as inspiration for others to do the same. However, the development and implementation of public policies to empower and facilitate the producer at the time of technological change (i.e. a change from a conventional production process for organic in combination with other producers) is necessary, especially in the state of São Paulo, where work is predominantly characterized by individualism.

References

Abramovay R (2003) "O Futuro das Regiões Rurais". Ed. da UFRGS, Porto Alegre, p 149

Bernardo EG, Ramos HR (2016) "Desenvolvimento Sustentavel na agricultural familiar". V International symposium on project management, innovation, and sustainability (SINGEP), Sao Paulo, p 16

Brito PRB de (2006) "O controle social no processo de certificação de grupo por auditoria externa: o caso APROVE pela AAOcert". São Paulo, p 62. Dissertação de Mestrado. Programa de Pós-Graduação em Ciência Ambiental da Universidade de São Paulo

Carvalho YMC, Dulley RD (1994) "O modelo de gestão da Fazenda Pirituba II, área 1, em São Paulo". Agricultura em São Paulo, São Paulo. 41(2):21–32

de Besi M, McCormick K (2015) Towards a bioeconomy in Europe: National, regional and industrial strategies. Sustain 7(8):10461–10478

Ehlers E (1999) "Agricultura Alternativa: origens e perspectivas de um novo paradigma". Guaiba: (ed) Agropecuaria, 2nd edn. p 157

Freire Paulo (1972) Pedagogia do oprimido. Editora Paz e Terra, Rio de Janeiro

Moreira RM, Carmo MSD (2004) "Agroecologia na construção do desenvolvimento rural sustentável". Agricultura em São Paulo, São Paulo 51(2):37–56, jul-dez

Normative Ruling No 7/99 (1999) Diário Oficial da União, 17 de maio de 1999

North DE (1990) Institutions, Institutional change and public performance

Oakley P, Clayton A (2003) "Monitoramento e avaliação de empoderamento". Intrac, Inglaterra. Junho. Instituto Polis, São Paulo, p 96

Organizacao para Cooperacao Economica e Desenvolvimento (1995) Sustainable agriculture: concepts, issues and policies. OECD, Paris

Romeiro AR (1998) "Meio ambiente e produção familiar na agricultura" In: Agricultura familiar: Desafios para a Sustentabilidade. Coletânea ed. Embrapa, Aracaju, vol.1, pp 255–276

Schmidt O, Padel S, Levidow L (2012) The bio-economy concept and knowledge base in a public goods and farmer perspective. Bio-based and Appl Econ 1(1):47–63
Sen A (1999) Desenvolvimento como Liberdade. Companhia das Letras, São Paulo
Guzman ES (2001) "As bases sociológicas da agroecologia". In: Encontro Internacional sobre Agroecologia e Desenvolvimento Rural Sustentável, Botucatu. Anais... CD-ROM. vol.1
Sevilla-Guzman E, de Molina MG (1993) Ecologia, campesinato e história. La Piqueta, Madrid

Part V
Bioeconomy: Advances on Bioenergy and Biofuels

Constructing a Sustainable Bioeconomy: Multi-scalar Perceptions of Sustainability

Aparajita Banerjee, Chelsea L. Schelly and Kathleen E. Halvorsen

Abstract Bioenergy holds significant promise to mitigate the climate-related problems associated with fossil fuel use in heat, electricity, and transportation fuel production. Many governments are encouraging bioeconomy growth with new policies. International trade between bioenergy producing and consuming nations has increased over the years. Developed countries with significant greenhouse gas emission (GHG) emission reduction goals are replacing fossil fuels with bioenergy, creating new export commodities for developing nations. However, increased bioeconomy development can put local social, economic, and environmental conditions in bioenergy producing areas at risk. To minimize the potentially adverse impacts of bioenergy development on existing socioeconomic and environmental conditions, several sustainability certification programs have recently been developed. However, there may be significant differences in how actors across multiple scales, including international non-governmental organizations, state and national governments, and local community members perceive a sustainable bioeconomy. In this chapter, we look specifically at two bioenergy development cases, one in the context of economic development in Latin America (jatropha-based bioenergy development in Yucatan, Mexico) and another in the context of a post-industrialized nation (wood-based bioenergy development in Wisconsin, USA) to understand how different actors view sustainability. Our conclusions suggest that, first, developing a sustainable bioeconomy requires addressing sustainability in all stages in the supply chain, and that, second, community perceptions matter in

A. Banerjee · C. L. Schelly (✉)
Department of Social Sciences, Environmental and Energy Policy Program,
Michigan Technological University, Houghton, MI, USA
e-mail: cschelly@mtu.edu

A. Banerjee
e-mail: abanerj1@mtu.edu

K. E. Halvorsen
Department of Social Sciences, Environmental and Energy Policy Program,
and School of Forest Resources and Environmental Science,
Michigan Technological University, Houghton, MI, USA
e-mail: kehalvor@mtu.edu

W. Leal Filho et al. (eds.), *Towards a Sustainable Bioeconomy: Principles, Challenges and Perspectives*, World Sustainability Series,
https://doi.org/10.1007/978-3-319-73028-8_19

developing a sustainable bioeconomy, thus there is value in a bottom-up approach to policymaking.

Keywords Bioenergy development · Bioenergy policy · Public perceptions of bioenergy

1 Introduction: Global Transitions to a Sustainable Bioeconomy

Bioenergy, a term describing the use of renewably-harvested biomass to produce energy for electricity, heat, and transportation fuel, can address issues related to both the contribution of fossil fuel production and greenhouse gas (GHG) emissions to climate change and global energy security (Halvorsen et al. 2009; Solomon et al. 2007). Given these benefits, many governments at both national and sub-national scales have introduced policies promoting bioenergy consumption and production. As some countries cannot meet ambitious bioenergy targets set by governmental policies domestically, international trade markets are developing (Junginger et al. 2013). Moreover, low biomass production costs in developing countries can provide supplies to the bioenergy-based transportation fuel market globally.

Sustainability is foundational to the use and production of bioenergy as it is promoted to reduce dependence on unsustainable fossil fuels. Consequently, all bioenergy production should be sustainable to maximize benefits. However, bioenergy production is not always sustainable (Halvorsen et al. 2011). Research suggests that implementing forest-based sustainability certification programs is difficult, specifically in developing countries, due to reasons like lack of demand for certified products, a gap between current and more sustainable management styles, weak adherence to policies by business entities, as well as a lack of implementation capabilities (Durst et al. 2006). However, international bioenergy trade has created a need for sustainability certifying bodies that evaluate the sustainability of bioenergy systems so that unsustainable production processes are not adopted, particularly in bioenergy-exporting developing countries. At times, bioenergy policy documents contain directives to ensure sustainability features prominently in the policy objectives like avoiding the use of food crop for biodiesel, rural employment generation, and greenhouse gas emission reduction (Eastmond et al. 2014; Solomon et al. 2015). Separately, some government agencies provide procedural guidelines for feedstock harvesting.

There are numerous international third-party certifying agencies that also verify sustainability processes in bioenergy production. Some of these sustainability certification programs are country-specific or product-specific. Most certification programs include environmental, economic and (to a lesser extent) social indicators to assess the sustainability of a bioenergy project (Solomon et al. 2015). In this chapter, we use a multi-scalar approach to examine two contemporary case studies of bioenergy development to understand what contributes to the sustainability of the emerging bioeconomy considering the challenges to creating a sustainable bioeconomy, and evaluating how local conditions are factored into national policies and international sustainability certification programs. Specifically, we question what it means for these bioenergy projects to be sustainable from multiple perspectives operating across social scales, from local community members to implementation of state and national policies to sustainability criteria in international certification programs. We first examine how community members perceive the impacts of local bioenergy projects regarding social sustainability, and we then evaluate the roles played by bioenergy policies in the development and outcomes of bioenergy transition projects. Finally, we explore how international bioenergy certification programs that impact our cases looks specifically into the social sustainability of bioenergy projects. Much can be learned from looking at these cases using a multi-scalar approach, examining both the policies and the perceptions of local community members to consider what sustainability means in the context of bioenergy development projects.

2 Understanding Sustainability: A Multi-scale Examination

New bioenergy production involves complex systems. The carbon neutrality of bioenergy is dependent on the type of feedstock used, how it is grown, the land-use changes associated with feedstock cultivation, and methods of bioenergy conversion and production (Ekardt and Von Bredow 2012; Solomon et al. 2007). There are real challenges associated with fuel sources that have competing uses, such as food crops (Sagar and Kartha 2007); sustainability requires multifaceted, multi-scalar examination. In addition to ecological sustainability, the social sustainability of bioenergy projects should also be an evaluation criterion, as bioenergy projects are embedded in local communities (Halvorsen et al. 2011). Negative externalities of the projects can flow into the local communities and impact community-level socioeconomic conditions (Manning et al. 2015). Moreover, social implications are not always straightforward, and an improved understanding of sustainability from multiple social scales within the same case studies of bioenergy development can help identify the complexities.

One way of assessing the social sustainability of a bioeconomy involves considering how community members perceive the socioeconomic impacts of the bioenergy proximate to their communities (Bain 2011). Any new energy project in a community entails new material realities in places where they are sited, and bioenergy projects are no different. Apart from the biomass processing plant, a new bioenergy project creates new ways or enhances existing ways of biomass extraction, storage, and transportation. All such activities that support a bioeconomy expose people in nearby communities to new risks and opportunities.

However, how people see the activities of the local bioeconomy, as risks or opportunities, depends on how they make sense of the world (Banerjee et al. 2017). Paying attention to differences in public perceptions and opinions, while based on a constructivist paradigm as it requires taking seriously constructions of reality, can contribute to understanding the real successes and failures in planning for a socially sustainable bioeconomy. Moreover, public interest or opposition to new technology projects in a community can enable or dissuade diffusion of new technologies (Wüstenhagen et al. 2007). Therefore, understanding how people perceive bioenergy projects can help reduce conflict and ensure smooth day-to-day operation of the projects and other supply chain stages critical to the projects. Additionally, moving towards sustainability requires public support, as sustainability concepts often challenge existing economic systems of production and consumption, making it necessary for sustainability projects to be backed by majority support to succeed in the long-run. Therefore, understanding how people perceive projects aimed at enabling transitions to sustainable bioenergy systems can inform future policy processes promoting sustainability.

Government policies play important roles in the transition toward new renewable energy (RE) systems and ultimately can shape the extent to which bioenergy development is sustainable (Delmas and Montes-Sancho 2011; Kivimaa and Mickwitz 2011). Many countries have public policies designed to promote national bioenergy development (Sorda et al. 2010). However, all countries with bioenergy policies have not been equally successful in developing a robust bioenergy sector (Solomon et al. 2015; Wiser et al. 2005). Since developing a bioeconomy requires the participation of a large number of investors, suppliers, and developers in the market who are vulnerable to different risks, governmental support for the sector is critical to encouraging their participation. Local, regional, and national policies can also mandate adherence to different sustainability programs, thus contributing to the development of a sustainable bioeconomy.

International trade in bioenergy markets is growing, ranging from forest-based bioenergy products to oil seeds. However, large-scale international trade can only result from a large-scale production of bioenergy, which can affect the social, economic, cultural and ecological sustainability of the producing areas. Thus, various certifying bodies have emerged, with different indicators that can be used for measuring the sustainability impacts of bioenergy projects. Sustainability certification programs often include some social and cultural sustainability indicators (Dale et al. 2013).

In the remainder of this chapter, we first introduce our two study cases and the methods used to conduct the research. We then present the results from our research on the two case studies one by one. In each case, we review the government policies and the sustainability programs applicable to each case before turning to the findings from our work focused on understanding the sustainability of bioenergy development from the perspective of local community members and the connections and conflicts in understandings of sustainability across local, regional, national, and international scales. Because neither of the projects discussed here (jatropha-based biodiesel projects in Yucatan, Mexico and wood-based electricity production in Wisconsin, USA) have resulted in sustained bioenergy production at the intended scale and scope, these projects can be deemed failures. While production of jatropha-based biodiesel has largely been abandoned in Yucatan, wood-based electricity production continues to take place in Wisconsin, although neither at the capacity nor from the sources that were originally targeted. In our conclusion section, we point out the key obstacles that affect the success of the two cases that we identify from our multi-scalar analysis.

3 The Case Studies

Our first case considers jatropha-based biodiesel production in Yucatan, Mexico. Jatropha, a euphorbia, bear seeds that have high oil content but little edible value, and has been widely promoted in countries by government organizations, aid agencies or local non-governmental organizations in Asia, Africa, and Latin America as a biodiesel feedstock. Jatropha was also adopted by Mexico as a biodiesel feedstock crop, and it was cultivated on land under different ownership patterns ranging from smallholder-operated farms to company-owned plantations. Our study focused on large plantations to understand their social sustainability impacts on local communities. We conducted semi-structured interviews in eight villages in Yucatan, Mexico, where three companies started jatropha plantation operations. These companies either rented communal or ejidal land or bought grazing land from cattle ranchers (Eastmond et al. 2014). Although the exact number of people employed on the plantations is unknown, local sources informed us that it was around 500 people in each of the projects.

Our Mexican research was divided into two stages. In the first stage, we analyzed all the government laws related to jatropha-based bioenergy development at local, regional, and national levels. We found that Mexico's bioenergy laws were all national-level policies with different federal agencies in charge of implementation. In some cases, like in Yucatan, state governments allowed private corporations to establish plantations by allowing them to buy or use local land. We also analyzed the policy documents and reviewed international bioenergy sustainability programs that applied to Yucatecan jatropha bioenergy development. After the first stage of documentary research, we interviewed 42 individuals (25 females, 17 males) who lived near the jatropha plantations or worked on a jatropha plantation or had a close relative who who worked in the plantations. As employee data from the jatropha companies were not available, we approached the village authorities (*comisarios municipal y ejidal*) for permission to do our research and help recruit interviewees. Using snowball sampling, each interviewee was asked to refer another of their colleagues for interview (Biernacki and Waldorf 1981). We focused on getting a roughly equal mix of male and female interviewees across a range of incomes and ages. Thus, our research methods ensured that we could approach aspects of the Yucatecan jatropha bioeconomy at local, national, and international scales.

The second case study involves the use of woody biomass in the state of Wisconsin, USA to produce electricity. There are two utility-scale power plants that use woody biomass in the northern part of this state. Like our Yucatecan case, we first researched relevant government policies and private sustainability certification programs that applied to the Wisconsin bioenergy case. In the second stage, this research involved interviewing local community members in the communities surrounding two bioenergy facilities in Northern Wisconsin. One of the facilities has intermittently used woody biomass like saw mill residues, railroad ties, logging waste, and house-building materials as biomass for electricity production since 1979. It was the first power plant in the United States to use biomass for electricity production. The second facility, a cogeneration plant, started operation in 2013. This plant was intended to use biomass to produce steam and electricity. The unit was constructed so that the electricity went to the grid and the steam produced from the plant could be used by an adjacent pulp and paper plant in the same premise. Per initial plans, the power plant was intended to use biomass obtained within a 75-mile radius to lower the cost of biomass acquisition. We interviewed 61 individuals in the communities surrounding these two facilities. Of our interviewees, 32 live in smaller, rural communities, and 29 live in the larger semi-urban communities. All the community members lived within a 75-miles radius of one of these two plants. We approached community members in public places like parks, restaurants, libraries, etc. to solicit an interview. We also asked interviewees to refer us to individuals they thought might be able to provide interesting insights, such as loggers, wood haulers, or people who use wood to heat their homes. Therefore, we used snowball sampling and purposive sampling methods (Biernacki and Waldorf 1981; Teddlie and Yu 2007; Tongco 2007).

A problem-centered interviewing method was used, which involved developing a standardized interview protocol that guided all interviewers, ensuring that all interviewees answered a short core set of questions that helped in comparing the answers (Witzel 2000). All interviews in both the cases were recorded and transcribed verbatim. Interviews in Yucatan were conducted, transcribed, and analyzed in Spanish; translation occurred post analysis. The transcripts were labeled, sorted by the response to each of our interview protocol questions, and coded using an open coding scheme. This open coding analysis allowed us to analyze the data for common patterns, and here we present findings specifically related to the theme of sustainability from the perspective of participants. A single researcher completed all coding and analysis; thus, no inter-rater reliability method was used to cross-validate the findings. None of the interviewees quoted below were actively working in the bioenergy sector; thus, their perspectives can be interpreted as based on their positions as community members rather than workers. The aim of presenting these results is to show the diverse ways people make sense of the local bioenergy projects, but given the sampling method and the interpretive process involved in analyzing qualitative data, these results should not be interpreted as empirically representative or generalizable, although we do believe that the conceptual themes provided by the interview data are theoretically generative and thus applicable beyond our particular case studies.

4 Results and Discussion

As mentioned earlier, in addition to interviewing local community members, this research project involved reviewing the past and current policies that shape the nascent bioeconomy in the two case studies discussed in this chapter. In our policy analysis, we used public policy documentation and were attentive to policies and policy gaps impacting the entire networked cycle of feedstock sourcing, processing, and transporting, energy production, and energy consumption. We also reviewed private international sustainability certification programs relevant to our cases, in order to address multi-scale dimensions across international, national, and state level policies. Below, we first present the results from the Mexican case study, including review of the bioenergy policies, sustainability certification programs, and findings from the interview data analysis. We then present the same three types of results for the US case.

5 Policies Shaping Jatropha Bioeconomy in Yucatan, Mexico

During Felipe Calderon's presidency (2006–2012), a series of new laws targeting renewable energy development in Mexico were enacted. The 2008, Ley de Promoción y Desarrollo de los Bioenergeticos (2008 Bioenergy Promotion and Development Act) was passed to encourage implementation of bioenergy projects nationally aiming simultaneously to increase rural economic development along with reducing fossil fuel use in the country. It is at the heart of Mexico's bioenergy program. The primary objective of the law was to promote bioenergy feedstock production from agricultural, forestry, algae, biotechnological and enzymatic activities and processes in the Mexican countryside. The idea was to usher revival of the rural economy through bioenergy by creating jobs where local community members could work in the bioenergy sector without having to migrate for economic opportunity.

The Interministerial Commission for Bioenergy Development (ICBD) was formed to write the law's rules and regulations. It included officials from the Ministries of Agriculture, Livestock, Rural Development, Fisheries, and Food (SAGARPA), Energy (SENER), and Environment and Natural Resources (SEMARNAT). ICBD was responsible for developing the Act's implementation plan to enable coordinated action between different level of government and was given a monitoring role so that it could oversee state, local, and industrial compliance.

Each ministry was entrusted with specific roles. SAGARPA was empowered to develop feedstock production and marketing programs, assess feedstock sustainability, monitor feedstock cultivation to avoid competition with food production, and promote scientific and technological bioenergy development. They were required to advise farmers on sustainable feedstock cultivation practices and create communication networks for bioenergy production actors. In 2008 and 2010, SAGARPA developed programs encouraging bioenergy feedstock cultivation and bioenergy production. It's 2008, Sustainable Production of Feedstock guidelines aimed to ensure that all stages of Mexican bioenergy production were sustainable were developed by SAGARPA (2009). SENER was charged with the developing industry supply chain processes. It was made responsible for the development of production operating guidelines, issuing bioenergy research and development grants, and imposing sanctions on bioenergy producers who failed to follow national production guidelines. SEMARNAT was made responsible for assessing biofuel supply chain stages for environmental compliance and enforcing requirements that all projects file Environmental Impact Statements. CONAFOR (the Mexican Forestry Agency) identified jatropha, a shrubby plant species that bears oil-rich seeds, as a potentially valuable bioenergy crop that could survive in degraded, dry soils in different parts of Mexico. SAGARPA and CONAFOR, therefore, developed incentive programs to encourage its establishment and processing, providing up to 6310 Mexican pesos (US$486) per hectare to jatropha cultivators under their ProArbol program. However, even with these policies in place, Mexico to date does not have either a domestic bioenergy market nor is an international supplier. As is noticeable in this review and will be discussed in more detail below, while bioenergy development policy in Mexico aimed to consider all aspects of the networked production to consumption system, in reality the focus was on boosting just feedstock production, with little attention to transportation, processing, or use of the cultivated feedstock supply.

6 Sustainability Certification Programs Applicable to Jatropha Bioeconomy in Yucatan, Mexico

The Mexican 2008 Bioenergy Promotion and Development Act applicable to bioenergy development in Yucatan, Mexico espoused the importance sustainability conditions in bioenergy production. We found repeated mentions of how sustainability should be an important criterion in the planning of a bioeconomy in the country. SEMARNAT was charged with Environmental Impact Assessment of the projects. However, no government level bioenergy sustainability guideline has been prepared to provide specific criteria for ensuring or assessing the economic, social, and environmental sustainability of the emerging bioeconomy (Eastmond et al. 2014). Instead, we found that third-party international guidelines emerged as the standard for assessment of sustainability as one of the three plantations we studied was certified by the Roundtable on Sustainable Biomaterials (RSB) (Eastmond et al. 2014). Moreover, use of these sustainability certifications was not consistently pursued as only one company pursued certification.

RSB certification assesses production on both environmental and community-level impacts. For example, some of the criteria require compliance on greenhouse gas emissions, human and labor rights, social development of rural areas, food security, wildlife conservation efforts, impacts on air, soil, and water quality, and land and water access rights. RSB certification can be obtained for supply chains or for specific production sources. In the single Yucatecan jatropha plantation case, only oilseed production was certified by RSB (as no jatropha biodiesel supply chain was established before the plantation closed its operation).

7 Sustainability of Jatropha Bioeconomy in Yucatan, Mexico: Community Perceptions

In Mexico, people with direct experience with the jatropha plantations (most either they had worked in the plantations or had a family member who did) were asked about perceptions of sustainable bioenergy development in their community. Seventy-one percent (27 of 38 interviewees asked the question) said that having the jatropha plantation companies was a good thing for their community. Fifty-one percent (19 of 37 interviewees asked the question) told us that the salaries paid by the companies were good, whereas 40% of the interviewees believed the salaries and perks of the companies were not good considering the hard-manual job they had to do in the jatropha fields.

Community members pointed out in details some of the benefits of having the plantations near their communities. These plantations provided local jobs, enabling people to stay in the village (rather than having to travel for work) and receive a regular paycheck. They described the jobs as having helped them improve their standard of living by, for example, allowing them to take out a loan for a vehicle because they had a steady source of income. Some also reported that one of the jatropha plantation companies provided additional benefits, such as providing breakfast to workers' children or helping them build a community basketball court. For example, one interviewee said,

> I think [the plantation company] was good because it meant people had work and had access to benefits such as low-interest credit. Many of the employees managed to get a loan from INFONAVIT (Instituto del Fondo Nacional de la Vivienda para los Trabajadores) and were able to buy a motorbike. We never thought that the company would close down (Interviewee 13).

However, though there were some social benefits from the nascent bioenergy sector in Yucatan, such benefits did not last long. After few years of operation, the plantations closed laying-off people. Our interviewees informed us that companies closed because the jatropha plants did not have enough yields that could make the plantations profitable. They also pointed out that yields suffered because the local climate was not suitable for the plants and problems like disease infestations and subsequent pesticide overuse was common. Though people lost their jobs, owning land and being able to continue subsistence agriculture helped most interviewees to survive. Seventy-two percent (28 of 39 asked the question) reported that either they or male members of their families had access to individual plots (formerly *ejidal* land) which was mainly used for subsistence agriculture (locally known as *milpa*) but also occasionally for raising cattle. Interviewees mainly cultivated corn, beans, squash and other vegetables. They often had fruit trees in their *solar* (back yard) where they might also keep pigs or an occasional cow and poultry mainly for home consumption. They also gathered wood from surrounding forested land for cooking. The jatropha companies operated on land that they bought from previous ranch owners and some community lands (*ejidal* lands). No community member reported renting their land to the jatropha plantations.

Talking about what according to them can help their community to achieve sustainability in the future, eighty-six percent of interviewees (32 of 37 asked the question) expressed desire for improved local economic opportunities. Some also said that the government should help companies like the jatropha plantations to open agricultural production units locally. They pointed out that working for companies was good because of the steady, regular paychecks. Others commented that rather than providing support through different welfare programs, the government should promote companies that provide local employment. One interviewee said,

> They (*government agencies*) send materials for your home… it certainly helps us and keeps us out of trouble, but then it gets used up. It's better to have a job, because you have money, even if it is only a little and you know the job will last (Interviewee 75).

Some of the interviewees said that the government should provide loans or some form of monetary assistance that would promote local, community-based entrepreneurship in the communities rather than opening or supporting new companies like the jatropha plantations. A few interviewees suggested that the government should provide financial benefits so that they could buy mechanized farm equipment that would help them increase their agricultural output.

8 Discussion of the Mexican Case Study

An analysis of the Mexican policies to support a transition to a bioeconomy points out policy gaps that we argue contributed to some extent in jatropha plantation failure. Understanding these gaps can throw light in how to develop a sustainable bioeconomy that have fewer chances to fail. The 2008 Bioenergy Promotion and Development Act, the core of the jatropha development in the country, provided few guidelines for project implementation (Eastmond et al. 2014). The Act mainly contained the goals, and most of the specific plans were left to federal agencies to implement. The federal agencies in the Yucatecan case incentivized private sector players to start jatropha plantations. Though the government did provide different support to private sector investors, private interest was limited. The lack of interest was mainly due to the lukewarm approach of the government evident in how very few policies were enacted to encourage the sector. Such inertia in government policy-making demotivated private sector players to invest a large sum of money in the different and interdependent supply chain stages necessary for a sustainable bioeconomy to grow. Not surprisingly, in the Yucatecan case, only three companies invested in the plantations and apart from feedstock establishment, no other steps in the supply chain were developed with much attention. Apart from government policies that incentivized feedstock establishment, no policies were in place that would incentivize investors to participate in other stages of the supply chain. The Mexican government also did not set any mandatory blending target for jatropha biodiesel that could have created demand for the product and helped in setting a foundation of a sustainable bioeconomy. Thus, the markets that would create demand for jatropha biodiesel never developed. Moreover, Mexico is a major oil producer and investors may have lacked interest in investing in an alternative to fossil fuel when domestic production of fossil fuel was a huge industry.

It is interesting to note that the 2008 Bioenergy Promotion and Development Act repeatedly states that Mexican bioenergy development should be monitored for sustainability outcomes. However, the government provided no producer guidelines that would help them follow sustainable processes (Solomon et al. 2015). With the lack of national level sustainability guidelines, one plantation in the area approached a third party international certifying agency for sustainability certification. Social sustainability is a part of RSB's sustainability indicators, and interviewees working for this plantation said that the plantation provided them with benefits like uniforms, protective gear when spraying pesticides, breakfast for the village children on some special days, and help building a basketball court in one of the villages. However, once certified by RSB, this plantation also fired half their staff. Further, the pesticide used by the plantation is known to have negative ecological impacts (Selfa et al. 2015). Thus, from the Mexican jatropha case, we find that international third party sustainability certification programs can certify a system, but that does not necessarily help sustain the projects alone.

Arguably, our results suggest that the main reason for high community-level acceptance of the jatropha projects was their role in the creation of hundreds of local jobs. Though the community members had mixed views on the quality of the jobs which were mainly low-skilled manual jobs not very different from the ones they did in their own fields or as day-laborers, the part that the jobs brought regular paychecks with benefits and perks were identified as a positive impact of the plantations. Localized jobs also helped plantation workers to live with their family and raise their children in a safe and known community. This helped in maintaining the social fabric of the community that is often affected in communities facing large-scale out-migration. Regular paychecks helped solve much of their worries in depending solely on subsistence agriculture increasingly becoming unpredictable due to changing weather patterns, disease infestations, and lack of irrigation. However, when the projects failed and the plantations closed, community members were somewhat protected from the dire conditions by the fact that they owned their land for subsistence use.

Access to land, which allows for diversification of strategies for subsistence and economic opportunity, is important in natural resource dependent communities where a singular focus on one development project, like the failed jatropha plantations, can be disastrous for community livelihoods. Also, access to land also helped them to supplement their income from the jatropha plantations. As the jatropha plantations were small and did not cover significant amount of land in the region, the people could continue to own land in the area. It was not surprising that when asked about what they think can help their communities to be sustainable in the future, our interviewees pointed out the importance of economic opportunities created by companies like the jatropha plantations as well as getting help to diversify their economic options.

Our results from our Mexican case study provide us with some key insights of developing a sustainable bioeconomy. The interviewees associated the economic opportunities as a social benefit of the bioeconomy. The new economic opportunity helped them to live in their communities that positively impacted their social life; living and raising a family in a safe place. We also found that having land was an important buffer that helped community members to absorb the shock of livelihood loss from plantation closures. This raises critical questions about how to ensure that projects like these do not fail, and how to have projects that ensures participation of local communities and can offer more than a steady paycheck for physically demanding and low skill work. It is also critical to understand at what level the feedstock plantations should expand so that community members' land access rights to both communal lands and personal lands are not compromised.

Another key insight from our jatropha bioenergy case that can inform our understanding to better develop future sustainable bioeconomies especially in developing countries is that in many cases, there is a lack of research and development that informs policy decisions. Though jatropha was identified as a feedstock crop by Mexican federal government, we did not come across any significant research that suggested the suitability of jatropha to Yucatecan weather conditions before 2008 when SEMARNAT adopted the jatropha as a feedstock crop. The government incentivized jatropha cultivation without ensuring the suitability of the crop thus spending money on ill-planned projects. Private investors seeing opportunity in the government incentives invested in establishing the plantations which they subsequently closed when the incentives programs stopped after a couple of years, and the yields were extremely low to generate any profit. The findings from Yucateca, Mexico case have interesting parallels with the case study in Wisconsin, USA, described next.

9 Policies Shaping Wood-Based Bioeconomy in Wisconsin, USA

The United States has had state and federal policies encouraging woody biomass power development since the 1970s (Laird and Stefes 2009). The federal 1978 Public Utility Regulatory Policy Act (PURPA) was the first policy to mandate that utilities purchase or generate power from renewable fuels, creating the first markets for utility scale bioenergy production (Aguilar et al. 2011; Lantiainen and Song 2015). The 1992 Energy Policy Act (EPACT) provided power production tax credits for use of woody biomass feedstock (Aguilar et al. 2011). The 2004 American Jobs Creation Act provided tax credits to utilities using biomass (Aguilar et al. 2011). The 2005 EPACT eliminated PURPA's mandatory purchase requirement (Aguilar et al. 2011), but also extended the eligibility period so that new woody biomass power plants could continue to receive PTCs and the 2009 American Recovery and Reinvestment Act (ARRA) provided additional tax credits (Aguilar et al. 2011).

Many federal policies in the United States provide grants and loan guarantees for new bioenergy projects (Becker et al. 2011; Halvorsen et al. 2009; Solomon et al. 2007). For example, the Renewable Energy Grant Program provides grants to projects established between 2010 and 2016, the 2003 Renewable Energy Systems and Energy Efficiency Improvement Program offers grants for biomass power feasibility studies, and the 2006 Residential Energy Efficiency Tax Credit Act, 2005 EPACT, and ARRA provide bonds to biomass power producers (Aguilar et al. 2011). At state-level, other policies also promote woody biomass electricity production through mandates or incentives, including in Wisconsin (Lantiainen and Song 2015). Since 1987, all biomass purchased for electricity production in Wisconsin is exempted from use and sales taxes by the state (Wis. Stat. §77.54). The 1993 Energy Priorities Act charged the Wisconsin Public Service Commission (PSCW) with promoting cost efficient, technologically feasible renewable energy development, prioritizing woody biomass electricity production (Wis. Stat. §196.378). Wisconsin's Renewable Portfolio Standards (RPS) was first enacted in 1999 and then amended in 2006. The RPS mandates that at least 10% of the state's energy production be from renewable energy sources, including from biomass, by 2015. The Woody Biomass Harvesting and Processing Tax Credit (Personal and Corporate) of 2010 provides a 10% tax credit toward the cost of woody biomass harvesting and processing equipment for power production (Wis. Stat § 93.547).

10 Sustainability Certification Programs and Woody-Biomass Based Bioeconomy in Wisconsin, USA

In the United States, forests have tremendous potential to supply biomass for wood-based utility-scale power production, and Wisconsin includes ample forested regions for bioenergy production. Typically, biomass feedstock includes the tops and branches after a timber harvest that mostly remains underutilized due to limited extraction (Becker et al. 2009). However, increased feedstock demand can create unstable conditions in the forest ecosystems. State-level laws regulate state and private forests, whereas national forests work under federal guidelines. While best management practices to protect water quality acts as a guide for following ecologically sustainable policies, some states have additional guidance for biomass harvests to eliminate negative impacts on forest health (Evans et al. 2013). Wisconsin's state specific biomass harvesting guidelines though do not address the social aspects of sustainable bioenergy production yet provides specific guidelines for ecologically sustainable biomass harvesting. However, most of Wisconsin's county forests (27 out of 29 of them) are additionally certified for sustainability by Forest Stewardship Council, or Sustainable Forestry Initiative, or both. These certification programs include social, economic, and ecological sustainability assessment criteria for evaluation. Private forests participating in state-level forest management programs (Managed Forest Law in Wisconsin) also participate in these certification programs.

11 Sustainability of Woody-Biomass Bioeconomy in Wisconsin, USA: Community Perceptions

Like our Yucatecan case study, we interviewed community members in Wisconsin to understand their perspectives of the impacts of woody-biomass based electricity production proximate to their communities. Sixty-one percent of participants in Wisconsin (37 of 61 interviewees asked the question) said that burning wood as a fuel source at an industrial scale is a good thing, but others (twenty-six percent; 16 out of 61) were concerned about utility-scale use of woody biomass. Another 15% (9 out of 61) expressed mixed opinions or could not form a clear opinion. Even with mixed support for utility-scale use of woody biomass, 82% (50 out of 61) said that the local biomass generation facility generated or can generate positive community impacts. For example, 26% (16 out of 61) of the interviewees said that bioenergy projects can be economically beneficial to their communities. They pointed out that such developments can bring in new jobs or provide new markets to local loggers. For example, one interviewee said: "it is adding that to the economy and to the loggers by giving them income" (Interviewee 717).

Forty-eight percent (29 out of 61) of the interviewees believed that using wood to produce electricity can be a good thing because of reasons related to locally available and renewable resources; these participants said that wood is locally available, it is a renewable resource, it is cleaner than coal or the use of scrap wood is a wise use of resources that would otherwise go to waste. For example, one interviewee said: "They are using the branches and things that are normally would be going to waste. So there is nothing wrong with that.... Anything that's renewable as far as I am concerned they can use it" (Interviewee 757).

Others, however, questioned whether the use of so-called "waste" wood was really a sustainable or ecologically beneficial practice. As one community member said,

> It's good for the economy because, you know… a lot of the guys that are loggers around here sell firewood.... I am not a big proponent of [woody biomass use in electricity

production] when you're taking everything out of the woods … you have no place for… bugs and snakes and… animals… for smaller animals or whatever because there's nothing there for them (Interviewee 709).

Around 46% (26 of 56 interviewees asked the question) talked about timber harvesting concerns when asked about potential negative impacts of utility scale wood use. Many opposed removing timber harvesting residues labeled as "waste" wood for use in the utility generation facilities because they saw it as an ecological benefit to leave them on site; as one interviewee said, some should be left "for the birds and the bees."

Some interviewees said that it was important to ensure that wood extraction is done with extreme care due to the different benefits to the community provided by the forest. For example, one interviewee said: "This would mean that you are cutting everything that supports life forms. What supports life for us here is the trees…. So we have to be very careful in how we're doing it, and… looking at the big picture instead of just seeing this small little picture" (Interviewee 702).

Around 25% (14 of 56 interviewees asked the question) of interviewees said that they were not concerned about negative impacts, as long as the facility continued to use only "tops and branches" as "cutting whole trees would not be good." In industry documentation, however, it is not clear that the promise of only using "waste" wood is actually being followed as practice.

Some interviewees pointed that wood is not the only available renewable energy resource, and felt that renewable energy portfolios should have carefully chosen mixes of energy sources. Some worried that large-scale mechanization of woody biomass use would bring an unsustainable level of extraction, while others wondered if Wisconsin timber harvesting regulations were adequate if woody bioenergy would incentivize greatly increased harvesting.

Thirty percent (17 of 56 interviewees asked the question) of interviewees identified other potential concerns about utility scale biomass generation facilities, many related to the negative impacts to property values, noise, and an unsightly plant in the community when the electricity generated is going to other communities. For example, one interviewee said: "My number one concern was the noise" (Interviewee 756). Another said: "It (the biomass plant during construction stage) was dirty, dusty, we couldn't have our windows open and, you know, it just was unpleasant" (Interviewee 763).

Other interviewees expressed dismay that, although there could be local economic benefits (see above), local loggers did not actually benefit economically from the utility-scale plant because they could not afford the grinder equipment necessary to turn their waste wood into wood chips for the facility. Others pointed out that the construction jobs that were created during the construction of the biomass plant went to specialized workers from outside the community. According to them, the money earned by these workers in these new jobs did not benefit the local economy as it was spent outside the local areas. Some interviewees expressed concerns about the health impacts for the local community and were concerned about the local pollution when the electricity was being used in other communities.

Some interviewees wondered if a holistic approach to understanding the plant - taking into consideration the pollution from truck traffic hauling the wood and the environmental and health impact of the ash resulting from burning - would change perceptions of biomass generation as a low pollution energy source.

Discussion of the Wisconsin, USA Case Study

The community members interviewed for this project demonstrated informed, thoughtful, and complex understandings of the benefits of pursuing bioenergy development for their community. While they recognized the bioenergy facility's potential economic value, they did not perceive that they or other community members were receiving actual economic benefits. Further, they were concerned about the sustainability of the bioenergy harvesting practices; even using so-called "waste" wood may not be a sustainable practice for those who value the ecosystem diversity and natural processes they can observe or experience in the forest. For those who live in wooded areas and who feel connected to the forest, the ecosystem services provided by "waste" wood in providing habitat and nutrients to the forest were worth questioning the sustainability of the bioenergy facility harvesting practices.

A policy analysis of relevant federal, state and local policies impacting bioenergy development in Wisconsin suggests that community members' perceptions of the lack of economic opportunity for loggers to benefit may be related to a policy failure. First of all, Wisconsin's Renewable Portfolio Standard (RPS), as amended in 2006, allows utilities to source renewable energy resources from out of state power producers (DSIRE 2016). This reduces demand for in-state energy production from renewable resources. Thus, opportunities for loggers to participate in the wood-based power sector remains limited.

While there are federal policies to promote use and manufacturing of bioenergy, there are very few policies for promoting economic development for those operating at the bioenergy harvesting stage in Wisconsin (Becker et al. 2011). The Woody Biomass Harvesting Processing Tax Credit program provides up to 10% tax credit for biomass harvesting equipment. However, from the perspective of community members, the cost of extracting the specialized biomass (mainly tops and branches) and grinding it for supplying to biomass plants can be prohibitive for small loggers. Moreover, wood-based power plants require clean biomass that has very few impurities, which requires special facilities (Dornburg et al. 2007). Another prohibitive factor is the expense of transporting chipped biomass from the place of processing to the biomass plants. Wood chips, as a material different from logs, require different containers for transportation; however, government incentives focus only on actual harvesting equipment, not on processing or transportation.

The Wisconsin's Forestland Woody Biomass Harvesting Guidelines were developed through participation from different stakeholders from the Wisconsin's wood-based industries to provide a comprehensive guideline for wood extraction rules for biomass harvest (Herrick et al. 2009). The majority of the state-owned forests and some private forests also participate in forest management programs and are also certified by third-party certifying bodies. These policies and certification schemes are focused on the ecological sustainability of bioenergy harvesting practices. However, by questioning the process of defining and using "waste" wood and wondering about the long-term ecological impacts of that, participants in this project questioned whether bioenergy harvesting practices matched their own understandings of ecological sustainability.

12 Conclusion: Moving Toward a Sustainable Bioeconomy

The proliferation of bioenergy policies and implementation projects with varying levels of success provides a rich set of cases to study the transition to a sustainable bioeconomy. National and international policy comparison provides an opportunity to consider this question across states and countries attempting to solve similar problems (Dodds 2012). National policies with the same goals tend to converge across countries (Drezner 2001; Heichel et al. 2005). Many have stressed the need to compare national and state-level public policies focused on renewable energy development (Ekins 2004; Lipp 2007). As bioenergy policies are becoming increasingly similar, important lessons can be learned from comparing policies regarding success and failure. This project identified policy gaps in both case studies that may explain the failures in each bioenergy project to fully promote a transition to a sustainable bioeconomy. Our research suggests that policy tools are not being used to consistently promote bioenergy development across all necessary stages from source production to end consumption. Inconsistent policy application hinders the development of fully successful bioenergy transitions, and bioenergy development projects may never provide optimal benefits to local communities without policies specifically focused on local economic opportunities.

Our findings from the policy analysis also suggest that policies should be designed with attention to all the stages of bioenergy supply chain, rather than concentrating only on one aspect, such as feedstock cultivation in the case of jatropha in Yucatan, Mexico, and electricity generation in the case of Wisconsin, USA. In addition to creating sustainable bioeconomies, public policies can also provide directives and guidelines for biomass production and feedstock harvest that requires adherence to sustainability conditions. Some research suggests that renewable energy policies should follow a bottom-up approach with public involvement (Dimitropoulos and Kontoleon 2009; Eltham et al. 2008), arguing that some of the challenges in implementing renewable energy policies can be reduced by allowing public participation in the decision-making stages (Zoellner et al. 2008). The community members' perspectives from both case studies presented here suggest that a sustainable bioeconomy should be developed in ways that aid in diversifying local economic capacities. A sustainable bioeconomy should be planned in ways that other forms of economic opportunities are not compromised, affecting the social and cultural foundations of the surrounding communities. From our study, we learned that community members who may be impacted economically, environmentally, or socially by bioenergy projects make sense of the bioenergy development close to their communities and their views can provide key information in defining a sustainable bioeconomy.

Thus, we offer two key insights for future developments attempting to pursue a sustainable bioeconomy. First, from the international to the local-level, multi-scalar policy development should address all aspects of the supply, production, processing, generation, and consumption chain. Policies that emphasize one element of this multi-stage process without holistic attention to all may create gaps in bioeconomy utilization, processes, and success. Second, understanding public perception is critical for conceptualizing a sustainable bioeconomy and can provide information critical for successful development for environmental, economic, and social sustainability in the future bioeconomy.

Acknowledgements This work was funded by the National Science Foundation's Partnerships in International Research and Education (PIRE) Program

References

Aguilar FX, Song N, Shifley S (2011) Review of consumption trends and public policies promoting woody biomass as an energy feedstock in the US. Biomass Bioenerg 35(8):3708–3718

Bain C (2011) Local ownership of ethanol plants: what are the effects on communities? Biomass Bioenergy 35(4):1400–1407

Banerjee A, Schelly C, Halvorsen KE (2017) Understanding public perceptions of wood-based electricity production in Wisconsin, United States: place-based dynamics and Social Representations. Environ Sociol 23:1–13

Becker DR, Larson D, Lowell EC (2009) Financial considerations of policy options to enhance biomass utilization for reducing wildfire hazards. Forest Policy and Economics 11(8):628–635

Becker DR, Moseley C, Lee C (2011) A supply chain analysis framework for assessing state-level forest biomass utilization policies in the United States. Biomass Bioenerg 35(4):1429–1439

Biernacki P, Waldorf D (1981) Snowball sampling: problems and techniques of chain referral sampling. Sociol Methods Res 10(2):141–163

Dale VH, Efroymson RA, Kline KL, Langholtz MH, Leiby PN, Oladosu GA, Davis MR, Downing ME, Hilliard MR (2013) Indicators for assessing socioeconomic sustainability of bioenergy systems: a short list of practical measures. Ecol Ind 26:87–102

Delmas MA, Montes-Sancho MJ (2011) US state policies for renewable energy: context and effectiveness. Energy Policy 39(5):2273–2288

Dimitropoulos A, Kontoleon A (2009) Assessing the determinants of local acceptability of wind-farm investment: a choice experiment in the Greek Aegean Islands. Energy Policy 37(5):1842–1854

Dodds A (2012) Comparative public policy. Palgrave Macmillan, New York

Dornburg V, van Dam J, Faaij A (2007) Estimating GHG emission mitigation supply curves of large-scale biomass use on a country level. Biomass Bioenerg 31(1):46–65

Drezner DW (2001) Globalization and policy convergence. Int Stud Rev 3(1):53–78

DSIRE; Database for State Incentives for Renewables and Efficiency (2016) Incentives/policies by State: federal: policies for renewables and efficiency. Available online 6 June 2016 http://programs.dsireusa.org/system/program?state=WI

Durst PB, McKenzie PJ, Brown CL, Appanah S (2006) Challenges facing certification and eco-labelling of forest products in developing countries. Int For Rev 8(2):193–200

Eastmond A, Garcia C, Fuentes A, Becerril-Garcia J (2014) Mexico. In: Solomon BD, Bailis R (eds) Sustainable development of biofuels in Latin America and the Caribbean. Springer, New York, pp 203–222

Ekardt F, Von Bredow H (2012) Extended emissions trading versus sustainability criteria: managing the ecological and social ambivalences of bioenergy. Renew Energy Law Policy Rev, 49–69

Ekins P (2004) Step changes for decarbonising the energy system: research needs for renewables, energy efficiency and nuclear power. Energy Policy 32(17):1891–1904

Eltham DC, Harrison GP, Allen SJ (2008) Change in public attitudes towards a Cornish wind farm: implications for planning. Energy Policy 36(1):23–33

Evans AM, Perschel RT, Kittler BA (2013) Overview of forest biomass harvesting guidelines. J Sustain For 32(1–2):89–107

Halvorsen KE, Barnes JR, Solomon BD (2009) Upper Midwestern USA ethanol potential from cellulosic materials. Soc Nat Resour 22(10):931–938

Halvorsen KE, Selfa TA, Becker DS, Hinrichs C (eds) (2011) Special issue: socioeconomic dimensions of U.S. bioenergy. Biomass Bioenergy 35(4)

Heichel S, Pape J, Sommerer T (2005) Is there convergence in convergence research? An overview of empirical studies on policy convergence. J Eur Publ Policy 12(5):817–840

Herrick S, Kovach J, Padley E, Wagner C, Zastrow D (2009) Wisconsin's forestland woody biomass harvesting guidelines. WI DNR Division of Forestry and Wisconsin Council on Forestry, Madison, WI

Junginger M, Goh CS, Faaij A (eds) (2013) International bioenergy trade: history, status & outlook on securing sustainable bioenergy supply, demand and markets, vol 52. Springer Science & Business Media

Kivimaa P, Mickwitz P (2011) Public policy as a part of transforming energy systems: framing bioenergy in Finnish energy policy. J Clean Prod 19(16):1812–1821

Laird FN, Stefes C (2009) The diverging paths of German and United States policies for renewable energy: sources of difference. Energy Policy 37(7):2619–2629

Lantiainen S, Song N (2015) Review and comparative analysis of US and EU public policies promoting wood energy. In: Dell G, Egger C (eds) World sustainable energy days next 2014. Springer Fachmedien Wiesbaden, pp 89–94

Lipp J (2007) Lessons for effective renewable electricity policy from Denmark, Germany and the United Kingdom. Energy Policy 35(11):5481–5495

Manning P, Taylor G, Hanley ME (2015) Bioenergy, food production and biodiversity—an unlikely alliance? GCB Bioenergy 7(4):570–576

Sagar AD, Kartha S (2007) Bioenergy and sustainable development? Annu Rev Environ Resour 32:131–167

Selfa T, Bain C, Moreno R, Eastmond A, Sweitz S, Bailey C, Pereira GS, Souza T, Medeiros R (2015) Interrogating social sustainability in the biofuels sector in Latin America: tensions between global standards and local experiences in Mexico, Brazil, and Colombia. Environ Manage 56(6):1315–1329

Solomon BD, Barnes JR, Halvorsen KE (2007) Grain and cellulosic ethanol: history, economics, and energy policy. Biomass Bioenerg 31(6):16–25

Solomon BD, Banerjee A, Acevedo A, Halvorsen KE, Eastmond A (2015) Policies for the sustainable development of biofuels in the pan American region: a review and synthesis of five countries. Environ Manage 56(6):1276–1294

Sorda G, Banse M, Kemfert C (2010) An overview of biofuel policies across the world. Energy Policy 38(11):6977–6988

Teddlie C, Yu F (2007) Mixed methods sampling a typology with examples. J Mixed Methods Res 1(1):77–100

Tongco MDC (2007) Purposive sampling as a tool for informant selection. Ethnobotany Res Appl 5:147–158

Wiser R, Porter K, Grace R (2005) Evaluating experience with renewables portfolio standards in the United States. Mitig Adapt Strat Glob Change 10(2):237–263

Witzel A (2000) The problem-centered interview. For Qual Soc Res 1(1)

Wüstenhagen R, Wolsink M, Bürer MJ (2007) Social acceptance of renewable energy innovation: an introduction to the concept. Energy Policy 35(5):2683–2691

Zoellner J, Schweizer-Ries P, Wemheuer C (2008) Public acceptance of renewable energies: Results from case studies in Germany. Energy Policy 36(11):4136–4141

Contributions of Public Policies to Greening Sugarcane Ethanol Production in Brazil

Issa Ibrahim Berchin, Sthefanie Aguiar da Silva, Pierre Bocquillon, Vitória Haendchen Fornasari, Larissa Pereira Cipoli Ribeiro, João Marcelo Pereira Ribeiro and José Baltazar Salgueirinho Osório de Andrade Guerra

Abstract The development of a green economy is key to achieve sustainable development. Investing in greening the energy sector contributes to mitigate greenhouse gas emissions while helping to develop a more resilient economy. Public policies have a role in driving the biofuel sector towards a more sustainable path, reducing its environmental footprint and enhancing its economic, environmental and social benefits. The aim of this study is to analyse how public policies can contribute to greening the ethanol sector in Brazil. To accomplish this objective, a review of the economic characteristics of the Brazilian ethanol sector was

I. I. Berchin (✉) · S. A. da Silva · V. H. Fornasari · L. P. C. Ribeiro
J. M. P. Ribeiro · J. B. S. O. de Andrade Guerra
Research Centre on Energy Efficiency and Sustainability (GREENS),
University of Southern Santa Catarina (UNISUL), Florianópolis, Brazil
e-mail: issaberchim@gmail.com

S. A. da Silva
e-mail: sthefanie.sads@hotmail.com

V. H. Fornasari
e-mail: vitoria.hfornasari@gmail.com

L. P. C. Ribeiro
e-mail: larissapcr@gmail.com

J. M. P. Ribeiro
e-mail: joaomarceloprdk@gmail.com

J. B. S. O. de Andrade Guerra
e-mail: baltazar.guerra@unisul.br

P. Bocquillon
University of East Anglia, Norwich, UK
e-mail: pmvb2@cam.ac.uk

P. Bocquillon · J. B. S. O. de Andrade Guerra
Cambridge Centre for Environment, Energy and Natural Resource Governance
(C-EENRG), University of Cambridge, Cambridge, UK

W. Leal Filho et al. (eds.), *Towards a Sustainable Bioeconomy: Principles, Challenges and Perspectives*, World Sustainability Series,
https://doi.org/10.1007/978-3-319-73028-8_20

conducted, based on official documents and reports, policies and on scientific studies. Then, a review of Brazilian public policies was conduced to analyse how public polices influence the sustainable production of the Brazilian ethanol sector. Brazilian bioenergy policies have contributed to establish a green biofuel sector, reducing greenhouse gas emissions, enhancing the efficiency in sugarcane production, creating jobs and improving air quality in cities.

Keywords Sustainable development · Biofuel · Energy policy
Green economy

1 Introduction

Considered as an important tool to achieve sustainable development and poverty eradication, the green economy agenda was widely discussed during the United Nations Conference on Sustainable Development—Rio +20 (United Nations 2012). There is no consensus about the green economy concept, but in its broadest sense, it is related to the economic dimension of sustainability (Ocampo 2015). One of the most accepted benefits is that the development of a green economy should result in an "improved human well-being and social equity, while significantly reducing environmental risks and ecological scarcities" (UNEP 2011a, p. 01).

According to the Final Report of the United Nations Conference on Sustainable Development, policies to foster a green economy must be based on sustainable principles raised and agreed on during international conferences on sustainable development (e.g. Rio 92, Agenda 21, Johannesburg, and Rio +20). In this regard, the United Nations Environment Programme (UNEP 2011a) highlights the importance of energy access to promote sustainable and equitable development worldwide.

Through green technologies and renewable energies, the development of a green economy has the potential to create jobs in new industries and emerging markets; to increase building capacity to address new challenges; to mitigate climate change through the replacement of carbon intensive industries, technologies and energies; and to favour inclusive and equitable development (UNEP 2011b; REN21 2015).

Renewable energies provided an estimated 19.1% of global final energy consumption in 2013 (representing an increase of 0.1% compared to 2012) (REN21 2015). According to REN21, in 2013, 9% of the global final energy consumption was based on traditional use of biomass (i.e. which refers to the burn of biomass for heat and cooking) while 'modern' renewable energy consumption increased by 10%. Liquid biofuel production was up by 9% in 2014, led by the United States (US) and Brazil, but with a significant increase in Asian productions. In 2013, 6.5 million people worked in jobs related to renewable energies, bioenergy being the sector that included the larger amount of jobs, with 2.5 million employees—especially liquid biofuels that employed 1.45 million workers (IRENA 2014).

Figure 1 illustrates that the Brazilian energy mix has remained relatively stable in the last six years, despite an increase in total energy supply. While energy supply from certain sources—e.g. petroleum and oil products, sugarcane products, natural gas, coal and coke and other renewables—has increased, other energy sources—e.g. hydraulic, firewood and charcoal—have seen their share in the Brazilian energy supply reduced.

For the Brazilian government, biofuels are defined by the Law No. 9.478, from August 6th, 1997, as a "substance derived from renewable biomass, such as biodiesel, ethanol and other substances established in regulations of the National Petroleum Agency, which can be directly used in internal combustion engines or for other types of power generation, and may partially or totally substitute fossil fuels" (BRAZIL 1997).

Bioenergy production is increasing globally, using food crops to produce energy (Gallardo and Bond 2011), consequently, concerns about the implication of biofuel production on land use and food insecurity, caused by rising food prices, are recurrent in international debates (Ferreira Filho and Horridge 2014). However, these concerns are alleviated by good policies towards the straightening of environmental concerns in sugarcane ethanol production (Gallardo and Bond 2011).

Regarding the transport sector, biofuels have great potential to stimulate a rapid substitution of oil products to renewable energy (Gallardo and Bond 2011; Hira and Oliveira 2009). Large biofuel production is mainly done by food crops that, if mismanaged, may increase food prices, increase CO_2 emissions and cause other environmental damages such as land degradation (Lucia 2010).

The increase of biofuel production worldwide has been led by governmental interventions in the sector (Sorda et al. 2010; Witcover et al. 2013; Enciso et al. 2016). Thus, policies regarding biofuel production might influence sustainability in

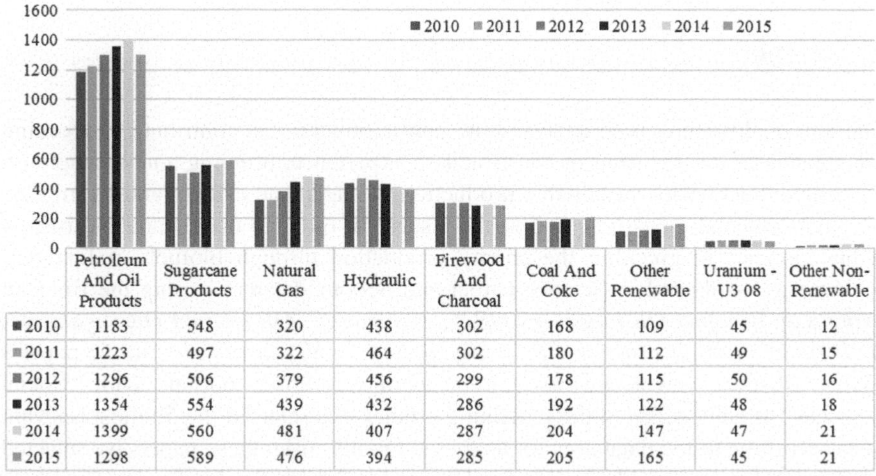

	Petroleum And Oil Products	Sugarcane Products	Natural Gas	Hydraulic	Firewood And Charcoal	Coal And Coke	Other Renewable	Uranium - U3 08	Other Non-Renewable
2010	1183	548	320	438	302	168	109	45	12
2011	1223	497	322	464	302	180	112	49	15
2012	1296	506	379	456	299	178	115	50	16
2013	1354	554	439	432	286	192	122	48	18
2014	1399	560	481	407	287	204	147	47	21
2015	1298	589	476	394	285	205	165	45	21

Fig. 1 Brazilian domestic energy supply (10 × 6 MWh). *Source* Created by the authors based on the Brazilian Energy Balance (EPE 2016)

all production chain, ensuring a green fuel that contributes to CO_2 mitigation, to increase employability, to enforce energy security and to promote rural development (Witcover et al. 2013; Enciso et al. 2016). In this regard, the sustainability patterns for biofuel production, asserted by the European Union have influenced its production worldwide (Lucia 2010).

According to Dincer (2002, pp. 138–145), "energy policies increasingly play an important role in addressing sustainability issues and a broad range of local, regional and global environmental concerns" because "a conscientious energy policy could speed up development towards efficient resource-saving technology". Basically, efficient and sustainable technologies find some barriers mainly due to the costs in technological and environmental issues, which could be avoided by governmental investments in sustainable policies, allowing new technologies to become reliable sources of energy (Tan et al. 2008). The success of those policies is guaranteed when innovation instruments and measures is considered using knowledge-intensive approaches (German et al. 2017; Hogarth 2016; Huttunen et al. 2014).

According to Pacini and Strapasson (2012, p. 388), "the Brazilian ethanol industry has shown itself to be highly innovative". Taking the country as an example, its "experience with bioethanol from sugarcane has shown the world the importance of government policy and support towards renewable energy" (Tan et al. 2008, p. 3364). Considering Brazil's dependence on sugarcane for ethanol production, Brazilian policies were essential factors to shape the production of sugarcane to become suitable and sustainable, making it the world's lower producer of CO_2 in sugarcane production (Tan et al. 2008; Galli and Wennersten 2013; Miyake et al. 2012).

2 Methods

The aim of this study is to analyse how public policies can contribute to greening the ethanol sector, focusing on the Brazilian experience, bringing a new insight on government action for sustainable production of sugarcane ethanol and its impacts on environmental, economic and social issues. In order to understand the role of public policies in greening the energy production through biofuels, a literature review was conducted in Science Direct and Scopus databases using the Boolean expression [(*biofuel OR bioenergy OR "green energy" OR ethanol OR sugarcane*) *AND* (*polic* OR governance OR "public polic*"*) *AND sustainab**] in the paper's "Title, Abstract and Keywords", in academic/scientific journals with "Social Sciences" as subject. From the remaining articles, we selected the 50 most relevant ones according to the databases for analyses. Finally, from these papers we selected a total of 24 papers, 17 from Science Direct and 7 from Scopus.

The procedures for selecting the most relevant papers for review in this research is illustrated in Table 1. These papers were used for the literature review.

Table 1 Procedures to select the literature used

Results	Science direct	Scopus	Total
Number of papers without the application of selection filters-only with the application of the Boolean expression: [(biofuel OR bioenergy OR "green energy" OR ethanol OR sugarcane) AND (polic* OR governance OR "public polic*") AND sustainab*]	20,484	37,615	58,099
Number of papers published in academic/scientific journals	18,301	25,445	43,746
Number of papers with the Boolean expression in the "Title, Abstract, keywords"	589	1,177	1,766
Number of papers restricted to Social Sciences	190	254	444
Selection of the 50 most relevant papers	50	50	100
Number of papers selected due to the complete paper revision	17	7	24
Papers selected	24		

Thus, to accomplish the aim of this study, analysing the Brazilian experience, a review of the economic characteristics of the Brazilian ethanol sector was made, based on official documents, reports, established policies, and on scientific studies. Then, a review of Brazilian public policies was made to analyse how they influence the sustainability and development of the Brazilian ethanol sector.

In order to develop the review on Brazilian public policies towards the development of the ethanol sector, we developed a search based on the connectors (ethanol) AND/OR (biofuels) AND/OR (energy) on Brazilian governmental laws and public policies. The search was made in Brazilian governmental websites, as ministries and the "Planalto" (Headquarters of the Federal Executive Branch, where the Presidential Office of Brazil is located)'s website, which contains all Brazilian public policies digitally available at http://www4.planalto.gov.br/legislacao). After the pre-selection of the policies, they were read in full and sixteen of them were selected for this review, according to their focus on sustainability or suitability to this research (see Table 2). Table 2 presents the description, the objectives and the benefits of each law, decree or government program selected.

3 Economic Characteristics of Ethanol Sector in Brazil

According to the Brazilian Energy Balance (EPE 2015), sugarcane production in Brazil reached 631.8 million tons in 2014, 2.5% less than 2013s production. Sugar production decreased by 5% in 2014 compared to 2013, while ethanol production increased by 3.3%, reaching 28,526 thousand m^3. Ethanol can be divided in two products, hydrous ethanol (used directly as fuel) and anhydrous ethanol (blended

Table 2 Brazilian policies towards biofuel

Title	Objectives and description	Benefits
Brazilian National Program on Ethanol (Decree No. 76593 of November 14, 1975)	Aims to reduce external dependence on oil, and seeks to meet the needs of domestic and foreign markets and the automotive fuel policy through the implementation of distilleries attached to sugar mills	Encourage domestic production
Legislation for Environmental Protection, CONAMA Resolution No. 01 of January 23, 1986	Imposes the need for Environmental Impact Studies and Report for obtaining periodical license for activities that may affect the environment in the agro-energy sector	Regulates the activities related to ethanol that can potentially change the environment
Decree No. 8723 of October 28, 1993	Aims to reduce GHG emissions from automobiles, imposing limits on vehicle emissions	CO_2 mitigation
Law No. 9478 of August 06, 1997	Reinforces that Brazilian national policies should increase the share of biofuels in the Brazilian energy mix, respecting economic, social and environmental basis	Regulates activities related to ethanol
Law No. 10.336 of December 19, 2001	Creates the Contribution for Intervention in the Economic Domain (CIDE), which put taxes on fossil fuels' imports and commercialization	Encourage the local production of ethanol
Decree No. 11241 of September 19, 2002	Aims to promote the gradual eradication of burning practices in sugarcane sector through mechanization of the harvesting process	CO_2 mitigation and sustainable production of sugarcane
Decree No. 5060 of April 30, 2004a	Reduces the rates of CIDE levied on imports and commercialization of oil and oil products, natural gas and their derivatives, and ethyl alcohol fuel	Encourage to the use of ethanol instead of gasoline
Social Fuel Seal (Decree No. 5297 of December 06, 2004b)	Seal granted by the Ministry of Agrarian Development to industrial producers who have business commitments with family farmers, increasing access to financing and tax benefits	Supports family farmers

(continued)

Table 2 (continued)

Title	Objectives and description	Benefits
Law No. 11.097 of January 11, 2005	Introduces the use of biofuel, stimulating the production and consumption of biodiesel. And focuses in the social inclusion and regional development	Job and income creation/ development
National Agroenergy Plan, 2005	Aims to organize and develop research proposals, development, innovation and technology transfer to ensure the sustainability and competitiveness of agroenergy chains	CO_2 mitigation, fossil fuels exchange to biofuels; contribution to social inclusion
Green Ethanol Strategic Environmental Project, 2007	Aims to recognize and reward good environmental practices in São Paulo's sugarcane industry with a certificate of conformity, renewed annually	Sustainable production of ethanol and environmental conservation
Agro-environmental Protocol, 2007	Develop actions that promote the sustainability of productive sugar chain, green ethanol and bioenergy, anticipating the deadlines to eradicate burning straw sugarcane	Stimulates the sustainable production of ethanol
Brazilian National Plan on Climate Change (Decree No. 6263 of November 21, 2007)	Encourages the sustainable increase in the share of biofuels in the national transport sector and work towards the structuring of an international market of sustainable biofuels	CO_2 mitigation and exchange from gasoline to ethanol
Sugarcane Agroecological Zoning, 2009	Aims to stimulate the sustainable production of sugarcane in Brazil, offering financial support to new producers of sugar and ethanol that respect sustainable principles	Stimulates the Brazilian sugarcane production
RenovAção Project, 2009	Aims to requalify the workers to adapt them to new job positions in São Paulo due to the eradication of burning practices	Improvement in the quality and adaptability of workers
Law No. 13033 of September 24, 2014	Stated that the federal government had the power to extend the blend of ethanol in gasoline by 27.5% and the blend of biodiesel in diesel oil by 15%	Encouragement to the use of alcohol instead of gasoline

with gasoline A to form gasoline C). Hydrous ethanol represented 57.1% of the total production of ethanol—16,296 thousand m^3 (EPE 2015).

In Brazil, ethanol from sugarcane is produced in more than 400 mills, covering an area of 4.8 million ha, many of the producers have the capacity to produce both sugar and ethanol, depending on market demands (Souza et al. 2015; Nogueira et al. 2015). The sector is dominated by large-scale production of sugarcane (60%), but also counts 70,000 small producers united in cooperatives, representing 40% of the national production (Diaz-Chavez et al. 2015). Thus, 85% of sugarcane producers in the southern region own small properties of 50 ha on average.

According to Shikida et al. (2011), the sugar market in Brazil started to be valued and encouraged in the 1930s, with the Institute of Sugar and Alcohol (IAA), and with the production quotas. In 1959, the Central Cooperative of Sugar Producers and Alcohol in São Paulo State (Copersucar) was created. On the chart, from the 1970s on, the evident increase in the production of both can be seen. This was mostly due to the creation of the National Alcohol Program (Proálcool) developed in 1975 to reduce external dependence on oil, and to meet the needs of domestic and foreign markets and the automotive fuel policy (BIODIESELBR 2006; Brazil 1975). Also aiming the increase in oil prices, with the first oil shock in 1973 and the consequent increased use of sugarcane.

In the 1980s the country came across with alcohol fuel shortages, demonstrating the weakness of the Proálcool, the program created in the previous decade. With the extinction of the IAA in the 1990s, the sector was deregulated, accompanied with the liberalization on exports and on the prices of sugar and alcohol. At the turn of the century the Brazilian Stock Exchange Alcohol Ltda (BBA) was created, aiming at a better control over the commercialization of the 170 associated plants.

From 2000 to 2014 a new rise in the price of oil generated the need to give better attention to alternative fuel sources. Thus, the creation of a new vehicle technology, the flex fuel vehicles, which makes the car able to operate with both gasoline and/or ethanol (Sorda et al. 2010). Although the flex fuel technology was created in 1980 in the United States, Brazil started to produce and sell vehicles with such technology only in 2003 (Stattman et al., 2013). In addition, the preference by the government for ethanol is evident, since in 2006, the tax rate applied to pure gasoline was higher than that applied to pure ethanol. Furthermore, the ethanol anhydrous used for blending with gasoline is not taxed, and gasoline blended with 13% of anhydrous or more, receives benefits from lower rates than hydrous ethanol (Martines-Filho et al. 2006).

Figure 2 presents the total amount of ethanol and sugar produced in Brazil since the 1960s. It also represents the total amount in Total Recoverable Sugar "which is the amount of sugar available in the raw material, minus the losses in the manufacturing process kept stable" (EPE 2015, p. 15).

In addition to the agricultural production of ethanol, the program also supported the development of technologies to enable the efficient blend of ethanol in gasoline for vehicles and the large production of vehicles running only on hydrous ethanol. From 1975 to 2000, about 5.6 million vehicles running 100% on hydrous ethanol were produced. The success of the program guaranteed savings of about US$11.5

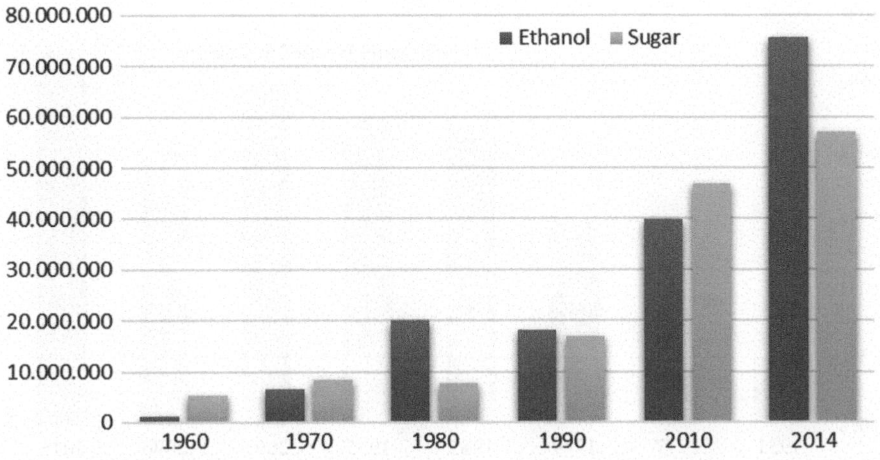

Fig. 2 Total sugar and ethanol production in Brazil in total recoverable sugar. *Source* Created by the authors based on EPE (2015)

billion due to reduction in oil imports and mitigated about 110 million tons of Carbon (BIODIESELBR 2006). As a result of the success of flex fuel cars, "in 2010, cars used nearly equal volumes of gasoline and ethanol" in Brazil (Ferreira Filho 2013, p. 85).

Between 2006 and 2010, biofuel production more than doubled worldwide (International Energy Agency 2013), fostered by supportive policies—such as blending obligations of biofuels in fossil fuels—in the three world leading markets: Brazil, the US and the European Union (Sorda et al. 2010; Witcover et al. 2013). Led by the US and Brazil, the International Energy Agency (2013) estimated that global demand for bioenergy will increase in all sectors from 1300 Mtoe in 2011 to 1850 Mtoe in 2035.

In Brazil, bioelectricity use almost doubled between 2010 and 2014, obviating the emission of 10.7 Mt of CO_2 and supplying energy equivalent to the consumption of 11 million houses (Souza 2015). The Brazilian productivity of ethanol increased 4% in the last decades, positioning sugarcane ethanol as an advanced biofuel, with a great potential to mitigate greenhouse gas emissions in the transport sector (Youngs et al. 2015).

Aiming to achieve greenhouse gas mitigation, the Decree No. 8723 of 28 October 1993, requires the mandatory addition of 22% of anhydrous ethanol in gasoline (Brazil, 1993). In 2014, the Law NO 13033 of 24 September stated that the federal government had the power to extend the blend of ethanol in gasoline up to 27.5% (Brazil 2014).

Figure 3 represents the percentage of ethanol added in gasoline in Brazil, from the first law in 1931 to the latest piece of legislation in 2014.

Brazil has one of the most advanced bioenergy programs in the world, both legally and technologically. Besides the blending obligation of ethanol in gasoline,

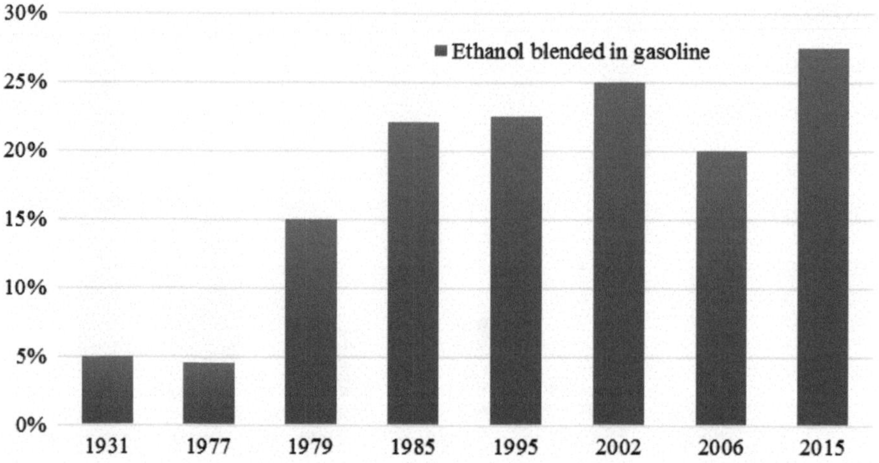

Fig. 3 Percentage of ethanol added in gasoline in Brazil. *Source* Created by the authors based on: Vieira (2011) and Brazil (2014)

the introduction of flex fuel cars in 2003 (cars that run both in hydrous ethanol and/ or in gasoline), lead to the success of biofuels in the country (Morgera et al. 2009). Currently, about 90% of vehicles (23.8 million cars) commercialized in Brazil are flex fuel, which represents 71% of the national light vehicle fleet (Souza et al. 2015). According to Joly et al. (2015), the use of ethanol improved air quality in Brazilian cities, especially in São Paulo.

4 Policies for Sustainable Production of Sugarcane for Ethanol in Brazil

In Brazil, ethanol is based mostly on so called 'first generation' biofuel from sugarcane, whose production originates from land that could be used to grow food crops. It can lead to displacing food production, and has often been accused of it, consequently, contributing to food insecurity. For this reason, it is especially important to invest in research and development to improve and develop 'second generation' biofuels that are not based on food crops (Nyko et al. 2010). Unlike most countries, Brazil has the advantage of having enough land to produce food, energy crops and to ensure forest conservation at the same time—if managed sustainably (Santos 2012).

Ethanol production is associated with both good and bad practices. The bad practices are often related to biodiversity loss through the expansion of sugarcane crops into conversation lands, high use of agrochemicals, intensive use of water and land degradation. On the other hand, the benefits are greenhouse gas mitigation in the transport sector and job creation (Joly et al. 2015). In this regard, Ferreira Filho

(2013) shows that sugarcane production in the Northeast Brazil is more labour intensive than the production in the Southeast region, due to the modernization level of the farms.

Hira and Oliveira (2009, p. 2456) indicates three phases of the Brazilian intervention in the biofuel market that ensured the success of its ethanol program. The first was "establishing and supporting the market in its infant industry phase and during its market crisis", the second was "investing in infrastructure and other long-term investments, including research and development", and the third was "weaning off state support once the market became viable".

Planning to boost the benefits and reduce the negative effects of ethanol production, since 1975 the Brazilian government through Proálcool has brought different national and regional policies to encourage and regulate the production, marketing and consumption biofuels, particularly ethanol (Stattman et al. 2013).

4.1 Historical Context of Brazilian Policies for Sustainable Production of Sugarcane for Ethanol

Intending to reduce the country's dependence on oil imports, due to the high cost per barrel with the oil crisis in the years of 1973 and 1975, the Brazilian government invested in hydroelectricity and bioenergy, to ensure energy security and sovereignty (Berchin et al. 2015). This resulted in the creation of Proálcool that was regulated by the Decree No. 76593 on 14 November 1975, aiming to encourage the replacement of oil derivate products to ethanol (Brazil 1975; BIODIESELBR 2006). Proálcool was one of the biggest projects for bioenergy production in the world and demonstrated the feasibility of sugarcane ethanol production and consumption (La Rovere et al. 2011), specially benefiting the State of São Paulo (Stattman et al. 2013).

The project was developed in two phases. The first was characterized by the production of anhydrous alcohol which is now used as an additive in gasoline; the second phase gives start to the production of hydrated alcohol for exclusive use as fuel in cars with the flex fuel vehicle technology, which was one of the project's inheritances. Thus, there was an increase of alcohol production from 600 million litres in 1975 to 12.3 billion in 1987 (BIODIESELBR 2006). Besides that, the project also aimed to improve environmental conditions, to create new varieties of sugarcane, to stimulate job creation, to increase skilled labour force, and to create, develop and improve vehicles powered by alcohol (IPEA 2010).

In 2001, the Brazilian government established a new policy that strengthens the objectives of Proálcool, establishing the Contribution for Intervention in the Economic Domain on the imports and commercialization of petroleum and its oil products, natural gas and its derivatives, and ethanol. In 2001 the taxation for petrol was US\$159.63 per m^3 and for ethanol was US\$9.30 per m^3, meaning a rate 1716% higher for gasoline. Within this context, it could be noticed the encouragement on production and commercialization of ethanol instead of fossil fuels (Brazil 2001). This law was regulated in 2004 by the Decree No. 5060 of 30 April, which reduces the rates established for gasolines to US$31.85 and for ethanol to zero, thus contributing again in encouraging the change from fossil fuels to biofuels (Brazil 2004a).

Another law from the Brazilian government encouraged the production of biofuels in the year of 2014. It was established concerning the progressive blend of biodiesel in oil diesel sold to the consumer (Brazil 2005, 2014). Still regarding these objectives, the Brazilian government created two high-impact programs within the industry, the National Agroenergy Plan and the National Plan on Climate Change.

The Brazilian National Agroenergy Plan is based on the Brazilian National Plan on Climate Change (Decree No. 6263 of November 21, 2007) which has seven objectives to mitigate climate change: land conservation; reduction in deforestation rates; increase renewable energies in the Brazilian energy mix; reduction of population vulnerabilities; adaptation to climate change; encouragement to the sustainable increase in the share of biofuels in the national transport mix; and working towards the development of a structure of an international market for sustainable biofuels (Brazil 2007).

The first three objectives are directly related to the improvement in sustainability of sugarcane and ethanol production, stimulating the use of bioenergy in Brazil's energy mix, phasing out the use of fire for clearing and cutting sugarcane in areas where mechanic harvesting can take place, increase electricity supply from cogeneration—mainly from sugarcane bagasse—and production of crops in the areas defined in the Sugarcane Agroecological Zoning Program with the aim to prevent the emission of 508 mt CO_2 by 2017 (Brazil 2007). The Sugarcane Agroecological Zoning was created in 2009 to encourage the sustainable production of sugarcane in Brazil, offering financial support to new producers of sugar and ethanol that respect sustainable principles (Brazil 2015).

The state of São Paulo, is the largest biofuel producer in Brazil (Ferreira Filho and Horridge 2014) giving it the biggest challenges in environmental concerns. The use of fire for clearing and cutting sugarcane reduces the sustainability of ethanol, causing high CO_2 emissions, environmental damages and health problems.

Aiming to improve the sustainability of ethanol production, the government of the State of São Paulo signed the Agro-environmental Protocol (Protocolo Agroambiental) which is part of the Green Ethanol Strategic Environmental Project (Projeto Ambiental Estratégico Etanol Verde) which anticipated deadlines, from 2021 to 2014, to eradicate the burning of sugarcane during the period of harvest. The project aims to develop actions to improve the sustainability of the sugar supply chain, ethanol and bioenergy production and awards the best practices in the sugarcane sector, mitigating their socio-environmental impacts and improving the sector both nationally and internationally (São Paulo 2015).

Producers participating in the project do it voluntarily and receive a Certificate of Environmental Protocol, which renewal is annual and monitored. The certificate influences the reputation of the power plants and associations in domestic and foreign markets, encouraging adherence to the design and adequacy of the actions of the actors of this sector plan (São Paulo 2015; UNICA 2009). In 2013/2014, harvesting without burning the sugarcane reached 90% of the total productive area, with a membership of 90% of São Paulo production, composed by more than 170 producers and 29 suppliers' associations (São Paulo 2014).

Also, aiming to mitigate the impacts of burning practices, in 2002 the government of São Paulo passed the Decree No. 11241 of 19 September, to promote the gradual eradication of burning practices in the sugarcane sector through the mechanization of harvesting processes (São Paulo 2002). According to Diaz-Chavez et al. (2015), the mechanization of harvesting process will improve sustainability, but will also reduce the employment rates in the sector.

To reduce negative impacts on employment, the Brazilian Sugarcane Industry Association (UNICA) in partnership with the Brazilian government and private institutions, developed the project RenovAção, in São Paulo, aiming at requalifying workers for new positions. To this end, more than 5700 rural workers were trained on 30 different courses (UNICA 2015). According to UNICA the rate of worker's reintegration in the job market reached 78%, a satisfactory number for UNICA. If addressed correctly, the policies applied by the State of São Paulo may be an example of benchmark for sustainability and productivity to other states and countries (Gallardo and Bond 2011), and "the Brazilian approach to biofuel and food security could be followed by other nations to provide a sustainable pathway to renewable energy" (Goldemberg et al. 2014, p. 14).

5 Discussion

Brazil is one of the key players on biofuel from sugarcane, with more than 35 years of research and technological developments in the sector (Nyko et al. 2010; Amaral et al. 2008). The success of the Brazilian bioenergy program can be attributed to investments in technology, supportive policies and the size of the market (Amaral et al. 2008). The Brazilian National Plan on Bioenergy acknowledges that Brazil's large consumer market has enabled economies of scale that reinforce bioenergy's competitiveness (Brazil 2006). Figure 4 represents the increase in Brazilian ethanol production and consumption from 1970 to 2014.

Besides Brazil's large internal market, whose ambitious national support policies have been key, ethanol exports have also increased since 1990, and so has the average price of Brazilian ethanol. Figure 5 shows the average price of ethanol exports in US\$/m^3. Sugarcane ethanol's value increased from 1990, but since 2011 it has followed a downward trend due to low oil prices.

The growing use of sustainability in ethanol production has contributed to mitigate greenhouse gas emissions and improve health quality in cities. Jank (2011) estimated that the use of ethanol in Brazil has saved about 36 million dollars per year in public expenditures for cardiorespiratory diseases, and has also contributed to prevent the release of about 600 Mt of CO_2 since 1970, when ethanol started being used in large-scale in Brazil.

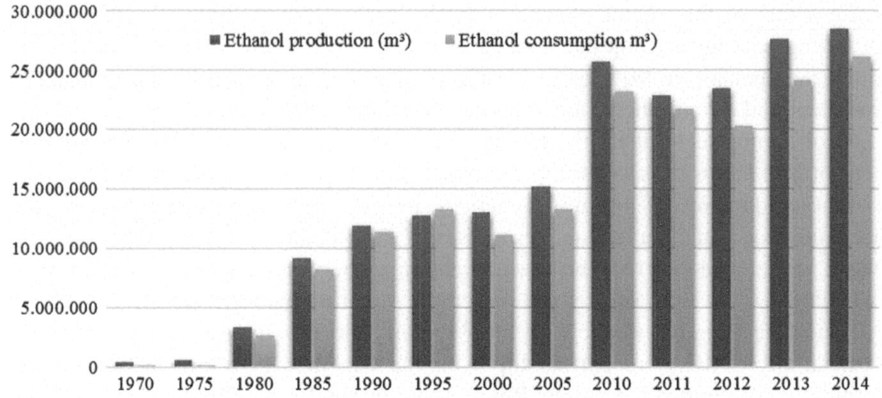

Fig. 4 Ethanol production and consumption in Brazil. *Source* Created by the authors based on: EPE (2015)

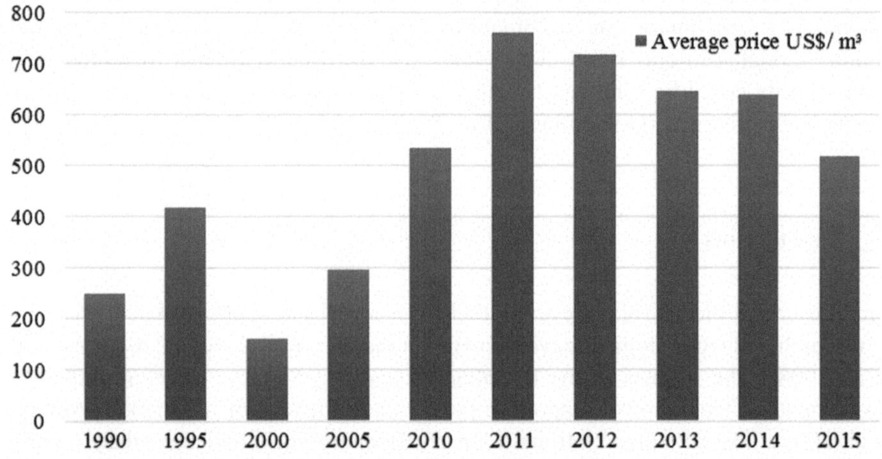

Fig. 5 Average price of Brazilian ethanol exports in US$/m^3. *Source* Created by the authors based on: EPE (2015) and Ministry of Agriculture, Livestock and Supply (2015)

Incentives to the mechanization of sugarcane harvesting and the ban on burning practices have also contributed to turn biofuels into a greener energy source (Youngs et al. 2015). According to Jank (2011), investments in new biofuel techniques and technologies have the potential to contribute in mitigating greenhouse gas emissions and reducing wastes, through the use of sugarcane bagasse and fibres, either for the production of second generation biofuels or for cogeneration.

Wastes from ethanol production could be used to improve soil efficiency by increasing the number of organic residues in the soil, by avoiding excessive water evaporation and reducing erosion processes (Jank 2011). Despite its low efficiency, the use of straw and bagasse to produce second generation biofuels and for cogeneration, contributes to the production of extra energy (Glithero et al. 2013; Costa 2013), that is used by Brazilian farms in a self-sufficient process, making them energy independent (Costa 2013). Youngs et al. (2015) claim that this extra energy could also be used for commercialization, complementing farmer's income.

6 Conclusion

Policies aiming to improve energy security and autonomy, through support to biofuels, have successfully bolstered the efficiency and competitiveness of the Brazilian biofuel sector, which relies on a large internal market and expanding export markets.

Public policies have played an important role in improving the sustainability of the ethanol supply chain, through prohibiting the use of fire for harvesting sugarcane, by supporting cogeneration of energy, by stimulating the production in appropriated areas through the Sugarcane Agroecological Zoning, and by stimulating the blend of biofuels in fossil fuels. Sustainability schemes implemented in export markets, notably in the EU and US, also encouraged the private sector to adopt more sustainable practices in order to guarantee the stability of exports. Thus, learning from the Brazilian experience, the ethanol sector can be a sustainable market, supported by governmental policies and regulations, reducing the amount of greenhouse gas emitted in the atmosphere in the production process, reusing the bagasse of sugarcane to produce extra-energy, also improving the quality of the air in cities due to mandatory blend of ethanol in gasoline.

Since biofuel production involves a broad range of sectors (e.g. energy, transport, agriculture, water, forest, wastes and cities), the promotion of green practices in the biofuel supply chain could foster the development of a green economy in Brazil. Incentives to increase the share of bioenergy in the Brazilian energy mix could contribute to greenhouse gas mitigation, reducing the dependence on fossil fuels, creating jobs and promoting welfare.

Although sugarcane production is mostly produced in large agro-industrial farms, small farmers also have an important role to play, and have received financial support from Brazilian government to engage in the biofuel market. By supporting small farmers, the government aims to increase family income in rural areas and improve welfare and social inclusion, reducing poverty and increasing rural resilience. Within this, future studies should investigate the implications of family farmers and smallholder farmers engaging in the biofuel sector to improve rural sustainable development.

Acknowledgements This study was conducted by the Research Centre on Sustainability (Greens) from the University of Southern Santa Catarina (Unisul), in the context of the project BRIDGE - Building Resilience in a Dynamic Global Economy: Complexity across scales in the Brazilian Food-Water—Energy Nexus, funded by the Newton Fund, Fundação de Amparo à Pesquisa e Inovação do Estado de Santa Catarina and the Research Councils United Kingdom (RCUK).

References

Amaral WAN, Marinho JP, Tarasantchi R, Beber A, Giuliani E (2008) Environmental sustainability of sugarcane ethanol in Brazil. In: Zuurbier P, Vooren J (eds) Sugarcane ethanol: contributions to climate change mitigation and the environment. Wageningen Academic Publishers, Netherlands. Available at http://sugarcane.org/resource-library/studies/Wageningen.pdf

Berchin II, Garcia J, Heerdt ML, Moreira AQ, Silveira ACM, Guerra JBSO (2015) Energy production and sustainability: a study of Belo Monte hydroelectric power plant. Nat Resour For 39:224–237

BIODIESELBR (2006) Proálcool—Programa Brazileiro de Álcool. Available at http://www.biodieselbr.com/proalcool/pro-alcool/programa-etanol.htm

Brazil (1975) Decree No. 76593 of November 14, 1975. Institui o Programa Nacional do Álcool e dá outras Providências. Available at http://www2.camara.leg.br/legin/fed/decret/1970-1979/decreto-76593-14-novembro-1975-425253-publicacaooriginal-1-pe.html

Brazil (1993) Law No. 8723 of October 28, 1993. Dispõe sobre a redução de emissão de poluentes por veículos automotores e dá outras providências. Available at http://www.planalto.gov.br/ccivil_03/leis/L8723.htm

Brazil (1997) Law no 9478 of August 06, 1997. Dispõe sobre a política energética nacional, as atividades relativas ao monopólio do petróleo, institui o Conselho Nacional de Política Energética e a Agência Nacional do Petróleo e dá outras providências. Available at http://www.planalto.gov.br/ccivil_03/leis/l9478.htm

Brazil (2001) Law No. 10336 of December 19, 2001. Institui Contribuição de Intervenção no Domínio Econômico incidente sobre a importação e a comercialização de petróleo e seus derivados, gás natural e seus derivados, e álcool etílico combustível (Cide), e dá outras providências. Available at http://www2.camara.leg.br/legin/fed/lei/2001/lei-10336-19-dezembro-2001-428889-publicacaooriginal-1-pl.html

Brazil (2004a) Decree No. 5060 of April 30, 2004. Reduz as alíquotas da Contribuição de Intervenção no Domínio Econômico incidente sobre a importação e a comercialização de petróleo e seus derivados, gás natural e seus derivados, e álcool etílico combustível (CIDE), instituída pela Lei no 10.336, de 19 de dezembro de 2001, e dá outras providências. Available at http://www.planalto.gov.br/ccivil_03/_ato2004-2006/2004/decreto/d5060.htm

Brazil (2004b) Decree No. 5297 of December 06, 2004. Dispõe sobre os coeficientes de redução das alíquotas da Contribuição para o PIS/PASEP e da COFINS incidentes na produção e na comercialização de biodiesel, sobre os termos e as condições para a utilização das alíquotas diferenciadas, e dá outras providências. Available at http://www.planalto.gov.br/ccivil_03/_ato2004-2006/2004/decreto/D5297.htm

Brazil. (2005). Law No. 11097 of January 11, 2005. Dispõe sobre a introdução do biodiesel na matriz energética brasileira; altera as Leis nos 9.478, de 6 de agosto de 1997, 9.847, de 26 de outubro de 1999 e 10.636, de 30 de dezembro de 2002; e dá outras providências. Available at http://www.planalto.gov.br/ccivil_03/_ato2004-2006/2005/lei/l11097.htm

Brazil (2006) Ministry of Agriculture, Livestock and Supply. Plano Nacional de Agroenergia 2006–2011 (National Agroenergy Plan 2006–2011). Embrapa Informação Tecnológica, Brasília. Available at http://www.agricultura.gov.br/arq_editor/file/Ministerio/planoseprogramas/PLANONACIONALDEAGROENERGIA.pdf

Brazil (2007) Decree No. 6263 of November 21, 2007. Executive summary: national plan on climate change. Available at http://www.mma.gov.br/estruturas/208/_arquivos/national_plan_208.pdf

Brazil (2014) Law No. 13033 of September 24, 2014. Dispõe sobre a adição obrigatória de biodiesel ao óleo diesel comercializado com o consumidor final; altera as Leis nos 9.478, de 6 de agosto de 1997, e 8.723, de 28 de outubro de 1993; revoga dispositivos da Lei no 11.097, de 13 de janeiro de 2005; e dá outras providências. Available at https://www.planalto.gov.br/ccivil_03/_ato2011-2014/2014/lei/l13033.htm

Brazil (2015) Ministry of Agriculture. Zoneamento Agroecológico da Cana-de-açúcar (ZAECana). Available at http://www.agricultura.gov.br/portal/page/portal/Internet-MAPA/pagina-inicial/carta-de-servico-ao-cidadao/politica-agroenergia/cana-de-acucar-e-etanol/zaecana

Costa AC (2013) Caso de sucesso—Produção de etanol, 2ª geração. (Case success—ethanol production, 2nd generation). Available at http://www.crq4.org.br/sms/files/file/caso_sucesso_aline_costa_etanol2.pdf

Diaz-Chavez R, Morese MM, Colangelli M, Fallot A, Moraes MAFD, Olényi S, Osseweijer P, Sibanda LM, Mapako M (2015) Social Considerations. In: Souza GM, Victoria RL, Joly CA, Verdade LM (eds) Bioenergy & sustainability: bridging the gaps. Áttema Editorial, São Paulo. Available at http://bioenfapesp.org/scopebioenergy/images/chapters/bioenergy_sustainability_scope.pdf

Dincer I (2002) The role of exergy in energy policy making. Energy Policy 30(2):137–149

EPE. Empresa de Pesquisa Energética (2015) Brazilian Energy Balance 2015: year 2014. EPE, Rio de Janeiro. Available at https://ben.epe.gov.br/downloads/Relatorio_Final_BEN_2015.pdf

EPE. Empresa de Pesquisa Energética (2016) Brazilian Energy Balance 2016: year 2015. EPE, Rio de Janeiro. Available at https://ben.epe.gov.br/downloads/Relatorio_Final_BEN_2016.pdf

Enciso SRA, Fellmann T, Dominguez IP, Santini F (2016) Abolishing biofuel policies: possible impacts on agricultural price levels, price variability and global food security. Food Policy 61:9–26

Ferreira Filho JBS (2013) Food security, the labor market, and poverty in the Brazilian bio-economy. Agric Econ 44:85–93

Ferreira Filho JBS, Horridge M (2014) Ethanol expansion and indirect land use change in Brazil. Land Use Policy 36:595–604

Gallardo ALCF, Bond A (2011) Capturing the implications of land use change in Brazil through environmental assessment: time for a strategic approach? Environ Impact Assess Rev 31:261–270

Galli E, Wennersten R (2013) Frame analysis of environmental conflicts in ethanol production in Brazil. Int J Sustain Soc 5(1):62–77

German L, Goetz A, Searchinger T, de Oliveira, GLT, Tomei J, Hunsberger C, Weigelt J (2017) Sine qua nons of sustainable biofuels: distilling implications of under-performance for national biofuel programs. Energy Policy 108:806–817

Glithero NJ, Ramsden SJ, Wilson P (2013) Barriers and incentives to the production of bioethanol from cereal straw: a farm business perspective. Energy Policy 59:161–171

Goldemberg J, Mello FFC, Cerri CEP, Davies CA, Cerri CC (2014) Meeting the global demand for biofuels in 2021 through sustainable land use change policy. Energy Policy 69:14–18

Hira A, Oliveira LG (2009) No substitute for oil? How Brazil developed its ethanol industry. Energy Policy 37:2450–2456

Hogarth, J. R. (2016). Evolutionary models of sustainable economic change in Brazil: no-till agriculture, reduced deforestation and ethanol biofuels. Environ Innov Soc Trans 24:130–141

Huttunen S, Kivimaa P, Virkamäki V (2014) The need for policy coherence to trigger a transition to biogas production. Environ Innov Soc Trans 12:14–30

IBAMA (1986) Resolução CONAMA nº 001, de 23 de janeiro de 1986. Available at http://www.mma.gov.br/port/conama/res/res86/res0186.html

International Energy Agency (2013) Renewable energy outlook. In: International energy agency. World energy outlook 2013. Available at http://www.worldenergyoutlook.org/media/weowebsite/2013/WEO2013_Ch06_Renewables.pdf

IPEA (2010) Comunicados do IPEA: biocombustíveis no Brazil, Etanol e Biodiesel (Announcements of the IPEA: biofuels in Brazil, Ethanol and Biodiesel). Available at http://www.ipea.gov.br/portal/images/stories/PDFs/100526_comunicadodoipea_53.pdf

IRENA. International Renewable Energy Agency (2014) Renewable energy and jobs: annual review. Available at http://www.irena.org/Publications/rejobs-annual-review-2014.pdf

Jank M (2011) Futuro verde: a cana-de-açúcar no centro da economia de baixo carbono (Green future: the cane sugar in the center of the low-carbon economy). Available at http://2011.ethanolsummit.com.br/arquivos/_Summit2011Portugues.pdf

Joly CA, Verdade LM, Huntley BJ, Dale VH, Mace G, Muok B, Ravindranath NH (2015) Biofuel impacts on biodiversity and ecosystem services. In: Souza GM, Victoria RL, Joly CA, Verdade LM (eds) Bioenergy & sustainability: bridging the gaps. Áttema Editorial, São Paulo. Available at http://bioenfapesp.org/scopebioenergy/images/chapters/bioenergy_sustainability_scope.pdf

La Rovere EL, Pereira AS, Simões AF (2011) Biofuels and sustainable energy development in Brazil. World Dev 39:1026–1036

Lucia L (2010) External governance and the EU policy for sustainable biofuels, the case of Mozambique. Energy Policy 38:7395–7403

Martines-Filho J, Burnquist HL, Vian CEF (2006) Bioenergy and the rise of sugarcane-based ethanol in Brazil. Choices. The Magazine of Food, Farm, and Resource Issues 21:91–96. Available at http://cepea.esalq.usp.br/pdf/artigo_heloisa.pdf

Ministry of Agriculture, Livestock and Supply (2015) Brazilian monthly ethanol exports. Available at http://www.udop.com.br/download/estatistica/etanol/14out15_exportacao_etanol_safra.pdf

Miyake S, Renouf M, Peterson A, McAlpine C, Smith C (2012) Land-use and environmental pressures resulting from current and future bioenergy crop expansion: a review. J Rural Stud 28 (4):650–658

Morgera E, Kulovesi K, Gobena A (2009) Case studies on bioenergy policy and law: options for sustainability. Food and Agriculture Organization of the United Nations, Roma. Available at http://www.fao.org/docrep/012/i1285e/i1285e00.pdf

Nogueira LAH, Leal MRLV, Fernandes E, Chum HL, Diaz-Chavez R, Endres J, Mahakhant A, Otto M, Seebaluck V, Wielen L (2015) Sustainable development and innovation. In: Souza GM, Victoria RL, Joly CA, Verdade LM (eds) Bioenergy & sustainability: bridging the gaps. Áttema Editorial, São Paulo. Available at http://bioenfapesp.org/scopebioenergy/images/chapters/bioenergy_sustainability_scope.pdf

Nyko D, Garcia JLF, Milanez AY, Dunham FB (2010) A corrida tecnológica pelos biocombustíveis de segunda geração: uma perspectiva comparada (The technological race for second generation biofuels: a comparative perspective). BNDES Setorial 32:5–48. Available at http://www.bndes.gov.br/SiteBNDES/export/sites/default/bndes_pt/Galerias/Arquivos/conhecimento/bnset/set32101.pdf

Ocampo JA (2015) The transition to a green economy: benefits, challenges and risks from a sustainable development perspective. United Nations (Summary of Background Papers). Available at http://www.unep.org/greeneconomy/Portals/88/documents/research_products/UN-DESA,UNCTADTransitionGE.pdf

Pacini H, Strapasson AB (2012) Innovation subject to sustainability: the European policy on biofuels and its effects on innovation in the Brazilian bioethanol industry. J Contemp Eur Res 8 (3)

REN21. Renewable Energy Policy Network for the 21st Century (2015) Renewables 2015: global status report. Ren21 Secretariat, Paris. Available at http://www.ren21.net/wp-content/uploads/2015/07/REN12-GSR2015_Onlinebook_low1.pdf

Santos FA (2012) Cogeração e Comercialização de Energia Elétrica nas Usinas de Açúcar e Etanol (Cogeneration and electricity trading in sugar and ethanol plants). Available at http://www.iee.usp.br/eventos/cogeracao_em_usinas.pdf

São Paulo (2002) State Law No. 11241 of September 19, 2002. Dispõe sobre a eliminação gradativa da queima da palha da cana-de-açúcar e dá providências correlatas. Available at http://www.iea.sp.gov.br/out/bioenergia/legislacao/2002_Lei_Est_11241.pdf

São Paulo (2014) Protocolo Agroambiental do Setor Sucroenergético Paulista: Dados consolidados das safras 2007/08 a 2013/14 (Agroambiental protocol of the São Paulo's sugar-energy sector: consolidated data from 2007/08 to 2013/14). Available at http://www.iea.sp.gov.br/Relat%C3%B3rioConsolidado1512.pdf

São Paulo (2015) Protocolo Agroambiental (Agroenvironmental protocol). Available ta http://www.ambiente.sp.gov.br/etanolverde/protocolo-agroambiental/o-protocolo/

Shikida PFA, Azevedo PF, Vian CEF (2011) Desafios da agroindústria canavieira no Brazil pós-desregulamentação: uma análise das capacidades tecnológicas (Challenges of the sugar cane industry in the post-deregulation Brazil: an analysis of technological capabilities). Revista de Economia e Sociologia Rural 49:599–628. Available at http://www.scielo.br/pdf/resr/v49n3/a04v49n3.pdf

Sorda G, Banse M, Kemfert C (2010) An overview of biofuel policies across the world. Energy Policy 38:6977–6988

Souza ZJ (2015) Programa de Certificação da Bioeletricidade (Bioelectricity certification program). Available at www.unica.com.br/download.php?idSecao=17&id=12472387

Souza GM, Victoria RL, Joly CA, Verdade LM (2015) Bioenergy numbers. In: Souza GM, Victoria RL, Joly CA, Verdade LM (eds) Bioenergy & sustainability: bridging the gaps. Áttema Editorial, São Paulo. Available at http://bioenfapesp.org/scopebioenergy/images/chapters/bioenergy_sustainability_scope.pdf

Stattman SL, Hospes O, Mol APJ (2013) Governing biofuels in Brazil: a comparison of ethanol and biodiesel policies. Energy Policy 61:22–30

Tan KT, Lee KT, Mohamed AR (2008) Role of energy policy in renewable energy accomplishment: the case of second-generation bioethanol. Energy Policy 36(9):3360–3365

UNEP. United Nations Environment Programme (2011a) Towards a green economy: pathways to sustainable development and poverty eradication: a synthesis for policy makers. United Nations, St-Martin-Bellevue. Available at http://www.unep.org/greeneconomy/Portals/88/documents/ger/GER_synthesis_en.pdf

UNEP (2011b) Towards a green economy: pathways to sustainable development and poverty eradication. Available at http://www.unep.org/greeneconomy/Portals/88/documents/ger/ger_final_dec_2011/Green%20EconomyReport_Final_Dec2011.pdf

UNICA (2009) Projeto etanol verde, que deu origem ao protocolo agroambiental, recebe prêmio Mário Covas (Green ethanol project, which gave rise to the agroenvironmental protocol, receiving Mario Covas award). Available at http://www.unica.com.br/noticia/12472387920338419546/projeto-etanol-verde-por-cento2C-que-deu-origem-ao-protocolo-agroambiental-por-cento2C-recebe-premio-mario-covas/

UNICA. União Da Indústria Da Cana-De-Açúcar (2015) Projeto RenovAção. Available at http://www.unica.com.br/projeto renovacao/

United Nations (2012) Outcome document of the United Nations conference on sustainable development: the future we want. United Nations, Rio de Janeiro. Available at https://sustainabledevelopment.un.org/content/documents/733FutureWeWant.pdf

Vieira RC (2011) Novo Marco Regulatório do Etanol Combustível no Brazil. Available at http://www.ie.ufrj.br/images/infosucro/Rita_Capra_Vieira.pdf

Witcover J, Yeh S, Sperling D (2013) Policy options to address global land use change from biofuels. Energy Policy 56:63–74

Youngs H, Nogueira LAH, Somerville CR, Goldemberg J (2015) Perspectives on bioenergy. In: Souza GM, Victoria RL, Joly CA, Verdade LM (eds) Bioenergy & sustainability: bridging the gaps. Áttema Editorial, São Paulo. Available at http://bioenfapesp.org/scopebioenergy/images/chapters/bioenergy_sustainability_scope.pdf

Sustainable Development: Implications for Energy Policy in Nigeria

A. V. Adejumo and O. O. Adejumo

Abstract As Nigeria's population grows significantly, it is expected that progress in physical infrastructures and environmental resources are utilized at a pace that can support population growth. Since energy resources, their uses and environmental impact are critical to the process of economic development, any quest for sustainability requires appropriate energy policy responses. In this regard, this paper explores the prospects for sustainable development given energy resource utilization in Nigeria. Specifically using the General Method of Moments and the Impulse-Response functions, the study articulated the environment-economic relationship with regard to energy resource utilization. The findings of the study revealed an asymmetric flow of energy resources to the Nigerian economy given the mixed results. Thus, the paper proffers policy instruments such as low cost energy and state-market approaches which can restore some sense of inter-temporal balance and advance the course of sustainability.

Keywords Energy policy · Sustainable development · Income
Population · Nigeria

The original version of this chapter was revised: Incorrect author names have been corrected. The erratum to this chapter is available at https://doi.org/10.1007/978-3-319-73028-8_30

A. V. Adejumo
Department of Economics, Obafemi Awolowo University, Ile-Ife, Nigeria
e-mail: adejumoakinvic@gmail.com; vadejumo@oauife.edu.ng

O. O. Adejumo (✉)
Institute for Entrepreneurship and Development Studies,
Obafemi Awolowo University, Ile-Ife, Osun State, Nigeria
e-mail: jumobum@gmail.com; adejumobum@oauife.edu.ng

W. Leal Filho et al. (eds.), *Towards a Sustainable Bioeconomy: Principles, Challenges and Perspectives*, World Sustainability Series,
https://doi.org/10.1007/978-3-319-73028-8_21

1 Introduction

The notion of sustainable development was said to be first used by the *World Conservation Strategy,* in 1980 at a programme referred to as International Union for the Conservation of Nature and Natural resources. In fact, the concept of sustainability became closely associated with the Brundtland report of 1987 in an article *our common future*, which was presented at the World Commission on Environment and Development. In this report, sustainable development was explained as meeting the needs of the present generation without compromising the needs of the future generation. Harris (2003) expanded the notion of sustainable development to include three broad areas which include economic, environmental and social issues.[1] In Harris (2000), he noted that:

> true sustainability means a major shift from existing techniques and organization of production (in areas as agriculture, energy, industry, renewable resource system) to newer techniques that will practically address the real issues without jeopardizing the future, but instead, preserve it.

Furthermore, other salient notions that emerged within sustainability include economic sustainability, ecosystem preservation and social equity Advocates of social development believe that institutional and infrastructural support mechanisms are essential aspects to achieving sustainable development. While areas of economic emphasis include Human Development index per capita, Gross National Product (GNP) or Gross Domestic Product (GDP), gender equality, and poverty. Also, the neoclassicals have analyzed sustainability as the maximization of welfare overtime. These include important elements of human welfare and social equity. Human welfare captures basic health and educational needs. But, social and infrastructural development especially with regard to the utilization of natural stocks for development purposes have been spotted differently. Thus, bringing to fore socio-economic features in relation to natural resource utilization, that have bearing with issues on environmental sustainability (Harris 2003).

Environmental sustainability addresses issues that has to do with conservation, utilization and posterity of natural capital. First, to achieve sustainability, per capita stock of natural capital is to be maintained through a stable level of human population.[2] Secondly, the minimization of the consumption of non-renewables as against the consumption of renewables must be ensued. This is possible when natural capital stock do not decrease which is known as *strong sustainability*; or where the total value of physical, human and natural capital do not decrease, which is referred to as *weak sustainability* (Harris 2003). The weak sustainability which is more advocated for hinges on the tenet that increases in other capital stocks will be

[1]Harris (2003) as stated in United Nations Report: *Report of the World Commission on Environment: Our Common Future.*

[2]Basic Principles of **Sustainable Development**. **Jonathan** M. **Harris**. June **2000**. Tufts University. Medford MA 02155, USA http://ase.tufts.edu/gdae/working_papers.

replaced with decreases in whatever natural capital stock that may be experienced. In essence, as natural stocks are being used up, it is expected that they can be replaced or used for the development of other stocks (renewables, physical and human). Thus, the sum of stocks within an economy will be maintained or even increased in varying forms; thereby having implications for posterity.

In an evaluation of the Brundtland report, development is expected to be a continual and evenly distributed to everyone as geographical disparity permits. Therefore, it has been argued theoretically that if development is to be sustained, it must be seen as contributing to the quality of life of humans and the natural environment (Jhingan 2013). This explains why some studies have connected the notion of sustainable development with a development path that ensures non-declining per capita well-being over some time horizon especially with an increase in population (Pearce and Atkinson 2002). Thus, to assess development and its prospects for sustainability in Nigeria, some of the major determinants of development must be considered. A cursory overview of the per capita income (PCI) in Nigeria as shown in Fig. 1 gives an insight to Nigeria's development stance.

The PCI[3] in Nigeria has revealed unstable positive and negative trends. For instance, the PCI in the post-independence era was between 10 and 20%; but by the mid-1980s, the PCI had a negative rate between −10 and −15%. In the early 1990s, the till the mid-1990s, the PCI was just a little above 2%. Although, in the early 2000s, there was a sharp increase in PCI between year 2004 and 2005, however, there was a decline again to about 4–6% in subsequent years. From the figure of an oscillating per capita, it appears Nigeria is struggling with development, thus raisings doubts on its ability to engineer, drive and sustain development process. Therefore, the study incorporates the PCI in relation to some other components of sustainable development such as population growth and energy consumption. This is with a view to examining the extent to which growth and development must be advanced in order to assure sustainability in Nigeria.

Population growth, as noted earlier has continuous effect in determining sustainable development in an economy. In view of the rate of population growth of Nigeria, and some of the challenges emanating from increased economic activities of the populace, the population growth pattern is considered vital in advancing the course of sustainability.

A quick examination of the previous censuses in Nigeria revealed that by inference, the population of Nigeria is growing at an exponential rate. For instance in Table 1, the population stood at 55 million in 1963, 89 million in 1991 and by 2006, it stood at an approximate of 140 million persons. Thus, the population statistics reveals that for the Nigerian economy to keep pace with the requirements

[3]PCI measured overtime is seen as an indicator of economic sustainability: Economic sustainability requires that the different kinds of capital that make economic production possible must be maintained or augmented. These include manufactured capital, natural capital, human capital, and social capital. Some substitutability may be possible among these kinds of capital, but in broad terms they are complementary, so that the maintenance of all four is essential over the long term (Harris 2003).

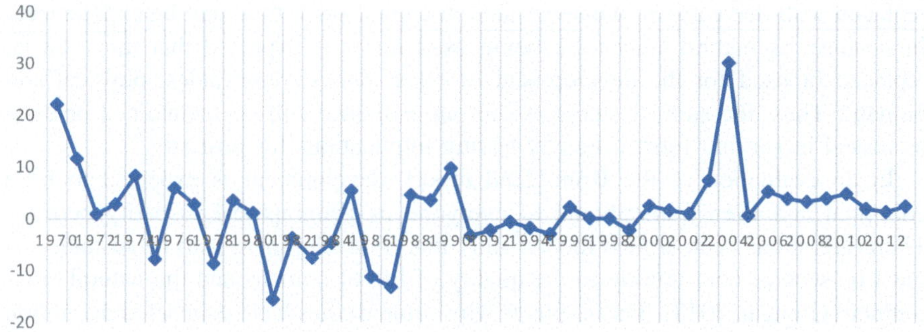

Fig. 1 Trend of per capita income in Nigeria, 1970–2013. *Source* World Development Indicators (2015)

Table 1 Population distribution of Nigeria by working and dependency ages in the 1963, 1991 and 2006 population censuses

Age-groups	1963	1963% distribution	1991	1991% distribution	2006	2006% distribution
0–14	23,925,586	42.98	40,088,028	45.04	58,736,297	41.83
15–64	29,593,340	53.16	45,996,457	51.69	77,158,742	54.94
65–85	2,151,129	3.86	2,907,735	3.27	4,536,751	3.23
Total	55,670,055	100	88,992,220	100	140,431,790	100

Source Nigeria Bureau of Statistics (NBS 2007) (formerly Federal office of Statistics) percentages computed by the author National Population Commission (1998, 2006)

for development industrially, materially, economically, and sustainably, there must be adequate social and infrastructural support mechanisms (energy, good governance and policies) which will serve as engine of growth. Amongst, the various in infrastructures, this study concentrates on the utilization of energy being a major input for development.

A relation as to how rapid population growth could be a major reason for environmental degradation is buttressed by a network in Fig. 2. As earlier noted, rapid population growth leads to rapid use of resources which creates pressure on the utilization of the resources of an economy. Theoretically, the neoclassicals believe that endowments of spatial resources, agro climates, animals and ecosystems are meant for human survival; nonetheless, the ecologist believe that despite the quest for economic growth, there is the need to check the overexploitation of resources for posterity sake (Hussen 2000). For instance, exploitation of forest and mineral wealth used for setting up industries, building roads and dams for electricity generation engender environmental degradation. These exploitative tendencies lead to increases in air and water pollution, loss of biodiversity, and soil degradation. Thus, explosive population growth could deplete resources and as such has the tendencies of threatening sustainable development (Jhingan 2012). However, if there is going to be a positive chain of sustainability, then policy instruments which may restore some sense of inter-temporal balance must be articulated.

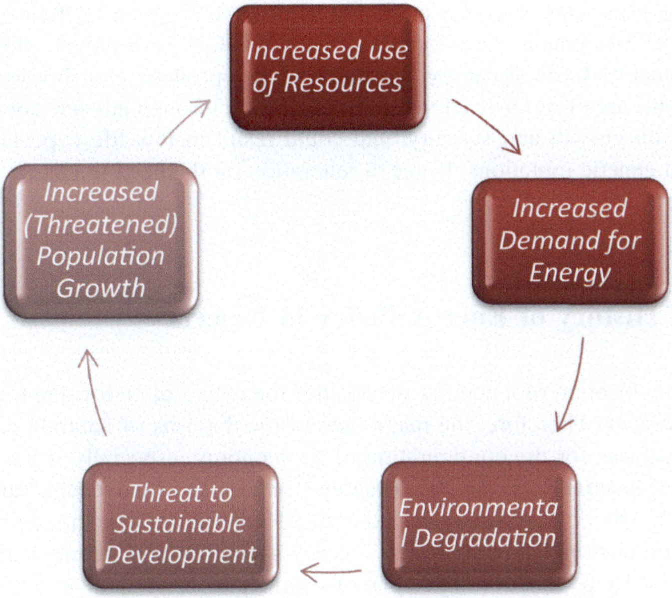

Fig. 2 Transmission mechanism between population growth, resource utilization and sustainable development

A classic example is the heavy reliance of the Nigerian economy on the export of primary energy products such as crude oil and the quest for economic growth and development; as well as the need to meet the national and international obligations has encouraged overutilization of natural resources (Oseni 2011). This is evident by the extraction of crude products which remains common sites in the Niger-Delta regions of Nigeria has led to the degradation of the soil. This degradation will in turn hamper other economic activities like agriculture; thus, threatening sustainable development. These threats to sustainability can in turn challenge social equity in a population with regard to deprivation resulting from overexploitation (Jhingan and Sharma 2008).

Furthermore, a review of the challenge of pollution through energy utilization is prominent in Nigeria. Gradually, industrial pollution is becoming acute in metropolitan areas like Warri, Kaduna, Port-Harcourt, Lagos of the Nigerian-state, where petroleum refineries, chemicals, Iron, and steel, non-metallic products, pulp and paper and textile industries, are concentrated. Even small scale industries like brick making factories, plastic and plastic products, chemical manufacturing, etc. are significant air pollutants. Some other causes of pollution range from energy sources like thermal power generation plant to shanty areas, slums and poorly ventilated regions where stoves, wood, or coal for cooking further increase air pollution. Thus, while energy is being produced or consumed, issues on sustainability emerge, which may be due to combustion, emissions, effluents, discharges, run-offs given the peculiarity of different activities. These issues identified explains why energy utilization vis-à-vis environmental impacts is one of the cardinal features of the Sustainable Development Goals (SDGs).[4] Moreover, the negative effects emanating through energy consumption also produce greenhouse gases and other carbons, are even far-reaching to human health through adverse consequences on population growth and stability, and could result in low life expectancy, birth defects and genetic mutations; hence, a reiteration on the need to chart a course on sustainable energy policies.[5]

2 Brief History of Energy Policy in Nigeria

Globally, the location of a country determines the extent of endowment in terms of natural resources. Therefore, the magnitude of the deposits of natural endowments have implications for the configuration of an economy, especially if the resources are annexed profitably. From the foregoing, it is a fact that energy and energy supplies (e.g. Oil, coal, natural gas, and hydroelectric power) are the foundations of the modern industrial economies. The 'energy shock' of 1970 dramatically proved the point. In 1970, the developed nations which comprised of one-third of the world's population was consuming annually almost 85% of the world energy production (UNEP 2012). But the non-oil producing third world countries also rely heavily on oil to fuel their growing Industrial and agricultural economies.

Since the early 1970s, Nigeria, though an oil-producing economy and a member of the (OPEC) equally relied heavily on the earnings from oil sale to embark on massive industrial development, provision of social utilities and general restructuring of the economy, Apart from earnings from crude oil sales, the Nigerian economy also required a considerable amount of energy for the growing economy. This was basically to help power machines and equipment meant for industrial activities, modern agricultural practices, as well as the provision of basic amenities, such as electricity for domestic and institutional purposes. Given the increases in domestic oil consumption and its attendant effect on the environment, an economically optimal strategy is to develop an alternative to oil with the obvious fuel available which is gas and gas derivatives. This will improve the quantum of oil for export, as well as promote the conservation of oil reserves. However, beyond the substitution of oil for gas (because gas fuels have been adduced to be more environmentally friendly than oil fuels) are also more environmentally friendly fuels

[4]The Sustainable Development Goals (SDGs), otherwise known as the Global Goals, became effective in January, 2016 and is deemed to end in 2030. It was enacted by the United Nations, which involves a universal call to action to end poverty, protect the planet and ensure that all people enjoy peace and prosperity.

[5]The 7th and 11th goals of the SDGs is majorly targeted at achieving clean and affordable energy; and sustainable cities and communities. http://www.un.org/sustainabledevelopment/news/communications-material/.

which are bio-fuel which are generated from biological products such as plants, animal and human residue.

In order to further explain the foregoing, a brief outlook of the Nigeria energy sector is reviewed. In Nigeria, major natural resources include natural gas, tin, limestone, niobium, lead, zinc and arable land; while conventional energy sources include crude oil, iron ore, coal, hydro, and gas.[6] Table 2 gives an overview of the major sources of energy used in Nigeria.

Coal production in Nigeria started in the early 20th century. Before, the discovery of oil, coal was the first fossil fuel to be discovered in Nigeria. Not until 47 years after the discovery of coal, was oil discovered in Nigeria. Table 3 presents an overview of coal production in Nigeria.[7] Coal production in Nigeria started in 1916 with an output of 24,500 t for that year. Production rose to a peak of 905,000 t between 1958 and 1959. Prior to independence in 1960, coal production was a major source of energy in Nigeria. Over 95% of coal production in Nigeria was consumed locally. This consumption was majorly for railway transportation, electricity production, industrial heating and production, as well as domestic consumption. Specifically, between 1952 and 1958, coal consumption by the Nigerian Railway Corporation accounted for about 60% of the overall consumption.

During the civil war, between 1967 and 1969, coal production stood between 20,400 t and 35,000 t. However, in 1972, coal production rose to 323,000 and by 1980, a progressive decline occurred to about 118,000 t as shown by Table 3. Statistics further revealed that in 1980, coal contributed less than 1% to commercial energy consumption between 1952 and 1966. This was due to the diesel conversion programme which started in 1966, which caused the share of coal consumption to fall by 30% in 1966. As a result, by 1986, the utilization of coal to drive productive activities within the Nigerian economy became insignificant. The cumulative effect of this decline was more reflective given the decline in the consumption of its two major consumers-railway and electricity. Besides, cost-effectiveness associated with the use of alternative energy sources—refined oil and hydro power—resulted in the edging out of coal in the national energy programme.

Presently in Nigeria, mining currently constitutes a central economic activity. The sector has been characterized by exploration, extraction and subsequent development of crude-oil, solid minerals and associated gas by trans-national companies (TNCs) in connection with the Nigerian Government. Crude oil which

---

[6]Oyedepo (2012) noted the Nigeria is richly endowed with a lot of the conventional energy resources which include petroleum, gas, coal and hydro. These resources are distributed in the various parts of the country. Coal deposits are found in large commercial quantities in places like Enugu and Okaba mines in Enugu and Kogi states respectively. Petroleum and its associated gas are found in commercial quantities in the southern part of the country, in places like Rivers, Bayelsa, Imo, Abia, and Delta states. Rivers suitable for hydropower are located in places like Kainji, Shiroro, Jebba, Makurdi, Mambilla, Lokoja.

[7]Ogunsola O. I. History of Energy Sources and their Utilization in Nigeria. Energy Sources 1990;12:181–98. The estimated cumulative production between 1916 and 1980 is about 25.3 million metric tonnes.

Table 2 Major sources of energy consumption in Nigeria

Indicators	Resources				
	Crude oil	Natural gas	Gas and lignite	Tar sands	Large hydropower
Reserves natural units	36 billion barrels	187 trillion SCF	2734 billion tonnes	31 billion barrels of equivalent	11,250 MW
Reserves energy units (Btoe)	4.76	4.32	1.92	4.22	1.11
Production	2.5 million barrels per day	7.1 billion SCF/day	Insignificant	–	1938 MW (167.4 million MWh/day)
Domestic utilization	445,000 barrels/day	3.4 billion SCF/day	Insignificant	–	167.4 million MWh/day

Source National Bureau of Statistics (2007)

Table 3 A summary profile of coal production in Nigeria

Period	Production (measured in tonnes)
1916	24,500
1958–1959	905,000
1967–1969	20,400–35,000
1972	323,000
1980	118,000
1983	52,700

Source Ogunsola (1990)

was discovered in 1956 in the Niger Delta region of Nigeria has since been rated to be the 4th largest in the world with almost 2 billion land-sized deposit. Also, Nigeria has one of the ten largest natural gas reserves in the world and roughly 50% of the deposits are discovered in association with oil. In addition, it possesses the largest deposits of natural gas in Africa, most of which are located in and around the Niger Delta region (OPEC 2017). Also, the oil discovered in the deep water terrain has projected that proven reserves will reach about 40 billion barrels by year 2020 and potentially 68 billion barrels by year 2030.

Initially, the mining of crude oil contributed so little to the National income of Nigeria. But since the oil boom in the mid-1970s, crude oil and gas have been important income sources as well as important energy sources (Ayodele and Falokun 2003). Given the commercial nature of this resource, the Nigerian oil and gas sector accounts for about 35% of its national income; while the exports of petroleum accounts for over 90% of export earnings. Furthermore, as against non-oil revenue, oil revenue consist of more than an average of 75% of the total income since 1985 till date (Central Bank of Nigeria (CBN) 2015). This statistics is

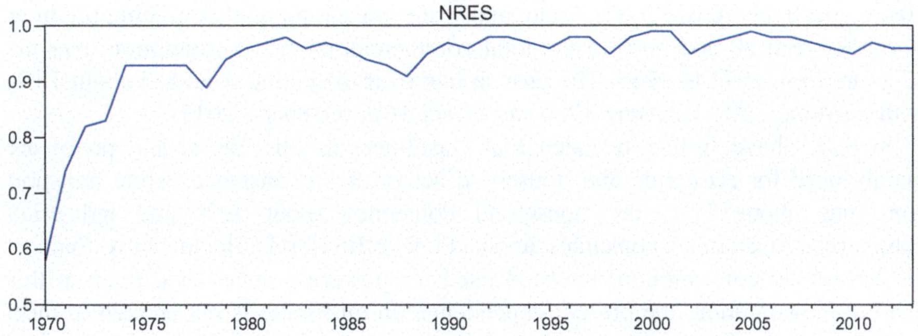

Fig. 3 Consumption intensity of natural resources (based on crude oil exports)

reflected in the intensity with which oil resources are utilized.[8] Using the share of Nigerian income from oil export in relation to total exports earned. Figure 3 shows that the intensity with which natural resources are being consumed have grown steadily; this is because the ratio of natural resource consumption/sales has gradually risen from a rate of 0.6 (60%) in 1970 to about 0.9 (90%) in 2010. This pattern of consumption may indicate some pattern of unsustainability in the depletion of unrenewable stock of natural capital/resources in Nigeria.

Furthermore, the CBN statistics revealed that the ratio of domestic production to exported products of crude oil is about 1:9. This ratio difference is an indication that most primary resources including crude oil are exported outside Nigeria. Also, it shows that local refineries produce at less than 15 million tonnes/year, indicating that most supply of petroleum and petroleum products have been augmented by imports of refined crude products. Nigeria's production capacity of crude oil before 2010 was about 2.4 million barrels per day. This level of production has been attributed partly to the problems in the Niger Delta and OPEC production restriction. However, average daily production has been projected to exceed to 4.0 million barrels per day by 2010 and potentially to over 5.0 million per day in year 2030.

The link with the Organization of Petroleum Exporting Countries (OPEC) in terms of oil prices caused a boom as regards mining activities in Nigeria—from about 0.5 million pb to $40 million pb in the 1980s, while in recent times, the sale of oil stood at an average of $80 million pb. In 1974, there was an announcement of a massive and an unprecedented 400% increase in the price of fuel by OPEC members; this in turn resulted in an enormous increase in the total export revenues of OPEC members from $14.5billion in 1972 to $110 in 1974. For Nigeria in particular, export in the mid-1980s stood at almost 500,000 barrels per day to about 600,000 barrels in the mid-1990s and further to 700,000 barrels in the early 2000; out of the total barrels produced, the magnitude of exports are over 88.9, 85, and

[8]Natural resource endowment (NRES) measures the intensity with which natural resources are consumed or used up. It is measured as the ratio of oil exports to total exports of Nigeria. Statistics are computed from Central Bank of Nigeria (CBN) Statistical Bulletin (2015).

80% respectively (CBN 2005). Incidentally, the consumption of petroleum products stood between 80 and 90% of the total commercial energy consumption over the 13 years from 1971 to 1984. The growth rate over the period averaged about 18%, with gasoline 22%, kerosene 17% and diesel 16% (Oyedepo 2012).

Statistics have further revealed that Gasoline, gas oils, diesel and petrol are mainly used for economic and household activities. For instance, while transport consumes about 77%, the household consumes about 12% and industrial/commercial operations consumes about 11% (NBS 2017). Incidentally, half of the household consumption has been used for powering generators. Even at this level of consumption, the present dependence on oil wealth is not enough to meet the energy needs of the country; hence, the interest in renewable energy development.

In order to achieve efficiency and sustainability of the petroleum sector, the Federal government of Nigeria deregulated the downstream oil sector. This is with a view to ensuring that the forces of demand and supply dictate the pace within the oil industry, thereby, leading to efficiency (Federal Ministry of Finance 2014). However, the challenges emanating from recent deregulation exercises which has led to increases in oil prices, the unavailability of electricity to majority of the population as well as high cost and energy losses associated with grid extension have increased the need for renewable energy. Also, government effort to foster a public-private partnership (PPP) has been hampered due to the technical and financial barriers; thus, reinforcing the need for alternative strategies.

Oyedepo (2012), noted that given the current reserves and rate of exploitation, the expected life-span of Nigerian crude oil is about 44 years, based on about 2 mb/day production, while that for natural gas is about 88 years, based on the 2005 production rate of 5.84 b scf/day. Given the prospects for the gas sector, investments in the gas sector was embarked upon by the Nigerian government to utilize the natural and associated gas deposits as a substitutes for oil both for domestic needs and foreign exchange earnings. As a result, in 1988, the Nigerian Gas Company (NGC) was created as a subsidiary and strategic business unit of the Nigerian National Petroleum Corporation (NNPC). NGC is responsible for the development of policies governing transmission, distribution, marketing, utilization, as well as pricing of Nigeria's natural gas and its derivatives to a domestic market and neighboring countries along the West African Coast.[9] NGC receives its natural gas supplies from the major International Oil Companies (IOCs) producers operating in Nigeria such as Shell, Chevron, ENI/Agip and Exxon Mobil.[10]

The natural gas reserves in Nigeria are estimated at 4.67×10^{12} m^3 at a mean specific volume of 1.56×10^3 m^3/kg. A mean gauge pressure of about 12 bar and a calorific value of 35 MJ/m^3. While the current production of associated gas stands at about 1.8×10^{10} $m^3/year$ (Energy Commission of Nigeria 2003). About 8.5 m^3

[9]National Planning Commission (2009).

[10]Tallapragada (2009).

of associated gas can be expected per barrel of crude petroleum. However, it has been observed that associated gas production and its utilization in Nigeria, is usually flared. Incidentally, the government policy permitted the flaring of gas; while its associated cost to environmental deterioration were rather monetized. For instance, in 1985, more than 80% was flared in 1985.

The NNPC statistics recorded that as at 2006 Nigeria was flaring 2.5 billion standard cubic feet (scf) of gas, while consuming only 300 mscf of gas per day. However, given the efforts by the government to monetize natural gas and the enactment of some environmental policies and agitations for the use of these gases, flaring has declined (CBN 2002). The Nigerian government in 2008 approved the Gas Master Plan together with Gas Infrastructure Blueprint and the Domestic Gas Supply Obligation. Specifically, the Gas Master Plan was geared towards addressing critical issues of energy security, gas availability, infrastructure, and commercialization framework and gas affordability. Efforts geared towards reduction in gas flaring since the late 1990s is gradually being actualized.

The Nigerian National Petroleum Corporation (NNPC) in conjunction with all oil companies in Nigeria has succeeded in reducing gas flaring in the Country by 26 percentage points (that is from 36 to 10%) thereby pushing Nigeria down from the second highest gas flaring nation in 2006 to the seventh position in 2016. The gas flare reduction was attributed to technological improvement and aggressive gas commercialization anchored on the Gas Master Plan.

Beyond the Gas Master Plan, in 2007, the Nigerian government launched a policy to enhance energy utilization in Nigeria—the Nigerian Biofuel policy and incentives. The NNPC was also mandated to enact a system that will cater for a domestic fuel ethanol industry. The aim of the programme was geared at reducing Nigeria's dependence on refined fossils, reduce pollution in the environment as well as create an industry that can bring about new economic opportunities. The Bio-fuel programme is a synergy between the agricultural sector and the downstream petroleum sector of the Nigerian economy (NBPI 2007). Since the enactment of this policies, the NNPC alongside with some supporting industries have been embarking on the production of Biofuel. However, this fuels have not been sufficiently commercialised within the Nigerian economy.

Energy history will not be complete in Nigeria, without an overview of the power sector which has witnessed series of transformation in recent times. Hydropower is a major source and form of energy in Nigeria. Nigeria built its first power plant in 1896 with a 20 MW power station at Ijora, near Lagos. This introduction of electricity came barely fifteen years after the introduction of Electricity in the UK (Adenikinju 2017). In 1951, the Electricity Corporation of Nigeria (ECN) was instituted to oversee the Electricity Sector. Later in 1960, the Niger Dam Authority (NDA) was set up to build and manage dams in Nigeria, with total installed generation capacity a little above 50 MW.

In 1972, Federal Government of Nigeria (FGN) approved the merger of NDA and ECN to form the National Electric Power Authority (NEPA), as a vertically integrated (monopoly) power utility, responsible for generation, transmission, distribution and trading of electricity in Nigeria (Amadi 2012).[11] The power sector in Nigeria has witnessed several reforms. For instance, in the year 2001, the National Electric Power Policy (NEPP) was developed, which set the pace for the Electricity Reform. As a follow up to this, an Electric Power Sector Reform Bill (EPSRB) was passed by the National assembly in March 2005. With the bill's enactment into law, an independent Electricity Regulatory Commission (NERC) was created, which commenced full operation in October, 2005. With the headquarters in Abuja, the commission was expected to engineer various regulations in its efforts to give direction to the industry operators and prepare the ground rules for public/private sector participation.

Meanwhile, in 2004, during the process of reforming the power sector, the National Electric Power Authority (NEPA) in January, 2004 commenced its decentralization process by creating 11 semi-autonomous business units from its former distribution sector. In April 2004, the transmission sector was decentralized into a semi-autonomous business unit charged mainly with transmission of electricity. Similarly, generation sector was subsequently decentralized into 6-semi-autonomous business units in November, 2004. The decentralization exercise was with a view to ensuring the eventual corporatization of the Business units on the enactment of the EPSRB. These acts were in anticipation of the expected to achieve the government's ultimate goal of divesting its interest in the Generation companies and Distribution Companies while the Transmission Companies was to remain state owned.

The reforms hinged on development of a wholesale electricity market through segmenting and licensing of each of Power Holding Company of Nigeria (PHCN) Successor Company (which are now CAC registered entities). The reforms also made for the establishment of a power Consumer Assistance Fund, and establishment of the Rural Electrification Agency (REA) to create and manage the Rural Electrification Fund. These efforts are targeted at opening up of the Power Sector for Private Sector Participation. At present, nearly 40 private entities have been licensed by the commission (to generate more than 12,000 MW) and many are already operational. At present, the nation has an installed generating capacity of 9000 MW (FGN and IPPs). The available useful capacity is about 4500 MW (as at October, 2011). The current national demand is estimated about 16,000 MW (ECN).

The government administration between 2009 and 2012 worked assiduously to increase power supply to 6000 MW by December 2012 and 14,000 by year 2013. This reaffirms the need for the completion of the Reform Program to enable private sector participation as government alone cannot fund the power sector (Amadi 2012). A cursory examination of Fig. 4 reveals a wide disparity is seen among electricity installation, generation, and distribution in Nigeria. For instance, in 1980,

[11]The merger of the Electricity Corporation of Nigeria (ECN) and the Nigerian Dam Authority (NDA) to form the National Electric Power Authority (NEPA) was found in the 1972 Decree No. 24.

31.2% of installed electricity capacity is generated, while 21.4% is being distributed. Incidentally, as the generating capacity increases via installation capacity, the distribution capacity remains low. For the year 1990, 2000 and 2010, while generation capacity stood at 33.7, 31.1; and 44% respectively, the distribution/consumption capacity stood at 19.5; 18.23 and 22.8% respectively.

As at 2015, Nigeria had 17 power generating stations with an installed capacity of 12.5 GW, but with an average daily generation capacity 3.9 GW (FGN 2017). This implies that out of the installed capacity, a larger percentage is not being utilized (about 68%); while about 30% is operational, barely 15% is distributed to end users. Thus, given the relationship between daily generation capacity and distributional operations, it can be inferred that electricity availability and consumption in Nigeria is facing internal management and delivery challenges, let alone sustainability challenges.

Several indicators of electricity consumption has been reflected by previous studies. For instance, the ability of the distribution of power system to perform its function given laid down principles for a period of time without failure is called Distribution System Reliability (DSR) (Warren 1996). Specifically, System Average Interruption Duration Index (SAIDI) is one of the DSR indices used in the assessment of distribution performance.[12] This reliability index is extremely high in Nigeria.

Ogujor and Orobor (2010), based on some statistics such as the Manufacturers' Association of Nigeria (MAN), have reported that Nigeria SAIDI stood at about 60,000 min. Similarly, Ogujor and Orobor (2010) in an independent study reported a SAIDI index of about 87,639 min as against other countries like France, USA and Singapore with 52, 88 and 15 min respectively. Besides the challenge of just 40% of the 160 million Nigerians that have access to electricity The high SAIDI also revealed that poor electricity has caused poor per capita electricity consumption of 125 kWh in Nigeria as against those of South Africa, Brazil and China estimated as 4500, 1934 and 1379 kWh respectively (Oyedepo 2012).

Despite the increase in new power generation capacity from the Power Holding Company of Nigeria (PHCN) and Independent Power Producers (IPPs), the effect of this decentralized power distribution system has been marginal. This situation is believed to be as a result of lack of adequate policies to drive private investment, particularly policies on renewable energy. In addition, there has been some unfinished policies that require inclusion and harmonization with regard to power generation and distribution implemented. For instance, the involvement of private sector in rural electrification have limited investment in new generation, transmission and distribution capacities. Besides, the partial use of renewable energies for both on-grid and off-grid electrification has resulted in poor access to electricity in rural areas. Other salient factors responsible for the interruption of electric power

[12]SAIDI—It represents in minutes, the annual average total duration of electric power interruption to a customer. Warren C. A.Distribution reliability—what is it? IEEE ApplMag 1996;2(4):32–7

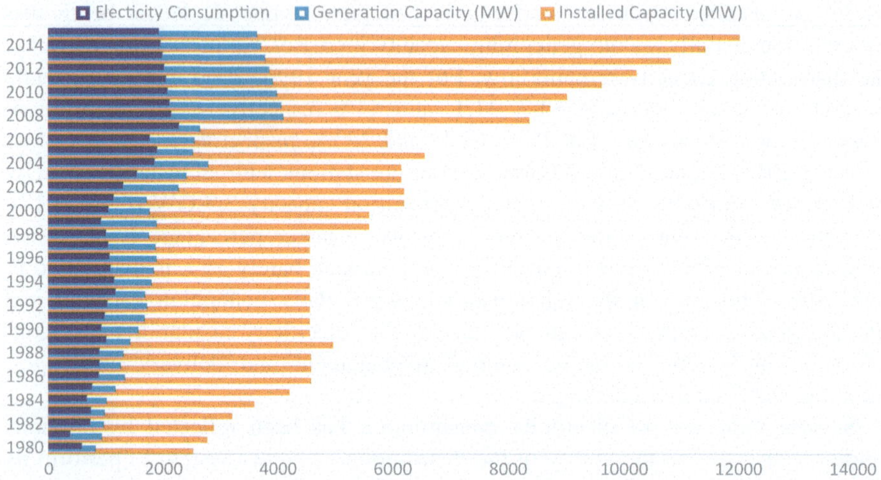

Fig. 4 Interactions among electricity installation, generation and consumption in Nigeria. *Source* NBS (2007), FGN (2017)

in Nigeria is load shedding as a result of inadequate generation, partial deregulation, corruption, obsolete infrastructures amongst others.

It has been clearly articulated that Nigeria is blessed with myriads of natural resources and specifically energy resources. However, these resources have not been properly managed to cater for the energy demands in Nigeria. This is reflected by the dependence of the Nigerian economy in the importation refined petroleum products. Furthermore, the paradox of economic dependence on exports of primary products, excessive fixation of oil prices and incessant power failure brings to fore issues on economic and environmental sustainability. Thus, if the drive towards encouraging investment in renewable energy is to be actualized in Nigeria, the government may have to adopt an approach that will aid and commercialize renewable energy development and production in Nigeria.

3 Literature Review

The assessment of energy on development has been on going in Nigeria. Specifically, the assessment of energy on development has been revealed through its effects on economic growth. For instance, Olomola (2006) discovered that energy shocks does not have any significant impact on the Nigerian economy. Similarly, Olomola and Adejumo (2006) assessed the effects of oil price shock on output, inflation, real exchange rate and money supply through regular channels through which price shocks transmit their impacts on the Nigerian economy. It was concluded that oil price shocks do not affect output. In consonance with these findings, some other studies that supported this null effects of oil price shocks (Iwayemi and Fowowe 2011; Babatunde and Adenikinju 2016).

Chuku et al. (2011) used the SVAR model to assess oil price shocks on current account balances in Nigeria, within the period 1970 and 2008. It was observed that more than 15% of variations in current account balances were influenced by oil price shocks. However, Kilian and Park (2009) using the threshold VAR examined the effects of oil shocks on macroeconomic indicators. The findings appeared to be consistent with the findings of Olomola and Adejumo (2006).

Corroborating the relationship between energy and economic growth in Nigeria, some other studies have also revealed that causal relations exist between energy and economic growth. For instance, while Aliyu (2009) found causal relationships between the prices of oil and real economic growth; Akinlo (2009) found a uni-directional relationship between electricity and economic growth, implying that the presence of electricity will enhance economic growth. However, a conflicting stance was reflected by the findings of Olubusoye et al. (2016) from a macro-economic model which assessed the impact of movement in global oil prices on the Nigerian economy. It was discovered that the Nigerian economy is oil dependent and as a result, it is extremely vulnerable to oil price shocks. As a midpoint between the debates on oil price shocks and economic growth, Olanrele (2017) noted that in the pre-1984 period, the Nigerian economy was moderately affected by crude oil price shocks; however, the post-1984 period showed more negligible effects of oil price shocks. Thus, he concluded that the longer the duration of the negative oil price shocks, the more it is likely to affect macroeconomic aggregates such as per capita income, GDP, investment and money supply.

Furthermore, Adeniyi et al. (2016), identified some other notable channels through which oil shocks could impact economic growth include real balance effects, international income transfers, monetary policy responses, potential output channels, Dutch disease channels sectoral shifts hypothesis and uncertainty and irreversible investments.

Beyond these, Olanrele and Adenikinju (2017) investigating the time varying effects of energy consumption (electricity) on economic growth, could not draw out any important relationship between the two variables. A plausible explanation they gave for the disconnect between electricity and economic growth is the heavy dependence on self-generation of electricity by individual economic agents; although, this is more expensive, however, these individual effects could have a multiplier effect on economic growth. Empirically, some other issues that could be adduced for the disconnect between the variables could also include non-linearity, asymmetric relationships, sample sizes and variable choices.

Adenikinju (2017) attempted to provide a relationship between energy and economic relationship. They are:

- A high correlation exist between energy consumption and levels of economic development.
- Energy ownership correlates more weakly with economic development than with energy consumption.

- The differential impact of energy production and consumption is more significant with oil production and consumption, than electricity production and consumption.
- A positive correlation exist between energy resource ownership and energy security.
- Energy security is weakly correlated with economic development.

Thus, having examined some existing studies, this study will lend credence to or refute some of these claims. By using a different approach, the Vector Autoregressive (VAR) approach is used to assess the performance the dynamic performance of energy on development in Nigeria. Besides, instead of using economic growth (GDP) directly, an index of sustainable development will be employed which is per capita income (PCI).

4 Theoretical Framework—Ecological Perspective on the Natural Environment and Human Economy

According to Hussen (2000), the ecological perspective of the relationship between the natural environment and the human economy is biocentric; this implies that the human beings cannot be viewed in isolation from the natural ecosystems. This is against the concept of the Classical and Neoclassical dominant approach to economic analysis since about the 1870s—which views of environmental economic relationship as anthropocentric. This means that the humans are treated as pre-eminent in the natural environment. Consequently, the human economy is rated above natural environment and humans are regarded as the universe's most important entity. The natural environment therefore is seen to exist to serve the human economy and environmental resources have no intrinsic value.

However, the basic objective of the ecological perspective is to establish a clear understanding of the basic principles governing the nature, structure and functions of the biosphere (and by extension, environmental resources) and the functional relationship between the biosphere and the human economy. From a purely ecological perspective, some basic tenets are as follow:

- Environmental resources are finite or scarce in absolute terms.
- In the biosphere, everything is interrelated. Hence, mutual interdependencies exist among all elements in nature.
- Physically and functionally, the biosphere is characterized by a continuous transformation of matter and energy which are governed by immutable natural laws.
- Material recycling is required for growth and revitalization of all the subsystems of the biosphere, including the human economy subsystem.

- The human economy is a subsystem of the biosphere therefore natural resources are not factors of production just lying outside the confines of the larger system.
- The natural tendency of human technology is overutilization of the natural system, eventually leading towards less stable, less resilient and less diverse ecological communities.

From Fig. 5, the clear demarcated circle symbolizes the earth and its finiteness within the biosphere. By locating it inside the circle, the human economy is perceived as a subsystem of biosphere. Also, the box inside the circle indicates that growth of the economic subsystem is 'bounded' by a non-growing and finite ecological sphere (and resources).

Figure 5 suggests that the human economy is dependent on the biosphere for its continuous withdrawal (extraction and harvest) of material inputs and as a repository for its waste (outputs) degraded matter and energy that are the eventual by products of the economic process. Furthermore, the biosphere (and hence the human economy) requires a continuous flow of external energy which comes from the sun. In addition, both the human economy and the biosphere are regarded as an "open system" with regard to energy (i.e. both systems require an external source of energy), the biosphere taken in its entirety is regarded as a "closed system" with respect to matter. Therefore, an ecological view as represented in Fig. 5 above appeared to incorporate the principle that the human economy is completely and unambiguously dependent on natural ecological systems for its material needs.

Fig. 5 Ecological perspective of the environment and the human economy

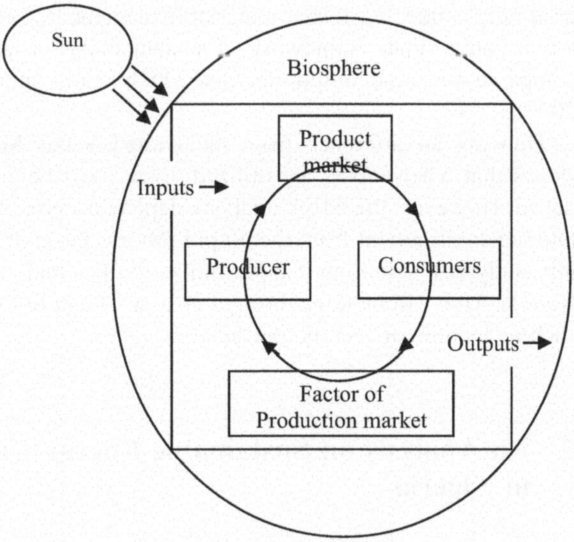

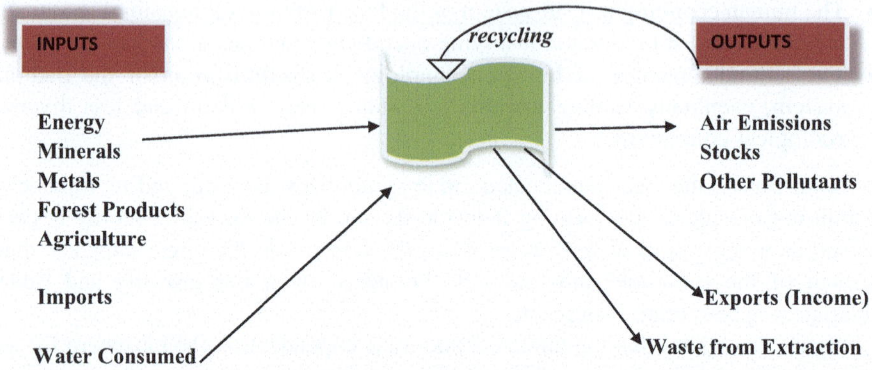

Fig. 6 Material flow analysis. *Source* Adapted from MIT OCW (Jeremy Gregory lecture series on Material Flow Analysis at the Massachusetts Institute of Technology, Department of Material Science and Engineering)

However, the perspective of biocentrism is that it does not explicitly recognize the main output of the economic system—non-material aspects of utility (enjoyment). It describes nature and the interactions that occur in nature between living and non-living matter in purely physical (energy and matter) terms. This is unlike the Material Flow Analysis (MFA), which considers both the input and output components in the environment. MFA is a systematic assessment of the flows and stocks of materials within a system defined in space and time. Specifically, MFA is used in resource, waste and material management. The MFA has been identified as a more appropriate framework of sustainability; especially when input factors are compared to some output factors which could indicate sustainability (Gregory 2009).

Drawing an inference from the United States Material flow of 1990, Fig. 6 shows that when inputs are utilized, there must be a reflection on the output produced. However, the MFA analysis depicts a necessary mechanism for feedbacks into the environment from the output process through recycling. But the concern of this study is to examine if input utilization has had any positive spillover effect in Nigeria. This will be done through an assessment of the effects of inputs on outputs such as income or income per head.

5 An Analysis on Sustainable Energy Consumption in Nigeria

In order to carry out an analysis on energy utilization in Nigeria vis-à-vis sustainability issues, the human economy is assessed. Quantitative method of analysis is employed using time series data between the historical periods of 1970–2015.

This period was specifically selected for assessment because it covers the oil boom years in Nigeria. Specifically, econometric techniques were employed for analysis (see Sect. 5.4).

Based on the classification of the Energy Commission of Nigeria (2003), energy consumption following usage pattern include sectors such as the industrial, transport, commercial, agriculture and household sectors. The household sector was noted to consume the largest share of energy in Nigeria which is estimated at about 65%. This is largely due to the low level of development in all the other sectors. However, the forms of energy consumed in Nigeria have increasingly diversified with innovations in science and technology. Compared to the primitive form of energy biomass in the form of wood fuel and solar energy, recent primary non-renewable energy forms which are have dominated the energy scene for several years; thus, this analysis will make use of the statistics available on these non-renewables. Also, per capita income is used as an index to assess sustainability within the Nigerian human economy vis-a-vis energy consumption.

5.1 Relationship Between Per Capita Income and Energy Consumption

A standard tool for macroeconomic analysis is the neoclassical production function which relates total output or per capita income as a function of some other determinants (inputs). To analyse the effects of the pattern of energy consumption on population or on per capita income, interaction among some indicators of energy consumption is modelled into a neoclassical growth model; this will lead to observing the dynamic interactions amongst these variables (Drabo 2011).[13]

A modified version of the inputs which included rate of investment in physical capital, and the rate of growth of labour force, environment, education and technology are replaced by some other determinants of per capita income (*PCI*). These determinants include variables that determine energy consumption (eng_t). Thus, based on the neoclassical augmented growth model, the effect of environment on economic growth can be specified as:

$$PCI_t = C + \alpha_1 PCI_{t-1} + \alpha_2 eng_t + \alpha_k X_{kt} + v_t \quad (1)$$

[13]The model that showed the relationship between environmental quality and economic growth adopted a variant for per capita income specified in equation: $\ln\left(\frac{Y_t^*}{L_t}\right) = lnA_0 + g_t - \frac{\alpha}{1-\alpha}\ln(\widehat{n}_t + \delta) + \frac{\alpha}{1-\alpha}\ln(s_k) + \theta_1 ln(Q_t) + \theta_2 \ln(E_t)$. $\left(\frac{Y_t^*}{L_t}\right)$ is expressed as per capita income (PCI_t); The model include explanatory variables inputs which are rate of investment in physical capital (s_k) and the rate of growth of labour force ($\widehat{n}_t + \delta$), environment (Q_t), education (E_t) and technology (A_0) are replaced by some other determinants of per capita income. See: Bassanini and Scarpetta (2001).

where PCI_t and eng_t represent the logarithmic form of PCI per capita and the energy consumption in Nigeria in period t; since it is a country-specific study. X is the matrix of the control variables introduced in the model and which have been used frequently in the empirical literature; while v_t is the error term. The coefficient α_1, measures the past effects of PCI on the present value, is expected to be superior to zero but less than 1 (i.e. $0 < \alpha_1 < 1$). While the relationship energy-PCI relations is determined by the coefficient of the α_2; which is expected to lie between 0 and 1 (i.e, $0 < \alpha_2 < 1$) (Bassanini and Scarpetta 2001). Also, a positive parameter (α_2) shows a complementary relationship; while a negative parameter (α_2) shows a conflicting relationship.

The hypothetical relationship between the energy consumption and PCI posits that an increase in energy consumption should improve PCI. Thus, the hypothesis is stated such that "*an increase in energy consumption should promote per capital income growth.*" The specific variables included to capture energy consumption are electricity consumption (*EEC*) and other growth determinants like natural resource endowment (*NRES*), carbon emissions (CO_2) and investment in physical capital (GCF). Investment in physical capital is introduced as a control variable; besides, it is a major determinant of per capita income. Thus, Eq. (1) is restated as:

$$PCI_t = C + \alpha_1 PCI_{t-1} + \alpha_{21} EEC_t + \alpha_{22} NRES_t + \alpha_{23} \mathrm{CO}_{2t} + \alpha_{1k} GCF_t + v_t \qquad (2)$$

From Eq. (2), while NRES, EEC, and GCF is expected to have a positive influence on PCI, CO_2, alone is expected to have a negative influence on PCI.[14]

5.2 *Dynamic Analysis Among Per Capita Income and Energy Consumption Variables*

In order to carry out a dynamic analysis, the Vector Autoregressive (VAR) model or Vector Error Correction (VEC) model[15] permits variables to interact in the regression with past values of one another without any theoretical structure on the

[14]All variables specified are in their log form or percentage; therefore, the coefficients are explained in terms of elasticities.

[15]The VAR-VEC models make all variables to interact in the regression with past values of one another without any theoretical structure on the estimators. The VAR models are dynamic and as such they can explain the dynamic structure of time series better than the static OLS estimation method. The VAR models make all variables to interact in the regression with past values of one another without any theoretical structure on the estimators. The VAR models are dynamic and as such they can explain the dynamic structure of time series better than the static OLS estimation method (Gujarati and Porter 2009).

The major distinction between VAR and VEC models is the incorporation of ECM term that attempts to give explanation as to the speed of convergence or divergence to a long-run equilibrium from a short-run equilibrium, since non-stationarity is experienced by the time series data to be used (that is, not converging to their mean).

estimators. The VAR models are dynamic and as such they can explain the dynamic structure of time series better than the static OLS estimation method. VAR-VEC models can be used to generate impulse-response functions/graphs. These functions/graphs can be used to analyze the performance of the response variable to the explanatory variables.

Specifically, the relationship among per capita income and energy consumption variables can be interacted. Thus, for this study, a Vector Error Correction Model (VECM) is used.[16] Based on the conventional VAR set-up, the coefficients of all other variables in Eq. (2) are restricted to zero; while only four major variables of interest will be considered (Hsiao and Hsiao 2006). These variables include per capita income (*PCI*), Electricity Consumption (*EEC*), Natural Resource Intensity (*NRES*), Carbon Emission density (CO_2).

VEC model applied for examining the dynamic interactions is specified as follows[17]:

$$\Delta LPCI_{t-j} = \alpha + \sum_{j=1}^{n} \beta_j \Delta LPCI_{t-j} + \sum_{j=1}^{n} \gamma_j \Delta LEEC_{t-j} + \sum_{j=1}^{n} \theta_j \Delta LNRES_{t-j} + \sum_{j=1}^{n} \partial_j \Delta LCO_{2t-j} + K_1 ecm_{t-j} + \varepsilon_{t-j}. \tag{3}$$

$$\Delta LEEC_{t-j} = \alpha + \sum_{j=1}^{n} \beta_j \Delta LPCI_{t-j} + \sum_{j=1}^{n} \gamma_j \Delta LEEC_{t-j} + \sum_{j=1}^{n} \theta_j \Delta LNRES_{t-j} + \sum_{j=1}^{n} \partial_j \Delta LCO_{2t-j} + K_2 ecm_{t-j} + \varepsilon_{t-j}. \tag{4}$$

[16]A natural progression from VAR specification is the VEC model. VEC models are used when the variables of interest are not stationary at their levels and are cointegrated. It has been widely posited that VEC can be used to determine short-run dynamic interaction among a set of variables from a set of long-run equations. Therefore, VECM leads to a better understanding of the nature of any non-stationarity among the different variables in the series, as well their long-run equilibrium. A VEC model improves long-term forecasting error over an unconstrained model (Sreedharm 2004) (see Appendix 1 for cointegration result).

The VEC specification to be applied takes the form:

$\Delta y_t = \alpha + \sum_{l=1}^{p=1} \emptyset \Delta y_t + \prod y_{t-1} + \varepsilon_t.$

where Δ is the differencing operator, such that $y_t - y_{t-1}$

where y_t is an (n × 1) column vector of endogenous variable; α is an (n × 1) vector of constant terms, $\emptyset$ and Π = coefficient matrices. The coefficient Π is known as the coefficient matrix, and it contains information about the long-run relationships.

[17]All variables specified in the VECM are specified in their natural logarithmic form.

$$\Delta LNRES_{t-j} = \alpha + \sum_{j=1}^{n} \beta_j \Delta LPCI_{t-j} + \sum_{j=1}^{n} \gamma_j \Delta LEEC_{t-j} + \sum_{j=1}^{n} \theta_j \Delta LNRES_{t-j} + \sum_{j=1}^{n} \partial_j \Delta LCO_{2t-j} + K_3 ecm_{t-j} + \varepsilon_{t-j}. \quad (5)$$

$$\Delta LCO_{2t-j} = \alpha + \sum_{j=1}^{n} \beta_j \Delta LPCI_{t-j} + \sum_{j=1}^{n} \gamma_j \Delta LEEC_{t-j} + \sum_{j=1}^{n} \theta_j \Delta LNRES_{t-j} + \sum_{j=1}^{n} \partial_j \Delta LCO_{2t-j} + K_4 ecm_{t-j} + \varepsilon_{t-j}. \quad (6)$$

The ecm_{t-j}, specified in Eqs. (3)–(6) is the error correction term and the ε_t which is the white noise residual are mutually uncorrelated. The size and statistical significance of the error correction term will measure the speed of convergence of each of the variables towards equilibrium. This implies that a significant coefficient will mean that past equilibrium errors play a significant role in determining the current outcomes.

5.3 Estimation Technique

This study is a quantitative econometric analysis covering the period between 1970 and 2015. The reason for the choice of this is due to data availability. Besides, the period is when the oil boom began in Nigeria. In addition, the study accounts for the period before and after the issue of sustainable development (during the 1980s) came to the fore. In order to assess the implication of energy consumption on sustainable development, the energy consumption variables will be regressed on per capita income. Specifically, the system Generalized Method of Moments (GMM) is adopted on Eq. (2).[18]

The system-GMM estimator which combines equation in level and in difference and then exploits additional moment conditions (Blundell and Bond 1998). Predetermined and endogenous variables are instrumented by both their lagged

[18]The GMM estimation which belongs to a class of estimators known as M-estimators are defined by minimizing some criterion function which is a robust estimator in that it does not require information of the exact distribution of the disturbances. The system-GMM estimator combines equation in level and in difference and then exploits additional moment conditions (Blundell and Bond 1998). The GMM estimation is based upon the assumption that the disturbances in the equations are uncorrelated with a set of instrumental variables. The GMM estimator selects parameter estimates such that the correlations between the instruments and disturbances are as close to zero as possible, as defined by a criterion function. Predetermined and endogenous variables are instrumented by both their lagged values in level and lagged value in difference (see Appendix II for result of stationarity).

values in level and lagged value in difference.[19] Two specification test check the validity of the instruments that will be used. The first is the standard Sargan/Hansen test of over identifying restrictions; but in this case the non-significance of the Prob (J-statistic) value explains if the issue of overidentification. The second test examines the hypothesis that there is no second-order serial correlation in the first-difference residuals which is tested by the Durbin-Watson result of being close to 2.

While the dynamic interactions among the variable employed Vector Autoregressive (VAR) and Vector Correction (VEC) Mechanism estimation (Adrangi and Allender 1998). VAR models are adjudged the best method for investigating dynamic interactions and shocks transmissions among variables because they provide information on impulse responses. The VAR models makes all variables to interact in the regression.

5.4 *Definitions, Measurements and Sources of Variables*

An overview of the variables selected, its apriori expectations and the variable sources are captured in Table 4:

5.5 *Results and Discussion of Analysis*

5.5.1 The Result of the Relationship Between Energy Consumption and Per Capita Income

The result in Table 5 revealed the relationship between energy consumption variables and per capita income in Nigeria. While the statistical properties show no serial correlation (Durbin-Watson = 1.974058) and no overidentification of the variables [Prob (J-statistic = 0.369734)]; the model can be adjudged as suitable to explaining the interactions between energy consumption/usage and per capita income growth.

From Table 5, according to prior expectations, it is seen that EEC had a significant positive effect on PCI in Nigeria ($t = 2.77202$; $P < 0.05$). It implies that a 1% increase in electricity consumption, will cause per capita income to rise by 1.98%. The implication of this is that electricity consumption or availability is capable of promoting sustainable development in Nigeria via increases in per capita income.

[19]Instrumental variables used included rank of eight (8); such that in addition to the lagged values of the variables specified in Eq. (2), other determinants of PCI such as inflation rate and human capital were included, and constant term.

Table 4 Variable definition and sources

Variables	Definition	Sources	Apriori expectation
Dependent variable Per capita income (PCI)	It is often used to measure a country's average income or it is often used to measure a country's standard of living. It is usually expressed in terms of a commonly used international currency such as the euro or United States dollar. It is calculated using GDP and population estimates	WDI[a] (2016)	–
Independent variables –Natural resource endowment	This is the ratio of oil exports to total exports of Nigeria, and it's a measure of natural resource intensity of the country	CBN Statistical Bulletin (2016)	Positive (direct)
–Electricity consumption	This is the amount of electricity made available out of total demand. An increase in electricity supply (use) is expected to improve per capita income	ECN (2014) and WDI (2017)	Positive (direct)
–Carbon emissions	It is used to assess environmental quality. These are carbon-dioxide emissions stemming from the burning fossil fuels and from other manufacturing processes. It includes carbon dioxide produced during consumption of solid, liquid, gas fuels and gas flaring	WDI (2016)	Negative (indirect)
–Investment in capital	This is measured by gross fixed capital formation. New investments in physical capital is expected to increase per capita income	WDI (2016)	Positive (direct)

Source Author's computation
[a]World Development Indicators (WDI)

Contrary to apriori expectations, it was discovered that the consumption of oil, which is measured by NRES had a significant negative effect on PCI in Nigeria (t = −2.716032; $P < 0.05$). This implies that a 1% increase in oil consumption or exports, will decrease per capita income by 6%. This makes it obvious that the earnings from crude oil production and consumption has not promoted sustainable development in Nigeria. Besides, it shows that despite the huge earning from crude oil production in Nigeria, there is no trickling down effect on the population of Nigeria, thus, challenging posterity. The significant effect of the energy utilization on the human economy (via the effects on the per capita income) reveals the dependence of the human economy within the bio-system; thus validating the biocentrism position of the ecologist view. However, the mixed effects of energy utilization on per capita income revealed that there is an asymmetric flow in the input-output relationship in Nigeria with regards to material flow; thus creating less prospects for recycling and thereby challenging sustainability.

Table 5 Relationship between energy consumption and per capita income

Variable	Coefficient	Std. error	t-Statistic	Prob.
C	−0.925253	2.509341	−0.368723	0.7143
PCI(1)	0.285201	0.164536	1.733362**	0.0909
EEC	1.982631	0.71523	2.77202*	0.0085
GCF	−0.674597	0.248376	−2.716032*	0.0098
NRES	−6.097525	3.25641	−1.872469**	0.0687
CO_2	−0.597708	1.38714	−0.430892	0.6689
R-squared	0.204225	Mean dependent var		1.183377
S.E. of regression	0.91759	Sum squared resid		32.83689
Durbin-Watson stat	1.974058	J-statistic		2.00000
Instrument rank	8	Prob(J-statistic)		0.369734

Source Author's computation using eviews 9.0. ** and * denotes significance at 10 and 5% respectively

In line with prior expectations, it was discovered that carbon emissions measured by CO_2, had negative effects on PCI (t = 0.430892; $P < 0.05$). Although these effects appeared insignificant, but the findings is consistent with previous studies on the negative effect of environmental pollution and income per capita (Kuznets 1955; Yang et al. 2010). The result revealed that a 1% increase in CO_2 emissions will cause per capita income to decrease by 0.5%. This position on the environment-income nexus is theoretically hinged on the Kuznet's hypothesis[20]—where he posits that decreases in environmental quality via emissions can only be addressed through increases in per capita income. Increases in per capita income will therefore lead to increases in the demand for clean environment and thus, create an environment for sustainability in development. But in Nigeria, although the effects from CO_2 emissions are not so significant, however, energy policies that will further reduce CO_2 emissions will improve per capita income, thereby, fostering sustainable development.

5.5.2 A Dynamic Assessment of the Relationship Between Energy Consumption and Per Capita Income

The impulse response function of the VEC model which shows the approach of tracing out the time path by which the variables respond to their shocks were generated and shown in Fig. 6. The Cholesky ordering followed for the study is based on the apriori principles that growth in per capita income (log PCI) will be influenced by electricity produced and available for use (log EEC), thus, influencing

[20]Kuznet's hypothesis which is also known as the Environmental Kuznets curve (EKC) hypothesis states that "as an economy's per capita income increases, the total amount of environmental impact of economic activities initially grows, reaches maximum and then falls.

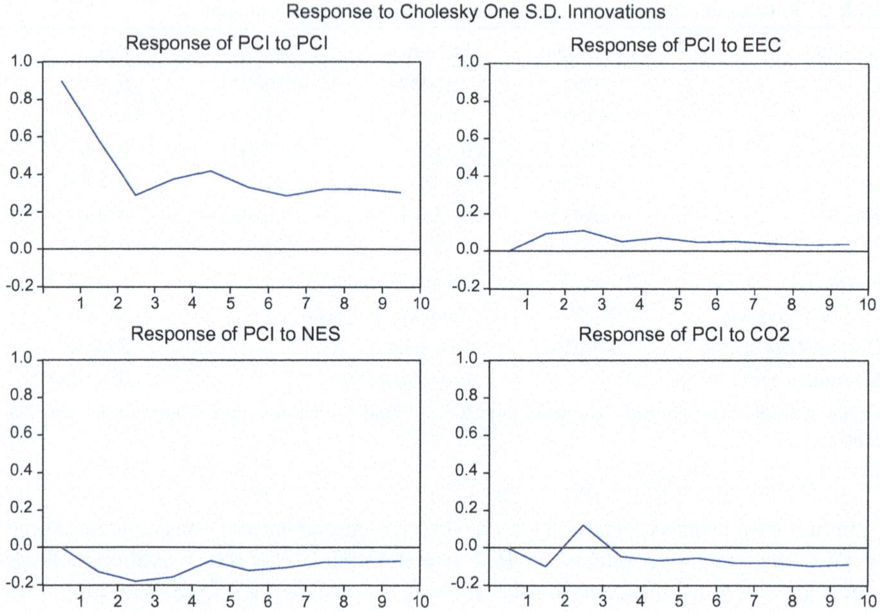

Fig. 7 Impulse-response function from VEC model

the intensity with which natural resources will be are consumed (log NRES), thereby leading to an increase in emissions from associated activity via production and consumption of transformed natural resources such as fossil fuels (log CO_2). Specifically, the Cholesky ordering follows some previous empirical studies; where it is expected that sustainable developments can be attained through an adoption of efficient energy scheme or policy, especially with regard to improved electricity availability and reduction in fossils as a primary source of energy and income (Jacob et al. 2012).

From Fig. 7, the impulse-response functions are generated showed the responses of PCI to short-term shocks from itself and other variables. A shock to PCI itself revealed that there will be an initial sharp positive response from which will gradually even out. This implies that a positive shock on per capita income, can cause a sharp positive response in the income per head of Nigerians; and such positive effects can be maintained overtime. Incidentally, a similar response is identified for a shock in electricity consumption (EEC). It is such that a shock to EEC causes PCI to improve positively and can be sustained over time. The implication of this is that any policy that improves electricity supply in Nigeria, will improve PCI positively and sustainably.

However, the responses of PCI to NES and CO_2 appeared to be negative. This pattern of negativity is corroborated by the result of the GMM analysis. This implies that any shock to the intensity with which natural resource is being consume or with regard to carbon emissions in Nigeria, there will be an adverse effect on per capita income; thus, threatening the stance of sustainability in development.

From the foregoing analysis, it can be inferred that policies that promote clean energy production and consumption, such as electricity generation via clean technologies will influence per capita income positively and promote sustainable development in Nigeria. However, overdependence on crude oil for income earnings in Nigeria and as a form of energy consumption in Nigeria (as depicted by carbon emissions) will retard per capita income and in turn retard the prospects for sustainable development in Nigeria; hence the need for alternative energy production and consumption policies that are more sustainable and that can facilitate sustainable development.

6 Sustainable Energy for Economic Development in Nigeria

Nigeria's overdependence and unreliability of supply in fossil fuel production, as well as the associated challenges of gradual environmental challenges has increased the need for alternative forms of energy. In addition to the need for this search are the challenges emanating from population growth and economic growth that appears to be lopsided in its distribution. Statistics have reported that the current trends of energy production is far from meeting local consumption.

Therefore, there is the need to increase energy production beyond the traditional means of energy generation, transmission and distribution. This is with a view to enhance productivity, social, economic and environmental performance through energy processes, production and consumption. In view of the foregoing search on an alternative to the traditional energy sources, renewable and sustainable energy may just be an option.

However, to promote policies on sustainable energies, a legal framework to regulate the industry may be required. In addition, clear roles and responsibilities of stakeholders must be clearly articulated. In addition, energy flow from supply to end- use are should form the bane of the energy policy framework needed to promote renewable energy technologies.[21] Although, efforts are being made to actualize the feat of renewable energy policies in Nigeria, these policies are at their primitive stage.

Oyedepo (2012) noted that effective policy and regulatory frameworks for advancing renewable energy in Nigeria are paramount to:

(i) Achieving long term reductions in carbon emissions;

[21]Renewable energy is energy that is generated from natural resources—such as sunlight, wind, tides, and geothermal heat—which are naturally replenished. The outlined energy sources are known as clean energy sources. This is because they introduce less or no greenhouse gases into the atmosphere when compared to fossil fuels.

(ii) Enhance the energy security of the country and establish a sustainable energy supply system;
(iii) Promote the policy of diversifying the energy supply so as to include renewable resources and technologies into the energy supply mix
(iv) Make electricity accessible to the rural dwellers through grid extension and mini-grids, considering that the level of electrification in the country is very low.

Energy production, given the fact that it cannot be recycled can be said to be a primary factor of production. This is unlike labor and capital which are intermediate factors of production and are therefore reproducible in nature. In the neo-classical school of thought, the quantity of energy available to an economy in any period can be determined. Although access to these energy resources may be restricted by biophysical and economic constraints, such as the amount of installed extraction, refining and generating capacity and the possible speed of efficiencies with which these processes can proceed. Nonetheless, the notion of sustainable energy use for economic development must be discussed with the framework of dynamic efficiency.

Generally, it has been observed that a major aspect of energy requirements is met by combustion of fuels such as wood, coal, kerosene, petroleum, diesel, natural gas, cooking gas, etc., which are believed to be formed from the remains of plants and animals. But in the past century, it has been seen that the consumption of non-renewable sources of energy has caused more environmental damages than any other human activity. Electricity generated from fossil fuel such as coal and crude oil has led to high concentration of harmful gases in the atmosphere. This has subsequently led to many problems being faced today in notable atmospheric aspects, such as ozone depletion and global warming. In addition to these, other challenges associated with energy generation and environmental issues include emissions from vehicles-carbon monoxide, chlorofluorocarbons, acid rains, oil spills, large-scale fossil fuel combustion and land reclamation (Jhingan and Sharma 2008).

As a result of these environmental threatening challenges, and the fear that the existing resources of fuels like coal, oil, natural gas and uranium are being used up, there is no doubt an imminent energy crisis as the future generation may be compromised (Harris 2003). Therefore, if energy is to by any means be sustained, the use of alternative forms of energy based on renewable recourses cannot be outmatched. The alternative possible energy which can be adopted in Nigeria will be discussed in turn briefly:

– Hydropower: This is basically power generated from water. This has been considered safe clean and cheap, as well as renewable. Nigeria already has a number of hydro power plants in Kainji Dam, Egbin thermal station, etc. Although there are the dangers such as dam failure that could cause loss of life, loss of farm area due to erosion loss of aquaculture due to thermal gradient, a controlled and continual exploration of this form of energy provision will be encouraged. This is because hydropower has some potential for pumped storage to reduce the peak demand and assist in electricity generation.

- Wind Power: It has been used to generate electricity desalinate sea water and pump water. A major challenge of wind power is tidal power, irregular nature of supply of wind and the accompanying necessity for energy storage. But the advantages are enormous; for instance, it is environmentally friendly, operation cost and maintenance are low, less noise unlike diesel powered engines. The Wind generating plant can then be located in remote areas to enable generation, transmission and distribution. Although the wind energy generation is capital intensive and requires highly skilled personnel, it is still exportable in Nigeria.
- Geothermal Energy: The most common method of derivation is the use of direct hot fluids from deep geothermal layers. While an alternative could be artificial pumping of water from the surface down through layers of hot rocks are being developed.

  Geothermal energy could be harmful to health by exposing people to toxic or potentially toxic-elements and non-nuclear agents. With each source having its own spectrum of pollutants and well identified, the potential effects on health has been said to be scanty, particularly in the long run and with limited exposure. In all, geothermal energy are useful additions to the source of energy and could be explored in Nigeria.
- Solar Power: The solar power is usually produced from small local sources or large central stations on land and on satellites. Unlike fossil fuel technology, solar technology does not make significant emissions to the environment when it is used. Its major challenge on human health lies in the setting up and discontinuation of the system. Besides, the solar voltaic energy which are land-based, will require large collection areas for effective functioning.
- Biomass Energy: Biomass production is a renewable form of energy that employs a far-reaching process of cultivating and harvesting residues of various types to produce bio-gas or bio-fuel. Therefore, in order to achieve efficiency in the use of bio-technology, improved means of production must be embarked on. A wide usage and distribution of bio-technology generation will require effective maintenance to avoid accidents and health hazards. For instance, the burning of wood may seem preferable to use of cooking stove due to the less carbon monoxide and mutagenic materials in the smoke from stove, but wood ash appears less toxic, but an increased use of wood could cause air pollution some condensable organic compounds which could affect ones health.

  In fact, a major form of bio-mass is biogas; this is where cattle dung, plant/crop residues and sewage are fermented, such that in the absence of air, certain combustible gases are produced-biogas. Common crops used as bio-energy sources include starch, sugar-based plants like sugar-cane or corn. Already, a lot of developing economies apply this technology such as the Northern region of Nigeria where animal waste are used as potential sources of energy, and in Songhai of Cotonu where human waste are recycled to get biogas. Hence the process of achieving effective utilization is paramount and as a result it can be improved upon to annex its optimal benefits.

A Case Study of the use of Bio Fuel by an Organisation: *LIFEBUILDERS*—located in Akinyele Local Government Area, Oyo State, Nigeria
In a communique designed by the Lifebuilders titled: "Climate Change and Fossil Use Reduction through Production of Moringa Bio-Gel as an Alternative Fuel for Domestic Use in Nkinyele LGA, Oyo State"; it stated its motto as …*combating climate change and its impact.* Lifebuilders is an organisation sponsored by the United Nations Development Programme (UNDP) and Global Environment Facility (GEF)
The project on biofuel production was borne out of the search for alternatives in cooking fuel. Instead of the use of firewood and kerosene which has caused diseases like pneumonia, cardiovascular diseases, increased infant mortality and lung cancer. According to the report off World Health Organisation (WHO 2013), almost 100,000 women die from the use of firewood. In addition, the report stated that a person who cooks with fire wood three times a day is similar to someone who smokes between 3 and 20 packets of cigarettes per day. In Nigeria, most rural dwellers, despite the knowledge of alternative fuels like gas and electric stove, most of them prefer fire wood due to its cost, thus leading to the falling of trees. However, this has compounded the issue of toxins, pollution and health problems
Lifebuilders are into the production of bio-gel for cooking. It is discovered that the sugar content of Moringa Oleifera stems could be exploited to produce ethanol which can be further processed to produce Moringa bio-gel for cooking. The organisation, which is located in the heart of Alore-Ijaye communities, through a grant scheme received has pursued the goal of: – establishing a bio-gel production center in its community – producing bio-gel using raw materials from Moringa Oleifera – building the capacity of women in bio-gel utilization in its community – increasing Moringa farmlands in its community for product and project and sustainability Having carried out the project, the following were discovered: – Moringa bio-gel burns with blue flame, heats up quickly and does not stain cooking pots – Moringa Oleifera stem is a viable source for ethanol production since the stems contain reducing and non-reducing sugars contents. – Moringa Oleifera plant has become widely cultivated in the host communities given its utility – the chaff, which is a bye-product from the stem of Moringa Oleifera has been used in as a valuable input for paper making industries. The implication of this project is that: – it creates cleaner and more friendly fuels; thereby creating sustainable energy consumption and promotes a healthy environment. – it reduces deforestation, thus promoting a sustainable environment – it reduces the risk of exposure to toxic gases, thus promoting sustainability in health – economically, it will create jobs for people and alleviate poverty. However, as promising as Moringa bio-gel appears, it faces the challenge of: – high cost of sales, when compared to fossil fuels and wood fuels. This is due to high cost of production – Inadequate knowledge about climate change challenges has led to low demand for bio-fuels – easy access to fossil fuels and wood fuel has discouraged the demand for fossil fuel. Nonetheless, the Moringa bio-gel is available and has been tested to be efficient. It may just require some policies such as tax holidays or reduction, subsidies and sensitization to recognize inculcate its use among the government and consumers in Nigeria. In addition, there is the need for ongoing intensive research to facilitate its production in commercial quantities

Source www.lifebuldersnigeria.org

7 Policy Instruments for Sustainable Energy Development

Environmental economics is basically concerned with the impact of economic activities on the environment, the significance of the ecosystem to the economy, and looks at ways in which cosmic balance can be achieved within a country. Environmental economist at every point in time bothers on the right volume of pollution a society can accommodate. As a result of this, it is advocated that if the goods that are environmental in nature should bear true values, then damages done to the environment are substituted by paying the right cost of depletion experienced. Thus, in a bid to achieve sustainable development, it is pertinent to understand certain critical issues that affect sustainability: that the earth does not exist because of humans alone, rather it requires a cosmic balance with all living and nonliving creature within the biosphere. Thus, the earth must be preserved at all cost while considering resources on earth meant for human consumption and the extent to which the earth replenishes itself.

Following this, an effective environmental policy that will regulate human activities to achieve sustainable development. In fact according to environmental economist, despite the intensity of human activities, and the wages or income they receive in return which enhances living conditions; a positive relationship is expected between income and environmental quality. With particular reference to developed economies, the more the demand for quality environment is high, but in bid to achieve a cleaner environment, human activities in form technological advancement will increase which will in turn lead to environmental degradation. But this is a less concern for less developed economies because, the primary needs of food clothes and shelter are still not met, hence there is less worry for what goes on in the environment. In addition, it is observed that in developing economies, hi-tech operations are scanty, as a result, there is less worry for what operates in the environment.

Although, in comparison with advanced economies of the world, the consumption of energy in developing countries, like Nigeria is minimal. Nonetheless, there is the need for policies that will ensure efficient use of current and future energy consumption. This, will in turn set a workable base for sustainable development, reduces environmental and economic cost, and build a good base for expanding energy services.

Before proposing some policy instruments for sustainable energy-environmental policies, some ten (10) commandments were postulated for Nigeria by Adenikinju (2017)

Ten Commandments to Promote Sustainable Energy Policies in Nigeria—By Adenikinju (2017)
– Avoid the violation of appropriate public choice rule: this involves a reinvestment of income from wasting assets like fossils into sustainable areas like human and physical capital development
– Avoid the creation of an enclave economy that is export dependent: that is a strong relationship is expected between oil and non-oil sector of the economy. Besides, avoid a situation where an insignificant portion of the population enjoys earnings from oil wealth
– Apply economic principles in efficient management of the energy sector: this involves an optimal exploitation of resources where proceeds from one sector is used to transform an economy after exhaustion of natural reserves; this is to foster self-sustainability
– Provide adequate and efficient investment to the energy sector
– Shun weak institutional governance that promotes rent-seeking and corruption activities
– Protect against opportunistic diseases (the Dutch disease and Resource Curse): This is where exports are narrowed to a few or single primary product to the exclusion of other products from other sector (like the manufacturing and agricultural sector) due to inability to compete on the international scene. Thus leading to import-dependence. In addition, it involves tackling the inability to turn wealth from natural resources to sustainable economic wealth
– Avoid the neglect of energy (oil) producing communities. Such as the Niger-Delta of Nigeria; there must be a design that gives back to the communities where resources that creates economic wealth are drawn
– Avoid foreign domination of the energy sector
– Avoid activities that lead to the destruction of the environment
– Avoid ignoring the first love of income generation. For instance, in Nigeria, the agricultural sector which is still largest employer of labour and highest contributor to the GDP must be vibrant. This will assist in complementing activities in other sectors such as the energy sector

Following the findings of this study, the policy instrument for sustainable development are discussed. Firstly, there must be the need to adopt the policies that will incorporate *newer technologies and low cost energy*. This is important, especially if there is a suspicion of future shortages of present energy demands. In addition to attaining energy efficiency through the use of new and least cost technology, there should be the incorporation of good operational and management practices that will help in minimizing energy waste. This implies that priorities should be given to renewable energies which can enhance energy generation and availability (Jhingan and Sharma 2008).

A policy mechanism that will ensure that the power requirements of Nigerians are met should be driven—That is a *population-power policy*. This has significant implications for industrial, social and domestic implications, as a result, the government of Nigeria needs to have a power policy that will be population driven. Already, it has been revealed that the 4500 MW of electricity consumed is far below what the Nigerian populace require, as the economy need over 16,000 MW to have a stable power supply. Although the presence of independent power producers is hoped to enhance performance, and the government intends to increase electricity generation to about 6000 MW, there is also a need to include in the policy, the need to explore alternative and cleaner methods of generating electricity for Nigeria.

A wide spread of the use of *clean technologies* could reduce local, regional and international challenges. For instance, in the non-commercial sector, improving the efficiency of traditional appliances such as wood, kerosene could help in ensuring cleaner environment. In recent times, fuel efficient stoves are already a common sight which could be advocated for by the Nigerian government, particularly for the rural dwellers and low income earners. Even support systems for transportation such as good roads, road worthy vehicles, fuel-efficient vehicles, and vehicle maintenance can help reduce emissions and carbons that are dangerous to the health. Hence, there should be greater reliance on advanced energy technology, particularly cleaner fossil fuels technologies.

A *balanced policy* approach should be pursued to achieve sustainable energy in Nigeria. This is a situation where in appropriate cases a good balance of state and capitalist policies should be implemented. Firstly, The Rio+20 outcome document, *The future we want*, which is set out as a mandate to establish an Open Working Group to developed a set of sustainable development goals considered and enacted by the General Assembly in its 68th session that the SDGs should be coherent with and integrated into the UN development agenda beyond 2015. In effect, government at every level should align themselves with these goals that include: Promote sustained, inclusive and sustainable economic growth, full and productive employment and decent work for all; Build resilient infrastructure, promote inclusive and sustainable industrialization and foster innovation; take urgent action to combat climate change and its impacts; conserve and sustainably use the oceans, seas and marine resources for sustainable development; Protect, restore and promote sustainable use of terrestrial ecosystems, sustainably manage forests, combat desertification, and halt and reverse land degradation and halt biodiversity loss.

Furthermore, in appropriate areas, *subsidies* should be phased out in areas that can inhibit sustainable development, while the government should equally pursue the functioning of national energy markets in ways that support sustainable development (Jhingan 2012). For instance government can assist in overcoming market barriers, help improve accessibility to cleaner energies, and provide a social environment suitable for the development of indigenous energy suitable for Nigeria as a country. This is evident in the deregulation of the downstream oil sector of the Nigerian economy where there is the need for fuel consumers to pay for scarce resources are subjected to the forces of demand and supply (Federal Ministry of Finance 2014). This is because subsidies for resources consumed by both private and public sector such as petrol diesel, electricity, gas, water, etc. will lead to wasteful use and environmental degradation.

With particular reference to issues that affect environmental degradation through oil spills and gas flaring, it is important to get prices right, internalize all environmental externalities, mobilization of scarce capital and increase the use of energy recycle or non-polluting energies (Oyeshola 2008). In terms of market-based approaches, there is urgent need for the protection of the environment. This is aimed at pointing to consumers and industries on the use of natural resources and the associated cost. These costs are therefore reflected in the prices of goods and services bought or sold. Other forms of market-based approaches will include

Fig. 8 Summary of policies that aid sustainable development

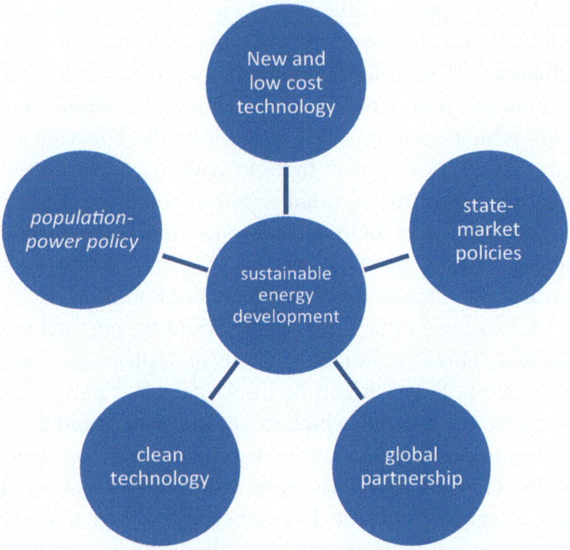

emission tax, differential tax systems for key energy users or suppliers, as well as depositor fund system.

Beyond these, there is the need to developing countries like Nigeria to put in place policies that will build its human capacity to advance hi-tech innovations—*technical know-how*. This can be achieved investing financial resources in training, re-training, research and innovations in areas the particularly affect Nigeria, as well improving the ability to implement diffused technologies.

Finally, the need for a *global participation* by the Nigerian government is important. These participation must include areas like the *Montreal Protocol*—which involves phasing ozone depletion, *Basel convention*—which involves trans-boundary movements and disposal of hazardous wastes and effluents, *Rio declaration and development*—which is strictly on sustainable development issues and the *GATT clauses on Environment*.

Figure 8 simply states an over-view of the policies that can fuel sustainable energy in Nigeria. No doubt, energy, which is a major component environmental sustainability, there is the need to advance positions that will put the Nigerian economy on the right perspective. There is a popular saying "little drops of water makes a mighty ocean". If every economy becomes concerned about a green environment, in no distant time the global economy will be better for it, especially in terms of human capital, life expectancy, and life sustainability (WWW.UN.Org/…/WSSD_PlanImpl).

Appendix I

Stationarity Test

Variables	ADF		*d*	PP		*d*
	Intercept and no trend	Intercept and trend		Intercept and no trend	Intercept and trend	
PCI	−0.15628	−0.26134	I(1)	−0.529462	−0.461986	I(1)
ΔPCI	−5.58412	−6.06767		−5.638035	−6.066968	
CO_2	−1.92355	−3.01153	I(1)	−1.85599	−3.033445	I(1)
ΔCO_2	−8.13129	−8.06662		−8.168986	−8.094495	
GCF	−1.24632	−1.85352	I(1)	−1.223757	−2.101440	I(1)
ΔGCF	−5.98611	−5.95246		−6.003219	−5.970621	
EEC	−1.738669	−3.476970	I(1)	−1.956525	−3.391263	I(1)
ΔEEC	−9.433786	−9508918		−9.504771	−9.508918	
INFL	−3.24025	−3.22163	I(0)	−3.199678	−3.199678	I(0)
ΔINFL	−6.65746	−6.62565	I(1)	−6.625646	6.313404	I(1)
NRES	−9.62187	−10.1620	I(0)	−10.16209	−14.06129	I(0)
ΔNRES	−7.00172	−7.08393	I(1)	−7.594651	−7.642427	I(1)
HUM	−1.44941	−1.96232	I(1)	−1.418720	−1.418720	I(1)
ΔHUM	−4.23421	−4.37414		−4.423885	−4.423885	
Mackinnon critical values						
Level						
1%	−3.61045	−4.21186		−3.610453	−4.211868	
5%	−2.93898	−3.52975		−2.938987	−3.529758	
10%	−2.60906	−3.19831		−2.607933	−3.196411	
1st difference						
1%	−3.61558	−4.21912		−3.615588	−4.219126	
5%	−2.94114	−3.53308		−2.941145	−3.533083	
10%	−2.60906	−3.19831		−2609066	−3.198312	

Source self computation using E view 9.0
Notes d denotes decision about the order of integration respectively

Appendix II

Cointegration Result

Trend assumption: Linear deterministic trend				
Series: PCI EEC NES CO_2				
Lags interval (in first differences): 1 to 1				
Unrestricted cointegration rank test (trace)				
Hypothesized		Trace*	0.05	
No. of CE(s)	Eigenvalue	Statistic	Critical value	Prob.***
None**	0.607703	74.18265	47.85613	0.0000
At most 1**	0.409427	33.01026	29.79707	0.0206
At most 2	0.183700	9.837133	15.49471	0.2935
At most 3	0.020388	0.906329	3.841466	0.3411
Unrestricted cointegration rank test (maximum eigenvalue)				
Hypothesized		Max-Eigen****	0.05	
No. of CE(s)	Eigenvalue	Statistic	Critical value	Prob.***
None**	0.607703	41.17239	27.58434	0.0005
At most 1**	0.409427	23.17313	21.13162	0.0255
At most 2	0.183700	8.930805	14.26460	0.2920
At most 3	0.020388	0.906329	3.841466	0.3411

*Trace test indicates 2 cointegrating equations) at the 0.05 level
**denotes rejection of the hypothesis at the 0.05 level
***MacKinnon-Haug-Michelis (1999) *p*-values
****Max-eigenvalue test indicates 2 cointegrating equations) at the 0.05 level

References

Adenikinju AF (2017) Energy and Nigeria's economic development: a troubled but indispensable marriage. An Inaugural Lecture Delivered at the University of Ibadan, Ibadan Nigeria. ISBN: 978-978-8529-07-1

Adeniyi O, Orekoya S, Oyinlola M, Omisakin M (2016) Oil and empirics: in the quest for development. Essays in Honour of Professor Akin Iwayemi at 70, A. Adenikinju, A. Jerome and O. Ogunkola, pp 331–367

Adrangi B, Allender M (1998) Budget and stock prices—International evidence. J Econ Financ 9 (22):57–66

Akinlo A (2009) Energy consumption and economic growth in Nigeria: evidence from cointegration and co-feature analysis. J Policy Model 15(3):681–693

Aliyu SU (2009) Impact of oil price shocks and economic growth in Nigeria: an empirical investigation. Available at: http://ideas.repec.org/p/pra/mprapa/16319.html

Amadi S (2012) The structure of the Nigerian Electricity Industry, major on-going projects; opportunities and challenges. Paper Presented to the Public and Private Development Centre, Procurement Monitors Training, 19th To 21st July

Ayodele AS, Falokun GO (2003) The Nigerian economy and pattern of development. Printoteque, 32, Shipeolu Street, Palmgroove, Lagos

Babatunde AM, Adenikinju A (2016) Energy consumption and economic growth in Nigeria: a time varying framework analysis: in the quest for development. Essays in Honour of Professor Akin Iwayemi at 70, A. Adenikinju, A. Jerome and O. Ogunkola, pp 331–367

Bassanini A, Scarpetta S (2001) Does human capital matter for growth in OECD countries? Evidence from Pooled Mean-Group Estimates. OECD Department Working Paper No. 282

Blundell R, Bond S (1998) Initial conditions and moment restrictions in dynamic panel data models. J Econom 87(1):115–143

Central Bank of Nigeria (CBN) (2002) Statistical bulletin. Central Bank of Nigeria P. M. B. 0187, Garki, Abuja

Central Bank of Nigeria (CBN) (2005) Major financial indicators. Central Bank of Nigeria P. M. B. 0187, Garki, Abuja

Central Bank of Nigeria (CBN) (2015) Statistical bulletin. Central Bank of Nigeria P. M. B. 0187, Garki, Abuja

Central Bank of Nigeria (CBN) (2016) Statistical bulletin. Central Bank of Nigeria P. M. B. 0187, Garki, Abuja

Chuku CA, Akpan UF, Sam NR, Effiong EL (2011) Oil price shocks and the dynamics of current account balances in Nigeria. OPEC Ener Rev 35:119–139. https://doi.org/10.1111/j.1753-0237.2011.00186.x

Drabo A (2011) Interrelationships between health, environment quality and economic activity: what consequences for economic convergence. Hal Archives Ouvertes, halsh-00552995

ECN (2014) National energy master plan. Retrieved from http://energy.gov.ng

Energy Commission of Nigeria (2003) National Energy Policy. Federal Republic of Nigeria, Energy Commission of Nigeria. Available online: http://www.energy.gov.ng/index.php?option=com_docman&task=cat_view&gid=34&Itemid=49

Federal Government Nigeria (2017) Economic recovery and growth plan 2017–2020. Ministry of Budget and National Planning, Abuja

Federal Ministry of Finance (2014) Frequently asked questions of the downstream petroleum sector and removal of oil subsidy. fmf.gov.ng

Gregory J (2009) Material flow analysis. Massachusetts Institute of Technology, Department of Material Science and Engineering, ESD. 123/3.560; Industrial Ecology-Systems Perspective Randolph Kirchain; Online: https://ocw.mit.edu/courses/engineering-systems-division/esd-123j-systems-perspectives-on-industrialecology-spring-2006/lecture-notes/lec14.pdf

Gujarati DN, Porter DC (2009) Basic econometrics. McGraw-Hill/Irwin Companies Inc

Harris JM (2000) Basic principles of sustainable development. Tufts University Medford MA 02155, USA. Global Development and Environmental institute Working paper 00-04

Harris JM (2003) Sustainability and sustainable development. Int Soc Ecol Econ 1(1):1–12

Hsiao F, Hsiao M (2006) FDI, exports and growth in East and South Asia: evidence from time-series and panel data causality analyses. Paper presented at 2006 international conference on Korea and the World Economy, July 7–8, 2006 at Korea University Seoul, Korea

Hussen MA (2000) Principles of environmental economics. Routledge, London and New York; ISBN:0-415-19570-5(hbk)

Iwayemi A, Fowowe B (2011) Impact of oil price shocks on selected macroeconomic variables in Nigeria: A multivariate cointegration approach. Energy Econ 32:71–85

Jacob M, Haller M, Marschinski R (2012) Will history repeat itself? Economic convergence and convergence in energy use patterns. Energy Econ 34:95–104

Jhingan ML (2013) The economics of development and planning. Vrind Publications, Mayur Vihar Phase-1, Delhi-110091

Jhingan ML, Sharma Chandar K (2008) Environmental economics-theory, management and policy. Vrinda Publications Limited, Mayur Vihar, Delhi

Jhingan ML (2012) The economics of development and planning. Vrinda Publications Limited, Mayur Vihar, Delhi

Kilian L, Park C (2009) The impact of oil price shocks on the US stock market. Int Econ Rev 50 (4):1267–1287
Kuznets S (1955) Economic growth and income inequality. Am Econ Rev 45(1)
National Bureau of Statistics (NBS) (2017) Nigeria data portal. Available Online: http://nigeria.opendataforafrica.org/xspplpb/nigeria-census
National Planning Commission (2009) Report of the Vision 2020 national technical working group on energy sector. Abuja, Nigeria
National Bureau of Statistics (2007). Nigerian data portal. http://nigeria.opendataforafrica.org/data#source=National+Bureau+of+Statistics+Nigeria
National Population Commission (1998, 2006) Population census of federal Republic of Nigeria. Analytical Report at the National Level Abuja
Nigerian Biofuel Policy and Incentives (2007) Federal Government Nigeria, Official Gazette of the Nigerian Bio-Fuel Policy and Incentives
Ogujor EA, Orobor NP (2010) SAIDI minimization through conventional energy resources: opportunities and challenges. J Econ Eng 54–56 (ISSN2078-0346)
Ogunsola OI (1990) History of energy sources and their utilization in Nigeria. Energy Sources 12:181–198
Olanrele AI (2017) Impact of oil prices on the Nigerian Economy using rolling impulse response approach. An Unpublished PhD Thesis submitted to CPEEL, University of Ibadan
Olanrele AI, Adenikinju A (2017) Structural breaks. Electricity generation and economic growth in Nigeria. NAEE Energy Forum
Olomola P (2006) Oil price shocks and aggregate economic activity in Nigeria. J Afr Econ Bus Rev 4(2)
Olomola P, Adejumo A (2006) Oil price shock and macroeconomic activities in Nigeria. Int Res J Financ Econ 3(1)
Olubusoye OS, Olofin S, Alade T, Oloko A, Ogboona and O Isah (2016) Forecasting the impact of global oil price movement on the Nigerian Economy. In the quest for development. Essays in Honour of Professor Akin Iwayemi at 70, A. Adenikinju, A. Jerome and O. Ogunkola pp 331–367
OPEC (2017) OPEC annual statistical bulletin. Organisation of Petroleum Exporting Countries (2017). Available Online: http://www.opec.org/opec_web/static_files_project/media/downloads/publications/ASB2017_13062017.pdf
Oseni MO (2011) An analysis of the power sector performance in Nigeria. Renew Sustain Energy Rev 2011(15):4765–4774
Oyedepo SO (2012) Towards achieving energy for sustainable development in Nigeria. Renew Sustain Energy Rev 34:255–272
Oyeshola D (2008) Sustainable development: issues and challenges for Nigeria. Daily Graphic
Pearce David and Giles Atkinson (2002) The concept of sustainable development. An evaluation of its usefulness. Ten years after Brundtland. Centre for Social and Economic Research on the Global Environment University College London and University of East Anglia: CSERGE Working paper PA 98-02
Plan of implementation of the world summit on sustainable development at the united Nation Conference. WWW.UN.Org/…/WSSD_PlanImpl-pdf
Sreedharm N (2004) Business cycle: a theoretical historical and statistical analysis. JEL Classification C32:1–38. https://doi.org/10.1007/978-3-319-73028-8_21
Tallapragada PSVN (2009) Nigeria's electricity sector—electricity and gas pricing barriers, pp 29–34. International Association for Energy Economics, First Quarter. http://www.iaee.org/en/publications/newsletterdl.aspx?id=72
United Nations Environment Programme (UNEP) (2012) Global environment outlook: environment for development. United Nations Environment Programme ISBN: 978-92-8072836-1
Warren CA (1996) Distribution reliability—what is it? IEEE Appl Mag 2(4):32–37
World Bank (2015) In: World Development Indicator, 2015. World Bank, Washington
World Bank (2016) In: World Development Indicator, 2014. World Bank, Washington
World Bank (2017) In: World Development Indicator, 2014. World Bank, Washington

World Health Organisation (2013) World Health Statistics. World Health Organisation. Available online @: http://www.who.int/gho/publications/world_health_statistics/WHS2013_IndicatorCompendium.pdf?ua=1

Yang H, Zhou Y, Abbaspour KC (2010) An analysis of economic growth and industrial wastewater pollution relations in China. Consilience (4):60–79

To Cultivate or Not to Cultivate? An Exploratory Analysis of What Influences Greek Farmers' Decisions Towards the Cultivation of Bioenergy Crops

Eugenia Petropoulou, Vasiliki Petousi and Irini Theodorakopoulou

Abstract Much emphasis is placed by the EU on bioenergy crops as a means of decreasing Europe's dependence on fossil fuels and environmental sustainability and in turn, promoting potential rural development benefits from biomass production and processing. Various reasons have been identified in the literature as influencing the future of bioenergy in the EU. Critical among them is the farmers' willingness to engage in bioenergy crops cultivation. While common concerns of farmers at the EU level have been identified in the relevant literature, evidence exists that such concerns differ at country and even local level. In this chapter, through an exploratory approach, we identify parameters which influence Greek farmers' decisions to engage in the cultivation of bioenergy crops. Thematic analysis was implemented for the exploration of data from a focus group, from central Greece, involving farmers and relevant stakeholders. General themes such as socio-economic, institutional/policy support, environmental and land-use considerations emerged. Numerous sub-thematics involving social and value considerations (e.g. generation gap), issues of social capital such as trust in government institutions and strategic planning complement the general themes.

This chapter is based upon work supported by the WATBIO research project under programme No. 311929 (EU's FP7 Programme). Any opinions, findings, and conclusions or recommendations expressed are those of the authors and do not necessarily reflect the views of the WATBIO programme.

E. Petropoulou (✉) · V. Petousi
Department of Sociology, University of Crete, Crete, Greece
e-mail: petrope@uoc.gr

V. Petousi
e-mail: petousiv@uoc.gr

I. Theodorakopoulou
Agricultural Economics Research Institute, Hellenic Agricultural Organisation—DEMETER, Athens, Greece
e-mail: theodorakopoulou@agreri.gr

W. Leal Filho et al. (eds.), *Towards a Sustainable Bioeconomy: Principles, Challenges and Perspectives*, World Sustainability Series,
https://doi.org/10.1007/978-3-319-73028-8_22

Keywords Bio-energy crops · Focus group · Farmers' and local stakeholders' perceptions · Trust · Greece

1 Introduction

The growing global interest in renewable energy observed in recent years has been driven by local and global environmental concerns, domestic energy security, land-use tradeoffs between food and energy, as well as potential rural development benefits from biomass production and processing. Renewable energy sources (RES) (wind, solar, hydro, biomass) for many countries (developed and developing) represent an important alternative source of energy with potentially positive prospects towards reducing energy dependency on finite energy sources (especially expensive imported oil) and strengthening energy supply security. Furthermore, potential positive effects on: (a) emissions of CO_2 and other pollutants, (b) the creation of new businesses and employment, and (c) rural development are identified in the literature and are suggested as pillars of sustainable development (Manolas 2007).

The European Union has taken major initiatives to promote Renewable Energy Sources (RES). Directive 2001/77 prioritises the increase in RES contribution, allocates obligations under the Kyoto Protocol on member states—and sets the Community framework. Member-States are required to comply with the above-mentioned framework at the national level and report at regular intervals on their progress towards these goals. Currently, at the EU level, a significant proportion of used energy is produced by renewable natural resources such as bioenergy, biomass in particular. The EU is the largest biodiesel producer worldwide (Flach et al. 2013) while Germany, France and Spain are the largest biodiesel producers in Europe followed closely by Italy that has aggressively invested in 19 biodiesel factories over the past few years (Palmieri et al. 2014).

Within the EU context, special attention has been given to the alignment of food production and energy crops production with a simultaneous consideration of the negative impact of a generalised economic recession on agriculture. Thus, under the framework of the recently reformed Common Agricultural Policy (CAP), the EU has developed a path aimed at revitalising rural development and consequently rural sustainability, while at the same time aligning food production with the conservation of finite resources and the natural environment. To that extent, regulations and directives related to transition of agricultural areas from food production towards energy production have been adopted and initiated (Arabatzis et al. 2006; Arabatzis 2008; Aggelopoulos et al. 2009). This orientation has also moved towards the development of economically viable alternative crop choices by farmers. Farmers can obtain support from recent agri-environmental schemes, since under the new CAP, payments for farmers are linked to the so-called "cross-compliance". This is the link between direct payments and the degree to which farmers respect (comply) with a set of EU rules concerning food safety, plant and animal health, the environment, the protection of natural resources and the condition in which farmland is maintained (European Commission 2011; Mattas et al. 2015). Consequently, at least at the planning and regulatory level, agricultural biomass production is placed in the spotlight of current rural development policies, bearing in mind the needs of the farmers within a context of general economic recession.

At country level, the Greek regulatory framework and guideline documents, such as the National Renewable Energy Plan (NEARP) issued in 2012 by the Department of Environment, Energy and Climate Change, and the National Plan for Strategic Rural Development in Greece (NPSRD) (2014–2020), issued by the Hellenic Ministry of Rural Development and Food in 2014, make the case for bio-energy development by asserting the potential national-level benefits. The following main goals of bio-energy development are stressed in the above documents: The first goal targets national energy supply, and includes benefits such as: A more diverse energy portfolio, energy security, energy independence, and a reduction in imported fossil fuels. The second goal focuses on national economic growth, and highlights the benefits in strengthening, diversifying, and providing new opportunities for more competitive biomass providers, biomass producers, and others in the extended supply chain. The third goal of bio-energy development emphasises environmental improvement, including benefits from reduced carbon emissions, enhanced water quality, biodiversity conservation, and improved air quality.

The Greek government's goals for bioenergy development however, assume a macro-scale focus, and stress the potential of accruing benefits at the national level rather than ascertaining rural revitalisation through agricultural biomass production, new "green jobs" or a good income for farmers. Sustainability concerns receive explicit albeit scant attention in subsequent policy statements. Consequently, farmers remain submerged in generic discussions of feedstock supply production and logistical issues (Panoutsou 2008; Soldatos et al. 2010). Nevertheless, as research has shown, in Greece, energy crops can be an attractive, alternative solution only if they are integrated properly into the existing agricultural activities which are not duly considered at the national regulatory framework (Mantziaris et al. 2017; Panoutsou 2008; Soldatos et al. 2010).

Despite the optimistic policy and planning promises and the initial support, RES projects have been frequently met with reactions stemming from local communities and environmental organisations (Wustenhagen et al. 2007). Thus, while until 2007, agricultural producers and environmental groups co-supported the development of bioenergy (Dauvergne and Neville 2009, 2010), the food-price crises of 2007 and 2011 led farmers to question the use of extensive land for fuel rather than food production, and many environmental activists withdrew their support (McMichael 2009). Concerns about the environmental and socio-economic impact of land use change, the competition between food and fuel production, and the fair distribution of economic risks and benefits have now become widespread in discussions about bioenergy. Bioenergy, like other forms of agricultural biotechnologies (e.g. GM foods) have developed into contested domains which merit careful attention.

Social and environmental contestation notwithstanding, the fact remains that the potential of relevant policies to be implemented widely depends on farmers' and consumers' acceptance of cultivation and use. However, policies and often research tend to limit their attention on farmers' and other stakeholders' instrumental role of producing and delivering energy feedstock supplies to address governments' renewable energy supply targets. This focus fails to consider how farmers and rural stakeholders formulate and express their views, not just about providing energy feedstocks, but about the wider, possibly more equivocal implications of agricultural bioenergy development for their localities and regions. In the case of Greece, for example, the above are further contextualised amidst rural decline evidenced by ageing farmers, uncertainty of farm succession and land ownership issues, price-cost squeeze, weak rural local economies and the trend towards increasingly individualistic values, thus, the challenges of bioenergy crop adoption.

Present research aims to address this knowledge gap by exploring the perceptions of farmers on bioenergy crops while taking into consideration the socio-economic, political and environmental advantages and disadvantages of their cultivation for local development and highlighting notions of institutional complexity as expressed by beliefs and emotions of insecurity between and among local stakeholders. In addition, land use, trust issues and their influence on local stakeholders' decisions and in turn, social, economic and environmental sustainability-viability are addressed. The chapter is based on the findings of a focus group session of farmers and relevant stakeholders in ASTRIDA (pseudonym) central Greece and is structured as follows: Firstly, factors influencing farmers' decision-making and interest in growing bioenergy crops as addressed in the literature are reviewed. Secondly, we present the implemented methodology and analysis. Thirdly, the results are presented in terms of the perception of respondents as regards environmental and socio-economic, political impacts etc. Finally, our findings are discussed and conclusions are provided.

2 Farmers and Rural Stakeholders in the Context of Agricultural Bioenergy: A Brief Literature Review

Current research addressing farmers' willingness to cultivate energy crops is usually based on and focuses on case studies—projects in a specific location with well-defined feedstock and conversion technologies–single projects (Wahlund et al. 2002), single countries (Foxon et al. 2005; Mantziaris et al. 2017; Negro et al. 2007; Panoutsou 2008), or multiple projects or countries (Soldatos 2015; McCormick and Kåberger 2007). This approach reveals that farmers' adoption of bioenergy crop cultivation depends on several factors and has the potential to account for the context, including social aspects the land use framework or the different stakeholders), economic aspects (energy prices or the demand structure), and environmental aspects (land use change and suitability of crops for the region) (Buchholz et al. 2009). The reasons for these differences lie in the distinct characteristics of the cases and the various disciplinary backgrounds of the researchers.

Among researchers, wide agreement appears to exist that it is "…non-economic and technical challenges that are hindering bioenergy, rather than economic and technical issues" (McCormick and Kåberger 2007, p. 447; Sherrington et al. 2008) and these are crucial determinants in the adoption of bioenergy crops (Altman and Johnson 2008; Panoutsou 2008; Paulrud and Laitila 2010; Soldatos 2015; Mantziaris et al. 2017). However, when it comes to the relative importance of these factors, there is little agreement in the literature although initial primary focus is being given to economic factors (Hipple and Duffy 2002; Jensen et al. 2007; Wiesenthal et al. 2009; Domac et al. 2009; Panoutsou 2008; Soldatos 2015; Mantziaris et al. 2017).

It is also worth mentioning that past and recent literature tends to agree that participation in environmental-sustainability policy measures depends on farmers' attitudes, perceptions and behavioural responses (Wilson 1996; Wynn et al. 2001), as well as RESs fitting the farming systems (Aylott et al. 2010; Jensen et al. 2007; Domac et al. 2009; Rossi and Hinrichs 2011; Arabatzis and Malesios 2013; Ostwald et al. 2013). Nevertheless, how and to what extent the above factors affect farmers' decisions on whether to adopt bioenergy crops cultivation or not, is not uniformly agreed upon. More sophisticated approaches have also been proposed, paying attention to a number of farmers' personal characteristics, such as motivations, values and attitudes. Wilson (1996) refers to 'external factors' and 'internal factors' while Wynn et al. (2001) incorporate farmers' behaviour into policy analysis, proposing the following classification model: (i) physical farm factors (size, type, etc.); (ii) farmer characteristics (age, education, succession etc.); (iii) business factors (tenure, etc.) and (iv) situational factors (interface between farmers and the policy: the amount of information received, advisors' negotiating skills supporting the entry decision, etc.).

Focusing on farmers' attitudes towards environmental protection and sustainability issues, Morris and Potter (1995) propose a 'participation spectrum' that classifies farmers into four groups: (i) active participants, who adopt voluntary policy measures for both environmental protection and financial reasons; (ii) passive adopters who enter environmental and sustainability measures mainly for financial reasons; (iii) conditional non-adopters who would participate under some circumstances (e.g. easier-to-fit measures and higher payments); and (iv) resistant non-adopters who do not participate in rural development schemes. This farmers' 'participation spectrum' has been further analysed by including some structural and household factors related to participation (Wilson 1996; Wilson and Hart 2001). Although criticised in the case of non-fixed maximum payment ceilings (Wilson 1996; Falconer 2000), this classification can provide suggestions to policy-makers as to how measures can be made more attractive, especially when there is low uptake of RES. In conclusion, the literature confirms that business factors, farm structures, farmers' characteristics, attitudes and situational factors, all interact in the farm response to bio-energy policies (Jensen et al. 2007; Panoutsou 2008; Sherrington et al. 2008; Aylott et al. 2010; Paulrud and Laitila 2010; Arabatzis and Malesios 2013; Ostwald et al. 2013), but gaps remain. Attitudes, perceptions and behavioural responses underlying RESs' uptake is therefore still an open question and a research challenge (Arabatzis and Malesios 2013; Chalikias et al. 2012; Kolovos et al. 2011; Chalikias 2010, 2013; Chalikias and Kolovos 2013; Mantziaris et al. 2017), especially in Greece where this field of research is rather new.

In Greece, previous studies have investigated biomass production and management and its impact upon local societies. Such studies have investigated issues of safety, optimum exploitation and rural revitalisation and have contextualised their findings in the current economic crisis (Arabatzis and Malesios 2013; Chalikias et al. 2012; Kolovos et al. 2011; Chalikias 2010, 2013; Chalikias and Kolovos 2013; Mantziaris et al. 2017). Our research considers the abovementioned factors but goes on to explore whether other factors may influence farmers' decisions as well.

Specifically, biomass policies are accompanied by significant social impacts and benefits which cannot always be estimated in monetary terms but lead to changes to the everyday habits of individuals/farmers (Jensen et al. 2007; Aylott et al. 2010; Rossi and Hinrichs 2011). According to Jones (2010), social impacts and benefits may be influenced by several factors including social capital parameters. Its main features are social trust, social norms, cultural perceptions and values, and the character of social networks (Field 2003; Jones 2010). Field (2003) distinguishes three types of social capital: Bonding, bridging, and linking. Bonding capital refers to relations within social groups, bridging refers to relations among different social groups, and linking refers to relations among individuals and groups that occupy different positions in social hierarchies. Trust, a subject of importance in the current research, can be linked to individuals (personal trust) or institutions and organisations (institutional trust).

Greek farmers comprise a special socio-economic category and played a key role in the country's processes of economic and political restructuring in the 19th and 20th century (Bika 2007; Papadopoulos 2008). In Greek rural areas, Common Agricultural Policies (CAP) and the Greek State in the 1980s generated price protectionism and a sense of income security to farmers. It also embedded the belief among farmers that social capital in the form of co-operation is unimportant, driving existing agricultural co-operations into decline (Iliopoulos and Valentinov 2012; Koutsou and Partalidou 2012). Nonetheless, Greek farmers nowadays, more than ever, are faced with a situation where they are imperatively obliged to become market-oriented and adapt to a globalised economy, under the influence of reduced CAP subsidies and meeting the targets of European Union energy directives such as 20% of energy requirements from renewable resources by the year 2020.

The rural development model which Greek society is currently experiencing is based on two premises: reliance upon the State, and EU protectionism and individualism. This has been a successful combination for the individual prosperity of many farmers for some decades, but has not always favoured sustainable development of Greek rural areas. Nowadays, with the reintroduction of protectionism and, individualism failing to guarantee farm survival, the search for a new development model for rural areas, based on renewable energy sources, may serve as an alluring alternative (European Commission 2012). Finally, it could be claimed that social costs and benefits of bioenergy crop cultivation may influence the decision of individuals to comply or cooperate with an environmental policy based on RES.

3 Methodology

The objective of the current research is to identify the factors that influence Greek farmers' and rural stakeholders' willingness to adopt bioenergy crops as farming practice. In addition to economic and financial considerations, the impact of socio-cultural and environmental factors has been assessed. Additionally, our group of farmer respondents was asked about their intentions regarding three bioenergy crops for which processing capacity was not under consideration at the time of the research: poplar, miscanthus and arundo. Farmers with significant production orientation towards energy crops and farmers with no such orientation were included in the sample.

Given the study's aim to understand the facilitators and constraints of the production and use of bio-energy crops by local stakeholders, focus group research has been implemented in the study because of this method's potential to better describe and understand complex and sometimes contradictory patterns of human perception, meaning and experience, particularly useful in research with marginalised or understudied social groups (Ragin 1994). Thematic analysis was our qualitative data analysis method.

3.1 Case—Study

ASTRIDA (pseudonym), a predominately rural area in central Greece with a population of approximately 113,000 (2011) and agriculture as the single most important sector (47.80% of the area's GDP; 38.41% of local employment) was selected as the locus of research for the following reasons: (1) the agricultural sector is one of the most important economic activities of the region which is lately facing economic difficulties; (2) decreasing employment opportunities, lack of alternative solutions for cropping and a constantly changing and complex policy situation has led to confusion among farmers, and conflicts with the government; (3) the climatic conditions favour the use of heat produced in a bioenergy scheme; (4) farmers had limited experience with energy crops (cynara) in the past; (5) the key informant in the region is the local development agency, ANASTRI (pseudonym) that has been prioritising energy plant cultivation among its activities for the past ten years. This is the reason why a new subsidiary has been established whose purpose is to organise the production and distribution of energy that results from biomass and biofuels.

3.2 Sample Selection and Description

Sample selection was based on the use of key informants in order to facilitate access to the field and assure sufficient sample size and wider possible stakeholder representation. A key informant, a person holding a position of responsibility in ANASTRI who has long and trusting relationships not only with potential participants but also members of the research team, was approached and provided with a list of potential participants' characteristics. The key informant was further given detailed instructions on how to approach (recruit) and inform potential participants about the goals of the study and the participants' role. ANASTRI's registry was the main pool of participants. Approval for the study has been provided by the Research Ethics Committee of the University of Crete.

A focus group meeting with a total of 14 participants was arranged on ANASTRI's premises in March 2014. The sample of the study included active farmers and livestock breeders (those that have farmed commercially the last two years), as well as other stakeholders, such as NGO members (i.e. environmental groups) and industrial stakeholders that use bio-energy crops to produce biomass and other forms of energy. Thus, the participants of the focus group involved several active farmers, a representative from a local environmental protection NGO, a representative of the industrial sector that uses bioenergy crops to produce energy and a local engineer, an expert in biomass.

A member of the research team moderated focus group discussion while assistant moderators monitored and taped the discussion, which lasted approximately two hours. Focus group discussion was based on a nine-theme discussion guide, as well as additional discussion probes.

The main goal of the focus group discussion was to investigate the different aspects of facilitators and the constraints regarding the production and use of bio-energy crops by local stakeholders. In order to avoid uninformed responses and to minimise the possibility that participants' views would be influenced by the views of other participants, the focus group meeting began with information about the economics of their production and questions about bio-energy crops in general, the relationship of producers and consumers with agriculture, food, nature, etc. The focal point of acceptance-rejection of production and use of bio-energy crops was assessed immediately afterwards.

Focus group discussions were transcribed and pseudoanonymised to safeguard against participant identification. Pseudononymised transcripts were used for the analysis. Data was analysed using a thematic analysis method with phrases as the unit of analysis. Phrases were then grouped into categories. Each researcher implemented this coding method individually. Final categories were discussed among researchers and agreed upon.

4 Findings

We identified various topics/frames and common arguments in the focus group discussion that might serve as possible barriers and drivers to the adoption of bio-energy crops in ASTRIDA. On the basis of this coding, the following themes emerged and are addressed here: (1) economic, (2) land-use (3) environmental (4) socio-cultural and (5) trust. A central focus was farmers' and local stakeholders' views on the potential and constraints for energy crops in their region, and/or nationally.

4.1 Economic Issues

Economic returns are clearly important to farmers. Whilst lifestyle and personal characteristics interact strongly with management decisions and a range of external factors (Damianakos 2002), profitability still remains a key factor underpinning and driving farm structure and structural change (Arabatzis and Malesios 2013) within the current context of the economic crisis. Still the solution, in this case the cultivation of bioenergy crops must fulfill two conditions: provide additional income without substituting current agricultural production. This is clearly indicated in the words of one of the participants: "…we are positive towards cultivating something different to what we already do if the new crop will add income to our main cultivation and not substitute it" (ASTRIDA, Focus Group, March 2014).

Despite their willingness to shift to something profitable, focus group participants were considerably cautious since their key restriction is shaped by economic uncertainty, lack of a target market, etc.:

> We are discussing energy producing cultivations without knowing where it will go, who will get it, which needs it will cover. It is what M…. said before, the question of distribution: who will get it and for what purpose… (ASTRIDA, Focus Group, March 2014).

Shortfalls in knowledge transfer and the technical support necessary for the adoption of bioenergy crop cultivation along with installation and machinery costs were identified by our participants as economic concerns. Limited backing for these concerns hinders farmers' shift from conventional cropping patterns for which they already have the equipment, experience and market outlet (Booth et al. 2009) to other types of farming, biomass farming in this case. In ASTRIDA, it appears that farmers have tried to 'make do' with what is available and 'innovate' in order to adapt to requirements of new farming practices:

> There is a deficit in knowledge and technical support. This does not mean that we cannot improve for example with the right equipment; because when you import new cultivations, you also need to import the equivalent equipment needed to carry out these new cultivations. We are trying, with existing methods, to obtain the best results… Well, we will see if we can succeed. In my opinion, we should first learn. We need to know that any plant we

choose, no matter which, has its own particular requirements regarding technical equipment, i.e. equipment for collection, alteration, transport… (ASTRIDA, Focus Group, March 2014).

4.2 Land Use Issues

As revealed in our focus group discussion, the farmers' decision on whether to adopt energy crop cultivation was closely linked to land use issues. Small size agricultural holdings–the average farm size in Greece is approximately 45 *strem-mata*[1] or 4.5 ha—and fragmentation of holdings characterise agricultural land in Greece. This pattern, for the most part, holds true for ASTRIDA as well, and is acknowledged by our research participants as a persistent restriction for the development of Greek agriculture. Small holdings and land fragmentation were viewed by our focus group participants as adversely impacting productivity and efficiency, further linked to wasted time and effort:

> Because if you rent 50 *stremmata* here and another 80 *stremmata* three kms away, you cannot do much… Therefore, anyone who has the disposition to do something tends to spend more time moving around than focusing on the productivity and efficiency of one farming place. He does not even have the time to discuss different pertinent issues with other farmers so that they could pursue common activities … (ASTRIDA, Focus Group, March 2014).

Various causes have been identified as linked to land fragmentation. First, there are land ownership policies which, for varying historical reasons, have been mostly circumstantial rather than the result of strategic spatial planning, and the current land registry system is deed-based rather than property-based. Second, in addition to the political economy of the agricultural sector in Greece, cultural practices which demand equal division of land between heirs, inheritance laws which facilitate or mandate and reaffirm such practices and past low prices of agricultural land, contribute to land fragmentation on the one hand and accumulation of land by vested individuals such as urban elites, in rural and urban areas (National Bank of Greece (NBG) 2014).

Land fragmentation as an exogenous imposition on farmers is summarised by our focus group participants as follows:

> It is bizarre for somebody to be a doctor or a lawyer living in Athens [the capital of Greece] and to have 500 *stremmata* of land either as an inheritance from his grandpa or cheaply bought…and of course, you have these politicians who try to implement policies that serve their own interest… It is time for the producer to be compensated accordingly… (ASTRIDA, Focus Group, March 2014).

[1]*Stremma* (plural: *stremmata*) is the area unit mostly used for agricultural land. A *stremma* equals 1000 m^2.

Within the context of land use consideration, however, additional issues emerged which are closely linked to themes which will be later addressed, namely, environmental and socio-cultural concerns. The first of these issues was indicative of a generation-gap, and the second was related to the notion of marginal land.

Younger focus group participants challenged many of the older farmers' practices of maintaining control of their land even after retirement–not renting it to younger, more active farmers, so they can increase their farming land—and simultaneously, using productive land to cultivate low farming intensity plants such as bioenergy crops. As one of the focus group participants put it:

> I think that older people remain 'bosses' of their agricultural holdings and persist with this till the end of their lives. But how may I convince these gentlemen here who are much older, to get involved with this? If you ask me to cultivate my best piece of land, as a pensioner, I wouldn't do that. The pensioner ought to go, to rest at home. This pensioner could get rent money out of his field better than the potential production return [wild artichoke]… (ASTRIDA, Focus Group, March 2014).

In ASTRIDA, energy crop production has enabled a small number of farmers to retire and thus, they have turned most of their land into energy crop production (i.e. wild artichoke); it is an attractive option in terms of grant support and limited working hours in the fields:

> I planted [wild artichoke] in good fields because it is not profitable to cultivate wheat or anything else apart from that … (ASTRIDA, Focus Group, March 2014).

Younger farmers however, and representatives of the local environmental and industry organisation oppose the use of productive land for energy crops. Although they do acknowledge that it may serve the needs of some people (mainly older farmers) they see negative implications of such cultivation:

> …The wild-artichoke is of certain interest; it suits some people who cannot cultivate cotton, wool or fruit and vegetables… they simply grow wild-artichoke for five years and they forget about any other agricultural activities. This particular cultivation is not sustainable for the product buyer nor the producer … (ASTRIDA, Focus Group, March 2014).

Thus far, our research has revealed land use considerations as inextricably linked to socio-political practices which hinder farm productivity and are particularly taxing for young active farmers. In relation to bioenergy crops, it appears that active farmers would consider it as a supplemental income generating activity. Nevertheless, land fragmentation and a generational decision-making conflict appear to negatively impact such decisions. Still, apart from the above-mentioned types of restrictions, many of the active farmers in our research voiced significant concern about the 'type of land' that should be dedicated to the cultivation of bioenergy crops:

> It seems that the idea of making good use of land where up to this day no productive plants were cultivated for food stuff, such as abandoned and/or arid land, is a good thing and very tempting. But it hides a few pitfalls. How do we define the border between different types of lands so that we can cultivate energy producing plants in areas of marginal productivity or

> in areas that need protection from land degradation? (ASTRIDA, Focus Group, March 2014).

Overall, focus group participants appear reluctant to use their most productive land for energy crops, and tend to agree that bioenergy crops should be grown on a farm's least productive land, including marginal arable land with low-yield expectations. Several more experienced farmers stressed the direct relationship between yield and land productivity, although crop management was also important:

> Whatever has no negative impact on the country should be adopted in order to help the agricultural sector. We should map the area before we try out a new plant. In this way, we can try and use every little piece of land to make it productive. But only for areas where nothing exists [that are abandoned] and not for those already cultivated… (ASTRIDA, Focus Group, March 2014).

It follows that farmers would consider bioenergy crop cultivation as a potential alternative with positive impact on agriculture, provided that agronomic conditions are suitable and they have confidence that energy crop production will be competitive in the long term through strategic planning and structural support available to farmers.

4.3 Environmental Issues

As aforementioned, farmers' concerns over the type of land which could be dedicated to bioenergy crops are also linked to their concerns over the environmental impact of biomass cultivation. The conversion from traditional food crops to bioenergy crops has been linked to several barriers whose impact intensified when farmers were asked to consider cultivation of perennial crops such as arundo, poplar and miscanthus, all of which require new machinery and must be further supported by research and agronomic information (Booth et al. 2009). Focus group participants arguing against bioenergy crops, referred to environmental frames suggesting awareness of emerging debates and conflicts over the environmental impact of bioenergy production. Specifically, participants questioned the argument of bioenergy crops' energy efficiency and pollution reduction. At least one of the participants explicitly excluded biomass from 'clean' energy forms and pointed to pollutants related with their production:

> In general, the biomass issue in Greece raises many questions regarding pellet quality. There are many locally produced as well as imported pellets of doubtful quality. A number of problems have started appearing regarding burners, boilers, pollutants and consequently, biomass gets a bad name. After all, biomass is not a clean energy source. Wind power, wind generators, and water power are clean energy sources. The issues concerning biomass are not clear yet. A lot of things should be sorted out first… (ASTRIDA, Focus Group, March 2014).

The perceived negative impact of perennial roots on field drainage was revealed as another important factor in the decision on whether or not to grow energy crops. Interestingly, these concerns were raised by potential growers rather than experienced ones, the latter group suggesting that careful site selection could safeguard against these problems. The uncertain cost of returning land to alternative production was yet another concern expressed by a number of farmers, along with field drainage and natural deterioration:

> It doesn't pay back. Now, why do we still keep going? Because we are like that [shows his hands tied together]. Now it is mud. You try to touch it and you touch mud. There is no way a single plant [wild artichoke] won't grow up to the ceiling. (ASTRIDA, Focus Group, March 2014).

In their discussion of biomass, the most common positive frames focused on the impact biomass might have on environmental quality. Many described biomass as an energy-efficient alternative that would lower fossil fuel consumption. This supports the view that the majority of participants shared a strong view that some of the more 'marginal' land within ASTRIDA could be used for biomass cropping, and there is some support for this perspective in the literature (Booth et al. 2009; Aylott et al. 2010).

4.4 Socio-cultural Issues

Among the socio-cultural issues linked to farmers' decisions to cultivate bioenergy crops, concerns over the lack of cooperation, increased individualism and low social prestige of farming, appeared as the most significant.

Cooperative organisation was perceived by focus group participants as a way to address some of the obstacles of becoming involved in the bioenergy sector:

> We don't have the knowledge of cooperation; of how one can help the other. In this country, there is a deficit of collectivity. We need to understand what collectivity means, to work together and make gains all together … (ASTRIDA, Focus Group, March 2014).

Co-operation between farmers was further seen as essential for the development of new and more- profitable crops. Sharing of experiences on the establishing, managing, harvesting, processing and marketing of crops, and collective purchasing of required machinery were some of the benefits of cooperation, according to our participants.

It was noted however, that historically, crop production has not resulted in significant levels of co-operation between farmers. On the contrary, longstanding *socio*-cultural values of individualism and beliefs about the importance of farming as a 'social condition' and not an occupation (Damianakos 2002), could make it difficult to form and sustain effective cooperative farming organisations (Iliopoulos and Valentinov 2012). Moreover, this type of behaviour may also illustrate the transition which the Greek rural society is currently experiencing: reliance upon

State and EU protectionism and individualism, and a lack of social capital (Koutsou and Partalidou 2012):

> … Individualism … holds very strong and that's the reason we have difficulties in establishing trustful relationships and agricultural co-operations. Everyone undermines each other… (ASTRIDA, Focus Group, March 2014).

Although not directly related to farmers' decisions to adopt bioenergy crop cultivation, another factor that was found to influence farmers' overall attitude towards development, risk taking, innovation etc. was the low social prestige and unattractiveness of farming as an occupation. Although the majority of farmers appeared to appreciate aspects of farming, such as autonomy, working outdoors etc., they emphatically portrayed it as an unattractive professional activity; something they would not want their children to do.

> When I told my wife to buy my son a cow, to become a stockbreeder, she reacted negatively. She didn't like this job … (ASTRIDA, Focus Group, March 2014).

For many of our focus group participants, farming was not a choice but rather the forced outcome of poor educational attainment, or lack of other occupational options:

> Can someone tell me of a person who was a good student, finished school, went on with further studies, and decided to become a farmer? Or of somebody who completed university studies and came back to farming, unless he is an agronomist… [and]…can exploit certain opportunities… (ASTRIDA, Focus Group, March 2014).

4.5 Issues of Trust

Focus group participants expressed strong pessimism about rural revitalisation. With regards to the potential of agricultural bioenergy production, this was shaped by a generalised lack of trust towards State Institutions at regional and local level, and companies as well.

Lack of trust towards State Institutions appears to be linked to a multitude of factors such as perceived and real inefficiencies, corruption, increased bureaucratisation, and the failing or disappearance of government extension services which had been the main avenue of information to farmers on this scheme (Charatsari et al. 2011). A particularly critical source of mistrust for farmers in our focus group was the CAP; since the '80s and on the basis of the CAP, subsidies were provided to farmers according to the magnitude of production, thus favouring specific geographic locations (i.e. the plains) and specific socio-economic categories (e.g. large-scale farms). As a result, focus group respondents asserted, large amounts of cash became available through informal networks and was spent mostly on urban dwellings and other amenities, intensifying thus, social distrust.

Today, the multifaceted crisis is impacting the rural sector financially through recession on jobs and production, as well as on scarcity of credit and cash

(Gkartzios 2013). The lack of adaptive capacity of farmers is due to past maladaptive policies and often linked to weak or corrupt government bodies and representatives:

> The 'system' has its own way because it is corrupt. It does not care about the weak but only supports the strong ones. Our politicians are crooked… (ASTRIDA, Focus Group, March 2014).

Farmers tended to express dissatisfaction and cynicism towards State Institutions based on their previous experiences, including lack of transparency, inappropriate handling of issues, inconsistency, bureaucracy. The following quote is typical of this sentiment:

> ...and think of how many 'papers' one should collect if you have to get involved with these kinds of transactions. A whole storeroom full of 'papers', in case the Bank decides to check it out. It is a pity for people to waste their lives in this way as if they don't have anything else more important to do… bureaucracy is…Well…here we are, trying to set up an alteration manufacture, and we need three years to get all the paperwork together (ASTRIDA, Focus Group, March 2014).

Corruption, however, and over-bureaucratisation are not the only sources of mistrust. Inefficiency and lack of commitment to the interests of the farmers and stockbreeders as a further indication of a generalised political crisis were additional factors of distrust:

> …and I would relate that to a more general crisis of our political system. These political elites do not want to have any relation with all this. Just think that in highly fertile meadows, they have installed solar cells (ASTRIDA, Focus Group, March 2014).

The government, however, was not the only entity in which farmers lacked trust. There was an important element of mistrust towards private companies:

> Personally, I got 'burned' a couple of times by companies. For example, there was this company with milk-thistle, the X Co, and another one, involved with cotton wool. We were taken for a ride as far as the milk-thistle was concerned (ASTRIDA, Focus Group, March 2014).
>
> …and there were certain companies which could not meet their financial obligations [bounced cheques]; Well…in this respect, I could call myself stupid. I lost 14 million (€) through these bad transactions… (ASTRIDA, Focus Group, March 2014).

Trust also had a major influence on communication and information sources regarding investment issues. Most farmers said they would like to receive more information about new crops and investment procedures, but this information must be delivered by a reliable government body, which of course they could not name, based on their past negative experiences. Although they show high levels of mistrust towards any State Institution, still, they opt for incentives, guidance and information from the State.

5 Discussion and Conclusions

Consistent with our expectations, our findings supported the view that almost none of the focus group respondents would consider growing energy crops without the support of a policy grant, due to the high upfront capital costs and uncertainties over resulting net income. As Mattison and Norris (2007) note, farmers are unlikely to grow bioenergy crops unless encouraging policy measures are in place. In addition, farmers have few incentives to grow energy crops without the existence of competitive markets, and potential users have little incentive to invest in the technologies necessary to develop these markets if supply is both limited and uncertain. Last but not least, the role of social capital is regarded as a significant parameter influencing farmer perceptions regarding the adoption of RES policies (Jones 2010). It would appear that the local stakeholders' rejection of social capital's premises, like bonding over subsidies and co-operation among farmers and other social groups, has hindered their ability to revitalise their rural area.

Some of our participants saw government programmes as a means of supporting specific socio-economic groups. These individuals cited their experiences with government agricultural programmes, which they saw as favouring retired farmers and absentee large-scale landlords and not helping young farmers like themselves. They suspected that State programmes supporting agricultural bioenergy would have similar outcomes, whether intentionally or not. Although skeptical about government support of agricultural bioenergy, focus group participants included the assertion that public funds need to move down the supply chain and focus more on (especially smaller) farmers as opposed to concentrating incentives at the refinery level: "They must involve us more," insisted one young full-time farmer in ASTRIDA (ASTRIDA, Focus Group, March 2014), making a plea for farmers' expressive contribution and meaningful participation in bio economy decisions.

The majority of respondents also perceived that without intervention to stimulate the bioenergy market, it may take a long time for this 'critical goal'—bioenergy production—to be achieved. In addition, farmers were concerned about how growing bioenergy crops would fit into their current farming operations. Similarly, Mattison and Norris (2007) found that the farmers' main concern was whether it would fit into their business plan. There was also broad recognition that producer groups and cooperatives could play a key role in establishing new markets, since end-users are unlikely to deal with individual growers.

Co-operation between farmers was another frame mentioned and was seen as essential if new and more profitable markets are to be developed. Areas where co-operation should bring benefits include the sharing of experiences and thus bonding capital on establishing, managing, harvesting, processing and marketing of the crops, and collective purchasing of required machinery. The importance of producer groups was acknowledged, but with awareness that they may not be the most impartial sources of information on energy crop production. The main concerns related to information on potential costs and returns, and the more problematic aspects of production.

These assertions accept the motivational factors (unknowns and uncertainties) of agricultural bioenergy, but also suggest questions about the trade-offs among capability, profitability, productivity and social concerns, as well as ecological consideration. Although there was optimism about the potential to address energy supply issues in contributing to national energy security, research participants did not see energy security as the primary goal for agricultural bioenergy development. Instead, the rural revitalisation benefits of bioenergy were the most desired, but also the least expected.

Finally, the study raises some questions regarding the adaptive capacity of the Greek agricultural sector, which has long been considered in crisis, at least with regard to structural factors such as employment and earnings. Strengthening this adaptive capacity calls for a further improved understanding of the involved motivational factors to develop effective policies at the national and European levels. On the other hand, increased global food prices, and the liberalisation of energy markets (leading to reduced subsidies for pilot projects) may be underlying driving forces hampering the continued development of energy crop production.

Although the economic or business incentives for changing production systems are strong, factors concerning values, knowledge, and land-use patterns are crucial for a change in production system. There has been no systematic investigation into the literature on why farmers decide to stay with a production system or change it, or what motivational factors serve as drivers of, or barriers to the cultivation of specific crops. In order to support farmers in expanding their production of energy crops, as stated by the Greek Government and EU,[2] this knowledge gap needs to be filled.

Moreover, focus group participants' concerns about the challenge of rural revitalisation were also shaped by their views on the possibilities of smaller-scale cooperative organisations within agricultural bioenergy. Many participants advocated for cooperative investment at various levels of agricultural bioenergy development as a way to revitalise the rural economy, and only a few expected State support to play a significant role in promoting the interests of farmers and rural economies. Since protectionism is being reduced and individualism is failing to guarantee farm livelihoods, this type of organisation, the co-operative, may serve as an alluring alternative for sustainable rural development. This affirms Levi et al. (2005) who claim that sometimes distrust can provide the foundation on which to build effective and efficient organisations.

Overall, project participants—farmers and non-farmers—demonstrated strong awareness of national and global economic and political trends shaping agricultural bioenergy and the ways in which their community and local economy might fit into, or potentially clash with, those developments. To a large extent, it is the clashes that participants highlighted most in their comments. They doubted that high levels of

---

[2]See more on this in Commissioner Phil Hogan's speech at the Greek parliament on October, 6, 2016. Available at:https://ec.europa.eu/agriculture/sites/agriculture/files/docs/speeches/hogan-2016-10-05-greek-parliament_en.pdf.

investment would create economically sustainable opportunities for farmers like themselves, and while they saw some need for government support, they also recognised its limitations. Their greatest hope for local rural revitalisation outcomes, through agricultural bioenergy, rested in locally-integrated cooperative enterprises and the devotion of bioenergy crops to marginal lands. Yet, they identified farm structures (i.e. size of land, stocking rates) and farmers' characteristics (i.e. young farmers, elderly farmers and farmers with no successors or education), as dominating the economic and business constraints (i.e. income, tenure) of bio-energy crop adoption.

The seemingly contradictory perspectives of young farmers, retired energy producers and environmental stakeholders reflect the complexity and trade-offs to be negotiated for a smooth transition to an energy future that includes bioenergy. These views help to counter the more instrumental renderings of the people and places expected to produce bioenergy crops by airing the attitudes of the farmers (Wilson 1996). In addition, evidence of mixed views on the benefits of an agricultural bio economy suggests some possible behavioural consequences. Most participants in this research had strong agrarian roots and histories, with work ties and social commitments to other facets of their rural communities. Their abiding concern about the need to revitalise their rural communities could mean that they, and farmers like them, will closely scrutinise proposed arrangements to produce arundo, miscanthus and poplar for commercial energy applications. For some rural stakeholders and farmers, individual incentives and technical assistance may be necessary, but not sufficient to compel production of arundo, miscanthus and poplar for bioenergy, given their skepticism about potential or durable gains for the local community. Their resignation about existing economic arrangements, which tend to favour the large-scale solutions of absentee landlords or retired farmers, however, may also suggest that some rural residents may be quietly critical, but will ultimately accept whatever structures are established by the more dominant actors in the bio economy. Downstream handlers, processors and energy consumers need to recognise these broader concerns of arable farmers and rural stakeholders. Such recognition has obvious instrumental implications as it may stimulate ideas for how to ensure necessary food production, especially when absentee landowners tie up the land. But such recognition could also put the design of economic and social relations for the bio economy on a sounder ethical footing than some past models for agriculture. Such design would attempt to more deliberately incorporate and balance both interests and influence up and down the bioenergy supply chain.

The findings of this research are based on a qualitative focus group study of a specific bioenergy project (WATBIO) and cannot be generalised to rural people and places everywhere. It can, however, offer insight for inquiry and practice into other settings, particularly in other European countries (Booth et al. 2009). Given the need to dedicate high-quality arable land to food production, governments may target perennial crops for energy uses, on marginal farmland. As in ASTRIDA, regions of relative agricultural marginalisation in Europe have often experienced economic and social marginalisation, as well (Booth et al. 2009). Further research should be conducted on how the objective conditions of rural marginalisation correspond to rural residents' subjective knowledge and experience of marginalisation. Comparative research in other settings could shed light on how social and cultural histories influence societies' abilities to redirect rural land use and economic development toward bioenergy end-goals.

References

Aggelopoulos S, Arabatzis G, Pauloudi A (2009) Agri-environmental policy of EU: the case of organic farming in Greece. Fresenius Environ Bull 18(6):1018–1024

Altman I, Johnson T (2008) The choice of organizational form as a non-technical barrier to agro-bioenergy industry development. Biomass Bioenerg 32:28–34

Arabatzis G (2008) The individual and social characteristics of poplar investors-cultivators and the factors that affect the size of poplar plantations according to the EU regulation 2080/92. Agricu Econ Rev 9(2):86–95

Arabatzis G, Malesios Ch (2013) Pro-environmental attitudes of users and not users of fuelwood in a rural area of Greece. Renew Sustain Energy Rev 22:621–630

Arabatzis G, Christopoulou O, Soutsas K (2006) The EEC regulation 2080/92 about forest measures in agriculture: the case of poplar plantations in Greece. Int J Ecodyn 1(3):245–257

Aylott M, Casella E, Farrall E, Taylor G (2010) Estimating the supply of biomass from short-rotation coppice in England, given social, economic and environmental constraints to land availability. Biofuels 1:719–727

Bika Z (2007) A survey of academic approaches to agrarian transformation in postwar Greece. J Peasant Stud 34:69–90

Booth E, Walker R, Bell J, McCracken D, Curry J (2009) An assessment of the potential impact on UK agriculture and the environment of meeting renewable feedstock Demands. Scottish Agricultural College, Aberdeen

Buchholz T, Rametsteiner E, Volk TA, Luzadis VA (2009) Multi criteria analysis for bioenergy systems assessments. Energy Policy 37:484–495

Chalikias MS (2010) Forest management and rural development in Northern Greece: the case of Pella prefecture. J Food Agric Environ 8(2):940–944

Chalikias MS (2013) Citizens' views in Southern Greece part I: the forests. J Environ Prot Ecol 14 (2):405–413

Chalikias MS, Kolovos KGK (2013) Citizens' views in Southern Greece part II. Contribution of forests to quality of life. J Environ Prot Ecol 14(2):629–637

Chalikias MS, Kyriakopoulos GL, Goulionis JE, Apostolidis GK (2012) Investigation of the parameters affecting fuelwoods' consumption in the Southern Greece region. J Food Agric Environ 10(1):885–889

Charatsari C, Papadaki-Klavdianou A, Michailidis A (2011) Farmers as consumers of agricultural education services: willingness to pay and spend time. J Agric Educ Ext 17(3):253–266

Damianakos S (2002) From peasant to farmer: Greek rural society towards globalization. Exantas and EKKE, Athens (in Greek)

Dauvergne P, Neville KJ (2009) The changing North-South and South-South political economy of biofuels. Third World Quart 30(6):1087–1102

Dauvergne P, Neville KJ (2010) Forests, food, and fuel in the tropics: the uneven social and ecological consequences of the emerging political economy of biofuels. J Peasant Stud 37(4): 631–660

Domac J, Richards K, Risovic S (2009) Socio-economic drivers in implementing bioenergy projects. Biomass Bioenerg 28(2):97–106

European Commission (2011) The future of rural development policy. In Agricultural policy perspectives briefs. Brief no. 4. Available at: http://ec.europa.eu/agriculture/policy-perspectives/policy-briefs/04en.pdf
European Commission (2012) Common agricultural policy: a business relationship between Europe and farmers. EU Publishing Agency, Luxembourg
Falconer K (2000) Farm-level constraints on agri-environmental scheme participation: a transactional perspective. J Rural Stud 16:379–394
Field J (2003) Social capital. Routledge, New York
Flach B, Bendz K, Krautgartner R, Lieberz S (2013) EU-27 biofuels annual report. Report no: NL3034. USDA Foreign Agricultural Service GAIN
Foxon TJ, Gross R, Chase A, Howes J, Arnall A, Anderson D (2005) UK innovation systems for new and renewable energy technologies: drivers, barriers and systems failures. Energy Policy 33:2123–2137
Gkartzios M (2013) 'Leaving Athens': narratives of counterurbanisation in times of crisis. J Rural Stud 32:158–167
Hipple PC, Duffy MD (2002) Farmer motivations for adoption of switchgrass. In: Janich J, Whipkey A (eds) Trends in new crops and new uses. ASHA Press, Alexandria, VA, pp 252–266
Iliopoulos C, Valentinov V (2012) Opportunism in agricultural cooperatives in Greece. Outlook Agric 41(1):15–19
Jensen K, Clark CD, Ellis P, English B, Menard J, Walsh M (2007) Farmer willingness to grow switchgrass for energy production. Biomass Bioenerg 31:773–781
Jones N (2010) Investigating the influence of social costs and benefits of environmental policies through social capital theory. Policy Sci 43(3):229–244. https://doi.org/10.1007/s11077-009-9107-1
Kolovos K, Kyriakopoulos G, Chalikias MS (2011) Coevaluation of basic woodfuel types used as alternative heating sources to existing energy network. J Environ Prot Ecol 12(2):733–742
Koutsou S, Partalidou M (2012) Pursuing knowledge and innovation through collective actions: the case of young farmers in Greece. J Agric Educ Ext 18(5):445–460
Levi M, Cook K, Hardi R (2005) Co-operation without trust, Russel Sage Foundation, New York
Manolas N (2007) The energy sector in Greece: trends and perspectives. CPER, Athens (in Greek)
Mantziaris S, Iliopoulos C, Theodorakopoulou I, Petropoulou E (2017) Perennial energy crops vs. durum wheat in low input lands: economic analysis of a Greek case study. Renew Sustain Energy Rev 80:789–800
Mattas K, Tsakiridou E, Michailidis A, Karelakis C (2015) Sunflower: a very high value crop? Aristotle University of Thessaloniki, School of Agriculture, Forestry and Environmental Sciences, Department of Agricultural Economics, Thessaloniki
Mattison EHA, Norris K (2007) Intentions of UK farmers toward biofuel crop production: implications for policy targets and land use change. Environ Sci Technol 41:5589–5594
McCormick K, Kåberger T (2007) Key barriers for bioenergy in Europe: economic conditions, know-how and institutional capacity, and supply chain co-ordination. Biomass Bioenerg 31:443–452
McMichael P (2009) A food regime genealogy. J Peasant Stud 36(1):139–169
Morris C, Potter C (1995) Recruiting the new conservationists: farmers' adoption of agri-environmental schemes in the UK. J Rural Stud 11:51–63
National Bank of Greece (2014) Land: an asset with great potential for Greece but with significant development challenges. Sectoral report. Available at: https://www.nbg.gr/greek/the-group/press-office/e-spot/reports/Documents/Land_and_Tourism.pdf
Negro SO, Hekkert MP, Smits RE (2007) Explaining the failure of the Dutch innovation system for biomass digestion—a functional analysis. Energy Policy 35:925–938
Ostwald M, Jonsson A, Wibeck V, Asplund T (2013) Mapping energy crop cultivation and identifying motivational factors among Swedish farmers'. Biomass Bioenerg 50:25–34
Palmieri N, Bonaventura-Forleo M, Suardi A, Coaloa D, Pari L (2014) Rapeseed for energy production: environmental impacts and cultivation methods. Biomass Bioenerg 69:1–11

Panoutsou K (2008) Bioenergy in Greece: policies, diffusion framework and stakeholder interactions. Energy Policy 36:3674–3685

Papadopoulos AG (2008) Aspects of social differentiation in the Greek countryside. Family farming, strategies, immigrants and flexibility. In Zakopoulou E, Kasimis C, Louloudis L (eds), *Agriculturalism, society and space.* Athens: Plethron (in Greek)

Paulrud S, Laitila T (2010) Farmers' attitudes about growing energy crops: a choice experiment approach. Biomass Bioenerg 34:1770–1779

Ragin CC (1994) Constructing social research. Pine Forge, Thousand Oaks, CA

Rossi M, Hinrichs C (2011) Hope and scepticism: farmer and local community views on socio-economic benefits of agricultural bioenergy. Biomass Bioenerg 35:1418–1428

Sherrington C, Bartley J, Moran D (2008) Farm-level constraints on the domestic supply of perennial energy crops in the UK. Energy Policy 36:2504–2512

Soldatos P (2015) Bioenerg Res 8:1562. https://doi.org/10.1007/s12155-015-9678-y

Soldatos P, Lychnaras V, Panoutsou C, Cosentino SL (2010) Economic viability of energy crops in the EU: the farmer's point of view. Biofuels Bioprod Bioref 4:637–657. View online at Wiley Online Library wileyonlinelibrary.com. https://doi.org/10.1002/bbb.257

Wahlund B, Yan J, Westermark M (2002) A total energy system of fuel upgrading by drying biomass feedstock for cogeneration: a case study of Skellefteå bioenergy combine. Biomass Bioenerg 23:271–281

Wiesenthal T, Leduc G, Christidis P, Schade B, Pelkmans L, Govaerts L, Georgopoulos P (2009) Biofuel support policies in Europe: lessons learnt for the long way ahead. Renew Sustain Energy Rev 13:789–800

Wilson GA (1996) Farmer environmental attitudes and ESA participation. Geoforum 27:115–131

Wilson GA, Hart K (2001) Farmer participation in agri-environmental schemes: towards conservation-oriented thinking? Sociol Rural 41:254–274

Wustenhagen R, Wolsink M, Burer MJ (2007) Social acceptance of renewable energy innovation: an introduction to the concept. Energy Policy 35(5):2683–2691

Wynn G, Crabtree B, Potts J (2001) Modeling farmer entry into the environmentally sensitive area schemes in Scotland'. J Agric Econ 52:65–82

Bioenergy Development in Sweden—Frameworks for Success

Anders Chr. Hansen and Anna Berlina

Abstract Sweden has gone through a transition substituting oil by bioenergy, in particular in the district heating and CHP sector. This paper focusses on the bioenergy-to-grid value chains and the institutional frameworks that have been essential for the development: Policy instruments addressing the competitive conditions of bioenergy-to-grid, multilevel governance enabling the entrepreneurship of local bioenergy communities and an institutional framework for controlling sustainability risks. We identify the policy instruments changing the competition conditions for bioenergy-to-grid and classify them according to the value chain links they address. We further asses the sustainability risks associated with bioenergy-to-grid value chains. We find that they have been manageable in Sweden, although the institutional framework is not flawless. We find that some degree of bottom-up governance has been essential in the multilevel governance. Empirically, the study is based on local examples from Jämtland and Västernorrland counties, government databases and legislative and planning documents.

Keywords Bioenergy · District heating · Sweden · Sustainability risk

1 The Transition to Bioenergy

The Swedish experience with development of bioenergy attracts increasing attention as similar transition processes would be welcomed in many other countries. It raises the question of which policies and framework conditions can enable a successful transition and according to which criteria it is successful. These questions

A. Chr.Hansen (✉)
Poppelvej 5B, 2630 Taastrup, Denmark
e-mail: anders@hanseco.dk

A. Berlina
Tartu, Estonia
e-mail: anna.berlina@nordregio.se

W. Leal Filho et al. (eds.), *Towards a Sustainable Bioeconomy: Principles, Challenges and Perspectives*, World Sustainability Series,
https://doi.org/10.1007/978-3-319-73028-8_23

are addressed in the present study.[1] We focus on the development of value chains transforming biomass, other waste, waste heat and electricity to heating, power and cooling, the bioenergy-to-grid (BtGR) value chains.

Previous research has described the policy instruments of the institutional framework (Regeringskansliet 2010), placed them in historical context (Di Lucia and Ericsson 2014; Brandel 2015) and in multiple levels of perspectives (Di Lucia and Ericsson 2014), related them to the technology innovation system (TIS) functions (Bergek et al. 2008; Energimyndigheten 2014) and in other ways (Ericsson et al. 2004; Nilsson et al. 2004; Ericsson 2005; Björheden 2006; Wang 2006; Hektor et al. 2011; OECD 2014; Andersson 2015; Ericsson and Werner 2016; Werner 2017).

We address the question by first, describing the conversion or remodelling of the physical structures delivering heating services through BtGR value chains. Then, we examine three sets of institutional frameworks enabling the development. First, the institutional frameworks enabling economically viable investments in BtGR value chains. Second, institutional frameworks allowing and empowering entrepreneurs to develop the value chains. Third, the institutional frameworks directing the development to follow a sustainable path. Finally, lessons learned and conclusions are summarized.

Substitution of oil was a strategic priority after the oil crisis in 1973 and a programme for oil substitution was introduced, initially motivated by security of supply concerns, but by 1980 with a clear ecological and renewable energy agenda (Andersson 2007).

From the 90s and onwards, bioenergy has been the expanding energy source, whereas oil consumption has contracted (Fig. 1).

Five bioenergy value chains are identified in Fig. 1. Firewood and pellets heating of residential and service sector buildings contributes with 3–4% of the total energy supply. Bioenergy in forest industry contributes with 11% by 2015. Bio-propellants (biofuels for transport) are just emerging and may be the expansive element in the 20s.

The most expansive bioenergy value chain in the recent decades has been the bioenergy-to-grid (BtGR) value chain. Its primary input is biomass and its primary output is district heating in combination with power. Other inputs including peat, fossil fuels and electricity also contribute and the grid suppliers also diversify into district cooling and biogas. Its biofuel consumption contributed with 1% to total energy consumption in 1983 and 10% by 2015.

This chapter focuses on the development of the BtGR value chain and the institutional frameworks that have allowed it to develop.

---

[1]The study is based on work preparing a comparative analysis of Sweden and Norway as part of the TRIBORN (Triple Bottom Line Outcomes for Bioenergy Development and Innovation in Rural Norway) project and the authors gratefully acknowledge the support for the project granted by the Norwegian Research Council.

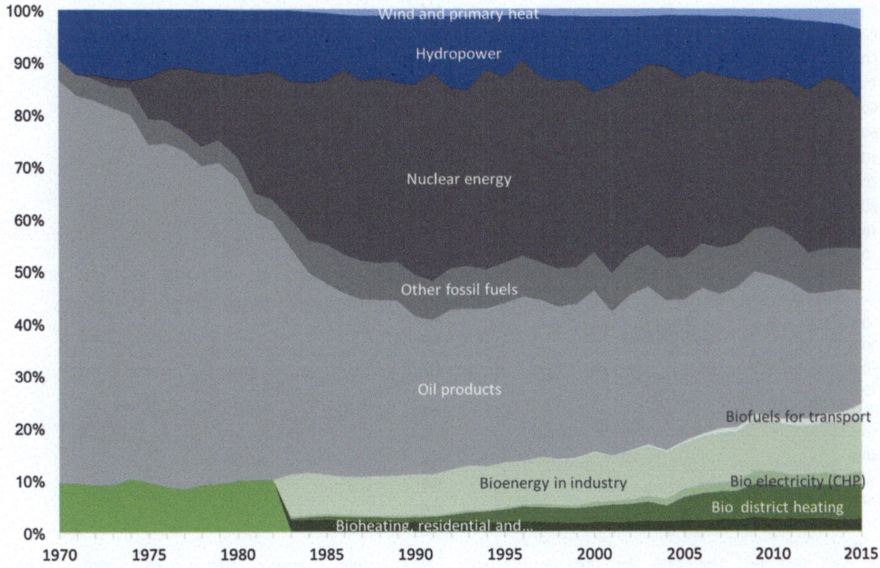

Fig. 1 Share of energy carriers in total energy supply in Sweden, 1970–2015 (%). *Note* The supply of electricity from nuclear power plants available for final consumption is approximately a third of the total supply of nuclear energy. *Source* Authors' calculations based on the Swedish Energy Agency (2017)

The consumption of biofuels in the BtGR chain is mainly a product of the heat demand, the share of district heating in the heat market and the share of biofuels in the market for input fuels.

District heating has increased market shares in multi-dwelling and non-residential buildings leaving around a fifth and a tenth respectively to electrical heating. Oil heating is expected to be competed out of the market by 2020. The substitution process is supported by increasing price trends for competing energy sources. The price trend was reversed by 2011 and it coincides with the stagnation in the substitution processes, a series of warmer winters and the approaching saturation point for district heating. The heated residential area is increasing (Swedish Energy Agency 2017), but the heating market is expected to decline due to higher energy standards of the building stock.

In district heating, biofuels have subsumed oil as the dominating heating fuel.

The district heating share of the heating market (Fig. 2) and the biomass share of the market for district heating fuels (Fig. 3) are both approaching their maximum. Future developments of the value chain are likely to take form as more valuable services rather than a continued quantitative expansion.

The reduction of fossil energy use in floor space heating was responsible for 46% of the Swedish reduction of GHG emissions since 1990. The reduction of fossil energy use for power and heat generation was responsible for another 10% (Authors' calculations' based on SCB 2017).

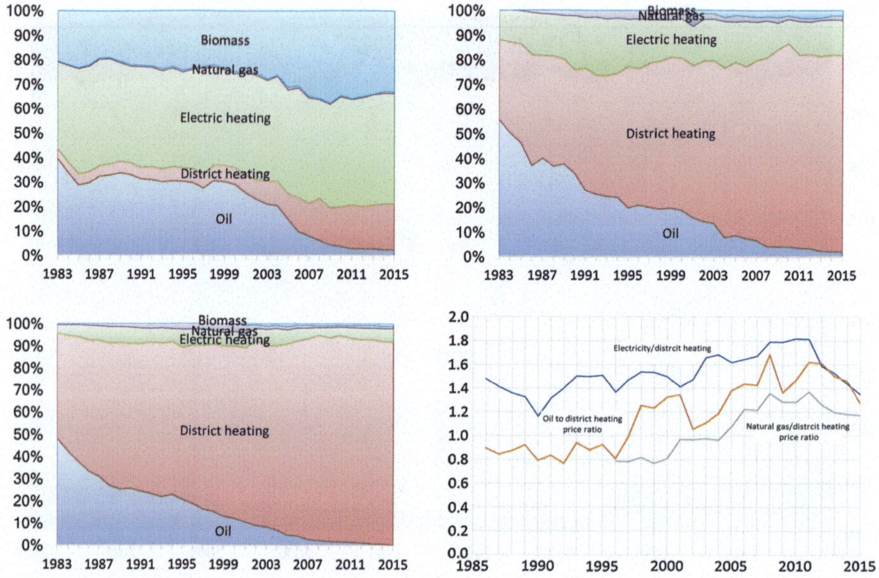

Fig. 2 Market shares in Swedish heating demand, 1970–2015 (%). Electricity, heating oil and natural gas price ratios, 1985–2015 (Ratio). *Source* Authors' calculations based on the Swedish Energy Agency (2017)

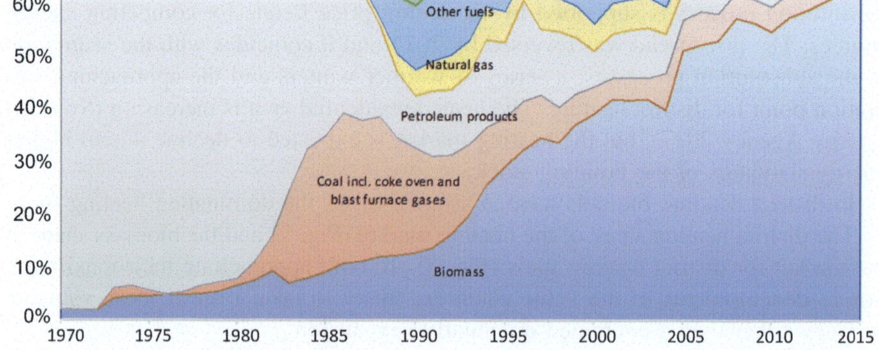

Fig. 3 Market shares of energy carriers for conversion to district heating in Sweden, 1970–2015 (%). *Source* Authors' calculations based on the Swedish Energy Agency (2017)

2 Bioenergy-to-Grid Chains Are Local

Nuclear and coal power plants are by nature centralised power sources. Bioenergy-to-grid chains are local as they convert low cost local biomass to heat satisfying local needs. The local availability of low-cost biomass sources (biomass with no valuable use) and other low-cost energy sources makes a business-case for BtGR. Fuels with a low energy density remain low-cost as long as they are "short travelled", which does not exclude the use of biomass from more distant sources to maintain the continuity of flows in the desired fuel-mix.

Thus, the intensity of bioenergy varies between regions. Contrary to the metropolitan regions, the forest rich Mid-Swedish regions are bioenergy intense, have lower incomes and are in decline.

Östersund, Härnösand and Sundsvall have a relatively high, 6%, primary sector contribution to the economy and lower per capita incomes. The level of employment has stagnated in Jämtland and declined in Västernorrland 2000–14 (Map 1b).

BtGR actors were visited and interviews were conducted with local and regional authorities, forest owners' associations, private forest companies as well as research and business networks. The focus of the interviews was on the role of bioenergy in producing positive social, economic and environmental outcomes in rural areas.

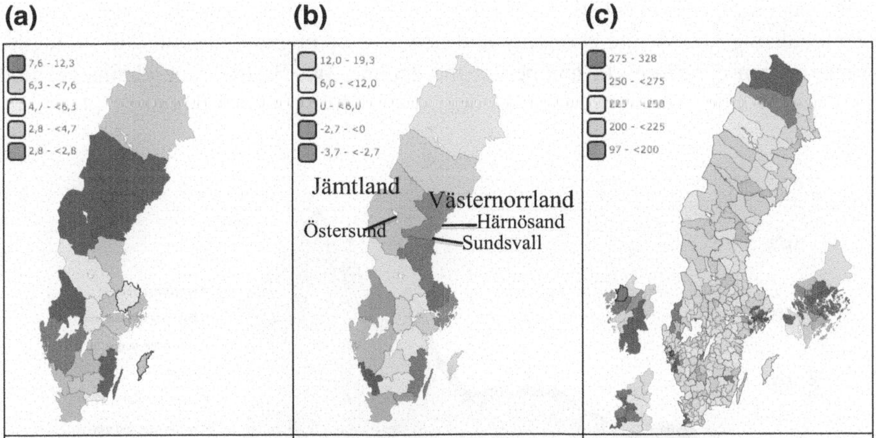

Intermediate use of biomass for generation of power and heat per capita by region. Average of 2009-14. MWh per capita.	Growth in employment 2000-14. Per cent.	Median income 2012 by municipality. 1000 SEK.

Map 1 Regional patterns in **a** bioenergy production, **b** employment growth and **c** median income. **b** shows the location of the case areas Östersund (Jämtland County) and Härnösand and Sundsvall (Västernorrland County). *Source* Author's calculations based on Statistics Sweden data (SCB 2016)

Jämtkraft AB is the leading force in the BtGR in Jämtland transforming primarily local biomass (forest residues, wood chips, recycled construction wood, peat and bark) to heat and power. At the supply side, the expansion of BtGR value chain has been primarily driven by the forest owners' association Skogsägarna Norrskog and in recent years by the forest company SCA Skog to some extent.

The public-private network of actors contributing to the development of bioenergy also includes the municipal and regional authorities, private entrepreneurs and "driven individuals" in developing of farm based biogas plants (Jämtkraft 2014).

Figure 4 shows the transition from fossil to bio-fuels over almost half a century at the local scale. Peat and coal were important substitutes in the beginning of the transition process, biomass and flue gas condensation at the end.

The publicly owned local utility HEMAB AB is the leading centre in Härnösand BtGR chain. At the supply side, the development has been driven by biomass and waste heat providers (SCA Bionorr, KBAB Ådalen, Norrskog, etc.) and residents contributing with food waste. A CHP was built in Härnösand in 2002 mainly fuelled with woodchips, peat and bark obtained from within a radius of about 100 km.

The pellets factory SCA Bionorr in Härnösand is part of the now international industry corporation SCA. Bioenergy has also become an important field of innovation for companies like SCA. It engages in production of biofuels, has converted fossil energy boilers to biofuel boilers and delivers waste heat to the Sundsval BtGR chain.

Peripheral regions in Sweden and other countries are marginalised by the trends of the globalising knowledge economy and negative demographic trends. In principle, bioenergy can contribute to hampering the decline, but BtGR value chains are like other energy value chains very capital intensive with little direct labour involved.

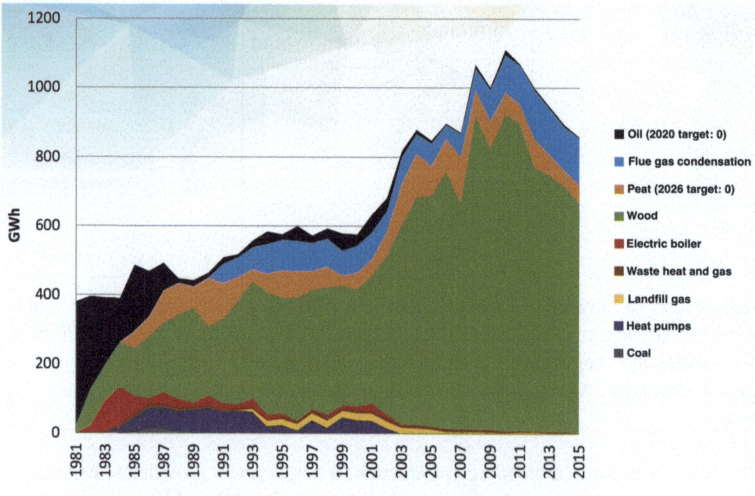

Fig. 4 The fuel switch from fossil to biomass in Jämtkraft, 1980–2015. *Source* Reprinted and adapted with permission from Ulf Lindqvist (Lindqvist, 2016)

The employment effects of substituting fossil energy by bioenergy are important because Sweden does not produce fossil fuel, but produces almost all the bioenergy it consumes. This effect of substituting import appears even at the regional level. HEMAB and Jämtkraft estimate that their bioenergy value chains generate around 130 and 400 local jobs, respectively.

3 Institutional Framework for BtGR Development

3.1 Policy Instruments in the Framework

The development of the bioenergy-to-grid value chains proceeded step-by-step rather than according to a master-plan. The independent Biofuel Commission (Jordbruksdepartementet 1992) did, however, provide a coherent and detailed analysis of challenges and solutions. In the following, we collect the various policy instruments that were actually applied to guide the development of BtGR chains (Table 1).

Table 1 Policy instruments addressing competition conditions at various links of the bioenergy-to-grid value chain

Final use of BtGR energy	Building code, municipal heat demand, mandatory heating grid connection, *CO_2-, energy, SO_2 and NO_X taxes, wholesale electricity and fuel prices, conversion subsidies and allowances*
Grid distribution	Heating and cooling, power, biogas
BtGR transformation[1]	*CO_2-, energy, SO_2 and NO_X taxes, wholesale electricity and fuel prices, green certificates, ETS, investment grants*
Biofuel processing	Chips, pellets, briquettes, municipal waste, sawdust, shavings, bark, wood pieces, black liqueur, other industry and construction waste, bio-based oil/coal/gas/waste heat/power
Low-cost biomass supply	Forestry residue collection standards, sorting, deposition ban, *deposition tax*, wastewater regulation, *perennial tree cultivation*, (Forestry: measures addressing afforestation, replanting, thinning, fertilizing)

Note: Main categories of biofuel and grid energy in parenthesis. Economic instruments in italics.

Note Main categories of biofuel and grid energy in parenthesis. Economic instruments in italics

[a]The terms "transformation" and "conversion" are often used synonymously to describe the generation of power or heat from fuels. Here, we use "transformation"

Markets fail to mediate the value of environmental qualities and of creating knowledge of how to use and adapt new technical solutions. This deprives solutions creating such values of competitiveness. A variety of policy instruments—economic and administrative, informative and R&D funding instruments—are applied to change the conditions for competition compensating the loss of competitiveness. Table 1 shows all of the policy instruments categorized according to the links in the value chain.

3.2 Final Use of BtGR Energy

The building code was reformed in 1994 instituting functional requirements rather than detailed technical prescriptions on solutions (Boverket 2011). Recent changes towards the near-zero energy standard pursue neutrality in the choice of technical solutions. Conversion to water borne heating installations allow district heating to compete against direct electrical heating in existing buildings. Economic incentives to conversion have been granted in the form of subsidies and tax allowances.

Heating grids beyond 20 km need a government concession. New buildings on municipality developed land can be required to connect to the grid (Nygårds 2011). Together with the heating demand of municipal buildings, the minimum scale required for competitiveness can be reached faster. Not all municipalities, however, use this option. Östersund municipality and Jämtkraft, for instance, prefer voluntary connection based on competitive offers (Jämtkraft 2016).

The economic instruments supporting the competitiveness of district heating on the heating market include taxes on CO_2-emissions, energy and sulphur. The taxes add to the cost of heating oil and direct electrical heating, whereas biofuels are carbon neutral and exempt from the energy tax.

They are incentives to convert to district heating and—outside the grid areas—to heat pumps and pellets burners. They also enable cost remuneration of biofuels upstream in the BtGR.

3.3 BtGR Transformation Link

CO_2- and energy, sulphur and NO_X taxes also make bioenergy solutions competitive in the link transforming biomass to grid energy, as biofuel use is neither taxed by CO_2- or energy taxes.

Within the ETS sector, also fossil and peat fuels have been exempt as the ETS emission allowance price would provide the same incentive if the price was equal to the tax rate. The ETS market, however, does not provide the intended incentives due to an oversupply of allowances. It has pressed the allowance price down to a level where it does not have a significant incentive effect.

Investment subsidies have been granted since the 80s under a succession of programmes. Pioneering initiatives in the 80s were followed by an investment grant programme from 1991 in addition to the CO_2-tax introduced in 1991. It was succeeded by the local investment programme (LIP) in 1998 providing subsidies to locally developed plans for investments in sustainable development with a focus on renewal of the built environment. In 2002 LIP was succeeded by the climate investment programme (KLIMP) focusing more on energy from waste based CHP plants (Naturvårdsverket 2013). From 2002 other electricity generating plants have received green certificates for the amount of renewable electricity delivered instead.

The municipality of Östersund realized 12 LIP and 23 Klimp projects, receiving approximately SEK 73 million of support. The projects stimulated the expansion of district heating network and CHP production, biogas production at Göviken sewerage plant etc. (Carbon Climate Registry 2012). Jämtkraft AB has benefited from investment grants as well as green certificates and CO_2-taxes through its long history. The green certificate scheme was essential for making its BtGR-switchover commercially viable (Jämtkraft 2016).

3.4 Low-Cost Biofuel Supply

The regulation of waste flows was also critical to the development of bioenergy. Prohibition of deposition of combustible waste and waste of biological origin and waste taxes was used to push the flow of biomass away from these destinations and towards the energy sector (e.g., Riksdagen 2001). The ban created a market of low or even negatively priced biofuels. It triggered the development of waste based BtGR in the 00s, which was also supported by the economic instruments including CO_2-and energy taxes and investment grants (KLIMP).

Forest residues are produced jointly with timber (if there is a market for it) and forest industry waste is produced jointly with forest industry products. Thus, the supply of low-cost biomass depends the harvested volume and volume processed by industry.

The Swedish framework does not include subsidies for biomass extraction. There are some policy instruments aiming at increasing the forest area and the production per area unit and they have indirect impact on the supply of low-cost biomass. The present supply of residues and dead wood for wood chips, for instance, is the result afforestation efforts through the 20th century as well as the naturally good conditions for forestry on Swedish land.

3.5 Competition and Competitiveness

It is customary to classify policy instruments in mandatory use of solutions (administratively prescribed or command-and-control solutions) or voluntary uses combining economic incentives with information and research funding. What is characteristic of the Swedish framework is, however, that the administrative instruments do not prescribe BtGR solutions, but ensure access to a market where they can compete with alternative solutions.

The competitiveness approach is essential because the development of the value chains is not merely a matter of quantitative multiplication of the technical solutions by 1980. It is rather a process of innovations in all links of the value chain. Combined heat and power generation and district heating was well known in the 90s, but how to design the system for biomass conversion and how to achieve the high transformation efficiency through flue gas condensation was not known at the time.

Economic instruments are almost exclusively applied in the final use and transformation links of the value chain, but not in the links collecting and processing biomass. There are few subsidies to the forest sector and they are granted to other purposes than energy, such as productivity, biodiversity and recreation.

The literature on the Swedish bioenergy experience converges towards the assessment that the introduction of CO_2 tax is the main key to the development of BtGR value chains so far (Nilsson et al. 2004; Lindblom and Rasmussen 2008; Ericsson 2009; International Energy Agency 2013; Andersson 2015).

Figure 5 shows the importance of energy and CO_2 taxes for the competitiveness of biofuels.

The heating oil prices before taxes shown in Fig. 5 are determined by the international oil price and the exchange rate. When the Swedish krona depreciates, imported goods like heating oil becomes more expensive and vice versa. These two factors have raised the costs of heating oil temporarily over a period from 2005 to 2014. In this period, the higher prices have been more important than the tax. This period is now over and without the CO_2-tax, there would be an incentive to slide back towards the unsustainable oil economy of the 70s.

In the future, however, it would be too optimistic to count on such periods with a helping hand from the international fossil fuel markets. As the transition proceeds, the demand for oil will be curbed and sooner or later the oil price will fall. There will be a need of more carbon taxing and carbon pricing, not less.

This is also the case for the ETS industries, which are exempt from the CO_2 tax. The EU allowance price is too small to be a significant incentive and it is not expected to rise to significant levels before late in the 20s. Alternatively, the exemption of ETS sector plants from the CO_2 tax can be ended, at least for plants that are not exposed to international competition. The Swedish government works on such measures and has appointed a commission to address the problem (Finansdepartementet 2017).

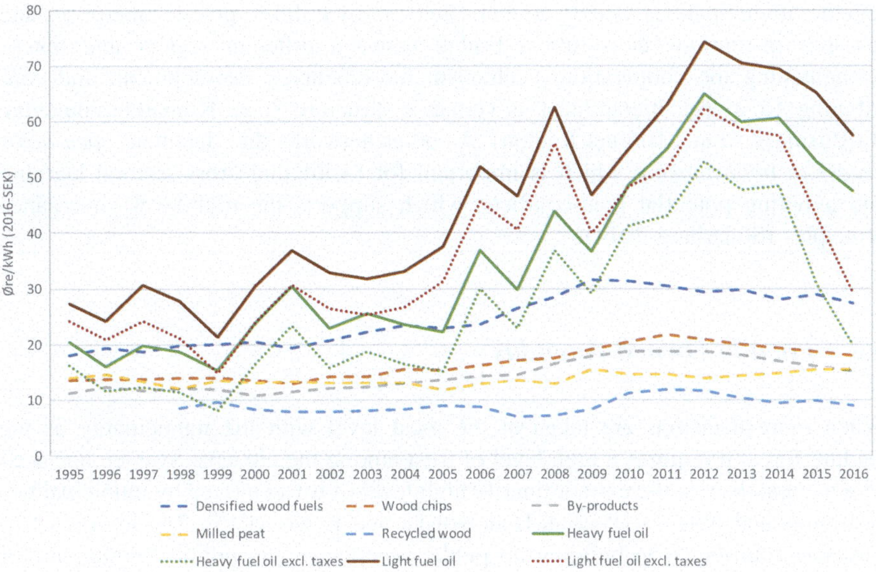

Fig. 5 Oil and solid biofuel prices for industrial and district heating use in Sweden 1995–2016, SEK (2016)/MWh (=10*øre/kWh). *Note* Deflated by the consumer price index. *Source* Author's calculations based on the Swedish Energy Agency database Energimyndigheten (2016)

4 Institutional Framework for Multi-level Governance

4.1 Bottom-Up Governance

Bioenergy in Sweden is developed in a multilevel governance setting, which both involves a "top-down" approach through agenda setting, policy goals and economic mechanisms and a more inclusive "bottom-up" approach, including the private sector, civil society and various groups of stakeholders. The change towards decentralisation shifts emphasis from top-down towards bottom-up.

Municipalities across Sweden have been responsible for energy planning and district heating at least since the large housing programmes of the 70s (Brandel 2015). The oil substitution policy introduced in the wake of the oil crises supported the switch from oil to domestic fuels and coal in plants such as Jämtkraft AB. As the direction of the switchover towards renewable energy became clearer, the municipalities were also invited to take a proactive role in it and in the 90s, the Agenda 21 processes combined bottom-up with local visions for sustainability. Investment grants supported local bioenergy CHP projects from 1991 and in another programme from 1997 (Naturvårdsverket 2013).

Multi-level governance of bioenergy development in Sweden is supported by collaboration arrangements between the municipalities/regions and multi-actor networks (Nykvist 2008; Söderberg 2014). Public-private networks bringing together municipalities, county councils and energy utilities, private enterprises and research institutions constitute a bridge between different sectors and levels, strengthening the cooperation mechanism for bioenergy development, and contributing to a less hierarchical governance structure (e.g. Klimatkommunerna, BioBusiness Arena, Biofuel Region). As other networks they facilitate *open actor access to policymaking* which is important for facilitating cross-sectoral learning and avoiding potential goal-conflicts, which supports the multilevel governance principles (Söderberg 2014).

4.2 Locally Owned Agendas

When more decisions are taken at the local level with the municipality as the leading force, it requires a high level of trust among the citizens. Sweden has been able to capitalise on the extraordinarily high level of trust enjoyed by municipalities and municipal officials (Węziak-Białowolska and Lewis 2015). The high level of trust has contributed to building up public acceptance and political legitimacy of bioenergy, as well as a high acceptance for community-wide technical solutions such as district heating (Ericsson 2009; McCormic et al. 2012).

However, the local agendas are not all about providing cheap heat. Jämtkraft perceives its role in society as improving the economic, ecological and social conditions for stakeholder groups including employees, suppliers, customers, partners, citizen groups, municipalities and society at large. It is in regular dialogue with the stakeholder groups (Jämtkraft 2014).

The open actor access to policymaking approach is likely to result in such agendas owned by the local actors rather than imposed by superior authorities. It reflects an active variety of trust unlike the more passive public acceptance many top-down governed agendas may hope for. The observations and interviews in Jämtland and Västernorrland gave the impression of a high level of active as well as passive trust in the local bioenergy agenda. Heat consumers stating a higher willingness to pay for energy from local sources and the presence of strongly engaged individuals in the development of biogas could be sign of such active trust. Sticking to biomass even when oil and coal prices decline could be another sign. Sustainability seems to be decisive for the public acceptance and local ownership of the bioenergy agenda. It is difficult to imagine the same local enthusiasm for an agenda of district heating based on heating oil. Swedish history and political culture also provides experience with similar local agendas: The traditional importance of forestry and forest industries for employment and livelihoods. The municipal responsibility for heat and electricity supply and the high level of trust. The generally high level of environmental awareness compared to other EU countries.

When municipalities and regions set more ambitious targets and goals than the national or EU targets, it is also a sign of local ownership to the agenda. The Climate Change Strategy for Jämtland 2014–2020 aims at a fossil free economy by 2030 (15 years ahead of the rest of Sweden) and at becoming a net-exporter of renewable energy (Jämtland County Administrative Board and Länsstyrelsen Jämtlands Län 2014). Östersund municipality is a member of the Swedish Klimatkommunerne and the EU Covenant of Mayors networks, committing to a faster pace of transition than Sweden and the EU, respectively.

The above examples indicate that the development of bioenergy value chains are motivated by more than the prices and costs and technical prescriptions controlled through the above institutional framework. The municipally owned utilities are formally limited companies, but they do not pursue profit maximisation as their main objective. Rather, they pursue cost minimization and maximization of their heat market share. Jämtkraft AB calls it *sustainable profitability* (Jämtkraft 2015).

5 Institutional Framework for Sustainability

5.1 Triple Bottom-Lines

Bioenergy is an indispensable part of a transition to a sustainable energy economy. Yet, it is also possible to develop unsustainable bioenergy value chains.

Preindustrial overexploitation of forest for energy and other purposes led to deforestation (McGrath et al. 2015). Europe is still struggling to restore the lost forest.

The IPCC concluded that bioenergy will play a critical role in climate change mitigation, but involves risks related to GHG emissions from land use, food security, water resources, biodiversity conservation and livelihoods (IPCC 2014).

The 2016 parliamentary consensus in the Swedish *Riksdagen* on a long-term framework for climate policy agrees on bioenergy as a key component of the low-emission economy, but also warns that "The expansion of biofuels is the factor that may lead to the deepest potential conflicts with other environmental goals…" (Government of Sweden 2016).

Poorly designed bioenergy value chains could entail economic, ecological and social sustainability risks. In such cases there would be a goal conflict between the primary ecological goal of bioenergy and other economic, ecological or social goals.

Sustainable development is now defined internationally as simultaneous progress in all three dimensions specified in 17 sustainable development goals (UN 2015). The development goals have also become a standard framework for corporate performance reporting labelled the triple bottom-line. In a addition to the ordinary financial bottom-line socially responsible corporations also require positive contributions to the sustainable development goals in the social and ecological dimensions. Following Elkington (1998), the standard ambition of a socially responsible corporation is to contribute to progress for people, planet and prosperity simultaneously. It is also applicable to BtGR chains.

Table 2 Bottom-lines suggested for bioenergy-to-grid value chains

Dimension	Risk	Budget (B)	Consumption (C)
Ecological (provision)	Deforestation or depletion of timber stock	Increment	Feelings
Ecological (regulating)	Carbon stock backsliding	Removals	Emissions
	Indirect emissions	9–13% of fossil alternative	Actual emissions
	NO_X and PM emissions	Rarefaction capacity	Emissions
	Nutrient deficit	Nutrient requirement	Nutrients in biomass outtake
	Pesticide concentration	Absorption capacity	Pesticides in hydrological flows
Ecological (cultural)	Loss of biodiversity and recreational services	Targets for biodiversity forest	Designation of forest areas for biodiversity
Economic	Market-power based pricing	Minimum cost pricing	Market power based pricing
	Misallocation of resources	Balance between uses	Unbalanced use
Social	Burden on low income budgets	Affordable burden of budget	Actual share of low income budgets

In practice, the triple bottom-line has more than three accounts. We have reviewed the literature and public debate and found ten risks that people raise concerns about (Table 2). They are grouped according to dimension—ecological, economic and social (or planet, prosperity and people). The ecological risks are grouped in sub-categories of ecosystem services—provisioning, regulating and cultural.

Many of the risks are actually risks related to forestry irrespective of the use of forest biomass. In a value chain perspective, however, the risks of losing forest ecosystem services are essential.

Forest ecosystems provide timber and overexploitation of this service leads to timber stock depletion or even deforestation. They also regulate the biogeochemical flows and these services are at risk of becoming disservices if forests are managed unsustainably. Finally, they provide cultural and recreational services.

We have also identified concerns that bioenergy value chains risk locking in sub-optimal resource use and expensive energy provision technologies. Finally, the distribution of any costs related to the transition involves risks in the social dimension.

The risks are taken seriously by Swedish government, cf. above, and has been so for generations. Therefore, it is instructive to examine the institutional frameworks set up for managing the risks.

Many of the risks are not specific for bioenergy. Air pollution, for instance, is a risk common to any plant or vehicle with combustion processes. Yet, it regularly appears as a challenge that must be solved in the development of new BtGR value chains.

Eight of the ten bottom-lines concern risks to ecosystem services. These risks are mitigated by environmental policies pursuing the intergenerational obligation: "... to pass on to the next generation a society in which the major environmental problems have been solved without causing increased environmental and health problems outside Sweden" (Author's translation). This obligation is specified in 16 goals defining ecological sustainability and 28 targets (SEPA 2017). The institutional framework for sustainable forestry includes among others the forestry act (Regeringskansliet 1979), its statutory instrument (Regeringskansliet 1993), administrative guidelines (Swedish Forestry Agency 2011), the Swedish Environmental Code (Regeringskansliet 1998) and industry standards including sustainability certificates (e.g., FSC 2017), ISO-standards (ISO 2015) and a variety of eco-labels. They have been developed since the forestry law of 1903 required replanting after felling. Other countries have a similar body of legislation concerning forestry and the environment and some of the provisions are EU law implemented in Swedish law. Sustainable bioenergy is also explicitly or implicitly defined in other long term strategies such as the energy and climate strategies (Regeringen 2008; Government of Sweden 2016).

5.2 *Timber Stock Depletion and Carbon Stock Backsliding*

A rough ecological bottom-line is the balance between the increment of standing timber and the felling of timber. This is the basic principle of sustainable forestry as defined three centuries ago (von Carlowitz 1732). Mandatory replanting was instituted in Swedish law by 1903 and it also prescribes a minimum tree age for harvesting. Afforestation programmes were carried out though the 20th century and the present resource wealth is the result of these efforts of past generations. It has allowed the harvested volume to increase from 60 Mcbm/yr around 1990 to more than 80 Mcbm/year in the 2010s, whereas the reserve of standing wood has increased simultaneously.

Gross felling grew by 1.1% annually while the reserve of standing wood and its carbon stock grew by 0.3–0.4% annually through 1988–2013 (Swedish Forest Agency 2014). Simultaneously with the development of the bioenergy value chains.

Several scenario analyses suggest that the generations of the present century can increase the harvested volume and the volume of standing timber, the carbon stock, the supply of biomass for energy conversion and the protected forest areas simultaneously through the 21st century (Egnell and Börjesson 2012; Claesson et al. 2015; Börjesson 2016; Pöyry 2016).

5.3 Direct and Indirect (Value Chain) Emissions

The renewable energy directive controls the risks of emissions upstream in the value chain by means of an ecological bottom-line. Maximum 9–13% of the emissions saved by replacing oil etc. can be emitted upstream in the collection, processing and transport of the biomass.

Local and regional forestry residues and industry waste easily complies with this limit. Jämtkraft estimates its value chain greenhouse gas (GHG) emissions in 2014 to be between 19 and 27 kg CO_2-equivalents per MWh energy (Jämtkraft 2014). The implied greenhouse gas savings are still about 90% of the emissions of an individual oil-fired burner.

The emissions are primarily due to the 7% peat in the fuel-mix, which is considerably higher than the 3% Swedish average. Peat resources are neither fossil nor biological. They emit more CO_2 by combustion than coal and entail upstream emissions from drainage of peat-lands.

The environmental protection agency finds that the bottom-line for peat is "in red" for biodiversity as well as cliamte reasons. Paradoxically, peat is exempt from energy tax and eligible for green certificates. The agency suggests to stop subsidising peat by certificates (Naturvårdsverket 2016).

The Swedish government has decided to maintain the subsidy. Arguments in favour of subsidies include its attractive properties in a fuel-mix with low-grade biomass fuels, making more low-grade biomass fuel competitive. The environmental conditions for supplying peat from a given bog will, however, be looked over.

Emissions of NO_X and particulate matter is a classical environmental risk, which is managed within the general framework for protecting the air quality.

5.4 Nutrient Cycles and Pesticides

New forest requires nutrients to grow and excessive outtake of biomass from the areas will deprive the land of its nutrients. Since 1998, the forest authority has issued guidelines for harvest of branches and tops, including a recommendation of leaving at least 20% for avoiding nutrients deficit and acidification and of returning ashes to the forest (Skogsstyrelsen 2010). Their outtake and the return of ashes are monitored, but not regulated. A critical report pointed out that ashes are only returned on a fraction of the area from which branches and tops are harvested, mainly due to lack of financing and delegation of responsibility (Aktörsrådet kring askåterföring 2014). Thus, locally there may be risks of nutrient deficit, but it still runs counter to the forestry interests and will hardly be a systematic risk.

Similarly, risks of pesticide use damaging ecosystems are being mitigated with the standard pesticide regulation and by innovation of chemicals free forest management methods. The FSC standard, covering about half of the Swedish forest is a driver of this innovation.

5.5 Loss of Biodiversity and Recreational Services

Sweden has since 1993 protected additionally 12–36 thousand hectares a year and continues to do so thereby increasing the protected area (SCB 2016). In addition to this, three quarters of the productive forest area are voluntarily certified[2] as sustainably managed forest (Lejon and Lidestav 2009; SCA Skog AB 2014; Skogsägarna Norrskog 2016).

The 2020 target is to add 150,000 ha protected forest, but the monitoring report concludes that despite considerable progress, it will be difficult to reach the target. This is, however, not because of increased logging activity, but rather because of the decision processes leading to protection (Naturvårdsverket 2017). Other measures include innovation in forest management methods without clear-cutting, that is, continuity forest.

The need of protection of areas of high value forest varies from region to region and the protection effort follows regional protection plans. In Jämtland, the regional plan states that it will be difficult to reach its 2020 targets. 84% of the forest area was continuity forest by 1957, only 21–32% was so in 2010 and only 5–16% will be in 2025.

The pressure from clearing on forest ecosystems, however, cannot be attributed to bioenergy. Forest biomass are primarily sold for processing in pulp and paper industry and building materials cf. above. Biomass for energy is a side-product of timber and wood products and a form of waste treatment after material use.

5.6 Market Power Based Pricing

The policy instruments reviewed above were generally designed to allow BtGR technologies to win market shares based on competitiveness. A market structure characterised by natural monopoly represents, however, a risk of pricing based on market power.

Until the 1996 reform of the energy market, this risk was mainly managed by organising the utilities as municipality owned non-profit corporations. The 1996 reform required that all power suppliers, including municipality owned CHP-plants, were "operated on a commercial basis and accounted for separately" (Riksdagen 1997, 8§2) (Authors' translation).

---

[2]The Programme for the Endorsement of Forest Certification (PEFC) or Forest Stewardship Council (FSC).

The liberalisation was, however, not accompanied by price control or other institutional frameworks, which has led to a debate on the impact on the actual prices. Several studies have found that the large heating companies supplying heating in several grids charge higher prices (Muren 2011; Abrahamsson et al. 2012; Holm 2013). Risk management measures proposed in the debate include price control or third-party access to the grids. Measures introduced include requirements of transparency, of consumer protection and of conducting cost-benefit analysis before investing (Energikommissionen 2017). It has led to the practice of a "price-dialogue" offering customers the right to cost information and to oppose price increases.

The question is whether this higher awareness on the risk of market power based pricing has had an effect on the actual price differences. We address this question by comparing the district heating price of the large heating corporations serving more than five municipalities with community based utilities serving their own municipality and maybe a neighbour municipality or two.

The seven largest district heating providers in Sweden include E. ON Värme Sverige, Solör Bioenergi Fjärrvärme AB, Rindi Energi AB, Vattenfall AB, Värmevärden AB, Neova AB, AB Fortum värme (co-ownership with Stockhom City). They supply between 6 and 20 municipalities each.[3] Rindi and Solör are part of the same corporation supplying a total of 26 municipalities.

We use data from the Nils Holgersson database, harmonizing prices and taxes to comparable building types.

The result is that these suppliers are still in the upper half of the price distribution (Fig. 6).

One of the seven large corporations supply at average prices, whereas the six others are all above the average price. Only nine of the 67 municipalities, they serve, pay heating prices below the average of all municipalities.

Comparing the average price of the two groups yields the results shown in Table 3.

The average heating price in municipalities supplied by large corporations (serving several municipalities) is 53 SEK/MWh or 6% higher than in the municipalities supplied by community based suppliers. There is no doubt about the significance of this finding as all municipalities and all suppliers are included in the dataset.

This finding indicates that the differences found in the above studies persist. Higher prices charged by larger companies are often seen in relation to the use of the larger volume as a basis for innovation, but this question is beyond the scope of this study.

---

[3]Skelletofteå Kraft also supplies 6 municipalities, but it is still, by the Authors' subjective judgement, a community company to such an extent that it is excluded from the group of large corporations.

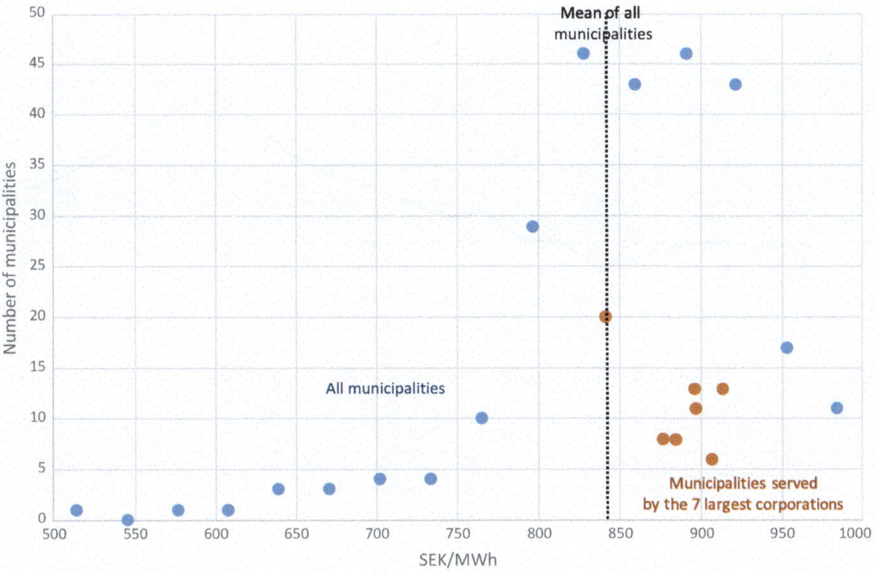

Fig. 6 Histogram of district heating price by municipality and by the seven largest corporations. SEK/MWh and number of municipalities. *Source* Authors' calculations based on the Nils Holgersson database 2016 (Nils Holgersson-gruppen 2016)

Table 3 Comparison of heating prices of large DH companies supplying many municipalities and community based suppliers, 2016

Seven large DH suppliers	882	SEK/MWh
Community based suppliers	829	SEK/MWh
Difference	53	SEK/MWh

SEK/MWh
Source Authors' calculations based on the Nils Holgersson database 2016 (Nils Holgersson-gruppen 2016)

5.7 *Misallocation of Forest Resources*

Industries processing resources from the forest often raise concerns about the risk that increased production and use of bioenergy will cause competition about the resources, they depend on. This could lead to higher prices and lower material use of wood (e.g., Söderholm et al. 2010).

The records of Swedish prices and production of wood since 1995 do, however, show a different pattern (Fig. 7).

The increasing production of bioenergy 1995–2016 did not lead to a decline in the use of the forest resource as construction materials and similar. It increased by 17%. Pulpwood used for pulp and paper increased by 28%.

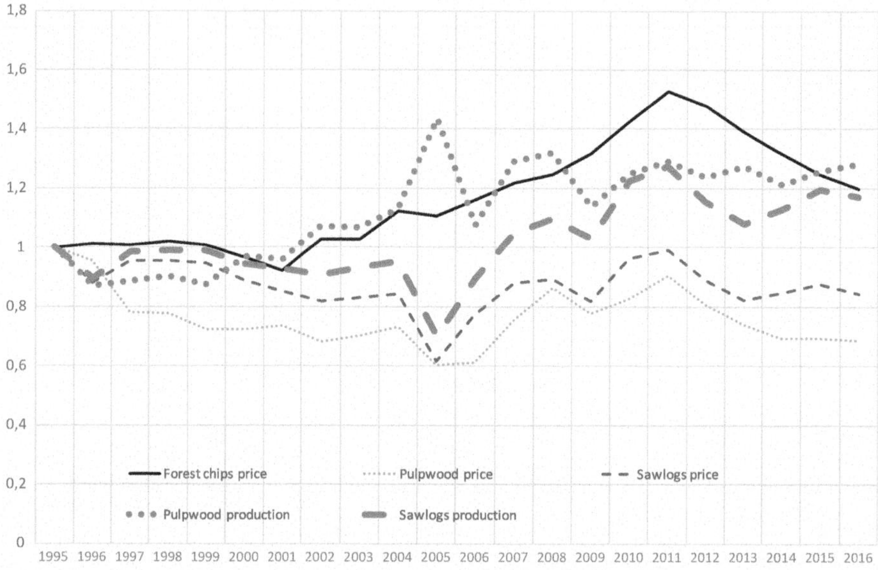

Fig. 7 Development of real prices and production of wood. Index 1995 = 1. *Note* Prices deflated with gross value added deflator. *Source* Author's calculations based on Swedish Forest Agency database (Skogsstyrelsen 2017) and Swedish national accounts (SCB 2017)

The risk of resource conflicts has been on the agenda in Sweden several times. The wood fibre act regulated the investments in bioenergy plants in 1987–93 to protect the supply of pulp wood to the forest industry, but it was abolished as the risk appeared to be limited (Jordbruksdepartementet 1992; Hektor and Hildebrand 2016). Since 1999, tall oil in energy use has been taxed in line with heating oil to secure supplies to the chemical industry. It has, however, not prevented the production of bio-fuels based on tall-oil in Sweden.

5.8 Burden on Low Income Budgets

Heating expenditures occupies a considerable share of low income household budgets. Innovation in the supply of heating may lead to temporarily higher heating costs, which again may place a heavy burden on low income household budgets. One indication on this risk is to compare the electricity and heating expenditure of households with small budgets (Fig. 8).

If looking at the unemployed, retired and other inactive persons, the heating and electricity share of the budgets of these groups are smaller in Sweden compared to other Nordic countries and the EU as a whole. Households in the northern parts of Sweden are exempt from taxes on electricity to avoid a heavy burden on household and service sector budgets. The downside is a risk of maintaing energy inefficient

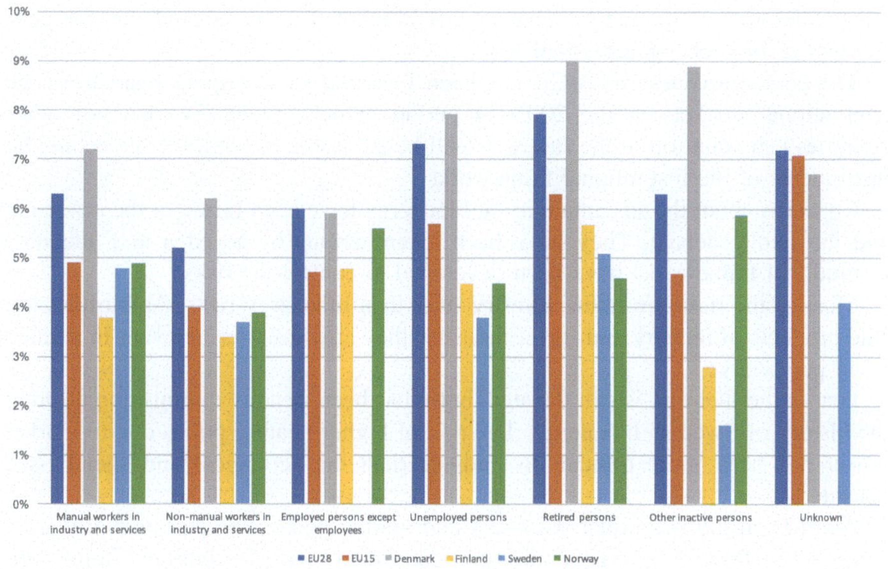

Fig. 8 Housing related energy share of all consumption expenditure in the EU and four Nordic countries, 2010 (%). *Sources* Author's calculations based on EUROSTAT (2017)

solutions. This risk is, however, manageable as households in the Northern regions can be compensated in many other ways.

6 Lessons Learned and Concluding Remarks

The progress of bioenergy in Sweden is the other side of the coin of nuclear phase-out and transition to a fossil free economy. The bioenergy development, which has been accomplished does not mount to a capacity that is close to matching the nuclear capacity, but a number of lessons can be drawn on the value chains feeding biomass to heat and power generation.

We identified three sets of institutional frameworks, which have been important for the successful development of the Swedish BtGR sector

- *Competition* conditions for BtGR competitiveness
- *Bottom-up* approaches in multilevel governance
- Conditions controlling the *sustainability risks.*

The policy instruments—administrative as well as economic—are designed to pave the way for *competitive* BtGR solutions. Economic instruments support the competitiveness of district heating against individual solutions and the competitiveness of biofuels against fossil fuels.[4]

The competitiveness of BtGR has been favoured by a helping hand from the international markets in the 2005–14 period, which neither Sweden nor other countries can count on in the future. It will be advisable to reinforce the economic instruments of the institutional frameworks.

Concerns about the sustainability of bioenergy have been raised in the academic and the public debate. They have been taken serious by Sweden and adequate institutional frameworks have been developed to control the risks.

Most of the risks are rather forestry risks than bioenergy risks. As biofuels are side-products of forestry and forest industry, they are, however, relevant in a value chain perspective.

Peat in the fuel-mix is still debated, but it has been steadily diminishing through the history of Swedish bioenergy. The risk of higher heating prices due to market power has been more difficult to mitigate than the ecological and social risks considered.

Apart from the risks related to peat and market power, the sustainability risks identified in Table 2 have generally been controlled by the institutional framework —in some cases through generations. These experience can be valuable for other countries.

Emphasis on bottom-up governance has enabled municipalities to seize their local opportunities and formulate locally owned agendas for development of sustainable bioenergy.

The entrepreneurs developing the BtGR value chains are municipalities with their utilities and networks of private suppliers and stakeholders. We call them *local bioenergy communities*. They are one of the secrets behind the Swedish success.

Policy instruments cannot be copied one-to-one by other countries, but the principles of competitive conditions, bottom-up governance and management of sustainability risks are probably useful in other countries aiming at making local bioenergy agendas realistic.

References

Aktörsrådet kring askåterföring (2014) Slutrapport från arbetet med aktörsrådet kring askåterföring

Abrahamsson K et al (2012) Uppvärmning i Sverige 2012. EI R2010:09, pp 1–74

Andersson L (2007) Bioenergi från jordbruket – en växande resurs, Statens offentliga utredningar (SOU). SOU 2007:36

Andersson K (2015) Bioenergy the Swedish experience

Bergek A et al (2008) Analyzing the functional dynamics of technological innovation systems: a scheme of analysis. Res Policy 37(3):407–429. https://doi.org/10.1016/j.respol.2007.12.003

[4]The support of competitiveness against peat is ambiguous, probably due to aspects of complementarity rather than substitutability in some uses.

Björheden R (2006) Drivers behind the development of forest energy in Sweden. Biomass Bioenerg 30(4):289–295. https://doi.org/10.1016/j.biombioe.2005.07.005
Börjesson P (2016) Potential för ökad tillförsel och avsättning av inhemsk biomassa i en växande svensk bioekonomi
Boverket (2011) Byggregler – en historisk översikt
Brandel M (2015) Översiktlig sammanställning/ analys av energipolitiska beslut mellan 1975 och 2009 i Sverige, pp 1–62 (December)
Carbon Climate Registry (2012) Östersund Municipality, Sweden
Claesson S et al (2015) *Skogliga konsekvensanalyser 2015 – SKA 15, Rapport 10/2015*
Di Lucia L, Ericsson K (2014) Low-carbon district heating in Sweden—examining a successful energy transition. Energy Res Social Sci 4(C):10–20. https://doi.org/10.1016/j.erss.2014.08.005
Egnell G, Börjesson P (2012) Theoretical versus market available supply of biomass for energy from long- rotation forestry and agriculture—Swedish experiences, p 23. IEA Bionergy Task 43, Report 2012: TR02
Elkington J (1998) Cannibals with forks: the triple bottom line of 21st century business. New Society Publishers, Gabriola Island
Energikommissionen (2017) Kraftsamling för framtidens energi
Energimyndigheten (2014) *Teknologiska innovationssystem inom energiområdet – En praktisk vägledning till identifiering av systemsvagheter som motiverar särskilda politiska åtaganden.* ER 2012:28. Available at: http://www.energimyndigheten.se/PageFiles/45332/2012-9091ER2014_23Teknologiskainnovationssystembilaga.pdf
Energimyndigheten (2016) Statistical database, Energipriser. Swedish Energy Agency. Available at: http://www.energimyndigheten.se/statistik/energipriser/?currentTab=1#mainheading. Accessed 22 Apr 2016
Ericsson K (2005) Corrigendum to? Bioenergy policy and market development in Finland and Sweden? [Energy Policy 32(2004): 1707–1721]. Energy Policy 33(1):129. https://doi.org/10.1016/j.enpol.2004.04.005
Ericsson K (2009) Introduction and development of the Swedish district heating systems. Critical factors and lessons learned. D5 of WP2 from the RES-H Policy project
Ericsson K, Werner S (2016) 'The introduction and expansion of biomass use in Swedish district heating systems. Biomass Bioenergy 94:57–65. https://doi.org/10.1016/j.biombioe.2016.08.011
Ericsson K et al (2004) Bioenergy policy and market development in Finland and Sweden. Energy Policy 32(15):1707–1721. https://doi.org/10.1016/S0301-4215(03)00161-7
EUROSTAT (2017) Eurostat database. Online statistical database. Available at: http://ec.europa.eu/eurostat/data/database. Accessed 27 Jan 2017
Finansdepartementet (2017) Vissa punktskattefrågor inför budget- propositionen för 2018
FSC (2017) FSC International. Forest Stewardship Council. Available at: https://ic.fsc.org/en. Accessed 6 Apr 2017
Government of Sweden (2016) Ett klimatpolitiskt ramverk för Sverige. 2016:21
Hektor B, Hildebrand D (2016) Cascading of woody biomass: definitions, policies and effects on international trade. (April)
Hektor B, Bruce L, Andersson K (2011) Country report Sweden. pp 1–9 (December)
Holm J (2013) Inför prisregleringen av fjärrvärme. Available at: http://www.konkurrensverket.se/globalassets/aktuellt/nyheter/infor-prisregleringen-av-fjarrvarme.pdf. Accessed 5 Oct 2017
International Energy Agency (2013) Energy policies of IEA countries 2013. Review Sweden
IPCC (2014) Climate Change 2014: Mitigation of Climate Change. https://doi.org/10.1017/CBO9781107415416
ISO (2015) Sustainability criteria for bioenergy. Available at: https://www.iso.org/obp/ui/#iso:std:iso:13065:ed-1:v1:en. Accessed 5 Apr 2017
Jämtkraft AB (2014) Jämtkrafts Års- och hållbarhetsredovisning 2014. Available at: https://issuu.com/jamtkraft/docs/ja____mtkraft_a____rs-_och_ha____ll/1?e=0/12704730
Jämtkraft (2015) Jämtkraft - förnybar jämtländsk energi

Jämtkraft AB (2016) Personal communication

Jämtland County Administrative Board and Länsstyrelsen Jämtlands Län (2014) Klimatstrategi för Jämtlands län, Diarienummer 423-1574-2014. Available at: http://www.lansstyrelsen.se/jamtland/SiteCollectionDocuments/Sv/publikationer/2014/Klimatstrategi-2014-2020.pdf

Jordbruksdepartementet (1992) Biobränsle för framtiden – slutbetänkanden av biobränslekommissionen. 1992:90

Lejon SB, Lidestav G (2009) Skogscertiering – vem, hur och varför? 11

Lindblom PG, Rasmussen RO (2008) Bioenergy and regional development in the Nordic countries. Nordregio Working Paper (5)

Lindqvist U (2016) Ta vara på energin – du kan värme andra, in presentation at the TRIBORN Seminar, Hadeland

McCormic K, Bomb C, Deurwaarder E (2012) Governance of biofuels for transport in Europe: lessons from Sweden and the UK. Biofuels 3(3):293–305

McGrath MJ et al (2015) Reconstructing European forest management from 1600 to 2010. Biogeosciences 12(14):4291–4316. https://doi.org/10.5194/bg-12-4291-2015

Muren A (2011) Exploatering eller reglering av naturliga monopol? Available at: http://www.ems.expertgrupp.se/Uploads/Documents/EMS-2011_2-till-webben.pdf. Accessed 5 Oct 2017

Naturvårdsverket (2013) Klimatinvesteringsprogrammen Klimp 2003–2012. Slutrapport

Naturvårdsverket (2016) Torvutvinningens och torvanvändningens klimat- och miljöpåverkan

Naturvårdsverket (2017) Etappmålen. Miljömålen – Årlig uppföljning av Sveriges nationella miljömål 2017. Available at: http://www.miljomal.se/Global/24_las_mer/rapporter/malansvariga_myndigheter/2017/etappmalen-2017.pdf. Accessed 5 Oct 2017

Nils Holgersson-gruppen (2016) Nils Holgersson| Rapport 2016. Available at: http://nilsholgersson.nu/rapporter/rapport-2016/. Accessed 5 Oct 2017

Nilsson LJ et al (2004) Seeing the wood for the trees: 25 years of renewable energy policy in Sweden. Energy Sustain Dev 8(1):67–81. https://doi.org/10.1016/S0973-0826(08)60392-0

Nygårds P (2011) Fjärrvärme i konkurrens. 2011:44

Nykvist B (2008) EPI in multi-level governance—a literature review

OECD (2014) OECD Environmental Performance Reviews: Sweden 2014. OECD Publishing

Pöyry (2016) 'Bioenergi från skog och skogsindustri'

Regeringen (2008) 'En sammanhållen klimat- och energipolitik', p Regeringens proposition 2008/09:162. Available at: http://www.regeringen.se/rattsdokument/proposition/2009/03/prop.-200809163/

Regeringskansliet (1979) 'Skogsvårdslag'. Svensk författningssamling 1979:429. Available at: http://www.riksdagen.se/sv/dokument-lagar/dokument/svensk-forfattningssamling/skogsvardslag-1979429_sfs-1979-429. Accessed 5 Oct 2017

Regeringskansliet (1993) 'Skogsvårdsförordning'. Svensk författningssamling 1993:1096. Available at: http://www.riksdagen.se/sv/dokument-lagar/dokument/svensk-forfattningssamling/skogsvardsforordning-19931096_sfs-1993-1096. Accessed 5 Oct 2017

Regeringskansliet (1998) 'Miljöbalk (1998:808)'. Svensk författningssamling 1998:808. Available at: http://www.riksdagen.se/sv/dokument-lagar/dokument/svensk-forfattningssamling/miljobalk-1998808_sfs-1998-808. Accessed 5 Oct 2017

Regeringskansliet (2010) The Swedish National Action Plan for the promotion of the use of renewable energy in accordance with Directive 2009/28/EC and the Commission Decision

Riksdagen (1997) 'Ellag (1997:857)'. Svensk författningssamling 1997:1997:857. Available at: https://www.riksdagen.se/sv/dokument-lagar/dokument/svensk-forfattningssamling/ellag-1997857_sfs-1997-857. Accessed 6 Oct 2017

Riksdagen (2001) Förordning (2001:512) om deponering av avfall Svensk författningssamling 2001:2001:512 t.o.m. SFS 2015:532. Available at: http://www.riksdagen.se/sv/dokument-lagar/dokument/svensk-forfattningssamling/forordning-2001512-om-deponering-av-avfall_sfs-2001-512. Accessed 26 Apr 2016

SCA Skog AB (2014) PEFC rapport certifieringsrevision 2014

SCB (2016) Statistikdatabasen. Available at: http://www.statistikdatabasen.scb.se/pxweb/sv/ssd/START__AM__AM0401__AM0401N/NAKUSysselAnkLAr/table/tableViewLayout1/?rxid=-f2a510a2-d407-4112-aa16-fd8f780b79ae#. Accessed 14 Apr 2016

SCB (2017) 'Statistikdatabasen'. Statistiska Centralbyrån. Available at: http://www.statistikdatabasen.scb.se/pxweb/sv/ssd/START__MI__MI0107/MI0107ELoFjarrN/table/tableViewLayout1/?rxid=bd097734-d7e3-4ca0-8725-613160c08d1c. Accessed 12 Apr 2017

SEPA (2017) Miljömål.se – om hur miljön mår och arbetet med Sveriges miljömål går. Swedish Environmental Protection Agency, Naturvårdsverket. Available at https://www.miljomal.se Accessed 17 Nov 2017

Skogsägarna Norrskog (2016) *Miljöcertifiering*

Skogsstyrelsen (2010) Skogsbränsle – Grenar och toppar. 1(2): 2010. Available at: http://www.skogsstyrelsen.se/Global/aga-och-bruka/Skogsbruk/Skogsbruksåtgärder/Skogsbränsle-Grenarochtoppar.pdf

Skogsstyrelsen (2017) Skogsstyrelsen - Statistikdatabas. Available at: http://pxweb.skogsstyrelsen.se/pxweb/sv/Skogsstyrelsensstatistikdatabas/?rxid=a6819262-3f5f-4ee9-b0f3-15984a277a41. Accessed 8 Oct 2017

Söderberg C (2014) What drives sub-national bioenergy development? Exploring cross-level implications of environmental policy integration in EU and Swedish bioenergy policy. Eur J Gov Econ 3(2): 119–137. http://www.ejge.org/index.php/ejge/issue/archive

Söderholm P, Brännlund R, Lundmark R (2010) *Kampen om skogen - koka, såga, bränna eller bevara?* SNS Förlag

Swedish Energy Agency (2017) Energy in Sweden 2017. Energimyndigheten. Available at: http://www.energimyndigheten.se/en/facts-and-figures/publications/. Accessed 1 Oct 2017

Swedish Forest Agency (2014) Skogsstatistisk årsbok 2014 Swedish Statistical Yearbook of Forestry

Swedish Forestry Agency (2011) 'Skogsstyrelsens föreskrifter och allmänna råd till Skogsvårdslagen 2011:17'. Skogstyrelsen, Skogsstyrelsens författningssamling. Available at: https://www.skogsstyrelsen.se/globalassets/lag-och-tillsyn/forfattningar/skogsvard/sksfs-2011-7-skogsstyrelsens-foreskrifter-och-allmanna-rad-till-skogsvardslagen.pdf. Accessed 5 Oct 2017

UN (2015) Transforming our world: the 2030 Agenda for sustainable development. Resolution adopted by the General Assembly on 25 Sept 2015. Available at: http://docbox.un.org/DocBox/docbox.nsf/GetFile?OpenAgent&DS=A/RES/70/1&Lang=E&Type=DOC

von Carlowitz HC (1732) Sylvicultura oeconomica – Anweisung zur wilden Baumzucht. 1st editi. Leipzig

Wang Y (2006) Renewable electricity in Sweden: an analysis of policy and regulations. Energy Policy 34:1209–1220. https://doi.org/10.1016/j.enpol.2004.10.018

Werner S (2017) District heating and cooling in Sweden. Energy 126:419–429. https://doi.org/10.1016/j.energy.2017.03.052

Węziak-Białowolska D, Lewis D (2015) Trust, local governance and quality of public service in EU regions and cities. JRC Sci Policy Rep. https://doi.org/10.2760/136878

Ecological Limits to Sustainable Use of Wood Fuels

Janis Abolins

Abstract A theoretical study of a simple analytical model of biomass accumulation to assess conditions of the neutrality of CO_2-emissions from burning wood (biomass) is reported. Conditions under sustainability defined with respect to harvesting are shown to satisfy requirements of CO_2-neutrality on local scale while burning wood under conditions of shrinking global forest area is not and should be taken into account in the balance of global emissions. Other ecological restraints—conservation of biodiversity in particular, are discussed concerning conditions imposed by the economic system and reflecting on the visions of bio-economy.

Keywords Bio-economy · Biofuels · Sustainability · CO_2-neutrality

1 Introduction

The shrinking reserves of oil, restraints of the available sources of wind and hydro energy, and CO_2 emissions of carbon released from fossil energy carriers contributing to the rise of the global temperature have stimulated corporate players to advance a bio-economy agenda (Mills 2015) promoting the search of renewable substitutes for the dwindling resources of materials and energy to support further economic growth and activity. However, environmental costs and social costs imposed by rising economic inequality mentioned as the two main factors of the disparity between the economic growth and the quality of life point to some kind of defect of the present system. A new industrial revolution is believed to solve the old and current problems by unlocking the potential of transformed economic relationship to materials and finite resources brought up by an innovated economic system based on principles of preserving natural capital, optimising the yields from consumed resources and eradicating externalities caused by pollution to foster

J. Abolins (✉)
Institute of Atomic Physics and Spectroscopy, University of Latvia,
19 Rainis Blvrd, Riga 1586, Latvia
e-mail: janis.abolins@lu.lv

W. Leal Filho et al. (eds.), *Towards a Sustainable Bioeconomy: Principles, Challenges and Perspectives*, World Sustainability Series,
https://doi.org/10.1007/978-3-319-73028-8_24

effectiveness, while maximising the productivity of assets or resources is regarded as a value driver of the new system—the circular economy (van Houten and Robbins 2016). Nonetheless, the critical remarks addressed to the bio-economy agenda being little more than expansion of the corporate-driven market system (Mills 2015; NOAH 2015) are neither answered in the vision of the circular economy explicitly focused on profiting from innovative technological improvements without discussing the basic economic issues concerning the purpose of production and distribution of the created wealth, not to mention the limits to growth.

Two contending visions of the economic background of a bio-based future knowledge society have developed within the EU research framework for a knowledge-based bio-economy, one of which to be achieved is oriented to modification of nature by genetic engineering while the other—to agro-ecological engineering to design agricultural systems sustained by minor inputs regulating and increasing productivity and crop protection (Levidow et al. 2012).

Wood and other products of photosynthesis are taken for granted being a renewable and CO_2-neutral source of energy (Omri and Khuong 2014; Demirbas 2004; Goldemberg and Coelho 2004) regardless to critical concerns (Agrofuels 2007), noticeable controversies (Brewer 2007; Searchinger et al. 2009; Greenpeace 2011; Redman and Tricarico 2013) and the evidence of growing atmospheric concentration of CO_2 after wood fuel started to replace coal in thermoelectric power stations (Greenpeace 2011; Zeebe et al. 2016). Emissions of CO_2 from burning wood and other biomass are still neglected and not taken into account (Upton 2015) regardless to the earlier arguments and common knowledge that CO_2 is released at burning any fuel containing carbon. Since carbon comprises about 50% of wood and other plant biomass (Smil 1999), 0.5 kg of carbon is released from each kilogram of wood burnt and not immediately sequestered from atmosphere into new biomass by photosynthesis. Large-scale industrial use of bio-energy is criticised for ignoring CO_2 emissions from burning biomass and the turning point at which a renewable resource becomes non-renewable for which reasons it is proposed to be excluded from the renewable energy definition (WRM 2015; Declaration 2016). Some authors suggest accounting for the carbon released at burning wood by "carbon debt", time of payback and carbon balance (Mitchell et al. 2012; Pingoud et al. 2015).

With regard to carbon storage and release, there is no essential difference between the carbon stored in fossil fuels and carbon stored in wood: standing forest and wood constructions (or products) are carbon deposits as well as coal, oil and natural gas. The difference is growing forest presenting a sink of CO_2 released in the atmosphere at burning fuels containing carbon and allowing for balance of atmospheric CO_2 (Hennigar et al. 2008; Ni et al. 2016).

Ecological limitations to sustainable use of wood as a renewable source of energy embrace CO_2-neutrality and along with socio-ethical concerns are vital pre-conditions for any realistic bio-economy. Hereafter conditions under which the energy of wood fuels is CO_2-neutral are considered within the theoretical concepts of biomass accumulation based on the empirical Richards equation (e.g. Zeide 2004) presented in normalised coordinates of the rate of growth and time (Abolins and Gravitis 2011). The model applicable to natural forest growth around the age of the maximum of the mean annual increment is chosen for simplicity of considerations.

2 The Basic Approach

The stock S of a natural forest stand, as considered by the theoretical model (Abolins and Gravitis 2011), is presented by

$$S(t) = \int dt \cdot R(t) \tag{1}$$

where t is time and $R(t) = {dS(t)}/{dt}$—the rate of growth (the current annual increment).

The rate of growth is assumed to be proportional to the rate of expansion of the active light-absorbing area L of the canopy. The active light-absorbing area as function of time (the age of the stand) is supposed exponentially approaching saturation limited by some finite value L_{∞} dependent on the land area occupied by the stand:

$$L(t) = L_{\infty}(1 - \mathrm{e}^{-at}) \tag{2}$$

the rate of its expansion being proportional to the actual active area expressed by Eq. 2 and the unsaturated area $(L_{\infty} - L(t)) = L_{\infty}\mathrm{e}^{-at}$:

$$\frac{dL(t)}{dt} = L_{\infty}^{2}(1 - \mathrm{e}^{-at}) \cdot \mathrm{e}^{-at} \tag{3}$$

providing expression for the rate of growth:

$$R(t) = const \cdot (1 - \mathrm{e}^{-at}) \cdot \mathrm{e}^{-at} \tag{4}$$

By putting derivative of function $R(t)$ equal to zero the age t_m at which the rate of growth reaches its maximum value is found from $at_m = \ln 2$ which substituted into Eq. 4 provides the maximum value of the rate of growth equal to $R(t_m) = const{\cdot}0.25$. Choosing the latter as the unit of the rate of growth, the current values of the rate of growth (Eq. 4) normalised with respect to $R(t_m)$ are presented by

$$R(t) = 4 \cdot (1 - \mathrm{e}^{-at}) \cdot \mathrm{e}^{-at} \tag{5}$$

By choosing a new, dimensionless time scale $x = \frac{t}{t_m}$ normalised against the value of t_m Eq. 5 is transformed as

$$\frac{dS(x)}{dx} \equiv R(x) = 4 \cdot \left(1 - e^{-x \cdot \ln 2}\right) \cdot e^{-x \cdot \ln 2} \tag{6}$$

Returning to Eq. 1 the stock at some cutting age x_c is presented by

$$S(x_c) = 4 \int_0^{x_c} dx \cdot \left(1 - e^{-x \cdot \ln 2}\right) \cdot e^{-x \cdot \ln 2} = \frac{1}{2 \ln 2} \left(1 - e^{-x_c \cdot \ln 2}\right)^2 \tag{7}$$

wherefrom the mean annual increment providing estimated land productivity at cutting age x_c

$$P(x_c) = \frac{S(x_c)}{x_c} = const \cdot \frac{\left(1 - e^{-x_c \cdot \ln 2}\right)^2}{x_c} \tag{8}$$

is found as a function of age having its maximum at $x_c \approx 1.81$.

The stock as function of the normalised time x is easily expressed in units of the asymptotic limit $S(x \rightarrow \infty) = S_\infty$ by rewriting Eq. 7:

$$\frac{S(x)}{S_\infty} = \left(1 - e^{-x \cdot \ln 2}\right)^2 \tag{9}$$

The normalised rate of growth (Eq. 6) and the mean annual increment (Eq. 8) are presented in Fig. 1 in units of the maximum of the rate of growth.

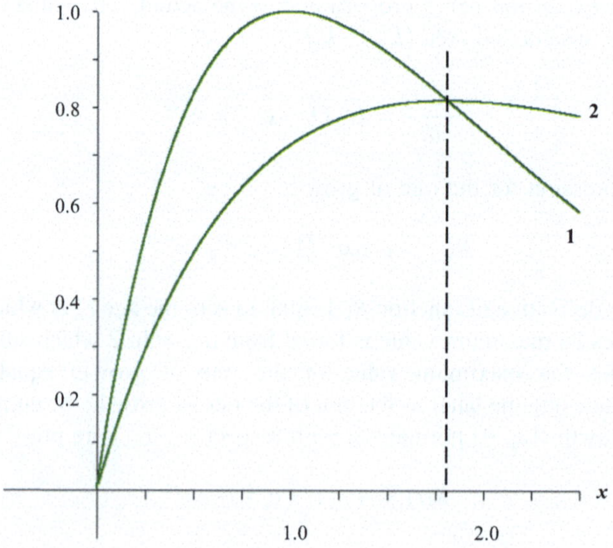

Fig. 1 Rate of growth (*1*) and productivity—the mean annual increment (*2*) in normalised coordinates

A				
A_0	A_0	A_0		A_0
1 year old forest stand	2 years old forest stand	3 years old forest stand	....	t_c years old forest stand

Fig. 2 Land area of a forest plantation necessary to provide sustainable annual yields of wood

3 Sustainable Harvest and CO_2-Neutrality

Sustainable harvests and CO_2-neutrality are two major concerns to be considered with regard to sustainability of using wood as a source of energy. Sustainable yields from forest lands being defined as permanent annual amounts of wood biomass harvested over unlimited span of time are secured by plantations (Fig. 2) comprised of stands of successive ages on land areas of equal size the number of which is determined by cutting age—period of rotation t_c (in years) chosen to provide maximum land productivity (Abolins et al. 2010):

$$P(t_c) = \frac{S(t_c)}{t_c} \tag{10}$$

Accounting for the time difference between emission of CO_2 at burning wood and being recaptured in new biomass by photosynthesis the CO_2-neutrality should be defined by identifying a reasonable finite time interval within which the released CO_2 must be recaptured. Since consumption of energy and accumulation of CO_2 in biomass have a pronounced seasonality in the temperate climate zone, duration of the natural seasonal cycles seems to be a rational choice confirmed by experimental data showing the growth of atmospheric CO_2 during winter and decline during summer. Thus, hereafter CO_2-neutrality is defined as balance between the amounts of carbon annually released in the atmosphere by burning wood and sequestered from the atmosphere by photosynthesis.

The total area A of the plantation comprises lots of equal size A_0 occupied by forest stands of sequential ages from one year old up to cutting age t_c: $A = t_c \times A_0$. The size of A_0 is the size of forest area to be felled for sustainable annual supply of wood.

4 Dynamics of Biomass and Carbon Accumulation

The function of the rate of growth is used to simulate the annual uptake of biomass or carbon the content of which in wood is proportional to the total amount of biomass comprising 50% or slightly more depending on the species. The optimum age t_o for felling is chosen as the age of maximum productivity at $x = 1.8$ in the normalised time scale being equal to all species and stands while the age t_m chosen as the time unit at which the stand reaches the maximum of the rate of growth depends on a number of factors apart from the biological potential of the species and may differ even between stands of the same species.

The annual uptake is estimated as proportion of the total amount of biomass or its carbon content accumulated by the age $x = 1.8$ or $t = t_o$ (Fig. 1, curve 2) the proportions being equal for both.

Dynamics of biomass uptake in a forest stand modelled by the rate of growth (the current annual increment) $R(x)$ is presented by a continuous function (Eq. 6 and curve 1 in Fig. 1) of the normalised time coordinate. The actual annual increment is a finite discrete amount of biomass accumulated during a particular year (represented by some time interval Δx) at a particular age. The normalised time interval Δx representing a year in the real time scale depends on the number of years (age) of the stand being considered, which, in the present case, is chosen as the optimum cutting age t_o—the number of years within the normalised time interval of 1.8:

$$\delta x = \frac{1.8}{t_o} \tag{11}$$

For further convenience the particular annual increment of stock ΔS_n is assessed in units of S_∞ from Eq. 9:

$$\left.\begin{array}{ll} \Delta S_n = S_n - S_{n-1} & n = 2, 3 \ldots n_c = t_o(\text{numberofyearsatcutting}); \\ \Delta S_1 = S_1 = \left(1 - e^{-\Delta x \cdot ln\, 2}\right)^2 & \\ S_n \equiv S(x_n) = \left(1 - e^{-x_n \cdot ln\, 2}\right)^2 & \text{where}\, x_n = n \cdot \Delta x, \;\; n = 2, 3 \ldots n_c = t_o \end{array}\right\} \tag{12}$$

Under assumptions inferred by Eqs. 12 equality

$$\sum_{n=1}^{n_c} (\Delta S)_n = \left(1 - e^{-x_c \cdot \ln 2}\right)^2 \tag{13}$$

where $x_c = n_c \cdot \Delta x$ is held for any arbitrary n_c.

By putting $n_c = t_o$ ($x_c = t_o \cdot \Delta x$) the stock S_o accumulated by the optimum cutting age is calculated from Eq. 13 the annual uptakes ΔS_n being found as proportions of S_o:

$$\sum_{n=1}^{n_c = t_o} \frac{(\Delta S)_n}{S_o} = 1 \quad \text{where} \quad S_o = \sum_{n=1}^{n_c = t_o} (\Delta S)_n \tag{14}$$

The entities of Eqs. 11–14 are presented in Fig. 3.

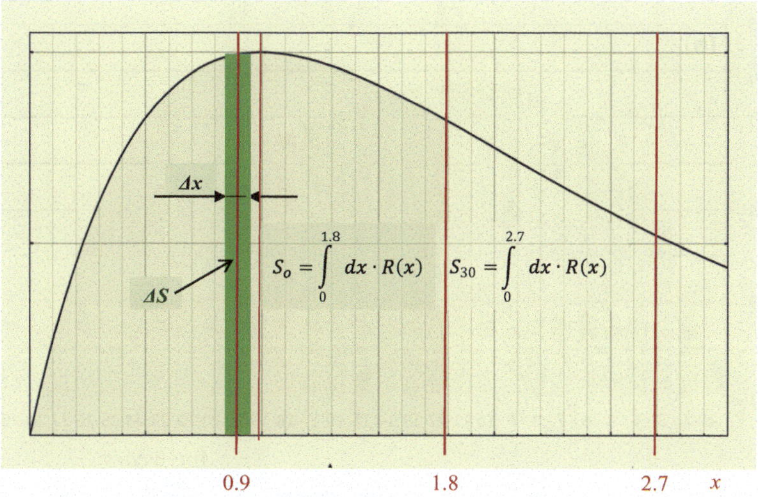

Fig. 3 Graphical description of the entities used to calculate proportions of annual uptakes presented in Figs. 4 and 5

The annual increments of biomass as proportions of stock accumulated at optimum cutting ages of 18 and 20 years old stands representing dynamics of biomass and carbon uptake are presented in Fig. 4.

As seen from Fig. 4, the proportion of an annual uptake depends on the age at which the stand reaches maturity—the maximum productivity with respect to the land area: the longer it takes to reach the optimum cutting age, the lower is the annual proportion (in other words, at slower growth the relative annual acquisition is smaller; the absolute values may be different though since, on the large, a slower growth is caused by poorer site fertility having a negative effect on its productivity). The effect of delayed harvesting (Fig. 5) is similar: the annual uptakes relative to the total amount of stock accumulated, e.g. by 30 years ($x = 2.7$) in a stand having reached maturity at the age of 20 years ($x = 1.8$) are smaller (Fig. 5).

Most effectively the biomass (and carbon) is accumulated by stands close to maximum of the rate of growth at $x = 1$ (Fig. 1) declining with age farther on. As seen from the results presented in Fig. 4, the total annual uptake by a plantation ensuring sustainable annual yields theoretically is equal to the harvested biomass the amount of fixed carbon being equal to that released at burning the harvested wood. From here it follows that the ecological "footprint" of burning wood measured by the total area of the plantation is equal to 18 ha or 20 ha per one ha burnt, as considered above. In case of delayed harvesting the "footprint" would be a 30 ha plantation of sequentially aged stands per one ha burnt.

The yield (and carbon content of it) from a ha of "overgrown" stand is higher compared with that the stand would have provided at the optimum cutting age the productivity of the extended plantation area being lower (in the case of the example considered above—by 6%). By practicing delayed harvesting on a regular basis to

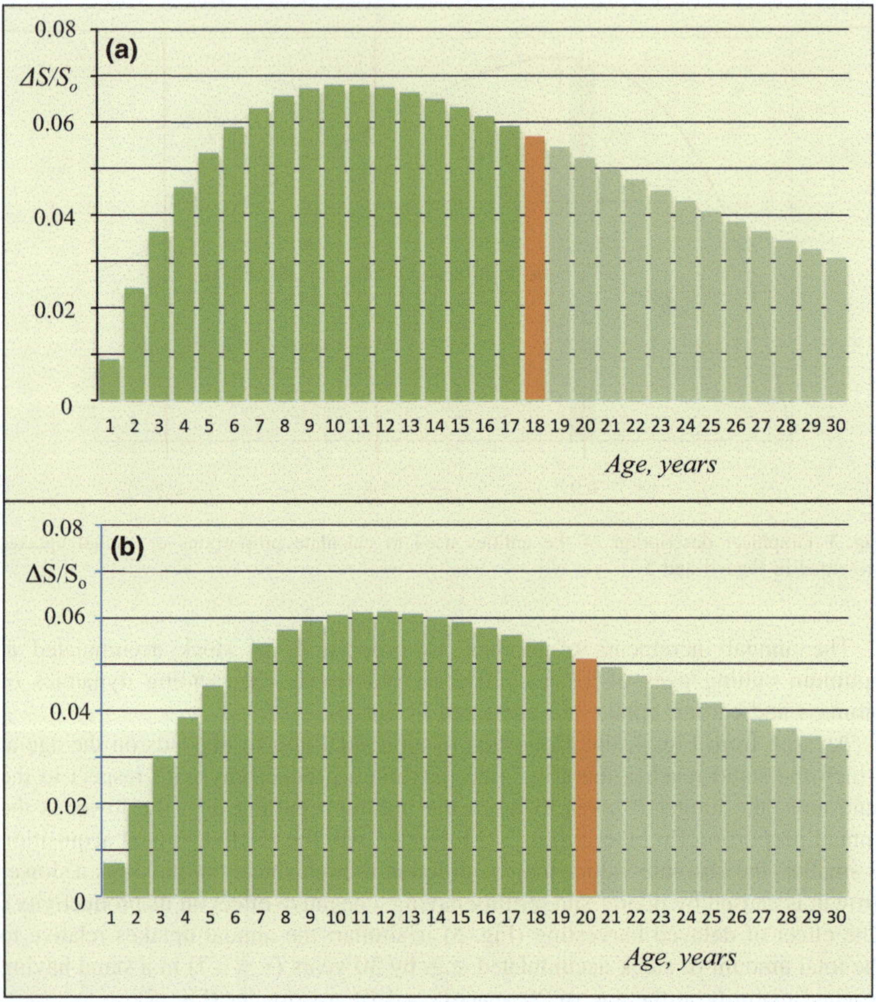

Fig. 4 Current annual uptakes of biomass and carbon as proportions of the amount accumulated by the optimum cutting age t_o of 18 (**a**) and 20 (**b**) years old stands

get higher sustainable current yield from hectare the area of the plantation would have to be extended by 50%—to 30 ha instead of 20. To absorb the surplus of carbon released at burning the wood harvested from one hectare of the overgrown 30 years old forest stand within a year would require nearly 7 ha of 10-year old stands or 3 years by one hectare of new forest added to the existing plantation.

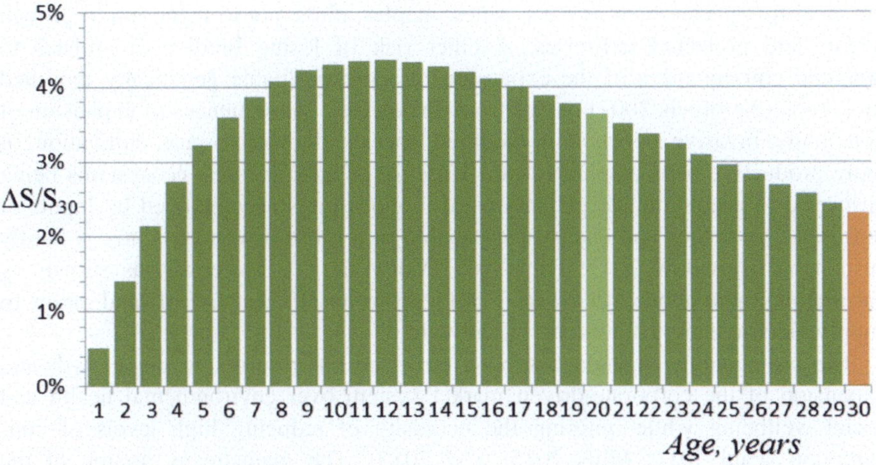

Fig. 5 Proportions of the annual biomass and carbon uptakes as *per cent* of the amount accumulated by an overgrown stand. S_{30}—the stock acquired by 30 years of a stand the optimum cutting age of which is 20 years

5 Discussion

Terrestrial photo-synthesisers absorb around 1.25% of the total energy flow of 1.373 kWm^{-2} (the solar constant) from the Sun reaching the upper boarder of Earth's atmosphere (e.g., Smil 1999). The average yields of dry biomass from short-rotation (3 years) willow being of 9 × 10^3 kg ha^{-1} (Aylott et al. 2008) retain 135 GJ ha^{-1} (37.5 MWh ha^{-1}) or 45 GJ (12.5 MWh) of energy per hectare of the plantation. Transforming this amount of primary energy to electricity at 35–40% efficiency because of slagging at higher temperature (BEDB 2011; Demirbas 2004) makes 4.375–5.000 MWh of electricity per ha of the plantation. The photoelectric solar panels at 10% efficiency and insolation of 1 kW m^{-2} would provide 8.76 GWh ha^{-1} per year. The price of the advantage of storing the converted solar energy in biomass under such circumstances is 1.72 ha of forest land per kWh electricity generated by burning the biomass of the energy crop. It means that under conditions of CO_2-neutrality and sustainable supply of wood for generating electricity the ecological footprint of a household consuming annually 1200 kWh is equal to slightly over 2000 ha of forest land. The calculations demonstrate that compared with wood the photovoltaic appliances are much more efficient transformers of solar energy into electricity and, therefore, a more efficient option to benefit from the land area.

Fast-growing species are more preferable and more efficient source of renewable energy for a growing demand by a profit-driven economy setting the available land assets under pressure to be used for cultivating the fast-growers at the expense of less profitable slow-growing species, which jeopardises biodiversity by shrinking the areal for species such as oaks, ashes, maples, elms, not to mention old-growth forests and protected territories. Another risk of losing biodiversity relates to potential consequences of the expansion of more profitable genetically modified fuel crops (Agrofuels 2007). Apart from unexpected consequences of expansion of potentially invasive genetically modified species, industrial-scale cultivation of more productive sorts of plants at the expense of natural forest ecosystems being destroyed to accommodate plantations of monocultures accompanied by inputs of fertilisers, pesticides and energy inescapable in a profit-driven economy is hardly sustainable (Levidow 2011; Paul 2013; WRM 2015). Moreover, as shown by calculations here-above, 45 GJ ha^{-1} is far from an efficient use of land assets to supply electric energy for economic growth.

The bio-economy agenda has been criticised for prioritising the growth and expansion of the corporate-driven market system over environmental health and social wellbeing while ignoring the necessity of reducing high levels of consumption (Paul 2013; Mills 2015; TNI 2015). The mainstream version of the bio-economy is likely becoming another "ecological modernisation" as a major driver of economic competition and an imperative of activities in maintaining and enhancing profits and dividends by stimulating technological and social innovation in more varied production-consumption networks (Kitchen and Marsden 2011). A critical examining of the bio-economy agenda (Agrofuels 2007; Levidow 2011; Paul 2013; Mills 2015) reveals the inconsistencies between the defining content and the chosen ambiguous name of "bio-economy". The prefix "bio" is strongly contrary to the content mainly oriented to technologies of industrial processing the allegedly renewable raw materials of biological origin and consuming "bio-fuels". The latter term has raised protests because the stuff is too far from conventional associations with living matter (the "*bio*") and long since suggested to be called "agro-fuels" regarding the agro-industrial dimension of the production of energy crops (Agrofuels 2007). The part of "*economy*" is presented by new, more efficient and cleaner ("greener") technologies and closed production cycles compatible with the zero emissions concept (Levidow et al. 2012; Paul 2013). Subjects specific to "*economy*" concerning the drivers and the socio-economic structure of the economic system, its compatibility with the *Declaration of Human Rights* and the natural environment, equality of distribution of the produced wealth, etc. are not considered as priorities[1]. The choice of the name reminds the "new-speech" in Orwell's *1984*—the same terms used in the contending visions have a different meaning (Levidow 2011). It seems more appropriate to speak about "*bio-economics*" instead of "*bio-economy*" with regard to the mainstream economic agenda.

[1]Russian researchers developing a complex program of biotechnologies for the Russian Federation having scrupulously studied the national bio-economy agendas of the EU member states and other countries have not found even a single outline of changing the economy as a system of productive social relations, exchange and distribution of wealth (http://greenevolution.ru/enc/wiki/bioekonomika/). See also: http://www.truth-out.org/news/item/27270-system-change-or-there-and-back-again-capitalism-socialism-fascism; http://www.paecon.net/PAEReview/issue53/Smith53.pdf.

More "*economy*"-oriented is the agro-ecological vision of agricultural systems of a knowledge-based bio-economy relying on the natural interactions between components of an ecosystem regulating themselves, increasing their own productivity, soil fertility and creating their own pest deterrents (Levidow et al. 2012). However, the agro-ecological projects do not get an equal financial funding to match that proposed by proponents of what Levidow calls the "life sciences perspective" oriented to technological modification of nature by genetic engineering and other "*bio*"-technologies to support the present corporate-driven market system (Birch et al. 2010). The reason of favouring such allocation of research funding is revealed in an article related to the Human Genome Project (Simitopoulou and Xirotiris 2000) the authors of which notice that free access to the results of the genetic research on the project caused a sharp drop of the NASDAQ index of the international stock markets indicating the loss of enthusiasm of the "investors" about the genetics industry and demonstrating the dominance of money over genetic research and the economics over science in general. A profit-driven economy provides funding to science, if the science can provide something to privatise for buying and selling—it is, for the existing market- and profit-dominated economic system. Unless the system is released from the obligation of making profit at any cost the research institutions will have to "give priority to complex, expensive and commercial science" for profit (Levidow 2011) to survive. The agro-ecological version of bio-economy promoting shorter value chains and higher gains for producers as first-hand suppliers does not comply with the globalised market allowing the army of traders to appropriate the major share of the value (and wealth) added by producers. It cannot expect being appreciated by the corporate players dominating the profit-driven system where independent producers are not wanted.

The dominance of finance and profit in a market economy obscures the reality that any kind of economy as a system of transforming materials into products actually is driven by energy in a similar way the energy sustains life, ecosystems and the whole biosphere. Meddling in the organisation of natural systems requires additional energy and more energy is needed to sustain the changes that are not sustained by the ecosystem itself. For that reason the "life sciences perspective" has a higher cost in terms of energy compared with the agro-ecological approach. Since any kind of renewable energy on the planet is limited by a finite energy flow from the Sun, either one of the two versions of bio-economy should take into account that energy from any available source has a limit and, therefore, in the reality unlimited economic growth is not and cannot be possible. In other words—unlimited growth of a profit-driven economy is neither sustainable, nor realistic. Attempts to sustain unrestricted demand for energy by any means are destined to turn the global forest into a non-renewable asset long before running out of land.

Presently, with a shrinking global forest area, burning wood for energy at best can be CO_2-neutral only locally, if the annual amount of wood biomass burnt does not exceed the local annual increment of the stock. On the global scale, as shown by the data on CO_2 emissions (Schulze et al. 2012; Schultz 2014), burning wood just cannot be CO_2-neutral, not while more trees are cut than replanted.

6 Conclusions

The land area to grow forest is the main asset ultimately limiting sustainable supply of wood including its use as a renewable source of primary energy.

CO_2-neutrality of burning wood depends on the standing global forest being a deposit of carbon along with coal, petrol, and natural gas reserves.

CO_2-neutrality of burning wood can be realised only on the local scale. On the global scale, under conditions of a shrinking area of the global forest, it is not CO_2-neutral, for which reason the emissions from the use of wood fuels should be accounted for.

With account for either the social or the physical evidence unlimited growth of a profit-driven economy is neither sustainable, nor realistic.

Acknowledgements The underlying studies drawn on in the paper have been made possible by the Latvian National Programme of Research and the Horizon 2020 project of grant agreement No 654371.

References

Abolins J, Gravitis J (2011) A simple analytical model for remote assessment of the dynamics of biomass accumulation. In Shahid Shaukat S (ed) Progress in biomass and bioenergy production. In Tech Open Access Publishers, pp 91–106. ISBN 978-953-307-491-7. Available at: http://www.intechweb.org/booksprocess/aboutthebook/chapter/14232/book/460

Abolins J, Gravitis J, Kosmacha J (2010) Optimising the yield of energy from biomass by analytical models of the rate of growth. Latv J Phys Tec Sci 47(5):25–31

Agrofuels (2007) Towards a reality check in nine key areas. http://www.econexus.info/sites/econexus/files/Agrofuels.pdf

Aylott et al (2008) Yield and spatial supply of bioenergy poplar and willow short-rotation coppice in the UK. New Phytol 178:358–370

BEDB (2011) Alkali content and slagging potential of various biofuels. http://cta.ornl.gov/bedb

Birch K, Levidow L, Papaioannou T (2010) Sustainable capital? The neoliberalization of nature and knowledge in the European "knowledge-based bio-economy". Sustainability 2 (2010):2898–2918

Brewer J (2007) The coming biofuels disaster. https://web.archive.org/web/20081025185709/. http://www.rockridgeinstitute.org/research/rockridge/coming-biofuels-disaster.html

Declaration (2016) Large-scale bioenergy must be excluded from the renewable energy definition. http://www.biofuelwatch.org.uk/2016/BioenergyOut-declaration/

Demirbas A (2004) Combustion characteristics of different biomass fuels. Prog Energy Combust Sci 30:219–230

Goldemberg J, Coelho ST (2004) Renewable energy—traditional biomass vs. modern biomass. Energy Policy 32:711–714

Greenpeace (2011) http://www.greenpeace.org/canada/en/recent/Burning-trees-for-energy-puts-Canadian-forests-and-climate-at-risk-Greenpeace/

Hennigar CR, MacLean DA, Amos-Binks LJ (2008) A novel approach to optimize management strategies for carbon stored in both forests and wood products. Forest Ecol Manag 256(4):786–797. https://doi.org/10.1016/j.foreco.2008.05.037

Kitchen L, Marsden T (2011) Constructing sustainable communities: a theoretical exploration of the bio-economy and eco-economy paradigms. Local Environ Int J Justice Sustain 16(8):753–769 (special issue: sustainable communities)

Levidow L (ed) (2011) Agricultural innovation: sustaining what agriculture? For what European bio-economy? http://www.crepeweb.net/wp-content/uploads/2011/02/crepe_final_report.pdf

Levidow L, Birch K, Papaioannou T (2012) EU agro-innovation policy: two contending visions of the bio-economy. Critical Policy Stud 6(1):40–65. https://doi.org/10.1080/19460171.2012.659881

Mills (2015) Bioeconomy: a primer. Transnational Institute. Available at: https://www.tni.org/en/publication/the-bioeconomy

Mitchell SR, Harmon ME, O'Connell KEB (2012) Carbon debt and carbon sequestration parity in forest bioenergy production. GCB Bioenergy 4:818–827

Ni Y, Eskeland GS, Giske J, Hansen J-P (2016) The global potential for carbon capture and storage from forestry. Carbon Bal Manag 11(1):Article number 3

NOAH, Biofuelwatch, Econexus, GFC, WRM, Rainforest Rescue, CEO (2015) Bioenergy out: why bioenergy should not be included in the next EU Renewable Energy Directive. http://wrm.org.uy/wp-content/uploads/2015/11/Why_bioenergy_should_not_be_included_in_the_next_EU_Renewable_Energy.pdf

Omri A, Khuong ND (2014) On the determinants of renewable energy consumption: International evidence. Energy 72:554–560

Paul H (2013) A foreseeable disaster: the European Union's agro-energy policies and the global land and water grab. Transnational Institute. 40p. Available at http://www.econexus.info/sites/econexus/files/A_Foreseeable_Disaster-HOTL-Agrofuels-July2013

Pingoud K, Ekholm T, Soimakallio S, Tuomas H (2015) Carbon balance indicator for forest bioenergy scenarios. GCB Bioenergy. https://doi.org/10.1111/gcbb.12253

Redman J, Tricarico A (2013) Wall street's climate finance Bonanza. Foreign Policy in Focus. Available at http://fpif.org/wall_streets_climate_finance_bonanza/

Schultz K (2014) Carbon dioxide emissions rise with rebounding economy. Scientific American. Available at http://www.scientificamerican.com/article/carbon-dioxide-emissions-rise-with-rebounding-economy/. 6 June 2014

Schulze E-D, Korner C, Law BE, Haberl H, Luyssaer S (2012) Large-scale bioenergy from additional harvest of forest biomass is neither sustainable nor greenhouse gas neutral. GCB Bioenergy 4:611–616

Searchinger TD et al (2009) Fixing a critical climate accounting error. Science 326:527–528

Simitopoulou K, Xirotiris NI (2000) The human genome project: the dominance of economy on science—ethical and social implications. Global Bioethics. https://doi.org/10.1080/11287462.2000.10800762

Smil V (1999) Energies. MIT Press, Cambridge

TNI (2015) Move to renewables jeopardized by EU corporate trade deals. Press release, 09 Dec 2015. Available at https://www.tni.org/en/article/move-to-renewables-jeopardized-by-eu-corporate-trade-deals

Upton J (2015) Pulp fiction. http://reports.climatecentral.org/pulp-fiction/1/

van Houten F, Robbins C (2016) Intelligent assets: unlocking the circular economy potential. Ellen MacArthur Foundation 2016. Available at: http://www.ellenmacarthurfoundation.org/assets/downloads/publications/EllenMacArthurFoundation_Intelligent_Assets_080216-AUDIO-E.pdf

WRM (2015) Bioenergy out: why bioenergy should not be included in the next EU Renewable Energy Directive. Available at http://wrm.org.uy/wp-content/uploads/2015/11/Why_bioenergy_should_not_be_included_in_the_next_EU_Renewable_Energy.pdf

Zeebe RE, Ridgwell A, Zachos JC (2016) Anthropogenic carbon release rate unprecedented during the past 66 million years. Nat Geosci. doi: https://doi.org/10.1038/ngeo2681. Available at http://www.nature.com/ngeo/journal/vaop/ncurrent/full/ngeo2681.html

Zeide B (2004) Intrinsic units in growth modelling. Ecol Model 175:249–259

Part VI
Bioeconomy: Advances on Bio-Based Forest Resources and Biomass

Sustainability-Driven New Business Models in Wood Construction Towards 2030

Anne Toppinen, Minna Autio, Miska Sauru and Sami Berghäll

Abstract In the transition towards a renewable material -based bioeconomy in Europe, growing interest is being directed towards wooden multistorey construction (WMC) as a sustainable housing solution. We analyse the changing WMC business, and the involved value networks towards 2030 based on service business model literature, with a focus also on consumer-driven models. Methodologically our study uses a three-round Delphi process focusing on Finland as a country where national bioeconomy strategy specifically acknowledges wood-based construction. Based on our results, the primary reasons for wood utilization are supporting the bioeconomy strategy with the use of renewable materials and addressing indoor air quality concerns. This happens instead of enhancing intrinsic motivation towards sustainable bioeconomy as such. Therefore, transforming business models towards sustainability calls for strengthening the positive image of the wood construction industry, especially among a largely neglected stakeholder group, i.e. residents. To achieve business model development, the industry needs to strengthen its orchestration of partner networks and capabilities, by including not only new co-creators as a part of the actor-to-actor network, but also residents as end-users.

Keywords Sustainability transition · New business models · Wood construction Finland · Europe

---

A. Toppinen (✉) · M. Sauru · S. Berghäll
Department of Forest Sciences, University of Helsinki, Helsinki, Finland
e-mail: anne.toppinen@helsinki.fi

M. Sauru
e-mail: miska.sauru@helsinki.fi

S. Berghäll
e-mail: sami.berghall@helsinki.fi

M. Autio
Department of Economics and Management, University of Helsinki, Helsinki, Finland
e-mail: Minna.Autio@helsinki.fi

W. Leal Filho et al. (eds.), *Towards a Sustainable Bioeconomy: Principles, Challenges and Perspectives*, World Sustainability Series,
https://doi.org/10.1007/978-3-319-73028-8_25

1 Introduction and Purpose

Sustainability is a key conceptual argument in favour of the general acceptability of bioeconomy as such. However, new sustainability-driven innovations are called for to renew the traditional sectors of economy into a truly new bioeconomy. The rise of wooden multistorey construction (WMC) is currently the most evident new business opportunity in the Nordic countries, especially as it has not faced sustainability-related criticism. Also, according to Bosman and Rotmans (2016), the Finnish national-level strategy towards bioeconomy is focusing on wooden buildings and construction materials. However, there is a lack of business model -related innovations (or business model research) in this dimension of the bioeconomy—especially from the service development viewpoint (Hansen 2016). This fact emphasizes the need to better understand both how and where *value* is created, but also the *roles* various parties have in this value creation within and outside the firm.

According to Bourdeau (1999, p. 364), the main challenge of a construction business is "to transform the demand for sustainable development into an opportunity, to create and access new markets, and to innovative responses which satisfy traditional industry demands and the new societal demands for sustainable development." Growing recognition of wood construction is a part of the rise of the green building concept. This also includes the use of hybrid structures (e.g. combinations of wood and steel; wood and concrete) (e.g. Wang et al. 2014). According to a recent systematic review on the drivers of green building, Darko et al. (2017) found that the demand and willingness of clients/customers determines the extent that green building systems are developed. Häkkinen and Belloni (2011) link this customer demand to issues such as knowledge of the issue, supply of solutions, methods, value and costs.

The construction business has also merely focused on cost-effectiveness when building competitive advantage for itself (Harris and Halkett 2007) and emphasized a view based on the importance of creating tangible objects—buildings. Overall, the construction industry is not perceived as an innovative business in meeting consumer expectations or transforming their core business models into consumer-driven ones. However, the industry has started to recognize consumers' diversifying preferences. For example, pro-environmental behaviour (Hu et al. 2013) has resulted in creating a more customer-centric culture (e.g. Killip 2013). Studies on consumer's sustainability-related choices towards green building have recently begun gaining more ground (e.g. Luo et al. 2017; Gold and Rubik 2009; Hoffman and Henn 2008; Holopainen et al. 2014; Hoibo et al. 2015).

Previous literature on business models in the construction sector is quite extensive (see a review by Mokhlesian and Holmen 2012). However, regarding the specific viewpoint of renewable wood-based construction, with the exception of Brege et al. (2014), Lessing and Brege (2015) and Höök et al. (2015), very little literature is available for combining the business model perspective to the wood building sector as a holistic view on how companies do business. To differentiate between strategic and operational effectiveness, Brege et al. (2014) focused on market position, system offering and operational platform as key business model blocks in the Swedish house-building sector. Using a case study approach, they found five business model elements to be important for WMC: prefabrication mode, changing actor roles in the building process, end-user segments, system augmentation and the use of complementary resources. In a follow-up study on two case companies, Lessing and Brege (2015) found that the use of end-customer knowledge to identify the target segment, development of an offering and sequentially increasing control over the production and value chain, is the recipe for success in business development. In a conceptual study by Pelli (2016), the use of a Service-Dominant Logic perspective (SDL, Vargo and Lusch 2016a, b) resulted in buildings being perceived as service platforms, but the built environment was also integrated with the natural environment. In Toppinen et al. (2017), the importance of the issues related to sustainable development was perceived to play an increasing role in the Nordic WMC market because future consumers and end-users are likely to be increasingly driven by the aims of finding sustainable lifestyle-based solutions for their housing.

In our paper, we combine the business model literature with the SDL as a platform for analysing the changing WMC business and the involved value networks towards 2030. Methodologically our study uses a three-round Delphi process conducted in 2016–2017 among construction value chain professionals in Finland, but we also draw from the expertise existing in the European pioneering WMC country of Sweden.

2 Conceptual Background

Teece (2010) describes a company's business model as a tool, describing how it converts resources and capabilities into economic output, i.e. it creates, captures and delivers value. A business model and company strategy are two constructs that bare close connection, and some scholars have even used the terms interchangeably (Magretta 2002). In a seminal paper Osterwalder et al. (2005) suggested that the conceptualization of a business model has nine elements under four pillars (product, customer interface, infrastructure management and financial aspects), and are approached using tools like a business model canvas in an empirical world. Furthermore, literature on consumer-driven business models (e.g. Anderson-Connell et al. 2002; Pynnönen 2008; Pynnönen et al. 2012; Pels and Sheth 2017) is currently scarce, yet certain studies have focused on consumer demand challenges within a business strategy (e.g. monitoring consumer needs, logistical solutions at the company level). However, scholarly literature has not paid much attention to seeing consumers as active agents interacting and communicating with businesses by contributing to business model development.

According to Vargo and Lusch (2008), SDL essentially builds upon "the application of competences for the benefit of customers; customers are operant resources, rather than operand resources, and they can contribute as value co-creators to the service process". SDL renewing traditional industries (e.g. construction or wood industry) would require users and consumers to be more actively involved in the development process of the products or services (see also Storbacka et al. 2012). To understand why, we need to delve more deeply into the concepts of value, value creation and the context of this creation (namely networks). For this we need a comprehensive view of institutional *service ecosystems*. The following describes the approach more thoroughly.

SDL (Vargo and Lusch 2004, 2016a, b) makes an effort to reframe the model of economic exchange. The core is that we should move away from seeing economic exchange as an exchange of products or product-like services.[1] Instead we should dive a little further into the economic process and ask: what is at the core of the "service" this exchange provides for various parties—i.e. what benefits the engagement provides (Vargo and Lusch 2016a, b). Vargo and Lusch (ibid.) argue that instead of focusing on searching for competitive advantage we should ask what "*strategic benefit*" the exchange provides to the "*actors*" involved (ibid). Thus, exchange in SDL is based on the intangible value that actors receive from the setting. Value is therefore not created by a producer of good, but co-created during the consumption of a service. Products are value carriers but actual value is "*phenomenologically*" defined by each of the partners for themselves (Vargo and Lusch 2004). However, while value is defined by each actor, the actor-to-actor (AtoA) relationship is the setting in which value is created (Vargo and Lusch 2016a, b). As value is created in an exchange setting, Vargo and Lusch (2004) see it as a co-creation process (as do other scholars e.g. Edvardsson et al. 2011; Payne et al. 2008). In line with the classic Hunt and Morgan (1995) Resource Advantage Theory of Competition, value is born out of the resources and capabilities of the exchange partners benefitting the other.

The previous view of value creation is therefore like a process of value facilitation. However, while value is co-created it is also framed by the context in which this exchange happens (Chandler and Vargo 2011). This is because resources and capabilities exist in a network that itself provides the context of this AtoA relationship. Thus also the values, norms and belief systems of the exchange setting have an impact on value co-creation. Vargo (2007) uses the term institutions (referring to social institutions) to describe the setting. Therefore, while value is co-created in AtoA exchange, it is conditioned by the network that it occurs in. Thus, as institutions are relatively stable in their mode of operation (as social institutions), Vargo and Lusch (2016a, b) see institutional changes as the only path to radical innovations. For the WMC setting, we see that radical systemic changes fit into this definition of innovations. The core result of this logical path is that new modes of operations often require not only new capabilities and resources, but also

[1]Vargo and Lusch (2016b) argue that due to Adam Smith's concentration on export as a means for national creation of wealth, economic models have excluded the value of products as instruments of service. This has given birth to the misconception of high value-added industrial products and low-productivity services. Thus, the SDL phenomenon discusses a far more fundamental phenomenon than a goods-dominant logic of thinking might at first view.

Table 1 Infrastructure management part of the business model by Osterwalder et al. (2005)

Business model block	Description
Value configuration	Describes the arrangement of activities and resources
Core competency/ capability	Outlines the competencies necessary to execute the company's business model
Partner network management	Portrays the network of cooperative agreements with other companies necessary to efficiently offer and commercialize value

new networks with social institutions different from the established ones, facilitating new modes of operations and the adoption of new ideas.

In a systematic literature review, Mokhlesian and Holmen (2012) analysed 38 studies on business models for the greening construction sector using the nine elements from Osterwalder et al. (2005). They found that the most important elements facilitating change towards green construction practices (falling under the 'infrastructure management' pillar) are capability, partner networks and value configuration. Based on this, we will also focus our empirical analysis on these elements. However, it is important to note that these elements are not only interconnected with each other, but also with the other six elements present in the Osterwalder et al. (2005) model (see Table 1).

3 Data and Methodology

The data were collected using a three-stage expert Delphi study during 2016–17. The Delphi was used to elicitate participant perceptions on sustainable business model innovations related to WMC. The core for using the Delphi method is bringing (geographically) dispersed experts together, while bringing up a personal viewpoint without group pressure due to anonymity is seen as an advantage of the Delphi method (see for example Landeta 2006). Therefore, the phenomenon being a new developing subject, the Delphi was seen as advantageous in gauging the subject with relative depth, but balanced by unrestricted modes of response possibilities for the respondents.

Our dissensus-based Delphi approach consisted of both thematic interviews (rounds 1 and 3, a total of 18 + 16 interviews) and an online survey (round 2, 17 respondents). Figure 1 depicts the Delphi process. The panelists involved in our study were of Finnish and Swedish origin, and were required to have in-depth knowledge of wood utilization in multistorey construction in the Nordic region. Due to emphasis on the market and business aspects, our study excluded for example consultants and architects with purely technical backgrounds or representatives from end-users. Omitting experts outside wood construction can be seen as a limitation, but in general qualitative studies such as this are limited in the number of respondents that can be included without the material growing too

Fig. 1 The Delphi process

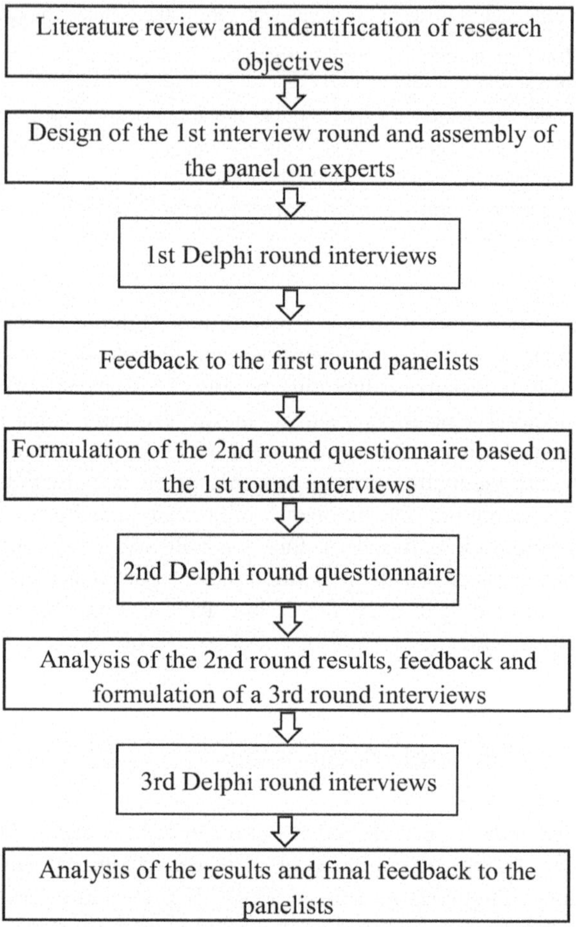

exponentially for analysis. Although we made an effort to ensure sufficient and diverse expertise on the panel, we faced some difficulties in finding experts for the personal interviews with the targeted high level of professional background. The Swedish dataset was smaller, as it was predominantly used to benchmark Finnish perspectives. The composition of the panel based on respondent backgrounds and professional expertise is given in an Appendix.

During the first round of data collection the study focused on the structure of the wood construction value chain and co-operation between the parties in this chain. We also focused on the more general market and institutional settings, with the interviews therefore covering the overall state of the wood industries, raw material and end-use market issues, but also the role of sustainable development as a concept. These themes were included in the interview guide based on a literature review and analysis conducted elsewhere (Toppinen et al. 2017). During the second Delphi round emphasis was given to the themes and topics considered the most thought-provoking or controversial in the first Delphi phase. The questionnaire consisted of 42 closed-ended questions, in which we again focused on business development-related items. Thus, the first-round interview material was used in round 2 for developing more concrete statements, the likelihood and desirability of which could be assessed by the panellists to explore areas where they believe future development could diverge. Data from round 3, which was built upon the previous rounds, acted to deepen the understanding of particularly the business model development issues. The third-round interview data were analysed using qualitative methods, namely thematization, building on the conceptualization of the business model by Osterwalder et al. (2005). All authors were involved in the data analysis during this stage, to ensure more reliable interpretation and identification of relevant aspects in a rather fuzzy setting.

Although we did not interview end-users (i.e. residents, consumers), we are able to analyse data of how experts position customers in business model thinking. For example, questions concerning the consumer market and how interviewees recognized consumer needs and changing consumer behaviour were included in both the first and third round interviews.

4 Results and Discussion

4.1 Partner Network

During the first-round interviews, it became obvious that the current cooperational schemes (and the future direction of this within WMC business) is a complex and difficult topic for the panellists. As a general observation, the panellists most experienced in prior cooperative efforts were respondents experienced in working in larger companies. Respondents from smaller wood industry companies were used to working with architects. However, the further upstream respondents were from the end-users, the less they appeared to actively search for cooperative relationships across the members of the construction value network.

It should also be noted that panellists from the building industry were interested in offering their expertise (and in some cases already utilized it) to the solution providers. Networked structures, cooperation with competitors, the standards, complexity and time-consuming nature of construction projects, large operators controlling a large share of the value chain, hybrid building, online marketplaces and even competing ecosystems within the WMC were all mentioned during the interviews. However, this occurred with little coherence between the respondents, as the following quotes show: "*It could be [that in the future] we will see competitors as partners. One manufactures the walls and the other manufactures the flooring.*" **Executive from wood industry, Finland**; "*It is more and more important to find strategic alliances.*" **Executive from wood industry, Sweden.**

Furthermore, respondents considered cooperation an important factor in the future, although their views differed widely on how and with whom this co-operation could happen. Smaller, local operators were more prone to discuss network models in which they co-operate with other small operators. The concept of hybrid material building solutions seemed particularly attractive to the panellists. However, the idea of actual co-operation between the wood construction sector and concrete solution providers was strongly challenged by some respondents. This was partly due to the on-going 'wood vs. concrete' dispute visible in the Nordic region. Examples from respondents illustrate this well: "*I don't think co-operation [with concrete] will happen—there is just too much competition*" **Executive from wood industry, Sweden**; "*The way construction has been developed by concrete builders for the past 50 years is not optimal for wood construction. We should optimize our material.*" **Executive from wood industry, Finland.**

Second-round data appeared to suggest a strong faith in organizational co-operation between various types of actors within the respondents. Instead, the topics that divided the opinions most were: "*Concrete builders will be significantly more interested in the opportunities offered by wooden building solutions than today*", *and* "*By 2030, the amount of companies involved in the wood construction value chain will be much higher than today*" (Table 2). Here a partnership model, in which (a) construction participants work together as an integrated collaborative team, (b) the team has a joint management structure under multi-party contractual arrangements of project partnering, (c) the network has an integrated delivery system, has gained growing interest, but has also been used to analyse the Finnish utility construction market (Lahdenperä 2012).

While the SDL perspective was not gauged per se, some of the Delphi round results support these arguments from a theoretical perspective, especially content related to (new) networks and related new partners (especially in the group of smaller operators). For example, the high likelihood and desirability of statement 38 support this conclusion and can be seen to offer support to a value network type of thinking. The same holds for statement 9, but to a lesser degree. These observations are in line with Vargo and Lusch (2016b), who argue that new (service) ecosystems demand new operators based on new social institutions.

The third round saw the panellists emphasizing the need to increase quality and availability of the new business solution to WMC. Respondents voiced hope for getting new actors to enter the market, but also a hope for more competition among the large-scale construction businesses. In the latter case respondents also voiced a wish to improve the quality of construction per se. It is notable that our expert panel did not recognize consumers as a part of the WMC business network: they only consider companies, NGOs (e.g. the Wood Industry Association) and other interest groups (e.g. civil servants) as key actors in the process of developing wood construction businesses.

However, they were aware of consumer needs concerning wooden housing, such as builders of one-family houses (using wood as a construction material), along with increasing pro-environmental attitudes among consumers (Holopainen et al. 2014; Luo et al. 2017). Respondents had observed this consumer value shift

Table 2 Delphi round 2 evaluations of statements related to the configuration of the partner network

Id	Statement		Likelihood (%)	Desirability (%)
9	By 2030, strong business networks within the industry will help us build competitive products more effectively and rapidly	Low	6	0
		Medium	18	18
		High	**76**	**82**
28	By 2030, concrete builders will be significantly more interested in the opportunities offered by wooden building solutions than today	Low	12	0
		Medium	35	24
		High	**53**	**76**
29	In 2030, the wood construction industry could be described as a network of specialized organizations of various sizes rather than a value chain consisting of only a few large companies	Low	0	0
		Medium	**65**	41
		High	35	**59**
38	In the future, my organization will co-create value with various types of actors, including customers and suppliers	Low	0	0
		Medium	6	6
		High	**94**	**94**
41	By 2030, the amount of companies involved in the wood construction value chain will be much higher than today	Low	12	6
		Medium	41	35
		High	**47**	**59**

Bold indicates the response group with highest share of respondents

towards green consumption: "*I think these [voices] valuing nature, decreasing consumption are important. I mean sustainable development and consumption, I would say these issues will guide consumer choices in the future*" **Field manager, Finland**; "*The needs of end-users and fulfilling them … I think these environmental, I mean the positive influences of wooden materials, should be communicate better [to consumers]*" **Research manager, Finland**; "*Wood has a positive image, [it is considered] green in values, good for the health, such things ... greenness is increasing*" **Technical manager, Finland.** Thus, based on these views, addressing end-user sustainability -related needs has been realized among industry actors, and in the future, underlining the sustainable image of wood could be emphasized even more as a competitive advantage. This is particularly relevant if combined with the earlier comments made by representatives from smaller companies.

Furthermore, participants articulated good examples of newly established alliance structures in Finland. A wood material -based new hospital project under construction in the northeastern region of Kainuu was a prime example. In general, the health benefits of utilizing natural building materials and the recently much discussed problems with indoor quality in public buildings, were seen as strong drivers towards the future diffusion of WMC (see also Toppinen et al. 2017). However, concurrently maintaining cost competitiveness was perceived as fundamentally important. Thus, addressing the health impacts in industrial construction business with the use of renewable materials appears to be more of an instrumental tool for increasing the use of wood than a derivative of an intrinsic motivation towards higher levels of (corporate) sustainability among the value network actors.

4.2 Capability Base

During the first round of interviews, the development of more sophisticated building solutions was seen as a process demanding financial capital. Respondents believed that only a few of the largest operators therefore have the necessary resources. Builders, especially ones that build with a range of different materials, were not interested in developing the solutions, and would rather remain interested in only bringing them to the markets once they exist. This is illustrated by the following quote: "*We will not do this validation [of wood-based building solutions], which needs to be undertaken by the wood industry itself.*" **Executive at a building company, Finland**. This attitude appears to indicate a limited interest in adopting the role of resource integrator [in line with Vargo and Lusch (2008)]. It also limits the propensity through which smaller contractors can be seen as game changers of WMC and overall uncertainty related to the future development of WMC.

Further, the differing timespans that operators in the value chain use to plan their actions, develop their resources, and especially building solutions based on wood, were cited numerous times as key issues for co-operation. Respondents also identified an issue with the core timespan being the length of a single project: "*[Building] processes can be very long, even unbearably so.*" **Research manager at a forestry organization, Finland**.

During the second round, the necessary core competencies were approached using the claims listed in Table 3. Based on this, highest conformity exists with item: "*...a building process from start to finish will be significantly shorter than today*". Statement "*Small companies will lack the money and know-how to develop more competitive and advanced wooden building solutions*" was perceived with a low desirability but a higher likelihood to happen. This appears to indicate continued scepticism in the ability of small and medium-sized enterprises (SMEs) to create a sufficient capability base (see the earlier remark on financial resources). Furthermore, the SDL line of thinking is evident in statement 33 despite divided reactions. Here the foreseen rise of prefabrication could benefit WMC, as the wooden elements are considerably lighter to transport than their concrete counterparts—especially if sustainability criteria are applied to the whole value chain.

During recent years the concrete-based construction industry has faced problems such as poor building quality in public buildings. Difficulties have particularly emerged in humidity control, which influences indoor air quality, but also makes structures prone to the growth of mould. As Luo et al. (2017) noted, indoor environmental quality (IEQ) has caused increasing public and scientific concern (e.g. eye, nose and throat irritation, headaches, breathing problems). Respondents emphasized the existence of these challenges in the construction business and

Table 3 Delphi round 2 evaluations of statements related to the capability base

Id	Statement		Likelihood (%)	Desirability (%)
8	By 2030, many experienced professionals will be both buying wooden building solutions as well as selling them	Low	12	6
		Medium	6	18
		High	**82**	**76**
31	Small companies will lack the finances and know-how to develop more competitive and advanced wooden building solutions	Low	18	**53**
		Medium	**59**	41
		High	24	6
33	By 2030, organizational cultures will be more prone to co-operation and strategic alliances between various organizations	Low	0	0
		Medium	29	24
		High	**71**	**76**
34	By 2030, a building project from start to finish will be significantly shorter than today	Low	0	0
		Medium	6	18
		High	**94**	**82**

Bold indicates the response group with highest share of respondents

believed wooden buildings to be a solution: "*The increasing needs of customers… especially these mould problems, people are increasingly aware of these issues, and there is need for healthier buildings*" **Executive from the building industry, Finland**; "*As a matter of fact, the healthiness of wood, what wood does to indoor air are the most influential aspects … if we [only] can prove these [health] effects, the better the positive influence of wood*" **Managing director from a wood industry association, Finland.**

As Forsythe (2006) noted, distrust in the construction industry originates from changes in construction culture. Previously the understanding of quality was handed down from master tradesman to apprentice, and either an on-site builder or possibly a leading hand carpenter generally supervised the work on a full-time basis (Forsythe 2006). Concerning the issues mentioned by Forsythe (2006), the availability of skilled personnel was also found to be a bottleneck among some respondents during the third Delphi round. However, others did not believe this to be an issue. Thus, further analysis is needed to connect this capability dimension with the network and business model issues of WMC.

4.3 Value Configuration

As the future of the value chain(s), especially how the actors view the operating logic, is the key point of interest in our study, the respondents were faced (Delphi round 2) with a bulk of statements that dig deeper into this logic. The statements and the reactions to the likelihood and desirability of related claims are provided in Table 4.

Table 4 Delphi round 2 evaluations of statements related to the value configuration

Id	Statement		Likelihood (%)	Desirability (%)
5	By 2030, we will sell and buy more wood construction products and services through open online platforms, such as web shops or professional digital networks	Low	0	0
		Medium	12	29
		High	**88**	**71**
10	By 2030, prefabrication will be the main operating logic, with less on-site building	Low	0	6
		Medium	0	6
		High	**100**	**88**
39	By 2030, we will have more standards, open-access platforms and public data banks for wood construction businesses to use	Low	0	0
		Medium	24	18
		High	**76**	**82**
40	In 2030, the best business model is to control a larger part of the value chain than today	Low	6	12
		Medium	**59**	**53**
		High	35	35

Bold indicates the response group with highest share of respondents

The panellists indicated highest conformity with item "*By 2030, prefabrication will be the main operating logic, with less on-site building*". Instead, item "*The best business model is to control a bigger part of the value chain than today*" divided opinions. This happened both in terms of likelihood and desirability, which suggests that there is scope for alternative value configuration arrangements, at least from the expert viewpoint.

Today's construction industry, with a distinctively project-based style of networking, could benefit from a service-led relationship (SLR) business model. Such a model would solve some of the problems of flawed construction at the company level, but also at the customer level. Razmdoost and Mills (2016) have observed that a "dark side of relationships" found in the project-based construction sector could be solved by emphasizing SDL instead of relationship marketing. Scholars suggest that process-based performance management increases the possibility of integrating the best available resources to meet project objectives, and SLR sees problem-solving as the main goal of a relationship. Thus, orchestrating a value network that includes end-users is essential for firms to build their business models on customer needs, to be able to recognize customer value and create a business model that will capture it (Pynnönen et al. 2012). As Vargo and Lusch (2008, 7) argue: "the customer is a co-creator of value, and value is co-produced in networks". Thus, considering for example homebuyers as one key partner in the construction business would bring new understanding on service offerings. Contrastingly, while the end-user is often closely connected to the building process itself in small-scale private housing projects, an almost total lack of this type of connection exists in multistorey construction.

To solve customer problems, and to offer end-users a valuable service (such as high quality indoor air, custom-built interiors, or design), businesses should strengthen the positive image of the wood construction industry. While transforming the business models towards sustainability, the respondents also emphasized price (cost-effectiveness), flat size, wood lightness and the swiftness of wooden construction. "*In my understanding* **…** ***the*** *average size of flats is decreasing. Today's flats are designed for one or two persons*" **Research manager at a forestry organization, Finland**; "*I think you are missing the two important competitive factors [of wood], lightness and ecological aspects*" **Wood industry sales executive, Finland**; "*I think reasonable size and reasonable price [matter most]. I would say in wooden construction [that if] we begin with so-called expensive and fancy architect houses, it does not serve the idea that we could build a wooden apartment house fast. We should have prefabricated units ready, storing them in a dry storage facility, and then quickly assemble the house*" **Executive from the building industry, Finland**.

5 Conclusions

In line with Brege et al. (2014), our study indicates that the key strategic aspects in the WMC business model are related to maintaining cost competitiveness, increasing construction speed with prefabrication and enhancing new hybrid material -based building solutions. Only to some extent the perceived role of wood is related to wooden buildings as a long-standing carbon stock in combatting climate change. Based on our findings, addressing end-user health concerns in the industrial construction business with the use of renewable materials appears to be a key channel for wood utilization. This also happens (possibly—further research is still needed) in the absence of intrinsic motivations towards sustainability among construction business actors.

Addressing health impacts can be aligned with social sustainability dimension for green/sustainable building, which is related to both physical and mental elements (e.g. Strobel et al. 2017; Burnard and Kutnar 2015). Wood has visual-aesthetic value as well. Furthermore, when companies develop sustainable-based business models towards more service and user-driven orientation, they could combine dimensions for green building into green consumption, and eventually provide their customers with offerings of green services (e.g. measurement of indoor air quality)—thus delivering value.

Issues, such as IEQ, and rapid and green building, could strengthen the image of sustainable and healthy wooden materials and the implementation of a national bioeconomy strategy. From the social equity viewpoint, the public sector's role can be decisive in terms of providing affordable and social housing, as well as in building better schools, kindergartens or service housing for the elderly. It could also be decisive in constructing a test-bed (via norms, regulations and innovation policy tools) for new business networks and ecosystem-based trials for new modes of operation.

From the capability viewpoint, our key observations highlight the complementarity of network partner resources as a basis for long-term partnerships among developers, architects and material providers alike. Therefore, the question arises: which of the network organizations should take the role of resource integrator within the WMC business? A Swedish study by Brege et al. (2014) painted a diverse picture: certain firms are capable of conducting designer planning by themselves, while new multi-skilled entrants to the business change the roles and responsibilities in the value network. Therefore, if the wooden construction business aims to engage consumers/end-users as a part of their network, firms need to better communicate with a wider group of stakeholders by offering new solutions to current problems. Thus, by applying a service-led relationship approach to their businesses, firms could develop sustainability-driven *services* for their end-users. Notably, as people do not necessarily only buy tangible objects—such as a flat—the key criteria are likely to be intangible in the broad sense of service: a good and healthy living solution at a reasonable price, the possibility of influencing the design and fulfilling the needs of various end-user groups.

Finally as a synthesis, our results appear to demonstrate an arising demand for altering business models towards wood material -based multistorey solutions towards 2030. However, as with any traditional industry sector, enhancing a stronger sustainability-driven culture sustainability-driven business models remains a challenge (see Toppinen et al. 2013). Furthermore, stimulating collaboration between agile companies, utilizing new market niches (start-ups and SMEs) and bringing in risk-aversive large-scale incumbent firms are needed to break free from existing silos—and especially needed to set free from the concrete-dominant regime (Prahalad and Bettis 1986)—in industrial construction business (Hemström et al. 2017).

A limitation of our study is that our expert panel did not include non-wood construction experts or representatives of end-users. Future research has an urgent need to analyse the types of changes needed in the value network to better ensure that consumers are truly embedded in the co-creation process. Competition from new actors entering the industrial construction market and collaborative networks between WMC and traditional concrete-based construction businesses also belong to the key topics that need more in-depth analysis. Similar argumentation goes for the formation of project-based alliance business models. Future research is therefore required to elicit especially project-level insights on actor values, norms and belief systems of the entire business ecosystem (e.g. Pulkka et al. 2016), and should be targeted to other European countries in which forests and wood materials have scope in promoting the implementation of national bioeconomy strategies.

Appendix

Composition of the Delphi panel, including participants' professional backgrounds and participation in the different stages of data collection.

Country	Gender	Years of prof. experience	Title	Organization type	Participation in rounds
Finland	Male	14	Senior vice president	Wood industry	1–3
Finland	Female	22	Director of CSR	Wood industry	1 and 2
Finland	Male	31	Managing director	Forestry	1–3
Finland	Female	1	Executive	Building industry	1–3
Finland	Male	16	Owner	Forestry	1 and 2
Finland	Male	15	Research manager	Forestry	1–3
Finland	Male	3	Field manager	Forestry	1–3
Finland	Male	26	Production director	Building industry	1–3
Finland	Male	5	Senior vice president	Wood industry	1 and 2
Finland	Male	22	Sales executive	Wood industry	1–3
Finland	Female	16	Planning executive	Building Industry	1–3
Finland	Male	23	Managing director	Wood ind. association	1, 3
Finland	Male	15	Senior advisor	Public sector	3
Finland	Male	12	Technical manager	Building industry	3
Sweden	Male	21	Senior advisor	Forestry	1 and 2
Sweden	Male	15	Managing director	Wood industry	1–3
Sweden	Male	11	Managing director	Wood industry	2–3
Sweden	Male	11	President	Wood industry	1
Sweden	Male	17	Vice pres. market dev.	Forestry	1 and 2
Sweden	Male	12	Academic expert	Building ind. expert	1–3

(continued)

(continued)

Country	Gender	Years of prof. experience	Title	Organization type	Participation in rounds
Sweden	Male	8	Sales manager	Wood industry	1 and 2
Sweden	Male	15	Academic expert	Building ind. expert	3
Sweden	Female	10	Managing director	Public sector	3

References

Anderson-Connell L, Ulrich P, Brannon E (2002) A consumer-driven model for mass customization in the apparel market. J Fashion Mark Manage: Int J 6(3):240–258

Bourdeau L (1999) Sustainable development and the future of construction: a comparison of visions from various countries. Build Res Inf 27(6):354–366

Bosman R, Rotmans J (2016) Transition governance towards a bioeconomy: a comparison of Finland and the Netherlands. Sustainability 8:1017

Brege S, Stehn L, Nord T (2014) Business models in industrialized building of multi-storey houses. Constr Manage Econ 32:208–226

Burnard M, Kutnar A (2015) Wood and human stress in the built indoor environment: a review. Wood Sci Technol 49(5):969–986

Chandler J, Vargo S (2011) Contextualization and value-in-context: how context frames exchange. Mark Theor 11(1):35–49

Darko A, Zhang C, Chan A (2017) Drivers for green building: a review of empirical studies. Habitat Int 60:34–49

Edvardsson B, Tronvol B, Gruber T (2011) Expanding the understanding of service exchange and value co-creation: a social construction approach. J Acad Mark Sci 39(2):327–329

Forsythe P (2006) Consumer-perceived appearance tolerances in construction quality management. Eng Constr Architectural Manage 13(3):307–318

Gold S, Rubik F (2009) Consumer attitudes towards timber as a construction material and towards timber frame houses—selected findings of a representative survey among the German population. J Clean Prod 17:303–309

Hansen E (2016) Responding to the bioeconomy: business model innovation in the forest sector. In: Kutnar A, Muthi S (eds) Environmental impacts of traditional and innovative forest-based bioproducts. Springer, pp 227–248

Harris M, Halkett R (2007) Hidden innovation: how innovation happens in six 'low innovation' sectors. National Endowment for Science, Technology and the Arts, London

Hemström K, Gustavsson L, Mahapatra K (2017) The sociotechnical regime and Swedish contractor perceptions of structural frames. Constr Manage Econ 35:184–195

Hoibo O, Hansen E, Nybakk E (2015) Building material preferences with a focus on wood in urban housing: durability and environmental impacts. Can J For Res 45(11):1617–1627

Hoffman AJ, Henn R (2008) Overcoming the social and psychological barriers to green building. Organ Environ 21(4):390–419

Holopainen J, Häyrinen L, Toppinen A (2014) Consumer value dimensions for sustainable wood products: results from the Finnish retail sector. Scand J For Res 29(4):378–385

Hu H, Geertman S, Hooimeijer P (2013) The willingness to pay for green apartments: the case of Nanjing, China. Urban Stud 51(16):3459–3478

Hunt SD, Morgan R (1995) The comparative advantage theory of competition. J Mark59:1–15
Häkkinen T, Belloni K (2011) Barriers and drivers for sustainable building. Build Res Inf 39 (3):239–255
Höök M, Stehn L, Brege S (2015) The development of a portfolio of business models: a longitudinal case study of a building material company. Constr Manage Econ 33(5–6):334–348
Killip G (2013) Products, practices and processes: exploring the innovation potential for low-carbon housing refurbishment among small and medium-sized enterprises (SMEs) in the UK construction industry. Energy Policy 62:522–530
Lahdenperä P (2012) Making sense of the multi-party contractual arrangements of project partnering, project alliancing and integrated project delivery. Constr Manage Econ 30(1):57–79
Landeta J (2006) Current validity of the Delphi method in social sciences. Technol Forecast Soc Chang 73(5):467–482
Lessing J, Brege S (2015) Business models for product-oriented house-building companies—experience from two Swedish case studies. Constr Innovation 15(4):449–472
Luo W, Kanzaki M, Matsushita K (2017) Promoting green buildings: do Chinese consumers care about green building enhancements? Int J Consum Stud. Forthcoming
Osterwalder A, Pigner Y, Tucci C (2005) Clarifying business models: origins, present, and future of the concept. Commun Assoc Inf Syst 16:1
Payne A, Storbacka K, Frow P (2008) Managing the co-creation of value. J Acad Mark Sci 36 (1):83–96
Pulkka L, Ristimäki M, Rajakallio K, Junnila S (2016) Applicability and benefits of the ecosystem concept in the construction industry. Constr Manage Econ 34(2):129–144
Magretta J (2002) Why business models matter. Harvard Bus Rev 80(5):86–92
Mokhlesian S, Holmen M (2012) Business model changes and green construction processes. Constr Manage Econ 30:761–775
Pelli P (2016) Service ecosystems as frameworks to elaborate sustainable futures. In: Proceedings of 26th RESER Conference. 8th–10th Sept 2016, Naples, Italy
Pels J, Sheth JN (2017) Business models to serve low-income consumers in emerging markets. Mark Theory. http://journals.sagepub.com/doi/abs/10.1177/1470593117704262
Prahalad CK, Bettis RA (1986) The dominant logic: a new linkage between diversity and performance. Strateg Manag J 7(6):485–501
Pynnönen M (2008) Customer driven business model–connecting customer value to firm resources in ICT value networks. Dissertation, Acta Universitatis Lappeenrantaensis. Lappeenranta University of Technology
Pynnönen M, Hallikas J, Ritala P (2012) Managing customer-driven business model innovation. Int J Innov Manag 16(4):1250022
Razmdoost K, Mills G (2016) Towards a service-led relationship in project-based firms. Constr Manage Econ 34(4–5):317–334. https://doi.org/10.1080/01446193.2016.1200106
Teece D (2010) Business model, business strategy and innovation. Long Range Plan 43(2–3): 172–194
Toppinen A, Wan M, Lähtinen K (2013) Strategic orientations in global forest industry. In: Hansen E et al (eds) Global forest industry: changes, practices and prospects (Book Chapter 17). Taylor & Francis Ltd
Toppinen A, Röhr A, Pätäri S, Lähtinen K, Toivonen R (2017) The future of wooden multistory construction in the forest bioeconomy—a Delphi study from Finland and Sweden. J Forest Econ (in press). http://www.sciencedirect.com/science/article/pii/S1104689916300952
Storbacka K, Frow P, Nenonen S, Payne A (2012) Designing business models for value co-creation. Rev Mark Res 9:51–78
Strobel K, Nyrud A, Bysheim K (2017) Interior wood use: linking user perceptions to physical properties. Scand J For Res. https://doi.org/10.1080/02827581.2017.1287299
Vargo S, Lusch R (2004) Evolving into a new dominant logic for marketing. J Mark 68:1–17
Vargo S, Lusch R (2008) Service-dominant logic: continuing the evolution. J Acad Mark Sci 36 (1):1–10

Vargo S, Lusch R (2016a) Institutions and axioms: an extension and update of service-dominant logic. J Acad Mark Sci 44:5–23
Vargo S, Lusch R (2016b) Service-dominant logic 2025. Int J Mark Res, in press. https://doi.org/10.1016/j.iresmar.2016.11.001
Vargo S (2007) On theory of markets and marketing: from positively normative to normatively positive. Australas Mark J 15(1):53–60
Wang L, Toppinen A, Juslin H (2014) Use of wood in green building: a study of expert perspectives from the UK. J Clean Prod 65:350–361

A Governance Framework for a Sustainable Bioeconomy: Insights from the Case of the German Wood-based Bioeconomy

Erik Gawel, Alexandra Purkus, Nadine Pannicke and Nina Hagemann

Abstract Increasing the sustainability of economic processes and products requires a path transition from the present, predominantly fossil resource-based "throughput economy" towards a renewable resource-based circular flow economy. The bioeconomy concept can contribute to such a transition. However, an adequate governance framework is necessary not only to overcome the current carbon lock-in and create fair competitive framework conditions for bioeconomy processes and products (*enabling function*), but also to ensure the sustainability of an increased use of bio-based resources (*limiting function*). At the same time, achieving a path transition is challenging due to, inter alia, interacting market failures which distort allocation decisions, and uncertainties about the economic, environmental and socio-economic impacts of different bio-based production pathways. Moreover, transitioning to a new "upper state" sustainability equilibrium requires a corresponding politico-economic equilibrium in markets for regulation that allows for the provision of necessary transition policies. In this chapter, we discuss the challenges of establishing an effective governance framework for the bioeconomy. Furthermore, focusing on the case of the German wood-based bioeconomy, we analyse how the enabling and limiting governance functions have been implemented in practice. Based on this, we identify scope for improvements. In particular, the case study highlights the important role that policies have to play in

E. Gawel (✉)
Department of Economics, Helmholtz Centre for Environmental Research—UFZ, and Institute for Infrastructure and Resources Management, University of Leipzig, Leipzig, Germany
e-mail: erik.gawel@ufz.de

A. Purkus · N. Pannicke · N. Hagemann
Department of Economics, Helmholtz Centre for Environmental Research—UFZ, Leipzig, Germany
e-mail: alexandra.purkus@ufz.de

N. Pannicke
e-mail: nadine.pannicke@ufz.de

N. Hagemann
e-mail: nina.hagemann@ufz.de

W. Leal Filho et al. (eds.), *Towards a Sustainable Bioeconomy: Principles, Challenges and Perspectives*, World Sustainability Series,
https://doi.org/10.1007/978-3-319-73028-8_26

establishing fair competitive framework conditions for bioeconomy applications, fostering innovation and safeguarding sustainability. While existing measures remain fragmented and insufficient to initiate a path transition, gradually developing them further may contribute to a dynamic that stimulates demand for more far-reaching transition policies on political markets.

Keywords Bioeconomy · Wood · Governance · Policies · Path dependencies Germany

1 The Bioeconomy as a Result of Sustainability Transformation

The concept of a renewable resource-based bioeconomy has gained increasing political attention in recent years. It is supported by strategies on an international level (e.g. OECD 2009; McCormick and Kautto 2013, German Bioeconomy Council 2015a, b), on the European level (EC 2012), by national-level strategies in member states such as Germany (BMEL 2014), and also by regional-level strategies (e.g. Landesregierung Nordrhein-Westfalen 2013; Government of Flanders 2013). While it is still a matter of discussion how the bioeconomy concept should be defined (see Staffas et al. 2013), the European Commission (EC 2012, p. 9) understands "Bioeconomy" as encompassing "the production of renewable biological resources and the conversion of these resources and waste streams into value added products, such as food, feed, bio-based products and bioenergy. Its sectors and industries have strong innovation potential due to their use of science, enabling and industrial technologies, along with local and tacit knowledge".

The bioeconomy is expected to increase the sustainability of economic processes and products, and is associated with a number of political aims—such as climate change mitigation, environmental protection, energy diversity and security, technological progress and innovation, growth and employment, or rural value creation (EC 2012; German Bioeconomy Council 2015a). However, merely substituting fossil resources for biogenic ones does not guarantee sustainability improvements. The controversial discussions surrounding bioenergy use have shown that replacing fossil fuels with biomass is not per se economically efficient or ecologically sustainable (WBGU 2008; Purkus 2016). As with bioenergy, there is a significant potential for conflicts between bioeconomy policy aims—in particular, expanding the bioeconomy will increase pressures on natural ecosystems, with possible adverse effects on environmental policy aims. As a result, there is a need to explicitly safeguard ecological and economic sustainability through adequate "governance" approaches. In particular, sustainability considerations suggest that the bioeconomy concept needs to be embedded into a wider path transition from the present, predominantly fossil resource-based "throughput economy" towards a renewable resource-based circular flow economy (Staffas et al. 2013; Richardson 2012;

BMEL 2014; Pannicke et al. 2015). This entails a closing of material cycles in industrial processes, and the realisation of significant improvements in resource use efficiencies (Carus et al. 2014a, b). At the same time, the sustainability of the (biogenic and non-biogenic) renewable resource base needs to be ensured.

Following this normative concept of a "sustainable bioeconomy" (see Pannicke et al. 2015), the question emerges how a transition towards a new "upper state" economic equilibrium can be achieved, which is more ecologically sustainable than the current equilibrium and also economically efficient, at least in the long term. The current equilibrium can be characterised as a "carbon lock-in" (Unruh 2000), which is much discussed for the energy context but also relevant for material resource uses. Initiating a path transition towards an "upper state" equilibrium requires an effective governance framework. We suggest that it needs to fulfil two basic functions:

- Enabling function, to overcome the carbon lock-in and create fair competitive framework conditions for bioeconomy processes and products, which allow for efficient allocation decisions;
- Limiting function, to safeguard the sustainability of an increased use of bio-based resources.

Importantly, besides incentivising the substitution of fossil resources and safeguarding the sustainability of bioeconomy pathways, the governance framework needs to generate incentives for strengthening innovation efforts regarding biogenic and non-biogenic renewable resources, to limit additional pressures on ecosystems (BMEL 2014; Carus et al. 2014a, b).

Designing an adequate governance framework, however, is a challenging task. Potential governance options encompass a wide range of coordination mechanisms, with markets and hierarchical government interventions as the end points of the spectrum. In-between, governance modes such as networks, associations, and private or public-private negotiation mechanisms can be found (Benz et al. 2007). An analysis of governance challenges and different governance options' performance in addressing them can contribute towards understanding what kind of framework is required to promote a path transition towards a sustainable bioeconomy. On a general level, a governance analysis can be understood as the analysis of the structure of rules which coordinate the actions of interdependent actors (Mayntz 2005; Benz et al. 2007).

In this chapter, we first discuss the role that a governance framework has to play in steering the development of the bioeconomy and the challenges it faces (Sect. 2). For analysing specific governance options, the breadth of the bioeconomy concept which encompasses diverse sectors and policy fields makes a narrower focus necessary. In Sect. 3, we therefore focus on a case study of the German wood-based bioeconomy, to examine how the enabling and limiting functions of bioeconomy governance have been implemented in praxis. The wood-based bioeconomy represents a sub-sector of the wider bioeconomy, and encompasses the material and/or energetic use of lignin-containing parts of plants such as trees and scrubs (Ollikainen 2014). This includes wood from forests (e.g. round timber, pulpwood, forest residues), wood from short rotation coppices (SRC) and landscape residues, as well as by-products, wood processing residues and recycled wood. As a case study, the wood-based bioeconomy is relevant because it does not directly compete with food production for biomass resources; given the food vs. fuel debate surrounding energy crops, interest in non-food feedstocks has increased within the last years (Ollikainen 2014; Carus and Dammer 2013). Whereas some conventional uses of wood materials (e.g. wood pellets, paper and furniture) are well established, the supply and demand for innovative wood-based applications still remains low despite the fact that political strategies aim at enhancing innovations and their market entry (De Besi and McCormick 2015).[1] Based on insights from the case study, Sect. 4 discusses central challenges and perspectives for the further development of the governance framework.

2 The Role of Bioeconomy Governance

2.1 Elements of a Bioeconomy Governance Framework

As a "structure of rules" (Mayntz 2005), a governance framework encompasses several elements (see Scharpf 1997; Benz et al. 2007; Mayntz 2005; Williamson 1985) which are also of interest to bioeconomy governance research:

- *Forms of interaction* (such as one-sided action, negotiation, majority decisions, hierarchical steering), which characterise certain governance structures (such as markets, relational contracts, networks, policy instruments which constitute regulatory contracts, hierarchies, etc.).
- *Institutional context of interactions*: Institutions can be understood as the formal and informal rules and contracts (including enforcement mechanisms), which—as an interacting, multi-layered system which has evolved over time—channel individual behaviour (see Furubotn and Richter 2005; North 1990). Several nested layers can be distinguished (Williamson 2000; Dixit 1996): informal institutions, such as customs, traditions and norms, which develop in an evolutionary process and change very slowly; the institutional environment which encompasses formal rules such as the constitution, the legal framework and property rights; transaction-specific governance structures; and incentives which actors face on markets, as expressed in prices and quantities.
- *Actors* (including individuals, organisations, subsystems of the society or states): these fulfil a double role—actors' behaviour is influenced by forms of interactions and the institutional context; but also, actors can try to actively influence the design of the rule structure.

[1]See Pannicke et al. (2015), Hagemann et al. (2016) for an analysis of the status quo of the wood-based bioeconomy in Germany.

In the literature, different concepts of governance can be distinguished. In a narrow sense, the term "governance" is sometimes used to describe only non-hierarchical coordination mechanisms; this understanding reflects a research focus on the role of non-governmental actors in governance processes (see e.g. Rhodes 1996). Here, a wider definition shall be adopted, which includes markets and hierarchies as the end points of a broad spectrum of possible governance structures (following Mayntz 2005; Benz et al. 2007; Williamson 1985). Depending on the analysed issue, governance modes which balance towards the hierarchical end of the spectrum can take the form of regulations, administrative bodies or companies (Benz et al. 2007). In governing transactions with different attributes (such as uncertainty, the frequency with which transactions recur, and asset specificity as central dimensions), different governance structures can prove advantageous, i.e. result in comparatively lower costs of coordinating transactions (Williamson 1985, 1996). For example, governance structures which balance towards the market end provide high-powered incentives, because actors directly bear the consequences of their actions. In adapting to changes in framework conditions and other disturbances, actors will autonomously search for cost-efficient solutions using the time- and space-dependent knowledge they possess (see also Hayek 1945). More hierarchical governance structures, on the other hand, provide stronger investment safeguards and coordinated responses to disturbances, but rely to a greater degree on centrally available knowledge. As gains and losses associated with actions are no longer the sole responsibility of each actor, incentive intensity decreases, and market incentives have to be replaced by administrative monitoring and enforcement mechanisms which give rise to bureaucratic costs.

Applied to the bioeconomy context, the governance framework has to be related to the societal goals associated with a transition to a sustainable bioeconomy (see Fig. 1). The institutional framework influences actors' production and consumption processes as "governance objects". On a systemic level, it can promote a transition to a more sustainable pathway of economic development, or reinforce the current, fossil resource-dominated pathway (see North 1990; Unruh 2000). From an economic theory perspective, the governance framework should enable allocative outcomes which are efficient (i.e. aims are implemented without using more resources than necessary), and sustainable also in an ecological and distributive sense. From these normative requirements, we derive the enabling and the limiting function as the two basic functions that bioeconomy transition governance has to fulfil. In its *enabling function*, the governance framework needs to support an efficient path transition away from the fossil resource-based throughput economy. The efficiency requirement implies that to be sustainable in an economic sense, the bioeconomy needs to generate products which are competitive at least in the long term and meet consumer demands. In its *limiting function*, the governance framework has to ensure that this path transition will in fact lead to greater ecological sustainability of economic activity, and that the transition is socially acceptable. Incentives for innovation are required as part of both functions, to reduce pressures on ecosystems as well as costs of bioeconomy processes and products.

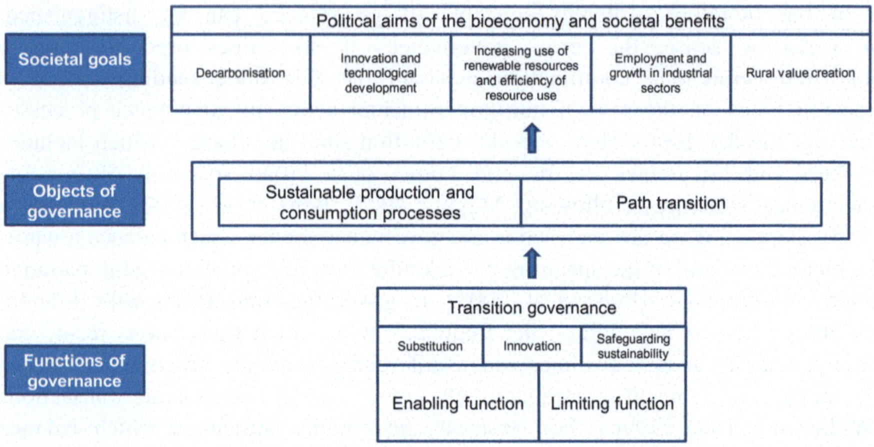

Fig. 1 The role of bioeconomy governance (*Source* own)

2.2 Challenges of Markets and Government Interventions in Fulfilling the Basic Functions of Bioeconomy Governance

In the following, we give a brief overview of central problems that markets and government interventions as two alternative archetypes of governance structures face when trying to implement the enabling and limiting functions of bioeconomy governance (based on Purkus 2016). Of course, the range of possible governance structures is much broader—however, basic challenges remain the same, with their relevance depending on the exact combination of market-based and hierarchical elements in hybrid governance structures.

In principle, renewable resource-based processes and products are marketable, private goods, for which both excludability and rivalry in consumption apply. As such, they are amenable to allocation by the market mechanism. However, a number of interacting market failures imply that markets alone will not suffice to fulfil the enabling and limiting functions of bioeconomy governance. First, greenhouse gas (GHG) emissions and other negative environmental impacts of fossil resource use (e.g. landscape degradation through mining, air pollution, waste) give rise to negative *environmental externalities*. In the absence of policy measures, these are insufficiently reflected in market prices, and distort competition between fossil resource-based and bio-based processes and products (Jenkins 2014; Lahl 2014). Moreover, environmental externalities also distort competition between different bioeconomy production pathways which may differ e.g. with regard to GHG balances or impacts on biodiversity, water and soil quality (Adler et al. 2015). Second, investments in innovation but also the deployment of innovative technologies give rise to *knowledge and learning spillovers*, which constitute a positive externality (Jaffe et al. 2005; Fischer and Newell 2008). Since investors are not able to appropriate the whole benefits of investments, this leads to an underinvestment in innovation and learning from a social perspective. Third, in competing with innovative technologies, established fossil resource-based technologies have the advantage that cost reductions from learning already took place in the past; moreover, they benefit from increasing returns to scale and network externalities. In interaction with the specialised nature of investments this creates a technological path dependency (Arthur 1989), which is further reinforced by the co-evolutionary development of infrastructures, consumption patterns, interdependent industries, and institutions. In combination, *technological and institutional path dependencies* can result in a persistent carbon lock-in (Unruh 2000). In the energy sector, the allocative outcome of market processes is further distorted by externalities resulting from the *public good nature of a secure energy supply*, and *market power* on the side of incumbents. Overall, the mentioned market failures imply that under a market governance framework, investments in innovative bioeconomy technologies would be lower than what would be efficient and consistent with a path transition towards a more sustainable economic equilibrium. As a result, the enabling governance function is not fulfilled. At the same time, interacting environmental externalities and knowledge and learning spillovers negatively impact the limiting governance function, because investments in bio-based processes and products and innovation efforts are not necessarily steered towards options which are particularly environmentally beneficial.

In principle, a voluntary internalisation of externalities through bargaining between affected parties would be possible; but particularly when global public goods such as the atmosphere as a sink for GHG emissions or biodiversity are concerned, private bargaining solutions are typically prevented by high transaction costs, ill-defined property rights and incentives for free-riding (see Coase 1960). Moreover, even if the market allocation was efficient, this would not ensure that the criterion of sustainability was met, because this also requires an equitable inter- and intragenerational distribution of resources (e.g. Daly 1992; Norgaard 1992). The limiting function of bioeconomy governance has to make sure that bio-based resource use does not exceed environmental carrying capacity (see Rockström et al. 2009), and it also has to address this distributive sustainability dimension.

Given the shortcomings of markets and voluntary bargaining solutions, government interventions are required to support a path transition. However, they too face a number of challenges which may give rise to "governance failures" (i.e. market or government failures). These challenges apply to enabling and limiting governance functions both. Of particular relevance are *information problems*, given the many uncertainties regarding the economic, environmental and social impacts of diverse and heterogeneous bioeconomy pathways, which compete for raw materials but also investments and human capital (McCormick and Kautto 2013; Edwards et al. 2008; Purkus 2016). Especially for innovative applications, costs and environmental impacts are often unclear, and will only become better understood with time as research and an upscaling to commercial production takes place. While these information problems also apply to market transactions, they prove more problematic for comparatively hierarchical policy interventions, because policy decisions affect the entire value chain, and costs of erroneous decisions can therefore be high (Hayek 1945). Another source of government failure can arise from *not establishing a consistent system of policy aims* (Jakubowski et al. 1997). Aims commonly associated with bioeconomy policy (see Sect. 1) show considerable potential for conflicts—with growth and rural value creation on the one hand and environmental protection on the other hand as an example. As with bioenergy pathways (Berndes and Hansson 2007; Purkus 2016), different bioeconomy pathways promise different contributions to the various relevant aims (Ober 2015). As a result, a prioritisation and transparent discussion of trade-offs becomes an important precondition for any cost-effective policy implementation. However, it can be politically expedient to leave the prioritisation of aims unclear, so that different aims can be emphasised when addressing different constituents (Kay and Ackrill 2012). Particularly, mixing efficiency rationales (i.e. the amelioration of market failures) and distributive rationales (which seek to enhance income opportunities for certain societal groups) can be a politically rational strategy to maximise support. From a *public choice perspective*, policy making is viewed not as a social welfare maximisation but as a political and social bargaining process, where politicians, bureaucrats, voters and organised interest groups attempt to maximise individual welfare (Dixit 1996; McCormick and Tollison 1981). This implies that policy makers may fail to optimally address market failures because of the political rationale inherent in the voting system, or because policies are captured by rent-seeking interest groups (see Helm 2010). Furthermore, government interventions are complicated by the *multi-level nature* of the bioeconomy governance problem, which arises from the transregional character of value chains and the fact that externalities (e.g. related to GHG emissions) and socioeconomic impacts (e.g. associated with land use changes) occur on different spatial scales. Building an effective multi-level governance framework—which may include both public and private actors—is challenging, with various problems related to its institutional design, its legitimation, and the coordination of governance levels (Benz 2009).

3 Governance of the Wood-based Bioeconomy in Germany

In this section, we analyse how the enabling and limiting functions of transition governance have been implemented in the case study of the German wood-based bioeconomy. The case study was carried out in the context of the accompanying research to the BioEconomy Cluster, funded by the German Federal Ministry of Education and Research. Findings are drawn from a review of scientific, industry and civil society publications, policy documents and legal texts, and discussions with stakeholders in cluster meetings and workshops. The following governance analysis builds on several studies which we conducted as part of the project, such as a detailed analysis of legal framework conditions (Ludwig et al. 2014, 2015), an analysis of political economy aspects (Pannicke et al. 2015), and a scenario analysis for the German wood-based bioeconomy (Hagemann et al. 2016).

Since an in-depth analysis of interaction forms, actors and the full institutional context of interactions is beyond the scope of this chapter, we focus on the role of governance structures such as markets, networks, and policy instruments, and those parts of the institutional environment which affect the German wood-based bioeconomy, specifically. Examples of the latter are forestry law or waste and recycling law. For the question of how to promote a path transition towards a sustainable bioeconomy, governance structures and specific legal framework regulations are of particular interest (see Purkus 2016), because they can be purposefully changed over manageable timescales (Williamson 1996). On the other hand, formal rules which affect economic development as a whole, such as constitutional rules or intellectual property rights, shall be taken as given in the context of this analysis. They are also an important framework condition for bioeconomy investments (e.g. North 1990; Sweet and Eterovic Maggio 2015), but take longer to change and result in part from evolutionary processes (Williamson 1996). Likewise, slowly evolving informal institutions can prove important—for example, the societal attitude towards technologies with high perceived uncertainties can play an important role in influencing the direction of technological change (e.g. Wynne 1983). In Germany, using genetically modified organisms in agriculture tends to be viewed critically (Nausch et al. 2015; Frewer et al. 2013), which is a relevant constraint for crop-based bioeconomy pathways. The use of genetically modified species for wood production is also being researched (Verwer et al. 2010), but plays a less prevalent role in the societal debate.

First, we provide a brief overview of relevant actor groups and their demand for active transition governance. Then, we examine what governance structures and bioeconomy-specific parts of the institutional environment play a central role in addressing the enabling and limiting governance functions. Lastly, we discuss evidence for governance failures and the role of policies, which emerge as particularly influential.

3.1 Identification and Characterisation of Relevant Actor Groups

The wood-based bioeconomy presents itself as a very heterogeneous field with diverse actors and interests (see Pannicke et al. 2015). The *business sector* alone encompasses conventional wood-related industries such as timber producers, saw mills, construction, woodwork and paper industries, but also consumer goods production and innovative applications in the chemicals industry (FNR 2014). Interests and attitudes towards the expansion of bioeconomy applications are equally varied. Especially actors whose production structure is predominantly based on fossil fuels, such as the chemical industry, lack interest in fostering a path transition (German Bioeconomy Council 2015c).

Domestic forestry actors do not face strong incentives to actively lobby for increasing the material use of wood, because market conditions are already favourable given demand from energy and conventional wood products sectors (FNR 2014). Further increases in wood demand might predominantly lead to increasing imports (Mantau 2012). Improving domestic wood availability is possible, but would need to be more actively pursued. Also, *agricultural actors*' interest in SRC plantations is limited so far, due to, inter alia, high initial investment costs and high uncertainties regarding the return on investments (Finger 2016).

Voters' support is essential for the credibility of bioeconomy policies. That said, it is rational for voters to prefer sustainability-oriented policies which do not impose additional costs on them (at least not perceivably) (Hansjürgens 2000). *Environmental interest groups* can influence public opinion on the wood-based bioeconomy either positively or negatively, but are still in the process of establishing their positions (McCormick 2011). Finally, on the side of *consumers,* awareness of bioeconomy products remains limited; often, they have at best similar characteristics as fossil resource-based products but are more expensive, making it challenging to communicate their advantages (Vandermeulen et al. 2012). Also, sustainability concerns can make consumers hesitant when it comes to choosing bio-based products, as demonstrated by the example of biofuels (see Pfau et al. 2014).

3.2 Institutional Implementation of the Enabling and Limiting Governance Functions

Table 1 provides an overview of relevant governance structures and bioeconomy-specific parts of the institutional environment which contribute either to the enabling function of bioeconomy transition governance, the limiting function or both. It builds on results from an inventory of institutions which impact the wood-based bioeconomy in Germany, which was undertaken as part of our research project (Ludwig et al. 2014, 2015; see also Pannicke et al. 2015). A combination of enabling and limiting functions can be found in institutions which explicitly try to enable bioeconomy applications which are particularly sustainable from an ecological and/or social perspective. As Table 1 demonstrates, such combinations of enabling and limiting elements in the same institution are a common characteristic of the case study's governance framework. To be able to reflect interactions between the two functions, we structure our discussion according to where in value chains governance measures are aimed at; that is, whether they focus on the bioeconomy resource base, bio-based processes and products, or whether they are aimed at reducing fossil resource use.

Table 1 Institutions with major implications for the wood-based bioeconomy in Germany

Governance functions	Focus on the bio-economy resource base	Focus on bio-based processes and products	Focus on reducing fossil resource use
Enabling function	– Associations of forest enterprises and owners – Financial support for SRC, e.g. Joint Task for the Improvement of Agricultural Structures and Coastal Protection – Support for wood recycling, e.g. waste and recycling law	– Energy Saving Ordinance – Incentives for energetic wood use in the electricity sector (feed-in tariffs/feed-in premiums) – Incentives for energetic wood use in the heating sector (mandatory minimum RES shares, grants and loans, reduced VAT on firewood) – Incentives for wood gasification in the electricity sector – Norms and standards, e.g. Bio-based Content (EN 15440); Wood-Polymer Composites (WPC, CEN/TS 15534), Compostability of plastics (EN 14995)	– EU Emissions Trading System (electricity sector) – Taxes, e.g. electricity tax, energy taxes for heating and transport fuels – Grants and loans for energy efficiency investments – Energy efficiency standards for products and buildings – Energy labelling for household appliances (EU Energy Labelling Directive) – Waste and recycling law (Waste Management Act) – Chemicals regulation (REACH)
Combination of enabling and limiting functions (enabling with explicit sustainability focus)	– R&D support – Financial support for e.g. afforestation, e.g. Joint Task for the Improvement of Agricultural Structures and Coastal Protection, agri-environmental schemes – Financial support for e.g. SRC under CAP (greening pillar)	– R&D support – Research networks and publicly promoted clusters, e.g. Cluster Forest and Wood, Leading Edge Cluster BioEconomy – Voluntary eco labels, e.g. Blue Angel Eco Label, Environmental Product Declarations (e.g. DIN EN ISO 14025:2011-10; DIN EN ISO 14040:2009-11) – Procurement law, e.g. Directive for the Procurement of Wood – Incentives for biomass to liquid through biofuels quota, with mandatory sustainability certification	
Limiting function	–Forestry law, e.g. German Federal Forest Act –Trade law for imports, e.g. Timber Trade Safeguarding Act, EU Timber Regulation (EC 995/2010) – Agricultural law (for SRC)		

Source Adapted from Pannicke et al. (2015)

3.2.1 Governance of the Bioeconomy Resource Base

Institutions governing allocation decisions in the biomass production sphere play an important role for the limiting function of bioeconomy governance. Wood production in domestic forests has to comply with the German National Forest Act and Federal Forest Acts, which implement sustainability as a guiding principle. Wood imports are governed by trade law, which seeks to safeguard against illegal logging. SRC plantations fall under agricultural law.

Resource flows of "used" wood are the domain of waste and recycling law. By setting framework conditions which are conducive to wood recycling, it contributes to the enabling governance function. Potentially, it could also contribute to the limiting function, because the cascading use of wood reduces pressure on forestry and agricultural ecosystems; to date, however, waste charges and recycling regulation fail to set effective incentives for cascade use concepts (Ludwig et al. 2015, 2016). Within these legal bounds, wood production decisions are mainly governed by markets. Besides supply contracts in value chains, associations of forest enterprises and owners are examples for private governance structures which influence production decisions, and can contribute to the enabling governance function. The focus here lies on reducing production and transaction costs through network and scale economies. However, there are also policy instruments which support the bioeconomy resource base. Financial support for projects that enhance the provision of wood is offered, for example, by the German Joint Task for the Improvement of Agricultural Structures and Coastal Protection, and agri-environmental schemes as part of the EU's Common Agricultural Policy (CAP). Under the CAP's greening pillar, SRC can only be supported if no mineral fertiliser and pesticides are used, whereas the "Joint Task" support implements no explicit sustainability requirements for SRC plantations.

3.2.2 Governance of Bio-based Processes and Products

In the governance of bio-based processes and products, a strong distinction is evident between material and energetic applications. Supply and demand for material applications is primarily governed by markets. Here, price premiums for sustainability advantages can be realised if consumers show a higher willingness to pay for "green" product characteristics. These can be signalled by using voluntary certification schemes, which contribute both to enabling and—by promoting compliance with sustainability requirements—limiting governance functions. However, research indicates that consumers' willingness to pay significant price premiums for "green" characteristics is limited (Carus et al. 2014a, b, Pacini et al. 2013; Schubert and Blasch 2010). Also, whether sustainability characteristics positively influence purchasing decisions may depend on product categories (Luchs et al. 2010). On a more general level, norms and standards for bio-products contribute to the enabling function, but are not tied to sustainability requirements (Pannicke et al. 2015).

Public support for material bio-based processes and products centres on research and development (R&D) and the promotion of public-private clusters and innovation networks. Again, enabling and limiting functions interact, as public support tends to be explicitly linked to the task of searching for sustainable solutions. This is complemented by selective support for niche applications. For example, environmental or social aspects can be included in public procurement, allowing for a promotion of wood use from verifiably sustainable sources. However, including such aspects is voluntary and allegedly increases the complicatedness of tendering processes; alongside information deficits, this limits the measure's effectiveness (Ludwig et al. 2014, 2015). In new buildings, the German Energy Saving Ordinance incentivises the use of energy-efficient wood constructions, without explicit sustainability requirements on wood supply.

On the other hand, there are a number of policy instruments which have proven highly effective in supporting the use of wood for electricity and heat production (Purkus 2016), thereby contributing to the enabling function. In the electricity sector, the Renewable Energy Sources Act has offered feed-in tariffs and later feed-in premiums for bioenergy use since 2000. Heat from renewable energy sources (RES) has been supported through the Market Incentive Programme's grants and loans since 2000, and since 2009 through the Renewable Energy Heating Act's mandatory minimum RES shares in new buildings. Moreover, a reduction applies for the value-added tax on firewood. So far, there are no binding sustainability requirements for wood and other solid and gaseous bioenergy carriers used in electricity and heating sectors, but the European Commission's 2016 proposal for a recast Renewable Energy Directive suggests their implementation for plants above certain capacity thresholds (EC 2016).

3.2.3 Governance of Fossil Resource Use

On the enabling side, markets set incentives for reducing fossil resource use when prices for such resources and associated processes and products increase. In guiding investment decisions, however, the volatility of price developments proves problematic, as the 2014/2015 oil price drop illustrates. In providing transition incentives, policy instruments which—at least partly—internalise the environmental costs of fossil resource use therefore play an important role. This role is fulfilled by climate policy instruments such as the European Emissions Trading System (EU ETS) and energy taxes. However, to date they focus primarily on reducing fossil resource use in the energy sector, rather than in materials sectors (see Rodi et al. 2011). Furthermore, emission allowance prices have been low and volatile in recent years, limiting the effectiveness of the EU ETS (Koch et al. 2014), while German energy tax rates are not closely aligned with the greenhouse gas emissions associated with diverse energy carriers (Gawel and Purkus 2015). Neither EU ETS nor tax regulations encompass sustainability requirements for wood, which would be problematic if these instruments triggered a significant wood demand.

In principle, chemicals regulation and waste and recycling regulation could incentivise reductions in fossil resource use in materials sectors (Ludwig et al. 2014, 2015). The European chemicals regulation REACH (Regulation (EC) No 1907/2006) aims at the replacement of substances which are hazardous for human health or the environment (Köck and Kern 2006). The search for substitutes may lead to innovation, but there is no preference for bio-based substances compared to fossil resource-based substances. In the German Waste Management Act, recycling requirements are subject to an assessment of "economic reasonableness"; here, it is the term's openness to interpretation which limits the instrument's effectiveness in promoting wood recycling and cascading uses (Herrmann et al. 2012; Ludwig et al. 2014, 2015).

3.3 Implications of the Case Study Analysis: Governance Failures and the Role of a Coordinated Bioeconomy Policy

In governing the transition towards a sustainable bioeconomy, policies emerge as a vital component of the rule structure. In the German case study, market price developments or private governance initiatives, such as voluntary sustainability certification, prove insufficient to fulfil the enabling function of bioeconomy governance beyond small niche applications. To comprehensively address environmental externalities, knowledge and learning spillovers, path dependencies and other market failures which distort allocation decisions to the detriment of bio-based processes and products, policy interventions are required. Moreover, effective policy instruments and legal framework conditions are necessary to implement the limiting function of bioeconomy governance, as markets and voluntary governance initiatives alone are not sufficient to safeguard the sustainability of an increased biomass demand.

However, the policy mix and legal framework conditions which govern bioeconomy allocation decisions prove very fragmented so far (see Ludwig et al. 2014, 2015; Pannicke et al. 2015) and fail to comprehensively address the enabling and limiting functions. To support a functioning innovation system, a coordinated mix of "supply push" and direct and indirect "demand pull" policies is required (Grubler et al. 2012; Foxon et al. 2005; see Fig. 2 and in more detail, Purkus et al. 2017; Pannicke et al. 2015). As highlighted in Fig. 2, such a policy mix needs to take differences in the commercial maturity of technologies into account. Market failures such as environmental externalities and path dependencies distort allocation decisions no matter whether conventional or innovative bioeconomy resources and applications are concerned, but knowledge and learning spillovers primarily apply to the latter. In the case of material sector pathways, the focus so far lies—apart from strategy development—on supply push policies which seek to enhance the supply of bio-based raw materials and technologies, for instance by providing

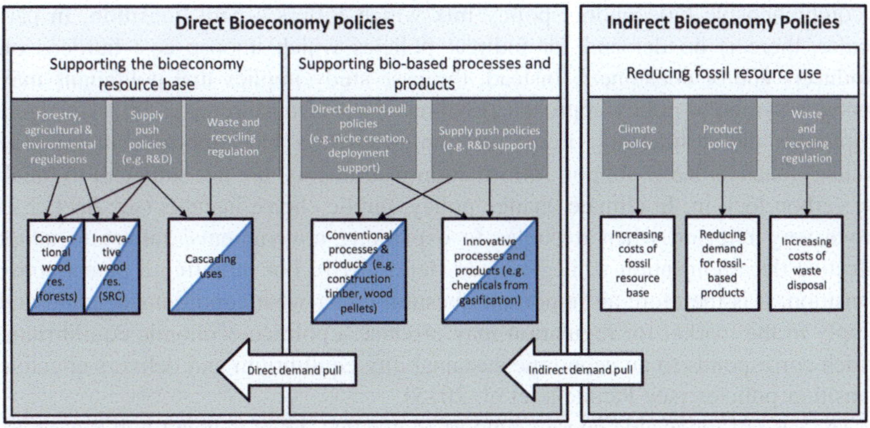

Fig. 2 Three pillars of wood-based bioeconomy policies (reproduced from Pannicke et al. 2015. *Note* blue areas indicate implementation of specific sustainability requirements)

funding for R&D and supporting innovative clusters. Direct demand pull policies, on the other hand, are only effective for energetic wood uses, which are to date predominantly based on comparatively established technologies (e.g. combined heat and power plants, wood and pellet stoves). When it comes to indirect demand pull policies, which increase demand for bio-based products, processes and raw materials by increasing the costs of fossil resource use, there is significant room for improvements—this applies to measures directed at the energy sector (e.g. EU ETS, energy taxes), as well as to measures directed at material applications (e.g. waste and recycling regulation, or the underdeveloped field of "material climate policy").

Moreover, where demand pull or supply push policies exist, they do not always explicitly take sustainability considerations into account—this is problematic, because there is uncertainty about whether existing land use regulations are sufficient to handle increasing pressures on ecosystems (Bringezu et al. 2014). In particular, there are many open questions with regard to safeguarding the sustainability of large-scale wood imports. If direct or indirect instruments implied significant increases in wood demand, sustainability constraints would need to be taken into account more explicitly in instrument design, e.g. through the implementation of mandatory sustainability certification schemes. Ideally, such schemes should be independent of the energetic or material end use of wood and coordinated on the EU level and beyond, because otherwise, distortions and leakage effects would arise (see Scarlat and Dallemand 2011). Meanwhile, the promotion of cascading uses avoids additional land use pressures, but this would require adjustments of the German waste regulation (Ludwig et al. 2014, 2015) and is limited by the amount of waste wood available (Mantau 2012).

However, from the perspective of politicians, it is only rational to supply transition policies if there is a political demand for such measures (see Pannicke et al. 2015). The overview of actor groups illustrates that so far, there is little demand for a comprehensive bioeconomy policy mix which fosters a path transition. In particular, there is no demand for indirect policies which impose cost burdens on producers and/or consumers. Instead, the case study implies that politicians may pursue a symbolic policy approach (Edelman 1964; Hansjürgens 2000): strategies emphasise the desirability of a transition towards a bioeconomy, but adopted policies (such as R&D support, information instruments) are not sufficient to break the carbon lock-in. In climate change policy, public choice insights (see Sect. 2.2) prove very relevant when it comes to explaining governments' failure to adopt effective decarbonisation strategies (see Helm 2010). For initiating a bioeconomy transition, it is therefore an important question under what conditions demand and supply in the market for regulation may produce a politico-economic equilibrium, which corresponds to an economic sustainability equilibrium and delivers effective transition policies (see Pannicke et al. 2015).

The case of renewable energy support shows that direct demand pull measures, such as feed-in tariffs, perform better in terms of garnering political support than indirect ones, such as the EU ETS or energy taxes (Jacobsson and Lauber 2006; Strunz et al. 2016). However, in the bioeconomy context, the design of direct demand pull measures is complicated by significant information problems (see Sect. 2.2). The design of European and German bioenergy support and associated sustainability and efficiency concerns can serve as an example for the problems of politically creating demand for bio-based production pathways with high uncertainties about costs, environmental and social impacts (WBGU 2008; Purkus 2016). The multi-level nature of the governance problem (particularly with regard to implementing effective sustainability safeguards) and the unclear hierarchy of conflicting political aims are further governance challenges which complicate the design of a comprehensive bioeconomy policy mix.

4 Outlook: Perspectives of Bioeconomy Governance

An efficient path transition towards a sustainable bioeconomy requires a clear governance framework with an active role of the state to provide fair competitive framework conditions for bio-based products, foster innovation and safeguard sustainability. The emergence of an effective governance framework is complicated by a number of relevant governance failures in the policy field, which arise when market or government interventions coordinate interactions between interdependent actors. Nonetheless, for addressing market failures and, in particular, the lock-in into fossil resource-based production and consumption structures, the case study analysis of the wood-based bioeconomy in Germany emphasises the important role of policies and the political system as rule giver.

The case study shows that there are already quite a few policies which impact the wood-based bioeconomy, but existing measures remain fragmented and are neither in their strength nor in their composition sufficient to initiate a path transition. At the same time, political demand for a comprehensive transition policy mix is low—the heterogeneity of the bioeconomy field and associated interests is an important factor in explaining this, alongside the comparative profitability of fossil resource-based production structures. Under these circumstances, there are a number of steps that can be identified to further develop the enabling function of bioeconomy governance. First, it is recommendable to gradually increase the stringency of existing instruments—this encompasses both policies with a direct focus on renewable resources and bio-based production, as well as policies aimed at increasing the costs of fossil resource use (such as strengthening the EU ETS and climate policy instruments for non-EU ETS sectors). At the same time, policy makers should communicate a long-term commitment to the path transition. Second, existing R&D support should be combined with targeted support for niches, e.g. improved green public procurement to promote the use of sustainable wood products. Stimulating public-private and private governance forms can further enhance niche creation and technology development; options are the use of information initiatives and labelling to increase consumer awareness of bio-based products with sustainability advantages, and by supporting knowledge exchange and networking between companies. Information instruments targeted at forest owners associations could be used to enhance the mobilisation of private forests.

In time, these first steps of an enabling bioeconomy governance may contribute to the forming of an "advocacy coalition" (Jacobsson and Lauber 2006) promoting more comprehensive changes in the institutional framework. On the other hand, given significant uncertainty regarding the sustainability impacts of different bio-based production pathways, policies which result in a large-scale direct demand pull for selected material wood uses should be avoided. Not only would policy makers face high information requirements when designing such interventions, but associated distortions in wood markets would also be significant.

Rather,—and this builds the bridge to the limiting function of bioeconomy governance—the focus should be on creating a selection environment that guides search processes towards sustainable wood-based resources, processes and products (e.g. supply chains based on recycling material and/or waste wood). This necessitates adjustments in waste and recycling regulations, to increase incentives for circular flow economy concepts, and a stronger focussing of support on sustainability-enhancing innovation, research and knowledge exchange. Also, particularly direct demand pull measures should be more strongly focussed on sustainable production and re-use, and innovative niches. Moreover, given relevant uncertainties, an adaptive approach to governance is required, with a gradual implementation of measures which leaves room for learning, and governance structures which support decentralised trial and error processes. Here, fostering the interaction of policies with private and public-private governance structures shows promise—for example, targeted support for clusters and forest owners associations could provide incentives to take sustainability concerns on board at an early stage of product and process development and resource management. Moreover, strengthening the limiting function of bioeconomy governance requires a re-evaluation of existing forestry, agricultural, environmental and trade policies, with regard to their ability to ensure sustainability in case of a significant increase in biomass demand.

References

Adler P, Budzinski M, Erdmann G, Majer S, Meisel K, Schock S, Thrän D (2015), Sachstandsbericht über vorhandene Grundlagen für ein Monitoring der Bioökonomie: Nachhaltigkeit und Ressourcenbasis der Bioökonomie, Deutsches Biomasseforschungszentrum (DBFZ), Leipzig

Arthur BW (1989) Competing technologies, increasing returns, and lock-in by historical events. Econ J 99(394):116–131

Benz A (2009) Politik in Mehrebenensystemen. VS Verlag für Sozialwissenschaften, Wiesbaden

Benz A, Lütz S, Schimank U, Simonis G (2007) Einleitung. In: Benz A, Lütz S, Schimank U, Simonis G (eds) Handbuch Governance. Theoretische Grundlagen und empirische Anwendungsfelder, VS Verlag für Sozialwissenschaften, Wiesbaden, pp 9–25

Berndes G, Hansson J (2007) Bioenergy expansion in the EU: cost-effective climate change mitigation, employment creation and reduced dependency on imported fuels. Energy Policy 35(12):5965–5979

BMEL (2014) National policy strategy on bioeconomy. Renewable resources and biotechnological processes as a basis for food, industry and energy, German Federal Ministry of Food and Agriculture (BMEL), Berlin

Bringezu S, Schütz H, Pengue W, O'Brien M, Garcia F, Sims R et al (2014) Assessing global land use: balancing consumption with sustainable supply. UNEP/International Resource Panel, Nairobi/Paris

Carus M, Dammer L (2013) Food or non-food: which agricultural feedstocks are best for industrial uses? Ind Biotechnol 9(4):171–176

Carus M, Eder A, Beckmann J (2014a) Industry report: GreenPremium prices along the value chain of biobased products. Ind Biotechnol 10(2):83–88

Carus M, Raschka A, Fehrenbach H, Rettenmaier N, Dammer L, Köppen S et al (2014) Ökologische Innovationspolitik – Mehr Ressourceneffizienz und Klimaschutz durch nachhaltige stoffliche Nutzungen von Biomasse, Umweltbundesamt, Dessau-Roßlau

Coase RH (1960) The problem of social cost. J Law Econ 3:1–44

Daly HE (1992) Allocation, distribution, and scale: towards an economics that is efficient, just, and sustainable. Ecol Econ 6(3):185–193

De Besi M, McCormick K (2015) Towards a bioeconomy in Europe: national, regional and industrial strategies. Sustainability 7(8):10461–10478

Dixit AK (1996) The making of economic policy: a transaction-cost politics perspective. MIT Press, Cambridge

EC (2012) Innovating for sustainable growth. A bioeconomy for Europe, European Commission (EC)/Directorate-General for Research and Innovation, Brussels

EC (2016), Proposal for a Directive of the European Parliament and of the Council on the promotion of the use of energy from renewable sources (recast). COM(2016) 767 final, European Commission (EC), Brussels

Edelman M (1964) The symbolic uses of politics. University of Illinois Press, Urbana

Edwards R, Szekeres S, Neuwahl F, Mahieu V (2008) Biofuels in the European context: facts and uncertainties. European Commission Joint Research Centre, Petten

Finger R (2016) Assessment of uncertain returns from investment in short rotation coppice using risk adjusted discount rates. Biomass Bioenerg 85:320–326

Fischer C, Newell RG (2008) Environmental and technology policies for climate mitigation. J Environ Econ Manage 55(2):142–162

FNR (2014) Marktanalyse Nachwachsende Rohstoffe. Fachagentur Nachwachsende Rohstoffe e. V. (FNR), Gülzow-Prüzen

Foxon TJ, Gross R, Chase A, Howes J, Arnall A, Anderson D (2005) UK innovation systems for new and renewable energy technologies: drivers, barriers and systems failures. Energy Policy 33(16):2123–2137

Frewer LJ, van der Lans IA, Fischer ARH, Reinders MJ, Menozzi D, Zhang X et al (2013) Public perceptions of agri-food applications of genetic modification—a systematic review and meta-analysis. Trends Food Sci Technol 30(2):142–152

Furubotn EG, Richter R (2005) Institutions and economic theory: the contribution of the new institutional economics, 2nd edn. University of Michigan Press, Ann Arbor

Gawel E, Purkus A (2015) The role of energy and electricity taxation in the context of the German energy transition. Zeitschrift für Energiewirtschaft 39(2):77–103

German Bioeconomy Council (2015a) Bioeconomy policy (Part I). Synopsis and analysis of strategies in the G7, Bioökonomierat, Berlin

German Bioeconomy Council (2015b) Bioeconomy policy (Part II). Synopsis of national strategies around the world, Bioökonomierat, Berlin

German Bioeconomy Council (2015c) Die deutsche Chemieindustrie – Wettbewerbsfähigkeit und Bioökonomie. BÖRMEMO 02, Bioökonomierat, Berlin

Government of Flanders (2013) Bioeconomy in Flanders. The vision and strategy of the Government of Flanders for a sustainable and competitive bioeconomy in 2030, Flemish Government, Environment, Nature and Energy Department, Brussels

Grubler A, Aguayo F, Gallagher K, Hekkert M, Jiang K, Mytelka L et al. (2012) Policies for the energy technology innovation system (ETIS), In GEA (ed), Global energy assessment—toward a sustainable future. Cambridge University Press/IIASA, Cambridge, pp 1665–1744

Hagemann N, Gawel E, Purkus A, Hauck J, Pannicke N (2016) Possible futures towards a wood-based bioeconomy—a scenario analysis for Germany. Sustainability 8(98):1–24

Hansjürgens B (2000) Symbolische Umweltpolitik – Eine Erklärung aus Sicht der Neuen Politischen Ökonomie. In: Hansjürgens B, Lübbe-Wolff G (eds) Symbolische Umweltpolitik, Suhrkamp, Frankfurt am Main, pp 144–182

Hayek FA (1945) The use of knowledge in society. Am Econ Rev 35(4):519–530

Helm D (2010) Government failure, rent-seeking, and capture: the design of climate change policy. Oxford Rev Econ Policy 26(2):182–196

Herrmann F, Sanden J, Schomerus T, Schulze F (2012) Ressourcenschutzrecht – Ziele, Herausforderungen, Regelungsvorschläge. Zeitschrift für Umweltrecht 23(10):523–526

Jacobsson S, Lauber V (2006) The politics and policy of energy system transformation—explaining the German diffusion of renewable energy technology. Energy Policy 34(3):256–276

Jaffe AB, Newell RG, Stavins RN (2005) A tale of two market failures: technology and environmental policy. Ecol Econ 54(2–3):164–174

Jakubowski P, Tegner H, Kotte S (1997) Strategien umweltpolitischer Zielfindung: eine ökonomische Perspektive. LIT, Münster

Jenkins JD (2014) Political economy constraints on carbon pricing policies: what are the implications for economic efficiency, environmental efficacy, and climate policy design? Energy Policy 69:467–477

Kay A, Ackrill R (2012) Governing the transition to a biofuels economy in the US and EU: accommodating value conflicts, implementing uncertainty. Policy and Society 31(4):295–306

Koch N, Fuss S, Grosjean G, Edenhofer O (2014) Causes of the EU ETS price drop: recession, CDM, renewable policies or a bit of everything?—New Evidence. Energy Policy 73:676–685

Köck W, Kern K (2006) Öffentlich-rechtliche Kontrolle von Umweltrisiken, insbesondere Probleme und Perspektiven der europäischen Chemikalienkontrolle. In: Vieweg K (ed) Risiko - Recht - Verantwortung, Heymanns Verlag, Köln, pp 279–320

Lahl U (2014) Bioökonomie für den Klima- und Ressourcenschutz – Regulative Handlungskorridore, BZL Kommunikation und Projektsteuerung/NABU (Nature And Biodiversity Conservation Union), Berlin/Oyten

Landesregierung Nordrhein-Westfalen (2013) Eckpunkte einer Bioökonomiestrategie für Nordrhein-Westfalen, Ministerium für Innovation, Wissenschaft und Forschung des Landes Nordrhein-Westfalen, Düsseldorf

Luchs MG, Naylor RW, Irwin JR, Raghunathan R (2010) The sustainability liability: potential negative effects of ethicality on product preference. J Mark 74(5):18–31

Ludwig G, Tronicke C, Köck W, Gawel E (2014) Rechtsrahmen der Bioökonomie in Mitteldeutschland – Bestandsaufnahme und Bewertung. UFZ Discussion Paper 22/2014, Helmholtz Centre for Environmental Research – UFZ, Leipzig
Ludwig G, Tronicke C, Köck W, Gawel E (2015) Der Rechtsrahmen für die Bioökonomie in Deutschland. Die Öffentliche Verwaltung 68(2):41–53
Ludwig G, Gawel E, Pannicke N (2016) Kreislaufwirtschaft im Bereich Holz – Rechtliche Bestandsaufnahme und Reformvorschläge für Kaskadennutzungen. Zeitschrift für das Recht der Abfallwirtschaft (AbfallR) 15(4):170–178
Mantau U (2012) Holzrohstoffbilanz Deutschland – Entwicklungen und Szenarien des Holzaufkommens und der Holzverwendung von 1987 bis 2015. Universität Hamburg, Hamburg
Mayntz R (2005) Governance Theory als fortentwickelte Steuerungstheorie? In: Schuppert GF (ed) Governance-Forschung. Vergewisserung über Stand und Entwicklungslinien, Nomos, Baden Baden, pp 11–20
McCormick K (2011) The emerging bio-economy in Europe: exploring the key governance challenges. World Renewable Energy Congress 2011, 8–13 May 2011, Linköping, Sweden
McCormick K, Kautto N (2013) The bioeconomy in Europe: an overview. Sustainability 5(6):2589–2608
McCormick RE, Tollison RD (1981) Politicians, legislation and the economy: an inquiry into the interest-group theory of government. Martinus-Nijhoff, Boston
Nausch H, Sautter C, Broer I, Schmidt K (2015) Public funded field trials with transgenic plants in Europe: A Comparison between Germany and Switzerland. Curr Opin Biotechnol 32:171–178
Norgaard RB (1992) Sustainability as intergenerational equity: economic theory and environmental planning. Environ Impact Assess Rev 12(1–2):85–124
North DC (1990) Institutions, institutional change and economic performance. Cambridge University Press, Cambridge
Ober S (2015) Noch ganz am Anfang. Viele offene Fragen beim Ausbau der Bioökonomie, Available online 18 April 2016: https://www.nabu.de/umwelt-und-ressourcen/ressourcenschonung/biooekonomie/19308.html
OECD (2009) The Bioeconomy to 2030: designing a policy agenda. OECD Publishing, Paris
Ollikainen M (2014) Forestry in Bioeconomy—smart green growth for the humankind. Scand J For Res 29(4):360–366
Pacini H, Assunção L, van Dam J, Toneto R Jr (2013) The price for biofuels sustainability. Energy Policy 59:898–903
Pannicke N, Gawel E, Hagemann N, Purkus A, Strunz S (2015) The political economy of fostering a wood-based bioeconomy in Germany. Ger J Agric Econ 64(4):224–243
Pfau SF, Hagens JE, Dankbaar B, Smits AJM (2014) Visions of sustainability in bioeconomy research. Sustainability 6(3):1222–1249
Purkus A (2016) Concepts and instruments for a rational bioenergy policy. A new institutional economics approach. Lecture notes in energy, vol 55, Springer International Publishing, Cham
Purkus A, Hagemann N, Bedtke N, Gawel E (2017) Towards a sustainable innovation system for the German wood-based bioeconomy: implications for policy design. J Cleaner Prod. online first: http://dx.doi.org/10.1016/j.jclepro.2017.04.146
Rhodes RAW (1996) The new governance: governing without government. Polit Stud 44(4): 652–667
Richardson B (2012) From a fossil-fuel to a bio-based economy: the politics of industrial biotechnology. Environ Plann C: Government and Policy 30(2):282–296
Rockström J, Steffen W, Noone K, Persson A, Chapin FS, Lambin EF et al (2009) A safe operating space for humanity. Nature 461(7263):472–475
Rodi M, Sina S, Görlach B, Gerstetter C, Bausch C, Neubauer A (2011) Das Klimaschutzrecht des Bundes – Analyse und Vorschläge zu seiner Weiterentwicklung. Umweltbundesamt, Dessau-Roßlau
Scarlat N, Dallemand J-F (2011) Recent developments of biofuels/bioenergy sustainability certification: a global overview. Energy Policy 39(3):1630–1646

Scharpf FW (1997) Games real actors play. Actor-centered institutionalism in policy research, Westview Press, Boulder, CO/Oxford

Schubert R, Blasch J (2010) Sustainability standards for bioenergy: a means to reduce climate change risks? Energy Policy 38(6):2797–2805

Staffas L, Gustavsson M, McCormick K (2013) Strategies and policies for the bioeconomy and bio-based economy: an analysis of official national approaches. Sustainability 5(6):2751–2769

Strunz S, Gawel E, Lehmann P (2016) The political economy of renewable energy policies in Germany and the EU. Utilities Policy 42:33–41

Sweet CM, Eterovic Maggio DS (2015) Do stronger intellectual property rights increase innovation? World Dev 66:665–677

Unruh GC (2000) Understanding carbon lock-in. Energy Policy 28(12):817–830

Vandermeulen V, Van der Steen M, Stevens CV, Van Huylenbroeck G (2012) Industry expectations regarding the transition towards a biobased economy. Biofuels, Bioprod Biorefin 6(4):453–464

Verwer CC, Buiteveld J, Koelewijn HP, Tolkamp GW, de Vries SMG, Meer PJ (2010) Genetically modified trees: status, trends and potential risks. Alterra Report 2039. Alterra Wageningen UR, Wageningen

WBGU (2008) Future bioenergy and sustainable land use, German Advisory Council on Global Change (WBGU), Berlin

Williamson OE (1985) The economic institutions of capitalism: firms, markets, relational contracting. Free Press, New York

Williamson OE (1996) The mechanisms of governance. Oxford University Press, New York

Williamson OE (2000) The new institutional economics: taking stock, looking ahead. J Econ Liter 38:595–613

Wynne B (1983) Redefining the issues of risk and public acceptance. Futures 15(1):13–32

Holistic Indicator for Optimizing Forest Governance

Mihail Hanzu

Abstract Forests have a key role in the wellbeing of the mankind. To fulfil sustainably all demands, forest governance must adapt to ever emerging needs and values of the society. The indicator proposed here is a criterion to optimize forest governance to ever changing social needs and development. It is based on an innovative mathematical theory, named holistic-integrative field theory, developed for this purpose. The theory uses linear algebra, statistics and discrete analysis, in order to integrate all forest outputs, perceived as important by at least one actor, into an indicator. Also some fractal and cybernetic principles are embedded in the logic and algorithms of the indicator. Problems raised by the heterogeneity of the outputs are solved using vector-based mathematics. Outputs are considered as vectors with an unknown number of dimensions but with known modules (lengths). Statistical methods and discrete analysis methods are used to compute the length of a resultant vector which represents the optimization criterion. The criterion measures the effects of change on forest outputs and is used as a feedback to improve forest governance. The indicator can integrate any available data, in an iterative manner. The holistic-integrative indicator has the potential to improve forests-society-science-policy-practice interface and to operationalize the concepts of natural capital and ecosystem services as well as to provide the means for a more sustainable, efficient and integrated usage of ecosystems. It also has the advantage that it uses general public as well as scientists, industry, political or any other actors' opinions in a continuous and integrated manner, thus it is promoting an inclusive and democratic approach in forest governance. It can also be updated for maintaining governance system connected with the development of society and to prevent unexpected negative side effects of governance changes. In the end of the chapter an example is provided.

M. Hanzu (✉)
Department of Forest Biometrics, National Institute for Research and Development in Silviculture" Marin Drăcea", 128, Eroilor, 077190 Voluntari, Romania
e-mail: mihail.hanzu@gmail.com; mihail.hanzu@icas.ro

W. Leal Filho et al. (eds.), *Towards a Sustainable Bioeconomy: Principles, Challenges and Perspectives*, World Sustainability Series,
https://doi.org/10.1007/978-3-319-73028-8_27

Keywords Optimal forest governance · General public acceptance
Inclusive forest policy · Equitability

1 Introduction

Constructing a sustainable bioeconomy is a challenge. The difficulty of the task is given by the constellation of actors involved—often with divergent or competing interests—and by the diversity and heterogeneity of the outputs provided by the ecosystems (Blagojević et al. 2016). Some of these outputs are the "raw material" of the bioeconomy, which is undoubtedly a sector that has the perspectives to increase in importance, if the natural resources are managed holistic and sustainable, under current dramatic climate changes (GISTEMP Team 2015).

Forests play a key role in the mankind's wellbeing and health (Gottdenker et al. 2014). Given the ever increasing uses of forest-based products and services, adapting forest governance to new challenges and demands becomes a must, in order to ensure a predictable and sustainable flow of positive outputs, from the forests, while maintaining all the other forest ecosystems functions and avoiding man-made natural disasters done in the name of bio-industries (Knudson 2009). Because the whole concept of bioeconomy is new and dynamic, the deciders tend to consider, in decision processes, only the outputs of the forest ecosystems with monetary or market value, as in neoliberal economics (Dimitrov 2006), omitting other benefits of equal or potentially higher importance. This lead to the creation of powerless institutions such as UNFF, just to ensure the general public that the deforestation and other forest-related subjects, perceived as a very important ones, and strictly related with bioeconomy, are on the political agenda (Dimitrov 2006). This is a worrying fact.

Considering the social and other implications of the forest ecosystems outputs (FAO 2015), in order to have a sustainable forest-based bioeconomy, the governance has to be a holistic, inclusive and democratic process. It has to include all actors and to be adaptable at any level such as global, continental, regional, national and local; simultaneously. It will have to provide fairness and equitability for all the actors by integrating all of them into a coherent, transparent and efficient framework. I want to emphasise that two actors identified as very important but whose role is often minor when decisions are taken are the local communities and the general public (Mustalahti 2015).

The general public concern regarding forest governance continuously increased, starting with the 1960s (Samuelson 1974). This concern was observed as a shift of the demands from provision of goods to provision of services and it lead to the emergence of new governace concepts ranging from discourses to institutions related to global forest governace (Arts and Buizer 2009).

The complexity of such a task, to design a functionating global forest governance framework might explain why despite the general concerns regarding global warming and the acknowledged importance of forests as carbon sinks, before the

Paris Agreement (UN 2015), there was no international-binding document or international framework or powerful institution to ensure sustainable use and protection of world's forests. Before Paris Agreement—which is still to be seen how effective will be on conserving and sustainable use of world's forests—only regional international regulations and organizations existed such as EUTR or ITTO and they seem to be unable to stop the global deforestation trend.

Because the criteria for optimizing forest governance were historically linked either with (i) sustainable timber either with (ii) maximum rent, antagonistic ideas appeared. The first criteria leaded to long rotation cycles while the second to short rotation cycles. However, they both neglected all other positive outputs of forests (Samuelson 1974).

Aiming for better governance, Decision Support Systems (DSS) were designed to improve the science-policy-practice interface, but recent studies show that, in some cases, levels of adoption have been lower than expected (Stewart et al. 2013).

Because modern forestry appeared in a timber-crisis, its governance structure continues to have a rather pyramidal shape in order to ensure the control on the system. But the needs of the human society are changing in time, therefore it is expected that forest governance process adapt itself to this social change and to new needs and threats.

The scope of this article is to present an indicator that is a democratic, equitable, inclusive and sustainable alternative to the pyramidal top-bottom model of forests governance. It can be used to avoid manipulation of forest governance by power networks operating independently of the interest of the general public, (Korvela (in Finnish) 2012 cited by Mustalahti 2015) possible because of a strictly neoliberal political thinking.

2 Method

Different governance has diverse effects on the structures of the forests at different spatial and temporal scales (Rametsteiner 2009). At their turn, each forest structure is determining distinct outputs of the forest ecosystem which can be measured, estimated or perceived. The levels where the outputs are externalized by the system are ranging from local to global. The outputs are assessed, more or less objectively, by different actors—including the general public. Considering the above, I hypothesise that forest structures are optimization criteria expressing indirectly—through their outputs—the efficiency of certain forest governance. The outputs can be evaluated, objectively or subjectively, but they are very heterogeneous therefore difficult to be integrated directly into an unique indicator. However the outputs are not independent.

In order to deal with the complexity of such task, there are two distinct challenges to consider in developing an indicator to be used when optimizing forest governance:

1. to design a structure of the indicator that will allow it to be adaptable in any particular context;
2. to integrate non-homogenous valuations of the forests outputs into an holistic ecosystem efficiency indicator.

For solving these challenges, a modular indicator was developed. A logical scheme of such a module of the indicator is presented in Fig. 1.

In the figure below is represented how each perceived output of the forests is evaluated, integrating all the actors' knowledge or perspective, in the final indicator, which is designed to be used to foster better forest governance. Each different output is considered from cybernetic perspective as a different dimension of the forest-society system, without omitting the perspective that they are interconnected.

Despite this heterogeneity, of the outputs they are not independent. For instance, a forest governance system that leads to substitution of European beech (*Fagus sylvatica* L.) with Norway spruce (*Picea abies* Karst.) might increase the timber production (Pretzsch 2009), expressed in cubic meters of timber, but such a structural change is increasing the acidity of the soil (Augusto et al. 1998), is decreasing the stability of the stand (Dobertin 2002), and is reducing the recreational effect of the stand (Hanzu 2009).

Because of the high heterogeneity of the outputs and due to the interconnectivity that exists between these outputs, I have chosen to model all the outputs using linear algebraic approach. Therefore any output is considered as a vector, whose length is equal with the estimated value of the output, no matter how subjective the evaluation is. According to this modelling approach, the efficiency of the system is given

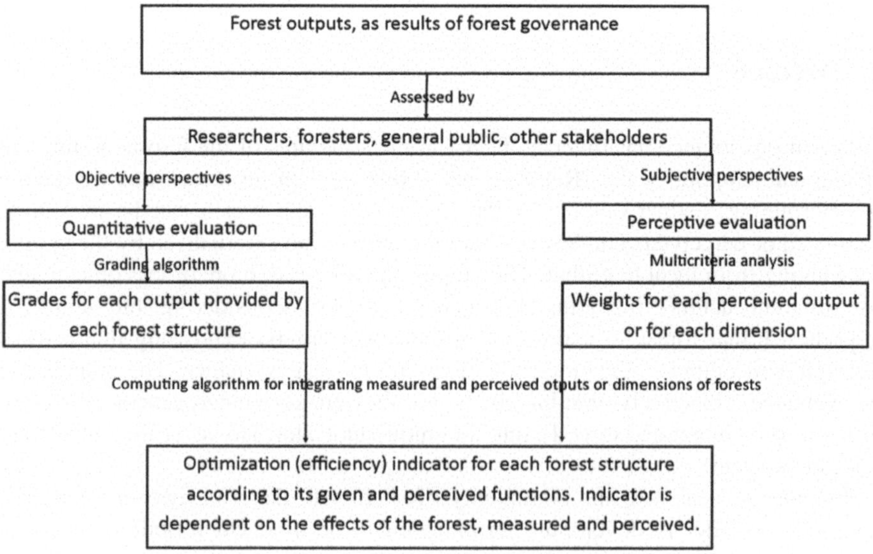

Fig. 1 The logical scheme of one module of the indicator

by the resultant of all the considered vectors. It is important to notice that the vectors, representing outputs, might be distinct ones, but this distinctiveness does not imply that they are linearly independent ones, as it is shown in Fig. 2. Their interdependency, which might be or not a causal one, should be somehow quantified, in order to compute the direction and the value of the resultant in the considered forest-society space.

For an easier understanding of the method's logic, let assume that the desired effect coincides with Output 2, as in Fig. 2, which depicts a 3D space. From the figure above, can be observed that in this vectorial modelling approach the participation of Output 1 to the desired effect of the forest equals to Output 1 * cos α. Therefore the desired effect is given by the following Eq. 1:

$$\textit{Desired Effect} = \textit{Output}\,2 + \cos\alpha * \textit{Output}\,1 \tag{1}$$

The equation above is identical with a linear model. However cos α cannot be estimated using this approach, not to mention that the dimensions of the forest-society system are, most likely, unknown entirely and they can appear or disappear in time, as result of the social progress. Therefore, in Eq. 1 cos α is substituted by the correlation coefficient between the two vectors. In this way the interdependency is quantified based on a correlation coefficient computed between the lengths of the vectors. By linking in this way linear algebra and statistics, a resultant of the output vectors can be calculated, even if not all the independent dimensions of the vectors are known. I named this innovative method of computing the length of the resultant output, considered as a vector, in a hyperspace with an unknown number of independent dimensions "holistic-integrative field theory", because it integrates statistical methods with linear algebra and allows one to integrate any new perceived output; therefore the method has the potential to holistically integrate all outputs. In this way a length of the resultant vector can be computed without knowing all the dimensions of the hyperspace. However, the

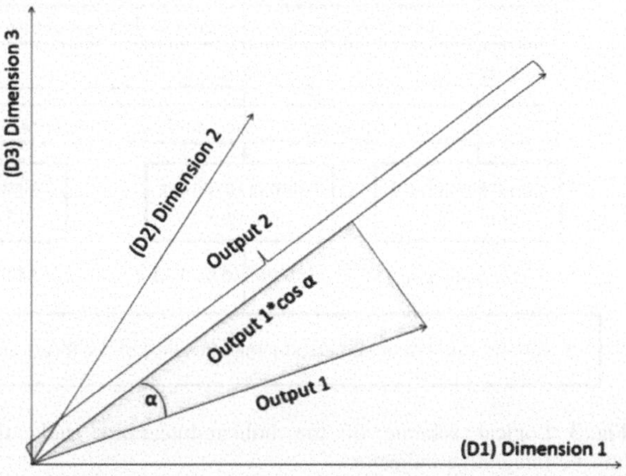

Fig. 2 Forests outputs, expressed as vectors, can be distinct and not linearly independent

vectors need to be expressed in a numerical manner, for computing the correlations, with no importance regarding the homogeneity of the used measurement units or of the quantification method.

This relative length of the resultant vector represents in fact the holistic-integrative indicator expressing the efficiency of the forest as result of a certain governance system. I give below, in Fig. 3, the logical scheme of the holistic-integrative field theory and of the computation of the indicator.

To better explain Fig. 3, let us consider a forest governance system which is shaping the forests in such manner that they are producing k outputs, named $Opt._1$; $Opt._2$… $Opt._K$. Their presence, absence or amount, are determining different perceptions, more or less objective, from different actors. The fact that one actor's opinion is less informed or less objective should not lead to inconsideration of that actor, because the actor might be a very important one by having decisional power or other kind of important impact on the system.

Let us now consider the outputs as vectors $Opt._1$; $Opt._2$… $Opt._k$ each of them with a number of "n" dimensions "d", noted $d_1 \ldots d_n$. Some of the dimensions of the space where the output vectors are externalized by the forests are unknown. Since the dimensions are not all known, integrating the outputs in a resultant is not possible by simple mathematical operators such as adition of the Opt_i dimension. However, a relative length of each vector can be known, even if they are heterogeneous, using each actor's view expressing the perceived importance of each output. The relative lengths of the output vectors can be computed by a discrete analysis method such as Frisco method for instance as described already in previous studies (Hanzu 2013).

In this way, relative lengths of the output vectors are computed. I use the word relative because it is a dimensionless (no measurement unit attached) value and because is dependent on the actors which are ordering the outputs when the discrete

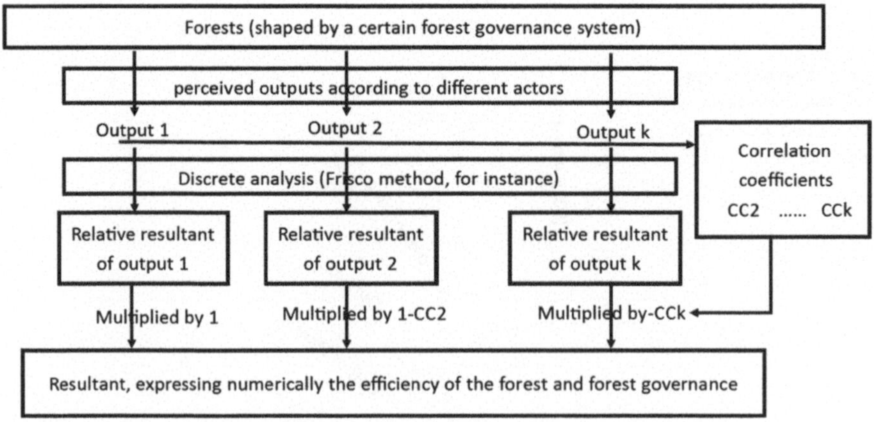

Fig. 3 Logical scheme of the holistic-integrative field theory used for computing the holistic-integrative indicator

analysis is performed (Hanzu 2013). Therefore it is relative to the number of actors and to the values the society holds.

I have to emphasise here that there is a very important difference between the output vector's lengths and output vector's relative lengths. The output vector's lengths have attached measurements units such as cubic meters of wood, tons of oxygen, values on a landscape preference scale, or any other imaginable measurement unit and are a result of any kind of measurement; while the output vector's relative lengths have no measurement unit and they express the importance given by the individual actors to one output relative to the others.

Since the lengths of the vectors are known, it is possible to compute a correlation coefficient between the vectors. This correlation coefficient is used in the following way. First, one output is considered, let say the most important one, and that vector's relative length in the resultant is adjusted with a coefficient equal to 1. After that, when a second vector' length is integrated, by addition for instance, its relative length is adjusted by 1-CC2, where CC2 is the correlation coefficient of 2-nd output, computed as relative to all the other outputs considered before. Similarly, when the i-th vector is integrated, its relative length is adjusted by 1-CCi, where CCi is the correlation coefficient of the i-th output, computed as relative to all the other outputs considered before. This process continues until all the percieved or considered outputs of the forests are used as inputs into the holistic integrative-indicator.

I want to emphasise now that the method is so flexible, and the mathematics behind the algorithms are so general that the indicator allows the integration of any kind of available data, without needing any particular data. Of course the results are more accurate if the data are as diverse, accurate and complete as possible.

In order to optimize forest governance, the holistic-integrative indicator should be used in an iterative manner, presented in Fig. 4. The steps are the following. First, one or few forest governance systems are evaluated. Second, once the value of the holistic-integrative indicator is known, it is compared with another value of the indicator computed either from a different governance system, either from the same governance system, but at a different moment in time. If possible the comparison can be done both between different states of the same governance system and between distinct governance systems. Third, knowing some of the differences between the governance systems or between the same governance system at different moments and having the values of the holistic-integrative indicator, the system which performs below the other or worse than in a previous state can be modified assuming that it will improve. After a certain time, dependent on the scale or level of governance, the evaluation is repeated and all the comparisons done again in order to orientate new actions.

It is important to notice here that if the indicator is to be used immediately, two different systems need to be compared, or ancient data should be used in order to see how changes of the forest governance system inflicted change in the forest outputs. It is important to notice also that the indicator changes in time, even if no changes are made on the forest governance system and on the forests in general, due to the changes in the outputs that are considered valuable by the society. It also has the capacity to integrate new outputs that are unknown at the initial moment of

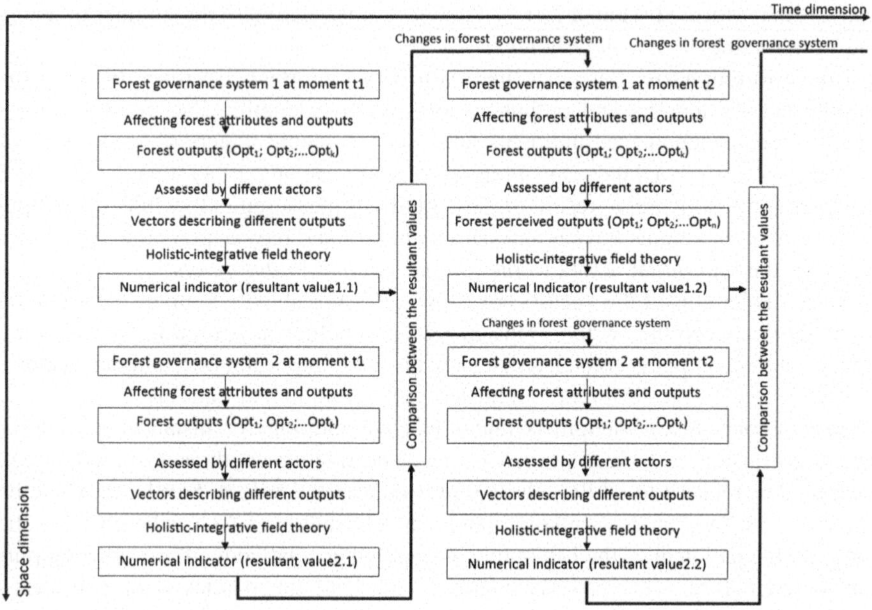

Fig. 4 Iterative usage of the holistic-integrative efficiency indicator for optimizing forest governance

usage. Thus the indicator is constantly adapting itself, and there is no danger to become archaic.

Below an example of using the algorithm presented in Fig. 1 and parts of the algorithm from Fig. 3 are used in order to establish optimal stand structures in sites appropriate for mixed forest stands located in Cindrel Mountains from Romania. This is the first step in optimizing forest governance. The iterative part of the presented method, presented in Fig. 4 could not be tested so far, because it is supposing repeated measurements and estimations, after a period of time that allows changes in the studied system, which was so far not carried on because I considered that not enough time passed since the first measurements were done.

3 An Example of Usage of the Holistic Indicator

The growing patterns of tree species that are forming mixed forest stands are different than the growing patterns of the same species growing in pure stands, even if they are located in the same sites. This natural reality might be explained by the dynamic equilibrium of the forest ecosystem. However, for practical purposes, the evaluation of the structures, the increment and therefore the dynamics of mixed forest stands is done by comparing the real stand structures with theoretical ones, obtained from models that were calibrated in pure, even age stands, not in mixed ones, and whom are known as yield tables (Giurgiu and Drăghiciu 2004).

It is known that one of the most sensitive structural characteristic of a forest, when the site conditions are varying, is the height of the trees. That is the main reason why the height, average or dominant, is used to establish the relative production classes of the species in a forest stand. This sensitivity of the tree height related to the site conditions makes it suitable to establish if there are differences in growth dynamic of tree species forming mixed or pure stands.

In order to evaluate the stand dynamic, the natural development series theory was accepted. According to this theory, the structure dynamic of a forest stand can be evaluated by determining static structures of forests that are growing in the same site conditions but are in different age classes. As studies have shown (Giurgiu and Drăghiciu 2004) this theory is suitable for building dynamic models for pure and even age stands, but it is disputable regarding mixed stands, mainly because of the natural change in time of the stands' composition. However, certain trends of the height dynamics for mixed stands can be predicted.

Further on a comparison is done between the height dynamics of pure spruce, beech and fir stands and mixed beech—spruce or spruce—beech—fir forest stands located in Cindrel Mountains, from Parâng Massif in Romania's Meridional Carpathians, known abroad as Transylvanian Alps. Some of the holistic integrative field theory algorithms are tested namely the ones depicted in Fig. 1 and partially in Fig. 3, in order to establish the efficiency of the stands.

The study was conducted in mixed forest stands that are natural forest types from Cindrel Mountains. The Cindrel Mountains are covering approximately 900 km^2 surface, with a maximum altitude of 2245 m above sea level; the tree line altitudinal limit is at 1800–1900 m, depending on the local conditions. Between 850 and 1300 m is the altitudinal belt of European beech-Silver fir-Norway spruce mixed stands. In the past, the human activity in the region was intense; leading to changes in the entire vegetation cover, therefore today only a surface of 4505 ha is covered by the described forest types.

Eighteen stands with an age varying from 55 to 140 years were selected for the studies. In each stand, plots of 0.25 hectares were sampled.

For evaluating forest structure efficiency and therefore forest governance efficiency, at this scale, a discrete analysis was done using the Frisco formula. Three criteria are taken into account, namely: (1) the productivity of the forest which is estimated by the average height; (2) the protective effect of the forest which is estimated by the stability of the forest stand for climatic hazards and (3) the aesthetic effect of the forest which is estimated by the people's perceptions and preferences for different forest structures.

The Frisco formula considered is:

$$Wk = (2Pk - Pmin + Sk + 0.5)/(0.5 * n + Pmax - Pk) \tag{2}$$

where: Wk—the absolute weight coefficient; k—the order number of the criterion; Pk—the global grade of criterion k; Pmin, Pmax—the maximum and minimum grades obtained by any of the criteria k; Sk—number of criteria whose global grades are smaller than the grade of current criterion.

The height measurements were carried out using a device with a precision is ±10 cm. The heights of all trees with a diameter of more than 7 cm were measured. Average heights were calculated for each stand and for each species. The average heights were compared with the heights from the yield tables. The used yield tables were developed for the entire Romanian Carpathian range.

The evaluation of the protective effect for each forest type can be done by the area affected by wind throws and wind breaks in a certain period of time divided by the total area of the forest, at the beginning of the time interval. The reason for this choice is the fact that any protective effect is conditioned by the very existence of the forest, therefore if the forest is more likely to be affected in a catastrophic way by a climatic hazard; the protective effect is less likely to exist. Such an analysis was not done for this area, but the literature suggests (Dobertin 2002) that for Romania's conditions the most sensitive to wind damage are the Norway spruce forests, followed by white fir forests. More stable to wind damage are the European beech forests; and the most stable are the mixed forests of resinous and beech forests.

The aesthetic effect of the forest was estimated through a perception and preference study that was carried out according to a methodology published by Santos (1998) and it is based on the so called cognitive psychological approach. According to a study on perceptions and preferences already done (Hanzu 2009) the most preferred scenery from the possible scenery in this vegetation belt is represented by mixed forest stands with high forest, followed by pure beech stands followed by spruce stands.

Regarding the productivity of the stands expressed by the average height the following results were obtained for the tree species that are forming mixed stands. As it can be seen from Fig. 5 the average heights for the tree species considered is constantly above the reference heights from the yield tables. The only exception is formed by the eighty-five years old beech that is bellow the reference limit of the yield tables.

Concerning the aesthetic effect of the stands, this characteristic was estimated using the cognitive-psychological approach. The method aims to estimate the reaction of the respondents using visual stimuli, namely pictures. Since it avoids verbal description it is assumed that it produces less biased results. Figure 6 depicts the results of such a clustering carried on in a similar research area. We can observe

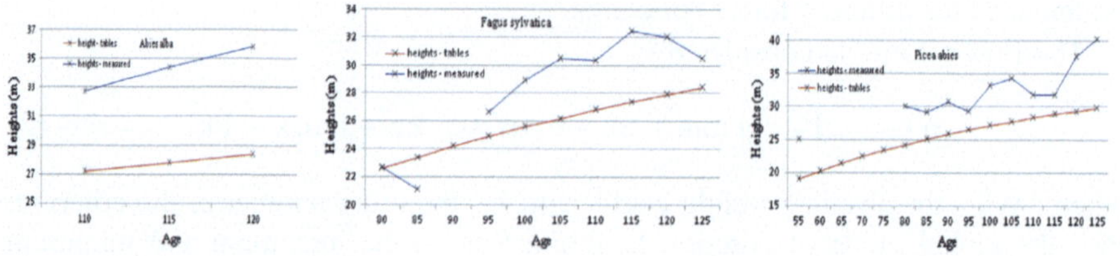

Fig. 5 The average heights for tree species growing in mixed stands in comparison with the heights from the yield tables

from this diagram that one main attribute considered by the general public when perceptions categories are formed is the age of the stands as well as the species composition of the stands.

The numbers below the clusters are the preference scores on a 5 point scale where higher scores indicate a more appreciated structure of the stand.

For the protective effects of the stands I extrapolated the known studies (Dobertin 2002). This assumption serves the demonstration purpose of this example. Therefore mixed stands are most stable stand followed by beech stands which are followed by the resinous stands.

Regarding the estimation of different forest structures integrated effect, the following analysis was done. The way how this integrated effect of the forest was

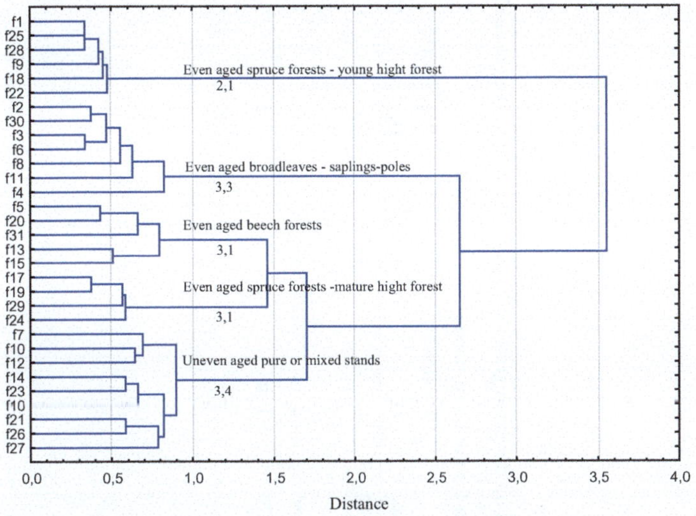

Fig. 6 The aesthetic effects of the stands

Table 1 Establishment of the relative coefficient of weight for the three criteria taken into account

k	Criterion	1	2	3	P_k	L_k	S_k	W_k	rw_k
1	1	0.5	0	0.5	1	2.5	1	0.833333	0.138889
2	2	1	0.5	1	2.5	1	2	4.333333	0.722222
3	3	0.5	0	0.5	1	2.5	1	0.833333	0.138889
Total								6	1

P_k—global grade of criterion k
L_k—the place of criterion k
S_k—number of criteria with global grades lower than criterion's k grade
W_k—the absolute coefficient of weight computed with Frisco formula
rw_k—the relative coefficient of weight computed as: $rw_k = W_k / \sum_{k=1}^{k=n} W_k$;
n—number of criteria

Table 2 Ordering of the possible compositions, based on the given grades G_k and on the relative weight coefficients rw_k

		Possible composition for the forest stand							
		Spruce stand		Fir stand		Beech stand		Mixed stand	
criterion	w_k	G_k	$G_k * rw_k$	G_k	$G_k * rw_k$	G_k	$G_k * rw_k$	G_k	$G_k * rw_k$
1	0.138889	8	1.111112	8	1.111112	6	0.833334	9	1.250001
2	0.722222	5	3.61111	6	4.333332	7	5.055554	9	6.499998
3	0.138889	4	0.555556	4	0.555556	8	1.111112	9	1.250001
		Total	5.277778	Total	6	Total	7	Total	9
		Rank	4	Rank	3	Rank	2	Rank	1

computed can be seen in Tables 1 and 2. The criterion assigned number is the same as in methodology section, namely 1—productivity, 2—stability, 3—aesthetics. As it can be seen from Table 2 the most efficient forest for these sites seems to be the mixed forest stands.

In the above example the obtaining of the rw_k corresponds to the obtaining of the relative resultants. However, due to the fact that protective effect was estimated based on existing literature instead of the area affected by natural hazards no correlation coefficients CCi, needed for the algorithm depicted in Fig. 3, were computed.

The height of the trees that are forming mixed forest stands are constantly above the reference model that is the yield tables. There are few possible explanations for this fact. One is that the yield tables were designed for pure stands and for all the area covered by spruce, fir and beech from Romania, while some local conditions, including that the species are forming mixed forest stands, are different. This is inducing a different dynamic of each species in the stand so therefore even if the production class was properly established, when the management plans were elaborated, the species forming the forest have different natural development curves than the ones considered. This leads, in time, either to overstocking or to under stocking of the stand, in both situations the possible amount of wood that is either unsustainable either not harvested entirely.

Regarding the possibility of estimating an integrated effect for the forests the holistic-integrative field theory seems to be quite an effective one for computing such an indicator. The result is in accordance with the previously known trends, that the mixed forests are more effective than the pure ones. However it might be difficult to apply such a method because of the subjectivity in the grade giving process. This disadvantage can be, at least partially reduced, by introducing a rigorous grading algorithm. The disadvantage can be totally avoided if some kind of numeric data is used. Due to this reason the result should be carefully interpreted.

Such a method can be used in the design of a sustainable and harmonized bioeconomy by integrating, hard to quantify social services of the forests and direct economic outputs of the forests, in order to estimate optimum structures to be promoted through human interventions and thus possible optimal forest governance.

4 Discussion

The aim of optimal forest governance, as part of the bio-economic concept, is sustainability and provision of forest goods, services and other perceived values needed by the society, in sufficient amount, using accepted methods. Therefore, in order to estimate the efficiency of the forest governance, the sustainability of the forest-society system and the perception concerning the provision of goods, services and other social values of any form is important. Perception of general public, concerning forest governance, has to be favourable. This is necessary because in democratic regimes no governance practice is a good one if it does not have the support or the acceptance of the majority of general public, no matter how efficient from technical perspective it is. Also it is very important that the actors which are providing inputs for the holistic-integrative indicator are expressing their opinion or knowledge in a free and uncensored manner, otherwise the results are less usable. This is why the method works best in democratic regimes, which are promoting freedom of expression as a social value.

As it is presented here the indicator is in its most simple structure, in order to make it as easy understandable as possible. Weights on the inputs of different actors can be further considered, using for instance the number of people affected by a certain decision, or the number of people in that actor's group. This is a sound use of the method, because all humans are equal and therefore should have the same importance when decisions affecting their future are taken. Therefore the indicator is somehow based on the human conservation instinct. This is a great advantage of the indicator because it allows that all actors at any level (international, national, subnational) are considered in a fair and proportional manner when the relative lengths of the output vectors are estimated, therefore the recentralization of forest governance, as observed by some authors, (Phelps et al. 2010) without the support of the majority of the stakeholders is avoided. Also unwanted, potentially inefficient or unsustainable decentralization is avoided. Such a multilevel perspective on forest outputs is necessary because some of the outputs appears only at certain scales, and otherwise might be not noticed.

The method works as a cybernetic system, because considering the value of the indicator as the feedback component, in correlation with different forest governance structures and practices, forest governance can be constantly and gradually improved over time, with no convulsions of the involved systems.

It is very important to emphasise here that the indicator considers as inputs not just measured or estimated data but also perceived information which, even if is subjective, is very important, especially in free-elective democracies. Therefore the policymakers are not put in the situation to take good decisions and obtaining bad results in elections, since the majority of the actors are agreeing with the way the governance systems are changing as result of policy measures.

5 Conclusions

The holistic-integrative indicator presented here comes to complete the existing methodology to assess forest governance (FAO 2011). It is proving an innovative way to foster better forest policies being both democratic and social-inclusive.

The mathematical apparatus used for the indicator, based on an innovative method which I named holistic-integrative field theory, is highly flexible and general, thus it gives the possibility to adapt the method to other kind of governance issues and to integrate new data as new research findings become available or as society is progressing.

The indicator can be applied for any output of the forests as result of a certain forest governance system, as long as it is perceived, due to the generality of the mathematics behind the computing algorithms. Even so, the results will need to be carefully interpreted, due to the fact that once the scale of the system changes, new properties of the system are appearing. In order to deal with this behaviour of the systems, which might escape from the perspective of the locals or of the high level governance bodies, international governance levels and actors as well as local actors should be always part of the process. Is very important to never disregard the local input (Hein et al. 2006), because some unexpected effects of the governance are first seen at local level.

The holistic-integrative efficiency indicator has the potential to express the outputs of the forests, as they are perceived by the whole society not just by some groups or some actors, thus it has the potential to operationalize and significantly improve the forest-society-science-policy-practice interface.

References

Arts B, Buizer M (2009) Forests, discourses, institutions. For. Policy Econ. 11:340–347

Augusto L, Bonnaud P, Ranger J (1998) Impact of trees species on forest soil. For Ecol Manage 105:67–78

Blagojević D, Martire S, Hendrickson CY, Hanzu M, Galante MV, Kähkönen T, Põllumäe P, Fontana V, Radtke A, Stojanovski V et al (2016) Making forest values work: enhancing multi dimensional perspectives towards sustainable forest management. South-east European for. [Internet]. [cited 19 Jan 2016];7. Available from: http://www.seefor.eu/vol-7-no-1-blagojevic-et-al-making-forest-values-work.html

COMM/CLIMA/0002: Forests and Agriculture. [Internet] 2010. Brussels: EU Commission Climate Action; [updated 2015 Dec 23; cited 1 Jan 2016]. Available from: http://ec.europa.eu/clima/policies/forests/index_en.htm

Dimitrov RS (2006) Hostage to Norms: States, Institutions and Global Forest Politics. MIT Press [Internet], [updated 2006 Mar 13; cited 25 Jan 2016]. Available from: http://www.mitpressjournals.org/doi/abs/10.1162/152638005774785499#.VqXW3ip95aQ

Dobertin M (2002) Influence of stand structure and site factors on wind damage comparing the storms Vivian and Lothar. For Snow Landsc Res 77:187–205

FAO (2011) Framework for assessing and monitoring forest governance, [Internet], [cited 12 Jan 2016]. Available from: http://www.fao.org/docrep/014/i2227e/i2227e00.pdf

FAO (2015) Foresters call for action: future land management needs better integration of sectors, [Internet], [cited 14 Jan 2016]. Available from http://www.fao.org/3/a-i5227e.pdf

GISTEMP Team (2015) GISS Surface Temperature Analysis (GISTEMP). NASA Goddard Institute for Space Studies. Available online at http://data.giss.nasa.gov/gistemp, updated on 7 Mar 2016, checked on 16 Mar 2016

Giurgiu V, Drăghiciu D (2004) Modele matematico-auxologice şi tabele de producţie pentru arborete, Bucureşti, (2004) (Mathematical—increment models and yield tables for forest stands) (in Romanian)

Gottdenker NL, Streicker DG, Faust CL, Carroll CR (2014) Anthropogenic land use change and infectious diseases. A review of the evidence. EcoHealth 11(4):619–632. https://doi.org/10.1007/s10393-014-0941-z

Hanzu M (2009) A method for evaluating the structure efficiency of the forest stands with recreational functions. In: Proceedings of the Conference Forest and Sustainable Development, (17–18 Oct 2008). Faculty of Silviculture and Forest Engineering, Braşov (RO).

Hanzu M (2013) Integrative indicator for optimizing decisions in multifunctional forestry. In: Sisak L, Dudik R, Hrib M (eds) Socio-economic Analyses of Sustainable Forest Management. Proceedings of the conference of IUFRO 4_05_00 division. 15–17 May 2013, Prague (CZ): Mendel University. Available from: www.iufro.org/download/file/…/40500-40501-40502-prague13_pdf/

Hein L, van Koppen K, de Groot RS, van Ierland EC (2006) Spatial scales, stakeholders and the valuation of ecosystem services. Ecol Econ 57:209–228

Knudson T (2009) The cost of the biofuel boom: destroying Indonesia's forests: yale environment 360. Available online at http://e360.yale.edu/feature/the_cost_of_the_biofuel_boom_destroying_indonesias_forests/2112/, checked on 16 Mar 2016

Mustalahti I (2015) Towards a sustainable bioeconomy innovative methods and solutions for the agriculture and forest sectors. Oral presentation. Barcelona

Phelps J, Webb EL, Agrawal A (2010) Land use. Does REDD + threaten to recentralize forest governance? Science 328:312–313

Pretzsch H (2009) Forest dynamics, growth and yield. Springer, Berlin

Rametsteiner E (2009) Governance concepts and their application in forest policy initiatives from global to local levels. Small-scale For. 8:143–158

Samuelson P (1974) Economics of forestry in an evolving society, [Internet] MIT. Economic Inquiry. [cited 5 Apr 2015]. Available from http://www.uio.no/studier/emner/sv/oekonomi/ECON4925/h11/Samuelsonj.1465-7295.1976.tb00437.x%5B1%5D.pdf, checked on 4/5/2015

Santos JM (1998) The economic valuation of landscape change. Theory and policies for land use and conservation, U.K

Stewart A, Edwards D, Lawrence A (2013) Improving the science–policy–practice interface. Decision support system uptake and use in the forestry sector in Great Britain. Scand J For Res 29:144–153

UN Framework Convention on Climate Change (2015) Adoption of the Paris agreement. Available online at: https://unfccc.int/resource/docs/2015/cop21/eng/l09r01.pdf

Qualitative and Quantitative Indicators of Coniferous in Boreal Zone After Care for a Forest

Dmitry Danilov

Abstract A distinctive feature of the Bioeconomy is the production and consumption of products from raw materials derived from natural biological processes, such as, for example, the growth of woody biomass used in the production of a wider range of products. The effects of intensive forestry and the intensive production of wood allow the use of renewable sources and efficient bioprocesses to stimulate sustainable production. On the basis of long-term observations on the permanent sample sites in the boreal zone obtained data on forest care regimen. The results allow us to give recommendations to intensify the cultivation of timber in natural and planted stands. Applications thinning together with fertilizers when caring for natural spaces allow getting to the age of 85 years, high-quality wood in large quantities than forestry plantations without care. In pure composition of pine and spruce stands after that increases the proportion of large-scale commercial timber. The timely regulating the composition of spaces produces a denser wood from pine and spruce plantations to a ripe age. It is now possible to analyze the results of growing plantations of pine and spruce intensive technology. It should be noted that the data on the modes of cultivation allow for consumers of woody biomass to reduce the time it was received. To create plantations for obtaining wood pulp can make recommendations based on the already acquired data. It should be noted that the density of pine and fir in the intensive cultivation to 40 years in the boreal zone to be at the level of the average of the region. Productivity indicators in 40-year-old plantation timber plantations exceed supply natural stands of the same age. The content of cellulose in the balance of raw materials derived from pine plantations exceeds that of wood older ages.

D. Danilov (✉)
Forestry Department, Institute of Forest and Environmental Sciences,
Saint Petersburg State Forest Technical University named after S. M. Kirov,
Institutskiy per 5, b.1, 194021 Saint Petersburg, Russia
e-mail: stown200@mail.ru

W. Leal Filho et al. (eds.), *Towards a Sustainable Bioeconomy: Principles, Challenges and Perspectives*, World Sustainability Series,
https://doi.org/10.1007/978-3-319-73028-8_28

Keywords Stock of wood biomass · Wood density · Mixed and pure stands Forest care · Complex care (cutting + fertilizer) · Plantations of spruce and pine

1 Introduction

Currently, the world still needs a larger amount of high-quality softwood timber due to the depletion of stocks of forest areas. The study of methods of forest management impacts on forest cenosis, provide a large amount of wood, with known quality parameters, and will be in demand now and in the future. Comprehensive care can improve the productivity of coniferous forest stands on all continents. According to the program JUFRO in the 1960s in many European countries were laid experiments with thinning and fertilization. In 1970—in Siverskoe forestry, Leningadsky region and in Karelia. All of these experiments proved the feasibility and prospect of complex care (cutting + fertilizer) reveal the features and possibilities of its application in different site conditions. At present, an opportunity to take stock of these experiments.

Plantation crops spruce and pine were created in 1976 in the same forestry. Fast growing softwood timber resources on plantations currently is of undoubted interest for forest practices, and wood for the needs of consumers. Getting in less time than making them profitable forest management activities in growing natural stands, the volume of wood required and the possibility of accommodation near the plantation wood consumers. Fast growing plantation requires minimize costs and maximize production from 1 ha of forest area. At present, the object of research (the culture of spruce and pine) have reached 40 years of age. Growing plantation pulpwood, while cutting back 40 years, these crops are of interest for the analysis of bulk and mass of wood indicators.

Effects of caring for the forest stand level lead to its changes at the level of the wood structure, affecting its important quality characteristics—density. The question remains practically unexplored. In different site conditions occurs variation of the basic wood density, which is related to a number of factors, dissimilar in nature. The different species composition conifer stands manifest their fundamental differences due to heredity and growth strategies of trees.

Different response of spruce and pine stands on the integrated care due to the difference in environmental and biological properties of these rocks and requires a special study. Little studied mixed pine-spruce stands. Research and evaluation of interaction between external and intrastructural changes in the comprehensive care facilities were incomplete and short term character. Changes in the density of the wood after forest management impacts on forest stand has its own characteristics, depending on the age and nature of the pattern. The stands have reached the age clear-cut after a comprehensive care, are an indicator of the effectiveness of this type of care for the forest. It is therefore a need to consider this topic in order to clarify and correct care regimen for these plantations.

2 Literature Review

The relationship between the degree and nature of the thinning of forest stands and the density of the resulting timber is an important but poorly understood aspect of the whole problem of thinning, since the density of the wood related performance spaces is defined by dry matter content (Poluboyarinov O.I.).

Obtained conflicting results on the effect of thinning on the density of softwood in different climatic conditions. According to Swedish researchers spruce wood density decreases to a greater extent than after thinning pine. In the United States, in experiments with loblolly pine was not found significant differences in wood density logging and passed unaffected and stands. Gildelbrant (Germany) noted that the site conditions determine changes in wood density in sparse spruce cuttings. Stronger rarefaction in the best growing conditions causes a decrease in the density of the wood. In Finland Viro pointed out that the reduction in wood density in softwood cuttings after slightly manifested.

Russia also has long conducted research on the subject and the results differ as well, depending on the location and age of the studied stands. After thinning, without making according fertilizer Isayeva R.P., Kurbatov G.V., under the conditions of the middle taiga, wood density of spruce increased. In Melekhova T.A., which investigated sparse spruce forests in the Arkhangelsk region, obtained similar results, i.e. increasing the density of the wood. In Chibisova A.G. in experiments on the effect of the corridor care in stands of spruce wood density increased by 20%. In his studies, Smirnov A.A. also showed no decrease in wood density after thinning in spruce forests of the Leningrad region. However, thinning sometimes lead to a decrease in wood density. According to Davydov A.V., Sennov S.N. and others a decrease in the density of spruce wood up to 10% of control without cutting. Increased density of pine after thinning noted Ryabukha E.V., Slyadnev A. P. Reduced density of pine after thinning observed Poluboyarinov O.I., Riabokon A.P., Sinkevich S.M. Influence of fertilization and thinning on wood density was considered in Aksenenkova Y.A., Babich N.I., Vyarbila V.V., Gelesi I. S., Kozlov V.A., Kisternoy M.Z., Korchagova S.A., Melehova V.I., Maksimov S. A., Poluboyarinov O.I., Riabokon A.P., Smirnov A.A., Stepanenko I.I., and many other. In Scandinavian countries, the impact of fertilizers on the density of the wood examined Viro P.G., Lundgren C., Mörling T., Valinger E., Lindkvist A., Nordlund C., Pikk J., Kask R., Peterson R. In America—Donald P. Hanley, H.N. Chappell, Ellen H. Nadelhoffer., C.S. (Cliff) Snyder, Allen T.J., Albaugh, R. Rubilar, C.A. Carlson, Allen H.L., in other regions-. P. Smethurst, G. M. Downes, E. Lowell, P.N Beets, K. Gilchrist, Jeffreys M.P., Zobel B.J., van Buijtenen J.P.

Analysis of the literature shows that depending on coniferous and natural conditions, the base density of the wood, after the application of various embodiments and doses of fertilizer on a background has a different gain increased performance. The type of fertilizer, depending on the basic elements—nitrogen, phosphorus, potassium, or their joint use in different degrees cause a change in the density of softwood. Experiments conducted in taiga zone showed that the use of nitrogen-phosphorus fertilizer causes a reduction of the density of pine wood (Gelesi, Stepanenko et al.), while the use of nitrogen has not revealed any trend.

However, as the analysis of recent literature data shows, the majority of domestic workers (Geles, Stepanenko, Poluboyarinov et al.) agree on the fact that nitrogen fertilizer and cutting increase the base density softwood, or at least do not reduce it (Korchagov S.A., Melekhov V.I., Smirnov A.A., Danilov D.A.) Many researchers in other countries have studied the effect of fertilizers on basic wood density also show different results. However, in general, there is a tendency of increase of basic wood density after comprehensive care in softwood, which is confirmed by long-term studies, regardless of region.

Thus, on the basis of the foregoing, it can be concluded that, depending on the species, growth conditions, care regimen, a particular type of timber structure is formed. The structure of wood is formed by growth factors, nutrition and development in trees peculiar to their conditions of growth. Differences in the structure of wood is shown in the boundaries of the basic features that a particular type of genetically fixed.

Long-term tests have shown that comprehensive care is the most rational view of maintenance of forests that maximizes the effect of thinning and fertilization.

Practice and research on thinning showed that improve the performance of pure composition stands alone logging can not, by what method they would have not been conducted. In the mixed two-tiered stands and it can happen, but there is the possibility of thinning limited. Fertilizer can solve this problem to obtain additional growth and improve the overall performance of the stand. The highest increase of growth achieved during the culmination of the growth of the stand, i.e., middle-aged forest stand, which is a natural reaction to the growing care.

Influence of thinning on the relationship of trees in pine and spruce stands on the basis of long-term experiments in conditions of a southern taiga are devoted Sennova S.N., Krankina O.N., Melnikov E.S. In have pointed to the greater stability of the stands, as compared with pure pine, to the adverse effects and thinning. A fuller use of their potential growth conditions and the possibility of controlling the composition according to the needs of forestry. Fir tree under the canopy of pine tier, despite the younger age, can go to the first tier of 50–60 years, in contrast to the deciduous-spruce stands. The need to develop a special system for thinning pine and spruce forests. Virtually no work dedicated to differentiation in pine and spruce stands after comprehensive care.

Plantation forest growing may become the main method of production of forest resources in regions with intensive forest management and infrastructure, including enterprises of timber industry complex. The need for extensive forest plantations has been recognized in the mid-20th century. So in the early 1990s, the area occupied by forest plantations was estimated at 100 million. ha worldwide.

The rate of increase in plantation area first 90 accounted for approximately 1–1.2 million ha per year. Dynamics persisted throughout the decade, and by 1999 created 8.5 million ha per year. By 2000 the area of forest plantations in the world amounted to 110–150 million ha (Bowyer 2001). Already in 2000, the world's 5%

of the forest land occupied by plantations have received 35% of round wood (Carle et al. 2001). This figure is expected to rise to 44% by 2020 ("Forest Loss" 2010).

Plantations are created and studied in Russia since the second half of the 19th century. Today, there are many areas of highly successful plantations of different tree species and different ages. The results of their study widely reported in scientific papers and monographs: Babich (1993), Redko (1994), Rubtsov (1997), Gavrilov (1969), and many other foresters.

A significant contribution to the establishment of forest plantations and their research workers have SPbNIILH and SPBGLTA: Shutov I.V, Maslakov E.L., Bel'kov V.P., Vyachkilev V.V., Markova I.A., Zhigunov A.V., Kovalev M.S., Redko G.I., and other generalizations of the results of many years of experimental work and the consideration of the comprehensive aspects of plantation forestry in relation to the conditions of the Northwest of Russia devoted to the book "Plantation Forestry" group of authors. SPbNIILH employees (Shutov et al. 2007).

Getting in less time required volume of wood than when grown natural forest stands, and the possibility of placing the plantation near the consumers of wood makes them profitable forest management activities. For the pulp and paper production is of considerable practical interest is to increase the volumetric parameters of timber and accompanied by an increase in mass. Low density pulp wood used in the pulp and paper industry, is highly undesirable because it leads to the production of quality pulp and less than the smaller its output, which in turn increases costs of preparation, delivery and processing additional feedstock volumes.

In addition to environmental and genetic factors that affect the density of wood, according to Poluboyarinov (1976), the leading role has forestry activities: the density and placement of crop thinning, fertilizing and draining. Poluboyarinov (1976) reported that an increase in the density of 34-year old pine tree crops increases and the density of large and medium-sized tree of categories. For small trees is characterized by feedback. For 40-year-old spruce plantation crops determining factor in the formation and structure of the width of the annual ring density stands stand.

In dense cultures, the proportion of late wood in the composition of tree ring after thinning is reduced and does not exceed one-third of its width, whereas in the absence of logging can be more than half (Glushkov 2011). With the increase of the initial planting density, an increase wood density (Ovodov 2010). Only a few papers discuss the quality of the wood grown on plantations of forest in Russia (Poluboyarinov and Fedorov 1991; Pekkoev 2010; Korchagov 2010a, b; Glushkov 2011).

3 Methodology

For experimental studies used stationary objects with complex care (cutting + fertilizer) for a period of 45–40 years of observations, as well as plantation crops spruce and pine with different types of experiments. All experienced objects located in the forest area of Gatchina district, Leningrad region, Russian North-West.

Table 1 Indicators of stands of pine (P. silvestris) and spruce (P. abies) myrtillus forest types of North-West Russia

Experiment variant	Compositions of forest stand (%)	Age, years	Average height (m) / Diameter (cm)	Stock wood (m^3)	Average base wood density (kg/m^3) / Variability (Cv%)	Stem woody biomass (ton)
Control	100 pine	85	$\frac{24}{24}$	338	$\frac{393}{16}$	132,834
Two-time thinning	100 pine		$\frac{26}{22}$	444	$\frac{386}{6}$	171,384
Two-time thinning + fertilizer	100 pine		$\frac{26}{28}$	469	$\frac{400}{2}$	187,600
Control	100 pine	85	$\frac{24}{26}$	326	$\frac{433}{4}$	141,158
Two-time thinning	100 pine		$\frac{24}{22}$	350	$\frac{444}{11}$	157,000
Two-time thinning + fertilizer	100 pine		$\frac{26}{28}$	355	$\frac{445}{6}$	157,530
Control	70 spruce	85	$\frac{24}{22}$	338	$\frac{435}{3}$	147,030
Two-time thinning	100 spruce		$\frac{24}{24}$	320	$\frac{456}{5}$	145,920
Three-time thinning	100 spruce		$\frac{26}{26}$	379	$\frac{458}{5}$	173,582
Control	50 pine	85	$\frac{25}{24}$	138	$\frac{535}{2}$	73,830
	50 spruce	80	$\frac{24}{20}$	156	$\frac{489}{5}$	76,284
Thinning + fertilizer	50 pine	85	$\frac{26}{28}$	166	$\frac{516}{6}$	85,656
	50 spruce	80	$\frac{24}{24}$	174	$\frac{491}{5}$	85,434
Thinning + fertilizer	60 pine	85	$\frac{27}{28}$	278	$\frac{526}{4}$	146,228
	40 spruce	80	$\frac{25}{26}$	244	$\frac{527}{12}$	128,588

The soil is loamy humus rough

The duration of the experiments allows keeping the methodological continuity in the collection and processing of information. Interpretation of the data is carried out within the framework of systems theory best explains the processes occurring in the natural ecosystem of the active economic impact.

We use the method of continuous enumeration based on permanent sample plots, traditional research on these objects. The period of repetition taxation—5 years. Measuring the diameter of the trees carried out with an accuracy of 1 mm metal caliper in two perpendicular directions at a height of 1.3 m from the root of the neck. The thickness of each stage (as species) was measured with altimeter height of at least 5 trees. The resulting equalized data graphically and used to determine the discharge heights on the diameter classes. Stock calculated according to the tables of heights and volumes of barrels (the cortex) stands for the Leningrad, Arkhangelsk and Vologda regions. The accuracy of the reserves—about 3%.

The selection of wood samples was carried out from the trunks of trees in the most presented levels of thickness growing incremental auger on H 1.3 m to the middle of the barrel. Depending on the variation in wood density were selected replications from up to 6–15 samples per stage thickness. Basic density of wood samples is determined by the method of maximum humidity. It calculates the average density of the samples for each diameter class. To determine the density of the entire trunk wood, the methodology proposed by Poluboyarinov (1976): at the height of the density H 1.3 m substituted into the correlation equation of communication density at breast height with the density of the entire trunk. We used the previously developed appropriate conversion equation for softwood phytocenoses with similar inventory indices—age, value class, type of wood (Smirnov 2007; Danilov 2011, 2012).

The resulting study data were processed in the computer environment and subjected to analysis to verify the validity of the changes after the maintenance of forests.

4 Results

Currently, the study of coniferous stands in the blueberry forest type under the age of 80–85 years old (the age of felling of mature stands) on the majority of objects were formed with the departure of high spaces. Figures shown in Table 1 data show that, depending on the species composition and the type of care form various final timber stock. In pure pine stands on the sections with logging (intensity of 30% of the stock of plantations) and fertilizer (N120 kg per ha) is now higher than in the timber stock control sections. Maintenance of forests has increased the performance of basic density of pine wood in comparison with the control variant without a care.

In spruce stands currently supply timber above targets only version of the comprehensive care of the forest. However, the density spruce up sections with leaves.

Of particular interest are mixed stands of pine and spruce with different interests in the species composition passed thinning and comprehensive care for the forest. At this experimental site was carried out low intensity (15% of the stock) Thinning forest and made ammonium nitrate (N 120 kg per ha) and superphosphate (P 80 kg per ha). Depending on the proportion of pine and spruce participation at sites with different goings formed timber reserves. However, with the predominance of pine in the stand density of wood and pine and fir higher than in sections with equal interests in the composition of plantations.

Fast growing plantation requires minimize costs and maximize production from 1 ha of forest area. At present, the object of research (the culture of spruce and pine) has reached the age of 40 years. Growing culture pulpwood, while cutting back 40 years, these crops are of interest for the analysis of bulk and mass of wood indicators for forest management and the development of intensive rearing conditions.

Plantation crops of spruce and pine, are grown on intensive technologies have now reached the wood stock indicators comparable with the stock of mature conifer plantations study region (Table 2). In 2000–2001, on sections with a density of 4000 and 2000, thousands of pieces of trees per hectare were carried out to decrease the density of 2000 pieces and 1500 pieces of trees. Carried out preventive treatments for a living ground cover (use of herbicides), and fertilization (N120 kg per ha) yielded so far to get the wood with a density below the average for the study region. The resulting stems wood biomass in the form above actually using herbicides.

5 Discussion

Analysis of the results of research of influence maintenance of forests shows that in natural pine stands of pine and spruce combined use thinning cuts and fertilizer in the myrtillus forest type gives the integral effect on the performance of wood. The stock of wood, and its density of pine and spruce up all the options with a complex care of the forest, than just after the thinning of forests. However, although the wood density and its variability deterministic sign after thinning often increases. After comprehensive care in pure coniferous plantations this rate does not increase significantly, in contrast to a mixed stand of pine and spruce. Apparently influenced by the quantitative distribution of the trees, the steps diameters planting cuttings obtained after decimation. In general, the resulting stem biomass higher at sites with leaves in mixed stands of pine and spruce.

For the pulp and paper production is of considerable practical interest, the fact that the increase of volume of wood was accompanied by an increase in weight and performance. The performed calculation of stem biomass of the wood on the basis of the obtained average basic wood density and plantation inventory data (Table 2) showed that, depending on the growing conditions, the initial density and types of care received various indicators.

Table 2 Indicators plantation cultures of spruce and pine on expe-imental objects

Experience version	Age (years)	$\frac{\text{Average height (m)}}{\text{Diameter (cm)}}$	Stock wood (m^3)	Average base wood density (kg/m^3)	Stem woody biomass (ton)
Spruce (P. abies)					
The primary density of 1000 per ha (950 per ha at this time)					
Control	38	$\frac{11}{18}$	200	412	82,400
Herbicides		$\frac{12}{18}$	269	414	111,366
Fertilizer + Herbicides		$\frac{13}{20}$	269	405	108,945
The primary density of 2000 per ha (1450 per ha at this time)					
Control	38	$\frac{14}{11}$	294	384	112,896
Herbicides		$\frac{18}{13}$	304	396	120,384
Fertilizer + Herbicides		$\frac{17}{13}$	272	390	106,080
The primary density of 4000 per ha (1900 per ha at this time)					
Control	38	$\frac{12}{11}$	299	384	114,816
Herbicides		$\frac{16}{14}$	356	412	146,672
Fertilizer + Herbicides		$\frac{16}{14}$	340	401	136,340
Pine (P. silvestris)					
The primary density of 4000 per ha (2000 per ha at this time)					
Control	38	$\frac{19}{19}$	280	397	111,160
Herbicides		$\frac{20}{20}$	300	356	106,800
Fertilizer + Herbicides		$\frac{20}{20}$	330	431	142,230

Soil type—sod podzolic

Processing sward herbicide gives the same silvicultural effect as a joint fertilizer and herbicides. The data on the basic density of wood of spruce and pine can be predictive of receiving the balance of raw materials to the age of 40–50 years in the growing crop with rates higher than the average growth for the region. The average basic wood density of pine and spruce cultures can be attributed to the balance sheets of their category I: pine—354–411 kg/m^3, spruce—329–358 kg/m^3 (Poluboyarinov and Fedorov 1991).

6 Conclusion

The results of the study showed that the thinning and fertilizing it possible to obtain a larger amount of wood with a density higher than the average for the study of the region, for pine and spruce. One thinning may increase in woody biomass only in pine plantations in the myrtillus forest type in the study area. The experimental data showed that to maximize the receive stem wood biomass to the age of mature plantings can be in mixed stands of pine and spruce.

In the studied site conditions, we can predict that when grown pine and spruce pulpwood will be received with great amount of raw material mass indexes for pulping, and when grown on logs with improved physical and mechanical properties of wood. For bioeconomics main factor is the production and consumption of foods from raw materials derived from natural biological processes, such as, for example, growth of woody biomass used in the production of a wide range of products. Conducting intensive forestry and intensive production of wood allow the use of renewable energy sources and efficient bioprocesses for the promotion of sustainable production.

References

Bowyer JL (2001) Environmental implications of wood production in intensively managed plantations. In: Society of wood science and technology state-of-the-art review http://www.forestinfo.org/files/u1/Plantations2.pdf

Babich NA (1993) Reforestation in the European North (generalization hundred silvicultural experience): dis.... Dr. Agricultural Sciences: Arkhangelsk Forestry Engineering Institute, Arkhangelsk, p 494

Carle J, Vuorinen P, Del Lungo A (2001) Status and trends in global forest plantation development. For Resour Div, 14 (English, FAO, Rome, Italy). http://www.fao.org/forestry/5010-0c773a78823b8b936c7f6c323919bd706.pdf

Danilov DA (2011) Patterns of structural change in the pine and spruce stands at the facilities of the complex care of the forest: the author's abstract dis. cand. Agricultural Sciences Saint-Petersburg State Forestry University, p 21

Danilov DA (2012) Indicators of wood structure in plantation crops of pine and spruce. In: Proceedings of the international conference "Renewable wood resources: innovative development in forestry" SPb GLTU, 6–8 June 2012, St. Petersburg, pp 53–56

"Forest Loss" (2010) United Nations system-wide Earthwatch. United Nations Environment Programme. Archived from the original on January 6, 2010. Retrieved 27 Oct 2011. http://web.archive.org/web/20100106001800/, http://earthwatch.unep.net/emergingissues/forests/forestloss.php
Gavrilov BI (1969) Plantations of fast growth. For J 4:14–16
Glushkov YP (2011) Impact of silvicultural departures on the growth of spruce plantation crops like. Author. Diss… for the Degree of Candidate of Agricultural Sciences, Yoshkar-Ola, p 20
Korchagov SA (2010a) Increase productivity qualitative silvicultural plantations on the basis of abstract: abstract dis…. Doctor. Agricultural Sciences, Arkhangelsk, p 40
Korchagov SA (2010b) Rationale cultivated breed at target growing pulpwood. In: Korchagov SA, Strebkov NN (eds) Bulletin of the Moscow State Forest University-Forestry Bulletin. MSFU 3:154–157
Ovodov AV (2010) Quality pine wood in plantations created by sowing and planting: Abstract. Dis. on soisk. Scientists. Step. cand. Agricultural Sciences, Arkhangelsk, p 16
Pekkoev AN (2010) Fast growing cultures of Scots pine in the middle taiga subzone of Karelia. Abstract. On the Competitor. Ouch. Ph.D. degree. Agricultural Sciences, St. Petersburg, p 20
Poluboyarinov OI (1976) Wood density: monograph. M.: Forest Industry, p 160
Poluboyarinov OI, Fedorov RB (1991) Quality wood pine plantation crops such as the North-West of the European part of the USSR. In: Forestry, forest crops and soil science. Mezhvuz. sbor. nauch. tr. LTA, pp 89–95
Redko GI (1994) In: Redko GI, Babich N (eds) Reforestation in the European North of Russia: monograph. Northwestern Publishing House, Archangel, p 188
Rubtsov MV (1997) In: Rubtsov MV, Nikolaev DK, Glazunov YB (eds) Silvicultural unique experience Porechye. All-Russian Research and Information Centre for Forest Resources. M.: Publishing House VNIITslesresurs, p 41
Shutov IV et al (eds) (2007) Plantation forestry: a monograph. St. Petersburg Research Institute of Forestry. Publishing House of STU, SPb, p 366
Smirnov AA (2007) Features of influence of the regular maintenance of forests on the productivity and quality of forest stands of spruce: Abstract. Diss. kand. Agricultural Sciences. Arkhangelsk, p 20

Bioeconomy Meets the Circular Economy: The RESYNTEX and FORCE Projects

Walter Leal Filho

Abstract This final chapter presents the projects RESYNTEX and FORCE, where bioeconomy meets the circular economy.

Keywords Bioeconomy · Waste · Circular economy · Resources Europe

1 Introduction

There were various definitions of bioeconomy used in the many chapters which compose this book. According to the Bioeconomy Science Center (BSC), "Bioeconomy summarizes all economic sectors that are involved in the production, processing and use of biological resources (plants, animals, microorganisms) for the production of food and feed, the provision of biomass as resources and the production of bio-based chemicals and materials and bioenergy" (BSC 2017).

But no matter which definition one favours, it is clear that the term "bioeconomy" entails all processes, sectors, activities and services which are based on, or use, biological resources.

As stated by Hodgson et al (2016), there is international recognition that developing a climate-smart bioeconomy is essential to the continuation of economic development, reduction of greenhouse gas emissions, and adaptation to climatic change. Bio-based products have an important role in making this transition happen.

Developing a sustainable bioeconomy has been part of European policy for a number of years, from 2002 with the launching of its life sciences and biotechnology manifesto (European Commission 2002) up to today. Internationally, the

W. Leal Filho (✉)
Research and Transfer Centre, Sustainable Development and Climate Change Management, Hamburg University of Applied Sciences, Ulmenliet 20, 21033 Hamburg, Germany
e-mail: walter.leal2@haw-hamburg.de

W. Leal Filho et al. (eds.), *Towards a Sustainable Bioeconomy: Principles, Challenges and Perspectives*, World Sustainability Series,
https://doi.org/10.1007/978-3-319-73028-8_29

Food and Agricultural Organisation of the United Nations (FAO) reported in 2008 that a bioeconomy approach may support cereal growth (FAO 2008).

Overall, progress in the field of bioeconomy has significant in some areas, and rather slow in others. Some of the areas where some substantial developments have been seen include:

- Forestry
- Agriculture
- Synthetic biology
- Systems biology
- Use of tools for the identification and utilisation of metabolic pathways
- Biotechnological approaches.

In respect of industrial biotechnology, some significant progress has been observed in the fields of sustainable production and conversion of different types of feedstocks and bioresources. On the other hand, there is a need for progress in areas such as the development of new biological based products, value-added products and supply services, and in terms of sustainable industrial processes.

In Germany, a country where the issue of bioeconomy is being intensively pursued, a national Bioeconomy Council was established in 2009, in order to support the development of bioeconomy in the country. In addition, the Federal German Government set-up the "National Research Strategy BioEconomy 2030", according to which the most pressing need seems to be in respect of breeding new crop plants and in the development of more efficient cultivation technologies to reduce harvesting losses. Also, the "National Research Strategy BioEconomy 2030" which is spearheaded by the Federal Ministry of Education and Research (BMBF), emphasizes the need to utilise soil as a resource on a more sustainable basis. As part of the strategy, five innovation areas of the federal government's high-tech strategy (HTS) and two of the four focal subjects of the HTS on climate protection are directly associated with the research for a future bioeconomy (BMBF 2011).

According to BMBF (2011), the research strategy lays out five priority fields of action for further development towards a knowledge-based, internationally competitive bioeconomy. These are: global food security, sustainable agricultural production, healthy and safe foods, the industrial application of renewable resources, and the development of biomass-based energy carriers. Thereby, food security always takes the highest priority (BMBF 2016). Substantial resources have also been invested in the area of biorefinery (Schieb et al. 2015).

Apart from current government efforts in Germany, a number of EU countries are also actively pursuing research in the field of bioeconomy. The UK's bioeconomy, for instance, has the potential to be world leading. However, political, economic, and social barriers are holding back innovation in this sector (Burns et al. 2016). In the Netherlands, RVO, a national agency, is responsible for the implementation of bioeconomy policies. The key objective is to foster sustainable biomass valorization ("value pyramid") or production of biobased materials and use residues for biofuels, electricity and heat ("co-production"). According to Langeveld et al. (2016), the realisation of the valorisation potential is done focusing mostly on biorefineries as key technologies.

According to the Organisation for Economic Cooperation and Development (OECD), the bioeconomy holds at least some of the cards to ensure long term economic and environmental sustainability. The OECD however believes that this potential will not become reality without attentive and active support from governments and the public at large. Innovative policy frameworks are needed to move forward to meet these global challenges, and these need strategic thinking by governments and citizen support (OECD 2009).

A recent publication has tried to present data, show opportunities, discuss R&D findings, analyze strategies, assess the wider economic impact, showcase achievements, criticize policies and propose solutions for the green revolution in biofuels, biochemicals and biomaterials' production and power generation (Sillanpää and Ncibi 2017).

2 How to Move Forward: Bioeconomy Meets Circular Economy

Whereas the field of sustainable bioeconomy offers huge opportunities in some areas (e.g. in agriculture or in the forest sector), there are still various challenges related to initiatives in streamlining waste, especially for energy generation. And even though it is widely acknowledged that there is a great potential in the utilisation of biomass to feed an expanding bioeconomy, this is not yet being realized. This state of affairs indicate that innovative technological and management approaches are needed. It also suggests that an integration of the bioeconomy with the principles of the **circular economy**.

For the purposes of definition, a circular economy can be described as an approach which goes over and above the traditional "use and dispose" of resources. It is a mind setting via which products and services are re-designed, maximizing their economic, natural and social capital. An integration of the principles of bioeconomy with the circular economy may be advantageous in a variety of ways. For instance:

- The circular economy acknowledges the need for regarding products not as simply as outputs of production processes, but as resources which may be re-used time after time again
- The use of resources such as biomass is maximized, in the sense that once the primary purposes are met (e.g. wood use for furniture or for construction), the materials can be re-used, for instance for energy generation
- The focus of the circular economy, which seeks to rebuild capital (e.g. natural, social, economic) is also on the natural/biological aspects of some products.

In simple words, once one product has been utilized, it may be used as a raw material for another product. This approach offers substantial opportunities for sustainability, since it reduces the pressure on the production of new, fresh resources, by re-using existing ones.

An example of how to link the bioeconomy with the sustainable economy comes from two HORIZON 2020 projects, namely RESYNTEX and FORCE. Due to their relevance, they will be presented in turn.

The project **RESYNTEX** is a research project which aims to create a new circular economy concept for the textile and chemical industries. Using industrial symbiosis, it aims to produce secondary raw materials from unwearable textile waste. The core aims of the project as stated on its web site (Fig. 1) are:

- To design a complete value chain from textile waste collection through to the generation of new feedstock for chemicals and textiles
- To improve collection approaches while increasing public awareness of textile waste and social involvement
- To enable traceability of waste using data aggregation. The collected data will evaluate the performance of the new value chains by means of a life cycle assessment (LCA) and life cycle costing (LCC)
- To develop innovative business models for the chemical and textile industries
- To demonstrate a complete reprocessing line for basic textile components, including liquid and solid waste treatment.

RESYNTEX has 20 project partners from across 10 different EU member states. Partners include industrial associations, businesses, SMEs and research institutes. Working together, the group creates an effective model for the whole value chain. The project is structured along 10 work packages:

Work package 1: lays out the strategic design and include scenarios for commercially successful and societally beneficial textile-chemistry partnerships.

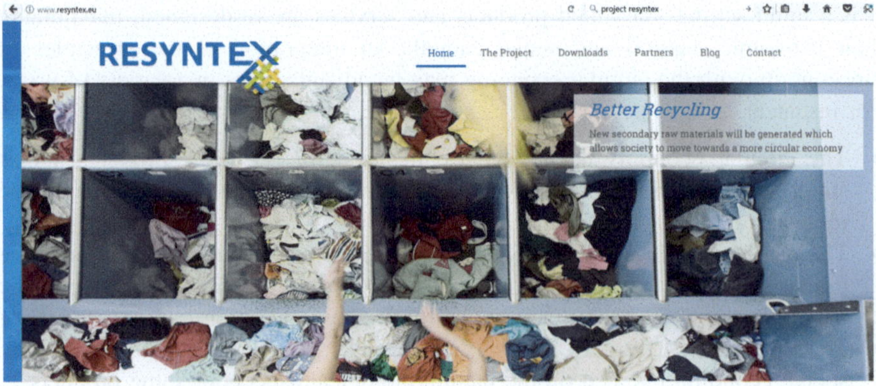

Fig. 1 Web site of the project RESYNTEX (http://www.resyntex.eu/)

Work package 2: aims to improve waste collection approaches, particularly for non-wearable textiles, by encouraging behavioural change. It will analyse stakeholders, promote recycling and enable greater social awareness and involvement.

Work package 3: looks to process textile waste as a suitable feedstock for recycling. This process must be suitable for a mechanical production line, sorting multicomponent waste and preparing it for recycling.

Work package 4: outlines an environmentally-friendly process for the transformation of natural and synthetic waste fibres into feedstock intermediates.

Work package 5: aims to develop industrial applications for the recovered feedstock from work package 4. Activities will involve end-users and experts, allowing effective exploitation of feedstock.

Work package 6: produces process designs and flow sheets for RESYNTEX at full scale. It will address all recovery stages needed for a single plant.

Work package 7: demonstrates the RESYNTEX concept in an industrial environment and aims to develop a demo-scale pilot plant.

Work package 8: uses state-of-the-art environmental life cycle assessment (LCA) and life cycle costing (LCC) to evaluate the performance of the new textile waste recycling value chains. It aims to guide the design and development of the value chains towards more sustainable solutions.

Work package 9: is the communications and dissemination programme, whereas Work Package 10 handles project management.

Figure 1 shows the web site of the project.

The fact that RESYNTEX emphasizes the re-use of textiles, hence reducing the pressure on cotton production on the one hand, and the energy consuming aspects related to the production of new textiles on the other—not to mention the chemicals used for dying clothes—means it is a good example of how the bioeconomy meets the circular economy.

The project **FORCE**

The FORCE project aims to minimise the leakage of materials from the linear economy and work towards a circular economy. The four cities involved in the project, Copenhagen, Hamburg, Lisbon and Genoa, engage enterprises, citizens and academia in 16 participatory value chain based partnerships to create and develop eco-innovative solutions.

Each city establishes a lead partnership for one of the four materials: plastic waste, strategic metals from electronic and electric equipment, surplus food and biowaste, and wood waste. Each city will also establish three local partnerships for the other materials. The work packages as specified at the projects's web site (Fig. 2) are as follows:

WP1 Ethics requirements: This work package sets out the 'ethics requirements' that the project must comply with.

WP2 Project Management, Project Monitoring and Coordination: the objective of WP2 is to provide the overall internal management of the project, to establish the administrative procedures for the project and to be the liaison between the project, its partners and the European Commission.

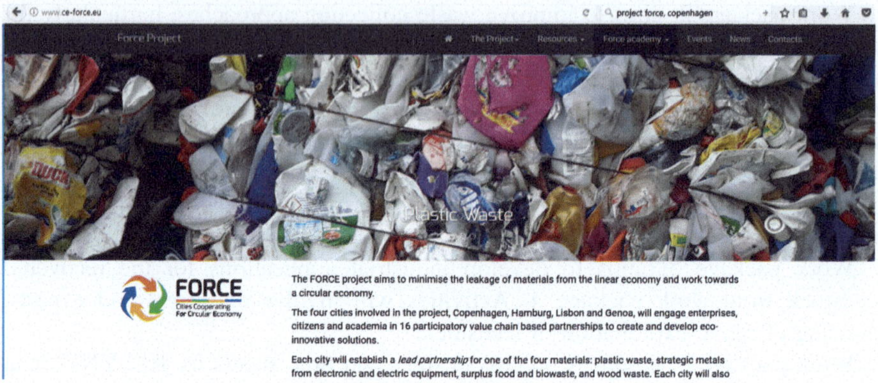

Fig. 2 Web site of the project FORCE (http://www.ce-force.eu/)

WP3 Plastic waste (from households and retail): it investigates and demonstrates possible applications of plastic waste otherwise considered as waste. The purpose is to exemplify new market opportunities for a variety of industries (plastics, packaging etc.), creating growth within Europe and contributing to the transition towards sustainable urban and peri-urban areas.

WP4 Strategic metals (from electronic and electrical equipment, EEE): the main objective is to identify the most effective approaches to increasing the re-use and continued use of electrical equipment and to govern the out dated or defect devices into effective resource recovery processes. The aim is to find the best routes to develop the market for used EEEs in such a way, that the individual citizen opts for a sustainable path when purchasing new electrical equipment or disposing of it.

WP5 Food waste prevention and bio-waste: the overall objective of this work package is to facilitate innovative ways to food management and prevent of foodwaste directly from where it is produced. With a simple way of referencing the food, day by day, that can be consumed; we will be preventing food of going to the waste bin.

WP6 Wood waste (from parks and forests, private households, companies, and 'other wood residues'): the aim is to establish an integrated wood waste management system that will increase the collection and recycling, ands put an innovative treatment system for residual wood waste to close the loop. In FORCE, a holistic and integrated action plan to deal with the urban wood chain in order to exploit synergies and business opportunities will be developed.

WP7 Governance and decision support: WP7 focuses on exploring the 'value chain partnership' as an innovative model of collaboration between public and private actors across the value chain. Within the FORCE project approach, the value chain partnership is the governance model for implementing the eco-innovative solutions outlined in WP3–6. WP7 analyses the cooperation processes and framework conditions of these value chain partnerships.

WP8 Dissemination and communication: the specific objectives of WP8 are:

- To provide targeted dissemination of FORCE results to the identified target audience.
- To coordinate the communication activities in WP3–6, and support the creation of exchange opportunities between partners, collect best practice, and replicate campaigns beyond cities.
- To set up the FORCE Academy that aims to: bring together stakeholders of partner cities, non-partner cities and other relevant stakeholders to stimulate mutual learning and exchange of experiences to support the development and dissemination of value chain based partnership governance models; to foster replication by training seminars (webinars, master classes, etc.).
- To create political awareness and acceptance as well as promoting the development of political framework conditions which support the implementation of eco-innovative solutions and continuous maintenance of value chain partnerships in European cities.

WP9 Exploitation, replication and market deployment activities: the main objectives of WP9 are to:

- Plan and address legal issues related to Intellectual Property Right among the partners, producing a "Plan for the Exploitation and Dissemination of Results" ensuring a smooth exploitation of the project results.
- Undertake a stakeholder analysis to better define and implement the strategy for the exploitation and market deployment of project outputs, to ensure the widest possible application.
- Inform key stakeholders and identified target groups at local, regional, national, European and international level on the project results and outcomes.
- Develop a business modelling which may open the way for organisations to expand their investment in circular economy, especially in direct re-use, repair, remanufacture, recycling and first market replication/manufacturing of new applications and products.
- Engage the business sector and run initiatives targeted to it, facilitating knowledge transfer and catalysing Investments in the fields of waste prevention and management.

In the area of wood waste, which is the most relevant one from a bioeconomy context, the specific objectives are to analyse, improve and integrate the existing wood waste management, design and test new, sustainable wood waste streams to increase their efficiency and to valorise wood matter during the processes phases. Another aim is to test the technical, operational and economics aspects of the integrated wood management system in an Urban Lab to verify its scalability at the city level and its replicability at European level.

3 Conclusions

Bioeconomy and circular economy should go hand in hand. According to the European Union, bio-based products are regarded as one of a set of six key markets. This means that a close cooperation between science, technology and policy making can yield substantial benefits when it comes to the conservation and long-term use of natural resources. By reducing and perhaps even avoiding the substantial wastes of resources being seen today-as the integration of bioeconomy and circular economy may do-, we can not only achieve a greater level of sustainability in various European industrial sectors, but also contribute to substantial reductions in CO_2 emissions, and assist efforts towards climate change mitigation. Some areas where action is needed, are:

- Supporting the transition towards a bioeconomy by catalyzing changes in production systems and consumption patterns
- Supporting climate change mitigation strategies, closely involving various sectors such as agriculture and forestry
- Supporting data collection to objectively measure the means to enhance sustainability in forestry, agriculture and related bioenergy systems
- Supporting optimisations in agriculture and forest operations as a means to ensure a more sustainable supply of biomass for industrial activities and minimizing wastage of raw materials
- Paying due attention to the various environmental challenges associated with the pursuance of the bioeconomy.

Last but not least, as shown by the examples provided by the projects FORCE and RESYNTEX, apart from the development of new technologies, there is a pressing need to take into account the economic and social challenges for the bioeconomy, ensuring that suitable policy mechanisms are in place in order to support such efforts, instead of inhibiting them.

In this context it should be pointed out that both basic and applied research and cross-cutting themes should be equally emphasized and supported, providing a solid basis for long term benefits.

References

Bioeconomy Science Center (BSC) (2017) Sustainable bioeconomy: concept and relevance. Available at: https://www.biosc.de/sustainable_bioeconomy_en

BMBF (2011) National research strategy bioeconomy 2030. BMBF, Berlin

BMBF (2016) Bioökonomie International. BMBF, Berlin

Burns C, Higson A, Hodgson E (2016) Five recommendations to kick-start bioeconomy innovation in the UK. Biofuels Bioprod Bioref 10:12–16. https://doi.org/10.1002/bbb.1633

European Commission (2002) Life sciences and biotechnology: a strategy for Europe. EC, Brussels

FAO (Food and Agricultural Organisation) (2008) Global cereal supply and demand brief. Crop prospects and food situation, Number 2. Apr 2008 www.fao.org/docrep/010/ai465e/ai465e04.htm
Hodgson E et al (2016) Horizon scanning the European bio-based economy: a novel approach to the identification of barriers and key policy interventions from stakeholders in multiple sectors and regions, Biofuels. Bioprod Biorefin 10(5):508
Langeveld JWA et al (2016) The bio-based economy and the bioeconomy in the Netherlands. Biomass Research report 1601, Ref. 59015257. Biomass Research, Wageningen
Organisation for Economic Cooperation and Development (OECD) (2009) The bioeconomy to 2030—designing a policy Agenda. OECD, Paris
Schieb P-A et al (2015) Biorefinery 2030: future prospects for the bioeconomy. Springer, Berlin
Sillanpää M, Ncibi C (2017) A sustainable bioeconomy—the green industrial revolution. Springer, Berlin

Erratum to: Towards a Sustainable Bioeconomy: Principles, Challenges and Perspectives

Walter Leal Filho, Diana Mihaela Pociovălişteanu, Paulo Roberto Borges de Brito and Ismar Borges de Lima

Erratum to:
W. Leal Filho et al. (eds.), *Towards a Sustainable Bioeconomy: Principles, Challenges and Perspectives*, World Sustainability Series, https://doi.org/10.1007/978-3-319-73028-8

The original version of the book was inadvertently published without incorporating the corresponding author's (F. S. Sosa-Rodríguez) corrections in chapter "Economic Assessment of Tourism Based on Shark-Seeing and Diving as a More Profitable Activity Than Commercial Fishing", which have been now incorporated and the incorrect author names "A. A. Victor" and "A. O. Oluwabunmi" have been corrected as "A. V. Adejumo" and "O. O. Adejumo" in chapter "Sustainable Development: Implications for Energy Policy in Nigeria". The erratum book has been updated with the changes.

The updated online version of these chapters can be found at
https://doi.org/10.1007/978-3-319-73028-8_12
https://doi.org/10.1007/978-3-319-73028-8_21
https://doi.org/10.1007/978-3-319-73028-8

W. Leal Filho et al. (eds.), *Towards a Sustainable Bioeconomy: Principles, Challenges and Perspectives*, World Sustainability Series,
https://doi.org/10.1007/978-3-319-73028-8_30

The manufacturer's authorised representative in the EU is Springer Nature Customer Service Centre GmbH, Europaplatz 3, 69115 Heidelberg, Germany. If you have any concerns regarding our products, please contact ProductSafety@springernature.com

Printed and bound by CPI Group (UK) Ltd, Croydon, CR0 4YY
30/09/2026
02225651-0019